Mrs Humphry Ward

John Sutherland has taught at the universities of Edinburgh and London. He now teaches literature at the California Institute of Technology. He is the author of *Victorian Novelists and Publishers* (1976), *Fiction and the Fiction Industry* (1980), and *The Longman Companion to Victorian Fiction* (1989). He has a longstanding interest in minor and unfairly forgotten Victorian fiction.

'John Sutherland's . . . account of her life is clearly the best yet'

Julia Briggs, *Observer*

'one of the most engaging aspects of the book is its author's unquenchable sense of amusement, indulgently exercised towards his subject and her world'

Jonathan Keates, *TLS*

'gripping . . . a story that unfolds with the fascination of high tragedy . . . we owe a debt to Mr Sutherland for making it so accessible'

Margaret Drabble, *New York Review of Books*

'model biography . . . brilliantly told'

Joan Smith, *Guardian*

Mrs Humphry Ward

Eminent Victorian
Pre-eminent Edwardian

JOHN SUTHERLAND

Oxford New York
OXFORD UNIVERSITY PRESS
1991

Oxford University Press, Walton Street, Oxford OX2 6DP
Oxford New York Toronto
Delhi Bombay Calcutta Madras Karachi
Petaling Jaya Singapore Hong Kong Tokyo
Nairobi Dar es Salaam Cape Town
Melbourne Auckland
and associated companies in
Berlin Ibadan

Oxford is a trade mark of Oxford University Press

First published 1990
First issued as an Oxford University Press paperback 1991

British Library Cataloguing in Publication Data
Data available
ISBN 0–19–282904–1

Library of Congress Cataloging in Publication Data
Data available

Printed in Great Britain by
Biddles Ltd.
Guildford and King's Lynn

Acknowledgements

I am grateful to Mrs Mary Moorman, the granddaughter of Mrs Humphry Ward, for permission to use manuscript material. The librarians and archivists at Pusey House Oxford, University College London, the Honnold Library Claremont, and the Houghton Library, Harvard, have been particularly helpful to me. I would like to thank the staff of Mary Ward House and its principal Patrick Freestone, for allowing me to examine their archive. Lieutenant-Colonel J. D. Sainsbury has been most useful in explaining details of Arnold Ward's military career. Others who have patiently and helpfully answered my queries and given assistance are: Rosemary Ashton, Sybille Bedford, James Bertram, Mrs Elizabeth Boardman, Morton N. Cohen, Jane Dietrich, Sarah J. Dodgson, Frances and Phil Edmonds, N. John Hall, Mrs P. Hatfield, Matthew Huxley, Enid Huws Jones, John Jones, Oscar Mandel, Mrs M. Trevelyan, Phyllis E. Wachter, Cindy Weinstein. I owe a special debt to Mac Pigman, Guilland Sutherland, John Watt, Ava Moeller, and Katie Ryde in helping prepare the book and to Kim Scott Walwyn for proposing the idea to me.

Contents

1

The Girlhood of Mary Arnold: 1851–1860

ON the morning of 5 April 1902, the famous novelist Mrs Humphry Ward made the following entry in her notebook:

> Last night for the first time since his death (Nov. 12th. 1900) I had a vivid dream of Papa—He was in bed I think—though not in the room where I had seen him last—and there was a sense of meeting again. I kissed him and laid my cheek against his—'Dearest Papa you do love me?'—'Yes, always.' Presently I said to him 'Papa I am so sorry to have argued so much about the [Boer] War. I am afraid I made you angry. Forgive me'. He smiled and said in the shy sly way he had sometimes when he was pleased—'Perhaps that was because I couldn't answer your arguments!' By which I knew he only meant to put me at ease—not any change of view. At this time he was standing, dressed, and looked like himself, only younger than he was at the last.[1]

It's a vivid glimpse of the author's inner mind. Mary Ward's father, Thomas Arnold, had been dead almost two years. She herself was 51 years old, and embarked on a furiously active career in public life—something that would make her, once an eminent Victorian, a pre-eminent Edwardian. And yet it is clearly not Mrs Humphry Ward—the grey-haired matriarch—in the dream. It is a young girl: Mary Arnold. And clearly this dutiful daughter who so wants to please an inscrutable father wistfully *knows* that for mysterious reasons she has failed. With the sharp intuition of the cartoonist, Max Beerbohm similarly discerned in the imposing bulk of Mrs Humphry Ward of 1904 the inner self of a little Arnold girl cringingly desperate for the good opinions of the imposing Arnoldian patriarch. Was she loved? Had she been a good child?

That she was an Arnold was the most important single fact in Mary's life. Printed references routinely identified her not as the wife of T. Humphry Ward (a middle-ranking *Times* journalist) but as the granddaughter of Thomas Arnold of Rugby and the niece of Matthew Arnold, Professor of Poetry at Oxford. In her autobiography, *A Writer's Recollections*, Mary Ward introduces herself like a Cruft's prize dog, inseparable from her pedigree and blood line: 'For me, the first point that stands out is the arrival of a little girl of five, in the year 1856, at a gray-stone house in a Westmorland valley, where, fourteen years earlier, the children of Arnold of Rugby, the "Doctor" of *Tom Brown's Schooldays*, had waited on a June day, to greet their father, expected from the South, only to hear, as the summer day died away, that two hours' sharp illness, that very morning, had taken him from them.'[2] It is a contorted opening to a novelist's life story. Apart from the name-dropping (an incurable vice with Mary Ward in her later years) what can one make of the syntactic skid by which—in the span of one sentence—her arrival at Fox How is blotted out by the morbid scene of her grandfather's departure by death, somewhere else, eight years before she was born?

As one reads on and the ritual references to 'the Doctor' multiply (including an extended description of his fatal attack of Angina Pectoris in June 1842) one recognizes it as a question of relative gravity. What her tortured opening sentence insinuates is that Mary was unimportant—measured, that is, by the scale of what mattered in the Arnold clan. For its members, all degrees of importance descend from the absolute importance of the patriarch, Thomas Arnold, whose dead but inextinguishable presence loomed over their subsequent lives like some deity in a Greek tragedy.

In fact, the god analogy was one which was commonplace during Arnold's lifetime. Lytton Strachey spins it out at malicious length in his iconoclastic *Eminent Victorians* (1918). Headmaster, 'Paternoster', patriarch, and father figure all merged in the awful form of Dr Arnold. Little Mary had been grounded in ritual obeisance to the Doctor from her cradle: almost the first words we know her to have said are 'grand-dada' addressed to the portrait of Arnold that dominated her family dining room in Tasmania. Through this icon, she drew 'the dignity of her race into her childish blood'.[3] Presumably the dead old man was confused in Mary's baby mind with that other 'our father' in heaven, whom she invoked in her bedtime prayers.

The Doctor had four sons, all of whom dutifully followed him into some form of educational administration or 'service'. The eldest, Matthew, Mary's 'Uncle Matt', is probably more famous to posterity than his headmaster father—something that would have rather surprised both of them. William Delafield Arnold, the fourth son, died prematurely in 1859, having reformed the Punjab education service and deflated the complacency of the East India Company with a powerful novel, *Oakfield* (1853). He was, at 31, burned out and sheer exhaustion played a major part in his death from fever as he returned invalided from the subcontinent. Mary never knew her uncle William. Edward Arnold, the third son, became an Inspector of British Schools. He too died prematurely, aged 52.

Thomas Arnold (1823–1900)—the second son, the Doctor's namesake, his favourite and Mary's father—was, by contrast, prodigally gifted; more so, arguably, even than Matthew. And he lived longer than any of the Arnold sons—seventy-seven years. But Tom's career would have been a singular disappointment—if not an outrage—to the Doctor had he lived to see it. It was certainly a gnawing disappointment to his eldest child, Mary, when in later years she came to review it.

The lives of his sons suggest the psychic brunt involved in being the offspring of a Victorian eminence like Arnold of Rugby. There was, of course, the silver spoon. Each of the Arnold boys had educational and social privileges assured them—a smoothed path from public school to Oxford; introductions to powerful friends of the family at the necessary moment. But each also had an essential freedom of choice—that of finding his vocation—removed from him. They were an educational caste. Overwork ('service') was mandatory—had not the Doctor killed himself by useful toil at 47? His distilled advice to Tom was the one word 'Work!', without any specification as to what the work should be. Effort alone would suffice. Working herself 'to death' would duly become Mary's fetish in a career which was one long theatre of heroic labour and invalidism.

Dr Arnold was a hard father for a boy to have since he was so immeasurably greater than any son could ever aspire to be. Matthew expresses a spectacularly traumatized filial reaction in his epic poem *Sohrab and Rustum*, published in 1853. It describes the man-to-man battle to the death of a father and son who do not know each other behind their armour when they fight as the champions of their respective Tartar and Persian armies. The contest ends with Sohrab's being slain by his more potent father, Rustum. Before he dies there is a pathetic recognition scene in which all is forgiven and Sohrab—in an extraordinary reversal of conventional family roles—instructs his father to live on and 'Do . . . the deeds I die too young to do.' *Sohrab and Rustum* expresses a rhetorical willingness to stand and fight the puissant father, even at the hazard of death or psychic extinction. In his actual day-to-day life Matthew's rebellion took the form of foppishness which he cultivated during his three years at Oxford. In the face of Rugby's arid 'earnestness' his popinjay mannerisms were resistance of a kind.

Tom's career, by contrast, took the better part of valour. It is marked not by fight but by 'truant flight'—physical, emotional, ideological—at every critical turn of his development. If Matthew chose the part of Sohrab, Tom was to be the scholar gypsy. Like Matthew (his elder by a year) Tom had the statutory six years at Rugby where, like most headmaster's sons, he bore the double burden of having to be a paragon while being regarded by his more subversive schoolfellows as a spy. (One refuses to believe that *all* the boys at Rugby were as idolatrous as Tom Brown and Scud East.) To his father's alarm, as early as his eighth birthday Tom made the extravagantly morbid claim that 'the years I have now lived will be the happiest of my life'.[4]

At 18, Tom won a scholarship to University College, Oxford. During the interval between school and university—always tense, emotionally—he witnessed his father's unexpected death. It was a scene of 'almost royal majesty', as immortalized in A. P. Stanley's hagiographic account. Early on 12 June 1842, Dr Arnold awoke with an agonizing pain in his chest. It was ominous; his own father had died of sudden angina, aged 53. A physician was sent for, and Mrs Arnold consoled her dying husband by reading aloud the fifty-first psalm. At this point, Tom came into the room. 'My son, thank God for me' said Dr Arnold; and as the boy did not at once catch his meaning, he added, 'Thank God, Tom, for giving me this pain: I have suffered so little pain in my life that I feel it is very good for me: now God has given it to me, and I do so thank Him for it.'[5] A few minutes later his family heard the death rattle and the Doctor was gone. Whether Tom in his prayers that night did humbly thank God for torturing his father with angina, we do not know. But the episode left an indelible wound. Even after fifty-eight years he could not bring himself to describe it in his autobiography. As one biographer records, 'from this time, the intermittent stammer that had been with [Tom] since childhood seems to have become a permanent handicap'.[6] Psychologically he was crippled by the sense that he could never attain his father's expectations for him. '[He] rated me much too highly,'[7] he wrote years later in his journal.

Nevertheless, at University College Thomas proved himself a brilliant student, and took his first in Greats effortlessly. He was thought 'the handsomest undergraduate at Oxford'. And theologically he appeared sound as a nut. He manfully resisted the seductions of the Tractarians and his father's great adversary Newman. Graduating in 1846, he impulsively proposed marriage to Henrietta Whately, the daughter of the Archbishop of Dublin. He was rejected, mainly because of his objectionably radical political and religious views. It upset him considerably. He thereafter read law in London in a desultory way for a few months. But the bar did not appeal to him. A career in the Colonial Office was duly arranged in April 1847, mainly by his mother's influence. His chiefs immediately took to Tom. His clear destiny was to be one of the senior managers of the English global estate, probably before he was 40. A career in politics would be a logical consummation. Or (were he prepared to curb his sexual appetites) he could eventually have taken up the fellowship at University College, which was earmarked for him in 1846.[8]

At this point, there occurred one of the first of Tom's quixotic defections from the fixed track of his Arnold destiny. It was the 'Hungry Forties' and, like many sensitive middle-class Victorians, he was deeply affected by the condition of the poor in London—a city which he 'loathed'. He was particularly haunted by the sense of belonging to the 'rich class'. 'I have *servants* to wait upon me,'[9] he lamented to his friend J. C. Shairp. He was steeping himself in George Sand, and pondering the 'sacred symbol' of 'Freedom, Equality, Brotherhood'.[10] In a spasm of 'young and democratic despair',[11] as his daughter later called it, Tom resolved to go to New Zealand, and embrace pioneer hardship on the land. In this 'colonist' condition of life there would be neither master nor servant, rich nor poor. He could, like others of his class, have purged his guilt less extravagantly by labouring as a Christian Socialist among the unfortunates of London's East End. His family fondly thought that a month or two in France or Germany might cure his wanderlust. But Tom's urge to flight was deeper than this. Geographically, New Zealand was as far as one could travel from Rugby, Oxford and Fox How and still remain on planet earth. He embarked for the other side of the world on an emigrant ship in November 1847.

Tom worked manually for four months, in mid-1848, clearing bush on his five-acre plot of land on the Porirua Road, living in a shack he called 'Fox How'. After this short exposure to the reality of colonist labour, he appreciated more clearly what an Arnold was cut out for and sensibly looked for a position school-teaching. Patronage and favours led in August 1849 to his being offered by the new Lieutenant Governor, Sir William Denison, the enviable position of inspector of schools in Van Diemen's Land (Tasmania), at a salary of £400 per annum. 'Our father's name', he wrote home, 'has been . . . more profitable than houses and land.'[12] He was 26. Other men would have been lucky to get an inspectorship in middle age.

Tom arrived in the small garrison town of Hobart in January 1850 to take up his new position. The island population was some 70,000, and was regularly supplemented by cargoes of transported convicts, whom Tom hated ('the worst among men'). In March he saw Julia Sorell for the first time. He recalled the episode, still awe-struck, fifty years later: 'I was at a small party . . . On a sofa sat a beautiful girl in a black silk dress, with a white lace *berthe* and red bows in the skirt of the dress. My friend Clarke presently introduced me to her. I remember that as we talked a strange feeling came over me of having met her before—of having always known her.'[13] Julia was both a belle and ostensibly eligible, the unattached granddaughter of a previous governor of Tasmania. There followed a whirlwind romance and the couple were married three months later—almost less time than it took the letter announcing the match to reach England.

It was not, for all the Sorell family's association with Government House, a match Tom could have contemplated, had his mother been at his side. The Sorells had some spectacular skeletons in their closet. Julia's mother was the daughter of a colonial merchant adventurer, 'Captain' Anthony Fenn Kemp (1773?–1868). In his early days a freebooting officer in New South Wales, Kemp subsequently became known as 'the Father of Tasmania', in recognition of his pioneer exploits in the country's wool, grazing, and banking industries. A quarrelsome man of violently republican political opinions and dubious commercial honesty, Kemp had eighteen children, one of whom was Elizabeth Julia. In 1825 she married William Sorell (1800–60), a son of Lieutenant-Governor William Sorell (1775–1848), who governed the colony from 1816 to 1823.

This match between Julia's parents cannot have begun auspiciously. Old Kemp had been partly responsible for old Sorell's dismissal in 1823 by publicizing details of the Governor's drunkenness and his having set up his mistress,

Mrs Kent, and his bastards as first family at Government House. This lady was in fact a junior officer's wife whom Sorell (then a colonel) had seduced in 1807 at the Cape of Good Hope. He had all the while a legal wife and six children languishing in England on a pittance which he irregularly sent them. In 1817, on assuming office in Van Diemen's Land, he was obliged to pay the aggrieved and cuckolded Lieutenant Kent £3,000 for criminal conversation with the gentleman's wife.

These rogues, then, were Julia's grandfathers. Her father was the oldest of the legitimate children of the Lieutenant Governor. In 1822, now an adult, young William Sorell wrote to the British authorities, indicating his intention to go to Van Diemen's Land 'to assert his claims on his father's attention in person'.[14] To effect some kind of reconciliation, he was given a free passage to the colony by the Secretary of State for War and Colonies, Lord Bathurst. Young William duly arrived in December 1823, only to discover that Bathurst had meanwhile dismissed old William to protect the moral tone of the colony. While his son was travelling out, the disgraced Sorell was preparing to travel back which he did in June 1824. He died in London in genteel poverty twenty-four years later.

Young William elected to remain at Hobart, where he was appointed registrar of the Supreme Court in Hobart in 1824, at a generous salary of £600. A series of lucrative promotions in the middle ranks of public service followed. In 1825, he married Miss Elizabeth Julia Kemp, and had five children. But the marriage proved eventually disastrous. As Janet Trevelyan puts it, 'the principal fact that is known about [Julia's mother] is that she deserted her three daughters after bringing them to England [in fact Brussels] for their education, and went off with an army officer'.[15] It seems that at least some of the children were taken off with her to Paris. The young Julia and her two sisters returned to Van Diemen's Land, disgraced and humiliated. They apparently never saw their mother again.

Julia Sorell was thus the daughter of a flagrant adulteress (after whom she was named) and the granddaughter of a flagrant adulterer. Nor—to compound the unsuitability further—was her own record entirely unblemished. Born in 1826, she was 24; long in the tooth by the standards of the colonial marriage market. As one account records, 'According to Tasmanian legend [Julia Sorell] was alleged to have seduced Lieut-Governor Eardley-Wilmot. Though this was improbable, it precipitated his recall [by Gladstone, in April 1846—on trumped-up charges of sexual delinquency] . . . she had been already engaged twice before she met Arnold.'[16] Her name, Tom learned, had also been coupled with Eardley-Wilmot's son Chester. It is probable that she was merely the victim of loose talk of the 'like mother, like daughter' kind, spawned by Mrs Sorell's lurid escapades. But such gossip was wildfire in Hobart, still seething from the Eardley-Wilmot scandal.

Echoes of the Kemp–Sorell scandals must also have impinged on the young Mary's consciousness in the form of strange silences about her maternal family. Later, when she was given garbled or malicious versions of her background, she must have felt burning shame. On the Arnold side, however, there was no connection of which a young lady could not be inordinately proud. Mary Arnold's early writing and her autobiography swarm with honorific references and literary dedications to 'my grandmother'—Mary Penrose Arnold. On the other side there were an assortment of the Kemp–Sorell family who could scarcely be mentioned in decent company, and certainly not little Mary's other grandmother.[17]

The Sorells supposedly drew on distant Spanish ancestry (about which Mary was later to romanticize) pre-dating their more recent colonial origins in the West Indies. And they were fiercely Protestant, having a direct paternal descent from persecuted Huguenot refugees. What is most prominently recorded about them, however, is the family's 'ungovernable temper'. The orthodox reading of the young Mary's character was that she inherited a Sorell 'dæmon', which was only gradually 'harnessed' by Arnoldian 'discipline' and civility. Thus—in her own person—she recapitulated the progress of colonial civilization, converting the wild pagan into the gentle Christian.

Tom and Julia married—with Arnoldian fidelity to such things—on the Doctor's birthday, 13 June 1850. The ceremony took place in Hobart's St David's Cathedral. The younger Sorell sisters Augusta and Ada were bridesmaids. Exactly a year after the marriage, the Arnolds' first child was born. She was a girl and missed the Doctor's birthday by forty-eight hours, being delivered on 11 June 1851. She was named Mary after her grandmother, the Doctor's wife, and Augusta after her aunt. Her godfather was Tom's friend Archdeacon Tom Reibey (a son of the wealthiest banking and business family in Tasmania). Her godmothers were her maternal aunt Gussie and her paternal aunt Jane (Arnold) Forster. Three sons followed: in 1852 (William), 1854 (Arthur, who died within twenty-four hours), and 1855 (Theodore).

As his first-born, Mary was an object of some curiosity to Tom, who noted her characteristics in his diary and in letters home to his mother. Two themes dominate: will and frailty. The child was, to his consternation, 'delicate and easily put out of order'[18] and had a perpetual cold and cough. Pain, it seemed, was of great interest to the little girl. On 15 February 1853 Tom noted that the 18-month-old clearly signalled her suffering 'in an odd way; yesterday morning she was playing about in the nursery when all at once she stood still, put her little hands on her stomach, began to cry and said "Poor, poor, poor!" in a doleful voice'.[19] Mary underwent agonies during teething (indeed was still suffering with her teeth sixty years later) and Tom described in detail the lancing of her gums, an operation which Julia was unable to watch: 'Dr Bedford brought the sweet little victim into the dining room, and while I held her fast on her back upon my lap, so that her head should be between the doctor's knees, he lanced the gums of her upper front teeth with the greatest ease.'[20] There was 'wonderful relief', but two weeks later the 'sweet little victim' was sick again, with fever and the inevitable toothache.

And yet this fragile little thing already had a will of iron, she was 'Sparta' to the 'Athens' of her docile brother Willie. When she was just a toddler, Tom recorded that 'by nature [Mary] requires very firm treatment'.[21] Just how he intended to be firm with her is not clear. She was, he later thought, 'domineering ... afraid of nothing, at least I cannot remember her ever showing fear if a dog runs up towards her and barks at her. She holds out her hand to him and says "Come little dog. Come you pretty little dog".'[22] She was less friendly to servants, and as a 3-year-old 'defied them all'. She was 'insensible to fear ... kind enough when she can patronize but her domineering spirit makes even her kindness partake of oppression'.[23]

Despite these small parental anxieties Tom seemed at last settled in life. After his New Zealand fling in the bush, he would quite likely end up being the director of education for all Australasia. He began licking the colony's educational system into shape despite opposition from his former patron, the reactionary Sir William Denison. Another respectably Arnoldian destiny beckoned. It clearly terrified him. There occurred the second of his great defections (more kindly to himself, he called them 'wanderings' in his autobiography). Geographical estrangement had clearly not worked, so now he tried ideological estrangement. He converted to Catholicism. Short of disembowelling the Colonial Secretary, there was scarcely a more shocking thing an Arnold could have done. The Doctor and Newman had been the irreconcilable champions of their respective Churches. The catastrophe would never have happened, Mary later asserted, if the Doctor had been there to stop it: 'Newman's subtle pervasive intellect ... now, reaching across the world, laid hold on Arnold's son, when Arnold himself was no longer there to fight it.'[24]

Tom was 31 in autumn 1854, when he first announced his intention to go over to Rome. And he was quite capable of raising his own arm in paternal severity, particularly where his wayward little daughter was concerned. In November 1854 he told his mother during the turmoil of his conversion that:

> Mary is distracting my attention very much, for she is very naughty this morning. A child more obstinately self-willed I certainly never came across. It is very painful to have to punish her (which I usually do by locking her up) for the resistance of her will, and the profusion of her tears and cries are wonderful; still it must be done, I suppose. 'What son is he whom his father chasteneth not?' The American system, under which I believe children of the age of 14 or 15 usually set their parents at defiance, so far as control is concerned, always disgusted me; and we see its fruits in the general character of the individuals of that most vulgar and disagreeable nation. And now, I am glad to say,

after the last locking up, Miss Mary has shown herself duly submissive and penitent, and after being seated in a chair a little while, to cool and collect herself after her crying fit, is happily employed in looking at the pictures in a volume of the Penny Magazine.[25]

In February 1855, Tom notes in another letter to his mother (who probably longed on her part to whip some sense into her erring son) that 'I have a regular pitched battle with [Mary] about once a day .... She showed at one time a slight tendency to story telling, but a timely whipping checked this.'[26] As at Rugby, the Arnold code enjoined corporal as well as moral discipline. But it does seem hard.

The other woman in his house was less easily subdued. Julia had exacted a six-month stay during which Tom was to take no decision on his religious future. But on his proving still obstinately inclined to Rome, she wrote a letter of ultimatum in late June 1855, telling him that if he went ahead with his plan to go over 'I will, I must, leave you.'[27] She went on to paint a picture of what their family life must be like, where she was an Anglican and he a Romanist and their children tugged between them. It was, in the event, a deadly accurate forecast. Julia (who was six months pregnant with Theodore) succeeded in extracting a promise that Tom would not convert until he had seen his mother. But he broke his word (blaming Julia's indiscretion in 'talking to everybody',[28] so forcing his hand) and was received into the Catholic faith on 18 January 1856. Julia did not, as she had earlier threatened, leave him. Two months after his conversion she was pregnant for the fifth time (another Arthur, later to be the black sheep of the family). The disgruntled but still spirited wife contented herself by throwing a brick through the window of St Joseph's pro-Cathedral in which her husband had been received into the Catholic faith, and by writing abusive letters to Newman ('that living saint',[29] as Tom called him in his belated letter of confession to his mother, 21 February 1856). During all this turmoil, the 4-year-old Mary was boarded with her godfather's staunchly Anglican family at Entally. 'Her temper it appears is greatly improved since she has been under Mrs Reibey's care,'[30] Tom complacently told his mother in September. He was apparently visiting her to teach her reading, and on 5 March told his sister 'she is by no means quick at that though so ready-witted in other things'.[31] It seems not to have occurred to him that this ready-witted child might have been distracted from her exercises by the adult passions swirling around her.

Thomas Arnold's apostasy made his position in Tasmania publicly difficult. The newspapers raised an outcry, calling for his resignation, as a pervert. The English middle classes were still paranoid about the Catholic 'Aggression' of 1851–2. It seems, incredibly, that Tom had hoped to keep the little affair of his conversion secret: a matter between himself and his spiritual adviser, Bishop Robert Willson. Again he blamed Julia 'for having talked so much about the matter'.[32] He still thought that he might keep his inspectorship—there was nothing specifically in the regulations about having to be a Protestant. Indeed, ten days after his going over, his salary was actually increased by the Legislative Council, presumably as a vote of confidence. But Tom was sick of Hobart. He loathed the convict system. He was disgusted by the genocide of the Tasmanian aborigines, who were freely shot as pests by English shepherds. He found the intellectual life of the colony impossibly stultifying and for some years had been confiding his yearning for England to his bosom friend Arthur Clough. When in April the Colonial Secretary offered him the option of resignation with eighteen months' half pay he took it.

The family duly sailed for England on 12 July 1856. It was a wretched trip for the five Arnolds and the couple of young cousins who were being shipped back to England under Julia's care. They were cramped for three-and-a-half months into a 400-ton rat-infested barque, the *William Brown*. Nor was it at all clear what they were travelling towards. Julia was very pregnant and constantly seasick. Tom nursed a sea-bird that died on him. The *William Brown* was a cargo vessel, with scant provisions for passengers. The genteel Arnolds were obliged to take their baths by day in two huge barrels of freezing sea water on the deck. They were destitute. Tom's half pay was required to pay departure expenses and he was reduced to begging a loan from his mother for their immediate needs in England. When they arrived in mid-October Mary claimed to remember 'being lifted—weak and miserable with toothache—in my father's arms to catch the first sight of English shores as we neared the mouth of the Thames; and then the dismal inn [the Vine] by [St Katharine's] docks where we first took shelter. The dreary room where we children slept the first night, its dingy ugliness and its barred windows, still come back to me as a vision of horror.'[33] Their distress and confusion were relieved next day by Tom's sister, the infinitely kind Jane Forster, who fed and housed them in more suitable lodgings to prepare for their train journey up to Mrs Arnold at Fox How.

In the best of worlds, Tom would now have taken some leisurely months to recuperate in the Cumbrian bosom of the Arnold family. But he dared not for doctrinal reasons stay any length of time at Fox How. As he pointed out to his mother, 'the Catholic Church forbids her members to hold religious communion with those who do not belong to her'.[34] This meant that it would be awkward for him to eat at a table where Anglican grace was spoken, and impossible to join in the morning and evening family prayers which were central events of the day at Fox How. There was also, although he was too tactful to mention it, the risk of being seduced back to the faith of the Arnolds. On the other hand, his wife was eight months pregnant, suicidally unhappy, had no close relatives in England and there were three children under 6. Tom could hardly now live the life of the scholar gypsy, wandering the Chiltern Hills with these dependants in tow.

On landing in England, Tom had written immediately to Newman in Dublin asking about the possibility of some tutoring in that city. It was frankly a begging letter. Newman, who genuinely liked Tom Arnold, replied by return offering him a position at the new—but already moribund—Catholic University, overriding the objections of his Irish colleagues. The appointment paid about half his starting salary as an inspector of schools in Tasmania. It was not a job Tom was fitted for by nationality or experience, nor did his 'hesitation' (as he called his stammer) qualify him as a lecturer. But it was in the circumstances a godsend. These, then, were the dismal circumstances of Mary's arrival at Fox How in October 1856. It was not the happy and auspicious 'homecoming' that she painted in *A Writer's Recollections*—more like a flimsy branch grasped on the slippery cliff's edge.

Tom moved directly to Dublin at the end of October, leaving Julia (who had never before met her Arnold relatives) to have her baby by herself at Fox How. The three toddlers, meanwhile, were looked after by Tom's mother and his younger sister Frances ('Aunt Fan'). All the Doctor's children had long since grown up, the nurse and governess had gone, and there was no equipment for three unhappy children at Fox How. It was winter and they were confined to the strange house, which had the musty air of a museum of childhood with its 'mysterious drawers full of cards and puzzles, and glass marbles and old-fashioned toys'.[35] Julia suspected the Arnolds disapproved of her. (They did. Mary Ward reconstructed the bleak tension of the situation in her late novel *Delia Blanchflower*, 1915.) All in all, it must have been a fraught three months. The next child (Arthur) duly arrived on 15 December, and soon after Julia departed for Dublin with her three boys, leaving the 5-year-old Mary behind at Fox How.

Fox How—'the Mecca of the Arnold family'—was the idyllic grey-stone house at Ambleside, under Loughrigg, built by the Wordsworth-loving Doctor (with Wordsworth's assistance) in 1834 as his retreat from the hurly burly of Rugby. 'The Fox How portion of our life', Tom recalled, 'was a time of unspeakable pleasure to us all.'[36] Mary herself never described Fox How without a flight of purple prose about 'the perfect outline of the mountain wall, the "pensive glooms" hovering in that deep breast of Fairfield,'[37] or whatever. And despite the never-ending wind and rain of the Lake District, she usually pictured it bathed in golden sunlight. After the Doctor's death in 1842, Fox How had become the Arnold stronghold. Mrs Mary Arnold retired there to finish bringing up her nine children. The beautifully landscaped house and the region, with its intimate links to English Romanticism, were to have a pervasive effect on all the young people of the family, especially those like Tom and Matthew with a literary bent. In 1856, when Mary arrived, Fox How was still under the superintendence of the Doctor's widow, now a matriarchal 65 and the 22-year-old maiden daughter, Frances. The younger mistress of the house was evidently something of a martinet. In one of the masochistic anecdotes testifying to her moral growth Mary is supposed to have declared, 'I like Aunt Fan—she's the master of me!'[38]

As was normal in mixed marriages, Tom Arnold's family was split along precisely defined sexual–religious lines. The boys would be brought up by the father as Catholics; the girl(s) would be brought up by Julia as Protestant. But in the upheaval of 1856 it was evidently not thought worth while to make separate provision at home for the single daughter who was left in England in the care of her aunt and grandmother. For the next ten years Mary Arnold was effectively orphaned, with her parents and siblings (including two later daughters) across the Irish Sea. Not that the larger Irish fraction of the family had much pleasure in their togetherness. Julia continued to keep her husband's bed. But, according to Newman, 'she used to nag, nag, nag [Tom], till he almost lost his senses'. The man was, Newman thought, 'weak and henpecked'.[39] She on her part was given to passionate outbursts of temper and extravagance in money matters. It was partly punitive; according to Newman, Julia delighted in taunting Tom with the poverty to which he had reduced his dependants. In their new Irish home, the Arnolds were to have five wretched years 'of straitened means and constant struggles, passed in dismal furnished houses in Rathmines or Kingstown'.[40] It would have strained a happy marriage. It made an unhappy marriage hell on earth.

According to Mary, she came to full consciousness on that morning in autumn 1856 when she arrived at Fox How. It was her first Wordsworthian 'spot of time'. Her previous five years of life existed for her only as vivid snapshots—either remembered or inserted into her memory as family lore. They included running through a field of Tasmanian lilies and of two of the last Tasmanian aborigines sitting behind her as she rode in a coach. In the memorial of her brother Willie she recalled another anecdote:

> A relation gave [Willie] on his fourth birthday, a little jacket, of which, as being no doubt a more masculine garment than he was accustomed to wear, he was vastly proud. A covetous elder sister of five [i.e. Mary] tried to coax it out of him, and when baffled, declared that selfish boys could not go to heaven. Willie protested that he was certainly going there, and then added, hugging his jacket to him, 'but I'll go with my jacket on though!'[41]

Someone as steeped in the Bible as Mary would appreciate the echo of the elder siblings' jealousy for Joseph's wonderful coat. And clearly it was a trial for Mary (until he failed at Oxford) that the clever and placid Willie was her father's favourite.

The newly conscious Mary found herself in October 1856 in a dilemma beyond any little girl's solving. She was abandoned, spiritually orphaned, and probably unwanted, though certainly not mistreated, at Fox How. She had arrived in England, the home of her fathers, but was effectively fatherless. She was among the Arnolds, but not of them. Her mother 'never wrote'[42] and her father rarely visited. And when he did it was not to see *her*; she merely happened to be there when he came to see other people. A recollection catches the loneliness of Mary which persisted even when her father was at Fox How: 'I can recall one or two golden days, at long intervals, when my father came for me, with "Mr Clough", and the two old friends, who, after nine years' separation, had recently met again, walked up the Sweden Bridge lane in the heart of Scandale Fell, while I, paying no more attention to them than they—after a first ten minutes—did to me, went wandering and skipping and dreaming by myself.'[43] It is likely that the family did not talk much about Tom in front of his daughter and there must have been some mysterious hushing when she was in the room. Above all, the little girl must have been excruciatingly lonely and confused.

The standard accounts of Mary Arnold's childhood are extremely skimpy. Her first sixteen years occupy only as many pages of Janet Trevelyan's 317-page biography. It seems that she had no childhood in the usual sense of growing up to independence from rooted domestic foundations. She was separated for ten years from her parents and siblings, without any adequate emotional substitutes. Her condition of life was to be nomadic: boarded out at a succession of schools and farmed out during the holidays (even Christmas holidays) to various relatives.

Mary Ward's own autobiographical account of her girlhood is vague, brief, and circuitous. Without giving any clear chronicle of her life from 5 to 16, it dwells instead at inordinate length on Fox How and the 'golden days' she spent there. This country house, we are to understand, was her 'second home',[44] where Mary Arnold had her seed-time. The five chapters which she devotes in *A Writer's Recollections* to Fox How cumulatively imply that she spent great stretches of her childhood in this 'house of paradise',[45] as had her father and Uncle Matt. She did not. In fact, she spent something over a year as a resident in the house, before being packed off as a boarder at the age of 7 to a small school for girls at Ambleside. Ambleside was very near Fox How; so near, in fact, that she could well have been a day pupil. There were around forty pupils at the school, and just one boarder apart from Mary Arnold. It is clear that Mary was not wanted at Fox How except at the weekends (Saturday only).

A minor encumbrance to her female relatives at Fox How, Mary was entirely invisible to important male Arnolds like Matthew, who at this period rarely came north anyhow. Uncle Matt seems not to have registered Mary's independent existence until many years later (and even then not as attentively as the hero-worshipping niece would have us believe). On her part, the grandmother—true to her Arnold values—gave priority to her eldest male grandchild, Willie, especially when he was at Rugby. He on his part was devoted to his grandmamma and as a privilege would 'say his hymn to her on Sunday afternoon'[46] when he was at Fox How.

Despite her father's early complaints at her slowness, Mary was precociously literate, and her early letters were clearly preserved in the family records for their impressive juvenile penmanship. Nevertheless, their rigid cadences often have a prophetic and invariably a poignant flavour to them. The earliest which survives is to her parents on 13 February 1857. She has been staying with her aunts Mary and Jane and has had a gift from a 'kind lady' of a box of bricks with which 'I've learned to build churches.'[47] Three days after, she wrote to Willie informing him about the wonderful bricks and adding: 'Yesterday I went to see a little baby which wouldn't come into my arms the mother said "Now you must not be naughty—no not naughty" but however he didn't come to me and I had to come away without him sitting on my knee.'[48] A later letter of 29 May 1859 from Eller How records the routines of her high and holy days at school:

My Dear Papa—I have got a Satisfactory this week. It has been very fine and very hot here. Mr Fell preached this morning. As we were in church a bird came in and flew about the Church and Mr Fell saw it. In the sermon he mentioned the death of Mrs Hornby, calling her a 'dear sister' and he said she was humble and devout in Church and that she was respected by all her friends and loved by the poor. On the Queen's birthday we had a half-holiday and we went over Loughrigg. We found a good many flowers. Among them was a pretty little flower called the Milkwort and the purple geranium. We went to a house and had some gingerbread and water then we walked home.—Love to all at home—I am your most affectionate child.—Mary.[49]

Another letter survives from 12 August 1860 in which the now 9-year-old Mary writes to her mother, asking with pathetic urgency to be allowed to do something, anything, for her new-born brother Frank 'My Dear mamma—will you please send me some [needle]work. I can do some for my little baby brother. Ellie Lewis came to tea on Thursday. Please Mamma will you write to me soon and send me work for I want it very much.'[50] In another, equally pathetic letter to Willie she asks him to teach 'baby' her name.

Reading between the lines, we can assume Mary to have been a problem to her Fox How guardians. The childhood anecdotes which are told about her dwell on her disruptive energy, passion, and downright naughtiness. She would 'get into great rages and scream, till everybody was quite tired out'.[51] Evidently the ladies of the house, nerves torn to shreds, would admit defeat and a burly footman would be called in to carry the 'wildcat', bodily and screaming, to her room, where she could scream herself out behind the locked door. Mary is described on one occasion at 'the top of a flight of stairs with a large plate full of bread-and-butter', flinging 'slice after slice smack in the face of the [school] governess standing at the foot, finally hurling the plate after'.[52] In 1859, on one of his rare trips, Tom tells his wife that he has impressed on her teacher the necessity of giving Polly [i.e. Mary] 'plenty of exercise',[53] doubtless thinking that exhaustion will usefully subdue her boisterousness. (He also in the same letter mentions her being constantly apprehensive of his leaving.) Visiting the little Ambleside school in later life, Mary Ward showed her companion where 'you can still see the damaged panel which I bashed in with my fists in my fury when I was locked into the cloakroom for punishment'.[54]

Any ambitious scheme of education for the young Mary Arnold was frustrated not just by her temper but by the problem of money and the fact that she was—when all was said and done—only a girl among a brood which included boys. Tom Arnold was still penurious and his family was inexorably growing as his salary from the ailing Catholic University remained obstinately stuck at around £300. Mrs Arnold was not rich and had needed the assistance of the Arnold Testimonial Fund to send her three sons to university. No help could be expected there, beyond occasional board for Mary.

The establishment in Ambleside which Mary Arnold attended from autumn 1858 was run by Anne Jemima Clough, later the first Principal of Newnham. Anne was the sister of Tom Arnold's particular Oxford friend Arthur Clough (who had earlier depicted Tom as the idealist Philip Hewson in *The Bothie*). It is possible that Mary's fees were reduced by Miss Clough, as they were at her next school. At 7 she can have known nothing of any such humiliating arrangement, if indeed it existed, but she must have missed the small fineries of dress that other better-off and less pious children had. Arnold girls were dressed in severe black.

Mary attended Anne Clough's school from 1858 to 1860. After Ward's death in 1920, her best friend there and fellow boarder, Sophie Bellasis ('only a year and two months older than I am'),[55] recalled in the *Cornhill Magazine* everyday details of their school life. Sophie particularly remembered a vivid anecdote of the 7-year-old Mary's queer destructiveness: 'I recollect she had a lovely doll, which her aunt, Mrs Forster, had given her, all made of wax. Once she was annoyed with this doll for some reason or other, and broke it up into little bits ... We put the bits into little saucepans and melted [the wax] and moulded it into dolls' puddings, and that was the last of her wax doll.'[56] A Victorian wax doll was no cheap thing. And the childless aunt, Jane Forster, who gave it to Mary Arnold, was also her godmother. Mary regularly spent holidays at the Forsters' house at Burley-in-Wharfedale. In later life she was to feel a strong affinity with Aunt Forster and admitted in 1894 that Jane constituted that part of Marcella that was not the author herself. All of which suggests some strong symbolical meaning in Mary Arnold's gleefully boiling down this particular doll into wax dumplings.

Elsewhere in her account, Bellasis recalls Mary's wild energy and mischief: 'out of doors she was never still, but was always running or jumping or playing'.[57] Indoors Clough's academy was evidently humane enough by Victorian standards (although it is recorded that refractory children—like Mary Arnold—were routinely locked in cupboards for punishment). But the overall atmosphere was Spartan. As Bellasis complained:

> We were never taken to amusements of any kind, such as a circus, for instance. [Miss Clough] seemed to think that amusements were unnecessary, that children must play because it was good exercise. We were sent out in the garden to play with hoops or at hide-and-seek, but always given to understand that it was a business, and that exercise was good for us, but otherwise a waste of time, which would far better have been spent in studying ... there was to be no coddling or self-indulgence of any sort whatever; she never indulged herself, and did not allow it in anybody in her house. There was no lying in bed, nor hot water in the mornings. We had to wash the winter through with lumps of ice in the basin, and we were never allowed to warm ourselves by the fire if we were cold. I had to practise on the piano in a cold room on bitter winter mornings by the light of a candle. The food too was of the plainest, and occasionally we protested ... Miss Clough told us we must not be fanciful about what we ate and drank.[58]

One need not be too sympathetic with these hardships. Unluckier children in Liverpool were begging in the streets, prostituting their bodies, or working ten hours in factories. But none the less, Miss Clough's school was surely oppressive to a boisterous tomboy like Mary, who must often have wished that she was Tom's boy, and with her parents in Ireland.

2

Schooldays: 1860–1867

AT nine-and-a-half, Mary Arnold was transferred to a more serious boarding establishment: the Rock Terrace School for Young Ladies at Shifnal, Shropshire. It was now an understood thing that she was not to return home for eight years. Understood, but not welcomed. In March 1861 she wrote to her mother, 'Oh! how I wish I was seventeen that I might go home and stay.'[1] Why she was thus exiled—unlike her three sisters—is mysterious. From hints in letters and elsewhere it seems that her 'temper' may still have been uncontrollable, particularly when she was close to her mother. Julia, Newman claimed, 'preached against Catholicism to her children, and made them unmanageable'.[2] The receptive Mary may have been the most unmanageable of all when at home.

The school at Shifnal was kept by a Miss Davies, whose titled sister was a friend of Tom Arnold's. Mary was taken in at this friend's charge—a charity girl. That Rock Terrace could be afforded was clearly a main if not the only consideration. The unhappy Arnold parents were still fruitfully multiplying in Dublin, and by 1860 had six children to support. All donations were gratefully received. Mary, however, was not grateful for Miss Davies's charity, then or later. She harboured her grudge thirty years and paid it out in the biting description of young Marcella Boyce's miseries at 'Miss Frederick's Cliff House School for Young Ladies'.

Mary's loneliness at Rock Terrace was virtually total. Her mother never wrote,[3] her father never visited. 'In contrast to every other girl in the school she had not a single "party frock".'[4] (The fat and Catholic German governess, Fräulein Gerecke, eventually made her one. Mary apparently reminded her of a dead sister.) By her own account, the discipline of the school transformed her 'for the time being, into a demon .... She hated her lessons, though, when she chose, she could do them in a hundredth part of the time taken by her companions; she hated getting up in the wintry dark, and her cold ablutions with some dozen others in the comfortless lavatory; she hated the meals in the long schoolroom .... The whole of her first year was one continual series of sulks, quarrels, and revolts.'[5]

She suffered 'perpetual colds',[6] toothaches, and headaches at this stage of her life. Miss Davies had no desire to pay doctors for a child who was not paying for her board, and Mary learned the art of stoic suffering. In November 1863 she reported to her father that 'My toothache nearly drove me mad, and

I plunged my head into a basin of cold water, and it really did some good for a time.'[7] For children who reported sick Miss Davies's treatment was routinely punitive:

> The rule of the house when any girl was ordered to bed with a cold was, in the first place, that she should not put her arms outside the bedclothes—for if you were allowed to read and amuse yourself in bed you might as well be up; that the housemaid should visit the patient in the early morning with a cup of senna-tea, and at long and regular intervals throughout the day with beef-tea and gruel; and that no one should come to see and talk with her.[8]

This was in direct imitation of the 'silent system', currently imposed on inmates of Victorian prisons—a system which, it was felt, stimulated 'conscience' and discouraged recidivism. Meanwhile, Mary's petted brother Willie was winning prizes for English literature at his civilized prep school and at 12 was coolly informing his mother that 'I can truly say with Pope that *Paradise Lost* has afforded me much pleasure.'[9] No gruel for him.

Mary Arnold's grinding sense of poverty and social humiliation among pampered fellow pupils of 'the tradesman class' is expressed in urgent letters to her father begging for tiny sums of money (i.e. the threepence she 'earned' by writing out lists for him). 'Do send me some more money,' she writes; 'it was so tantalizing this morning, a woman came to the door with twopenny baskets, so nice, and many of the other girls got them and I couldn't.'[10] In *Marcella* the visit of the cake-woman on Saturday is described as one of the schoolgirl heroine's 'only pleasures' (the other two being 'mad games of tig' and the kindness of 'fat Mademoiselle Rénier, Miss Frederick's partner',[11] i.e. Fräulein Gerecke). The tone and the devious psychology of the begging letter would recur thirty years later in her interminably wheedling requests for subs from her publisher George Smith. Mrs Humphry Ward was to be an author always asking for more money, more advances, just another few hundreds.

The secretarial services she performed for her father were not, of course, principally valued for the pittances they earned and that he was slow in disbursing. More important was the sense that she was needed by her father—that she could be his handmaiden. In June 1862 she asks her mother, 'will you tell me when you write how Papa has got on with his index now I am not there to help him'.[12] The fact was, he seems not to have much registered her presence or absence. Nor to have responded to the barrage of letters she sent him, showing off her French and German, her precociously fine essay-writing and her pietistic poetry, all interspersed with abject apologies for wrongdoing ('Dear Papa, I *do* think I did very wrong when I was at home . . . . I will try to be better').[13] Privately she kept a picture of the Madonna in her dormitory, and evidently fantasized conversion and the veil to please him. Deferring to Tom as her *real* teacher she would make humble requests for advice on reading: 'will you advise me on my choice of books when I return at Christmas [1863]?

I shall spend all my money on books.'[14] One curt response from Tom survives, in which he makes some frigid corrections to the scansion of her latest effort in verse and signs himself—with incredible priggishness—'T. Arnold'.[15]

In 1862, Tom made one of his rare visits and she wrote to tell him that 'I was very unhappy for some time after you left me but Miss Davies made me go into the schoolroom, so that cheered me up a little.'[16] Miss Davies seems here less than sympathetic and elsewhere positively sadistic. In May 1862, she was making the girl (always prone to colds and chills) take a cold bath every morning, to remedy her torpor (another treatment usefully borrowed from the British penal system). At this period, Mary Arnold began to take a morbid interest in death, finishing her letter of 24 May 1862 to her father with the news that 'Mr Slaney the member for Shrewsbury fell down in the exhibition and hurt his side and it brought on eresipellis [sic] and afterwards mortification from which he died. It seems most strange that Mr Slaney was but last week in very good health and now I can hear the bell toling [sic] for his funeral. Is it not sudden?—Goodbye dearest Papa.—And with best love to Mama and the Children—Believe me *ever* your affectionate child,—Mary.'[17] Mary was evidently meditating her own death—either by suicide or by martyrdom at the hands of the school disciplinarians of the aptly named Rock Terrace.

Mary Arnold remained three-and-a-half years at Shropshire and, although never happy, calmed down somewhat after her passionately rebellious first terms. Towards the end of her time at the school, she was strongly affected by the vicar of Shifnal, the Reverend Cunliffe. Inspired by his sermons she had her first significant religious experiences, accompanied by a romantic fixation on the blameless vicar and his wife, to whom she secretly wrote poems ('I love thee as the thirsty flourets turn to worship the heaven on high.').[18] She recalls the girlish crush dispassionately enough in *Marcella*, where the Cunliffes are portrayed as 'Mr and Mrs Ellerton'.

It was in her third year at Rock Terrace—when she was between 12 and 13—that Mary records the growth of a taste for 'reading of the adventurous and poetical kind'.[19] She pored over such hoary historical romances as Bulwer's *Rienzi*, Jane Porter's *The Scottish Chiefs*, and Scott's *Marmion*. As she recalls, she 'laboriously wrote a long poem on the death of Rienzi', which in fact survives in an early 1864 notebook. It opens with a fine clang:

> Hark to that deep sound! Hark.
> See yon heroic form
> And on his brow the mark
> Of one to rule the storm
> Far-famed Rienzi.[20]

In October 1864, aged 13, she completed her first surviving story, the eleven-chapter 'A Tale of the Moors'. A steamy Spanish romance, it borrows heavily from one of her favourite novels at the time, Grace Aguilar's *The Vale of Cedars*. The first sentence indicates not only Miss Mary Arnold's impeccable grammar but her trick—even at this juvenile stage—of opening her stories with a lush scenic description: 'It was a hot summer's day but the sun's burning ray could not penetrate through the thick foliage of a beautiful avenue of acacia trees under which the coolness was delicious after the hot and dusty high road.'[21] (Compare the famous opening of *Robert Elsmere*: 'It was a brilliant afternoon towards the end of May . . .'.)

As it continues, 'A Tale of the Moors' is a fascinating and strikingly precocious piece of work, enjoyable for such charming Daisy Ashfordisms as: 'It was midnight. The mosque clock had chimed the hour of twelve.' Baldly summarized, it is about a young heroine, Inez—'a girl beautiful to excess'—who lives in Granada in the thirteenth century. Inez supposes she is the daughter of an unpleasant Moor, Don Pietro. While on her casement 'under the silvery radiance of the majestic moon', she overhears conversation which reveals that her supposed father is a spy for the Castilian King. She upbraids him with a volley of thees and thys: 'My father, Oh! why did that hateful spy ever come near thee to stain thy patriotism and to alienate thy love from thy daughter!' But he repulses her attempt to save him. Subsequently denounced, he in turn denounces Inez to the Alcalde as in fact not his but the daughter of a Spanish grandee and an enemy Christian. Arrested for treason, Inez shames her 'craven' father at her trial, and protests to her judges, 'My lords, never till this moment did I know that I was Spanish.'

In the climax of 'A Tale of the Moors', Inez chooses martyrdom for her new faith and is told that in deference to her youth and virginity she will be allowed to meet her end in a nun's white veil. Her last night in the death cell is the occasion for some intense authorial piety in which punctuation rather gets swept away:

> Inez our heroine is occupied preparing for the execution on the morrow. She kneels in earnest prayer yes prayer to the God of the Christians for in the recesses of her dungeon she had found some scattered leaves of the Bible precious indeed were these to her and the God of the Christians became her father her saviour.

The next morning at dawn, Inez serenely awaits beheading on Granada's eastern ramparts when suddenly 'the wall is breached' by the forces of the King. 'St Iago of Castile, the King and Our Lady!' shouts a young cavalier 'impetuously leading his men on the enemy'. Inez falls into the caballero's arms, saved. Her false father dies, 'tended by the being he had so wronged'. The young cavalier, Don Vencino, recognizing noble birth in Inez (with whom, of course, he has fallen in love) carries her off to Cordoba, the capital of Spain. At court, she is reunited with her true father, the Marquis Loyola, who blesses his new-found daughter's marriage to the gallant Vencino.

'A Tale of the Moors' is a charming and revealing performance. And it helps answer the question why Mary Arnold (unlike any other Arnold girl) became a novelist; and why indeed she so persistently wrote novels from her thirteenth to her seventieth year. She wrote them even when, as it seemed for the first two decades, they would never be published; or if published never sell; or if sell never be admired by those judges whose opinion she most valued. Like Anthony Trollope, Mary's story-telling virus originated in childhood loneliness, playground wretchedness, and compensatory day-dreaming. As he recalled in his *Autobiography*,

> As a boy, even as a child, I was thrown much upon myself ... other boys would not play with me. I was therefore alone, and had to form my plays within myself. Play of some kind was necessary to me then, as it has always been. Study was not my bent, and I could not please myself by being all idle. Thus it came to pass that I was always going about with some castle in the air firmly built within my mind .... For weeks, for months, if I remember rightly, from year to year, I would carry on the same tale .... I myself was of course my own hero. Such is the necessity of castle-building .... There can, I imagine, hardly be a more dangerous mental practice; but I have often doubted whether, had it not been my practice, I should ever have written a novel.[22]

Mary Ward in *Marcella* offers a strikingly similar aetiology of her novel-writing addiction. It began in the silent incarceration of the school sick-room, where the heroine 'learnt, under dire stress of boredom, to amuse herself a good deal by developing a natural capacity for dreaming awake. Hour by hour she followed out an endless story of which she was always the heroine.'[23] For young Trollope, fiction was a defiant, self-preserving response to being sent to Coventry by his schoolfellows. For Mary, it was the self-preserving response to the sensory deprivation and punitive immobility of Miss Davies's 'sanatorium'. Brutal schools were for these two, as for others, the nurseries of Victorian fiction.

Mary Arnold's release from Rock Terrace finally came on 10 December 1864. It is the only diary entry that survives from this most wretched period of her life. She started from the school at just past noon. She was to travel alone, and called first at the vicarage, where kind Mrs Cunliffe was waiting for her 'in brown silk' and refreshed the little traveller with 'wine and biscuits'.[24] She then caught the train for the Christmas holiday for which she had been counting days since summer. She was going home. Or, as she must have felt, she was at last to be joined with her father.

Mary Ward was to claim in later life that she, of all the Arnold children, 'knew my father best'.[25] But during her childhood her connection with him was at best perfunctory. She rarely saw him and he rarely wrote to her directly. But Tom Arnold was, at the time, paternally close to his growing boys, in whose schooling and Catholic instruction he took a direct interest. Thus while Mary was penning 'A Tale of the Moors' the favoured William accompanied his father on a trip to Ireland in August 1864. According to *Marcella*'s account, Mary Arnold was visited at Rock Terrace by her father only once, after she had been there the best part of four years. In the novel, he immediately perceives that the school is no good and arranges to have his daughter removed. Tom was still alive at the time of the novel's publication (Miss Davies clearly wasn't) and *Marcella*'s version was a flattering fiction designed to save his feelings. Unlike Marcella's father, Tom Arnold actually visited Mary at school in 1862, when she was at her most miserable, and did absolutely nothing to rescue her beyond bidding her to 'be brave'.

If remote and self-engrossed, Tom was not a cruel man and there remains the mystery as to why Mary was not then or two years later in 1864 united with her family and why she continued to be educated apart from them. There were now six children, all the girls at home except her. She could surely at 14 have assisted her mother in managing the household, the usual subaltern chore for the eldest daughter. And some decently Protestant day school could easily have been found for her. The family were now no longer in alien Dublin. In January 1862, Tom had moved to Birmingham, to work directly under Newman as senior classics master at the small but select Oratory School, where Mary's luckier young brothers were enjoying a useful preparation for Rugby. (Mrs Mary Arnold's reaction to hearing that Willie had won a £5 prize in English literature was typically appalled: 'Oh! to think of *his* grandson ... being examined by Dr Newman!')[26] The Arnolds now had a little more money and a large house in ultra-respectable Edgbaston.

Tom doubtless felt some belated anxiety as to what was happening to his clever daughter. But she was still not to come home to stay, whether home meant Birmingham or Fox How. After a couple of months at Birmingham in the winter of 1864–5 she was sent instead to a smaller, 'much sought-after', and more expensive finishing school at Clifton (Solesby in *Marcella*), a watering place at the coast, near Bristol. There is no direct evidence, but it seems likely that the presence of Mary and the tug of war she provoked aggravated the always tense relationship between her parents. She recalls seeing Newman walking about the streets of Edgbaston and 'shrinking from him in dumb, childish resentment as from some one whom I understood to be the author of our family misfortunes'.[27]

Tom had spent some time with a cousin in Bristol in January 1864, during a protracted breakdown in his health following an attack of scarlet fever. While convalescing by himself he wrote a learned article on Bristol's churches, which was published in *Fraser's Magazine* (February 1865). With his usual engagingness, while surveying the city's ecclesiastical architecture, Tom had evidently managed to make some useful contacts which enabled him to place his daughter in a school which was—after the brutality of Rock Terrace—the saving of her. She was pathetically grateful. Her letters of early summer

1865 are a cascade of enthusiasm, dutifulness, and contrition for former waywardness. 'Mother dear,' she wrote shortly after arriving at Clifton, 'I could not bear to be here as I was at Rock Terrace.'[28] She now actually *enjoyed* her lessons: 'Our German class is such fun, only the worst of it is I generally get so excited that the poor declensions go completely out of my head .... I am quite tired of laughing.'[29]

Mary flowered intellectually under what was clearly not just a happier but a better educational regime in which modern languages, singing, drawing, and deportment were the main subjects. She was soon doing well in her exams and after a year was the 'undisputed head'[30] of the school. More significant, she had evidently made heroic efforts to correct her demon. On 20 June 1865 she wrote to her mother that her headmistress, 'darling Miss May ... did not let my fourteenth birthday [11 June] pass without some serious thoughts. You and Papa have indeed had little cause to bless the day I was given to you. My whole life short as it has been has been nothing but selfishness and conceit. He [i.e. God] can forgive, I know he can give us strength though we have none of ourselves. It will be very hard at first dearest Mother, but you will help me and bear with me, won't you?'[31]

Mary Arnold had meanwhile fallen headlong in love with Miss May. It was, as she herself later realized, a full-blown sexual passion. The adoration is recalled—with some retrospective irony—in *Marcella*. In that novel Miss May appears as 'Miss Pemberton': 'A tall slender woman, with brown, grey-besprinkled hair falling in light curls after the fashion of our grandmothers on either cheek, and braided into a classic knot behind—the face of a saint, an enthusiast—eyes overflowing with feeling above a thin firm mouth—the mouth of the obstinate saint, yet sweet also.'[32] At this point in her growing up, Mary's temper problems drastically subsided. Puberty played its part. By 14 Mary Arnold had the figure of a young woman, and was mistaken in the street for 17.[33] More-stimulating lessons were another civilizing factor. Mary concedes that she was 'well taught [at Miss May's] and developed quickly from the troublesome child into the young lady duly broken in to all social proprieties'.[34] Evangelical moral discipline was a third element in the 'breaking' process, mixed as it now was with a repressed charge of sexuality. Miss May was extravagantly pious, and Mary records treasuring the moralistic letters she sent her pupil during the holidays, letters which were reread until they fairly dissolved with the wear, tear, and tears spilled on them.

Mary Arnold's socialization was accelerated by another momentous event. Since his illness of 1864, Tom Arnold had been inexorably backsliding (did he ever slide in any other direction?) towards Anglicanism. The formal reason was his increasingly 'liberal' scepticism about miracles and resistance to papal authority. But for some time Tom had in his vague way been yearning towards an academic career which would be easier on his health than school-teaching.

In October 1864 Newman refused him a rise in salary for his work at the Oratory School, which Tom took very ill. His great work, *A Manual of English Literature* (first published in 1862) was under constant revision and needed time spent on it. He had other scholarly ventures in mind that required leisure and Bodley. A retired donnish life must have been irresistible.

In November 1864 Tom Arnold gave Newman six months' notice that he would resign his teaching post. In June 1865 he formally returned to the religion of his fathers. The ever-reliable Arnold network—led by Dean Stanley—rallied round the returned prodigal and eased his return to academic life. In the autumn of 1865, the family temporarily rented a house in St Giles, Oxford, where Tom could take in pupils. But, strangely, Mary was even now not brought back. Thus, in winter 1865, she wrote to Mrs Cunliffe (the vicar's wife who had been so kind to her at Rock Terrace) in the awkward terms of someone in but not of the Tom Arnold household: 'Do you know that we are now living at Oxford? My father takes pupils and has a history lectureship. We are happier there than we have ever been before, I think. My father revels in the libraries, and so do I *when I am at home*'.[35] Mary was actually to remain at Miss May's until summer 1867, her sixteenth birthday. Nevertheless, her first visit to Oxford in the summer holiday of 1865 remained with her as one of her most treasured memories. Her father brought her from the station in a hansom and as the two of them drove through the empty streets of the long vac. she looked with a 'thrill of excitement' at the colleges while he pointed out Balliol, where the 'arch-heretic' Jowett lived. [36]

Mary returned to Oxford again in December 1865, when she evidently heard her uncle Matt lecture as Professor of Poetry, but was too shy to introduce herself to him. In August 1866, during her second long holiday at 'home' in Oxford, Mary Arnold made another stab at writing a story. The resulting 'Lansdale Manor' was significantly different from the escapist romance of 'A Tale of the Moors'. Subtitled 'A Children's Story', the new work was designed as part of a larger fiction sequence, provisionally called 'Alford Rectory'.[37] The model now copied by Mary was Charlotte Yonge's saga of the May family, *The Daisy Chain* (1856), and its proliferating sequels. She was also clearly influenced by Elizabeth Sewell's pious saga of Anglicanism and useful spinsterhood *Laneton Parsonage* (1846).

There was now no solitary foreign heroine. Mary Arnold's new story centred on the solidly English Victorian domestic unit with its paterfamilias, invalid mother and complement of unmarried daughters and boisterous boys. Sewell and Yonge were—famously—spinsters who wrote from deep within the bosom of their families. (Yonge, for instance, was largely educated by her father—an Anglican churchman—who 'believed in higher education for women but deprecated any liberty for them.'[38]) 'Lansdale Manor' was never finished although, significantly, it was dedicated in manuscript to her 'dearest

Grandmamma'. As ever with Mary, the novel was conceived principally as an 'offering' to her elders. The portion of the story which reached written form concerns the domestication of a wild and book-loving heroine, Edith—transparently a self-portrait of Mary Arnold. The main episode in the narrative is Edith's being so distracted by reading books that she lets her young brother Percy fall over a cliff. He is rescued from the 'fatal precipice' by their father (an Anglican vicar) with the help of the 'the short iron-pronged Alpenstock which was his constant companion'. In a paroxysm of hysterical remorse, Edith confesses her guilt to her invalid mother, while upstairs Percy teeters between life and death. An avalanche of homily descends on the young culprit. While her mother reassures her that 'the Lord is *very* pitiful dearie' her father sternly outlines her 'duties to the home in which God has placed you. Learn to watch against the beginnings of evil my child, the neglect of some little duty, the temptation to excuse and overlook in yourself some act of self-indulgence which would appear to you blamable in another. Those are what you have to fight against and pray against, Edith.' The young invalid recovers, but 'the days of Percy's sickness and convalescence were the turning point in Edith Lansdale's life'.

Turning points were much on Mary's mind in August 1866. She was leaving the limbo of boarding school for permanent residence at home, at last. And it was, miraculously, a home without religious division. Rock-like stability is the overriding domestic feature of 'Lansdale Manor'. The ethos is Anglican and Devonshire rural (homage to Yonge—although Mary could not resist giving her manor a Westmorland name). Above all, the Lansdale household is settled. The taming of headstrong Edith, and the voluntary sacrifice of her intellectual ambitions for the good of the family has something of the pledge about it. This was the price that Mary was prepared to pay to rejoin her family.

As William Peterson has noted, Mary Arnold's diary entries that have survived from this 1866 period similarly reflect a severe self-negation, framed in the conventional rhetoric of evangelism. The following, for instance, was written on 14 January 1866, at the conclusion of her first Christmas holiday at home: 'Oh how bitterly bitterly have I failed this holiday with regard to my brothers. If in one thing the struggle is successful in another it is as weak as ever if not more so. I am writing in miserable pain.'[39] On another occasion Mary upbraids herself in her diary for a failure to extinguish self: 'These last three days I have not served Christ at all. It has been nothing but self from beginning to end.'[40] Her father's return to Protestantism had followed Mary Arnold's new lease of ostentatiously good behaviour, in early 1865. For a melodramatically inclined girl the temptation to see cause and effect was irresistible. Her reformed conduct was what had reunited the family.

The 67-year-old Mary Ward sums up her decade of school education bleakly in her autobiography: 'On the whole they were starved and rather unhappy years; through no one's fault. My parents were very poor and perpetually in movement. Everybody did the best he could.'[41] Even after sixty years, she sounds bitter and unconvinced by her own forgiveness. According to Janet Trevelyan, she 'often' recalled her hardships at Rock Terrace to her own luckier daughters, in later life.[42] Many Victorian novelists have their Warren's blacking warehouse or their Lowood, and this was evidently hers. Two features stand out from what we know of these years. The first is the quantity of sheer physical separation from her mother and father. Her relationship with them was scarcely more remote than if they had remained in the antipodes and she had been sent back alone (as often happened with the children of colonial families). Thus she found out about her father's reconversion to Anglicanism in June 1865 by reading about it in the newspapers. She was obliged to convey her hysterical joy at the miracle by letter to her mother:

> My precious Mother—I have indeed seen the paragraphs about Papa. The Lancasters [distant relatives] showed them to me on Saturday. . . . You can imagine the excitement I was in on Saturday night not knowing whether it was true or not. Miss May has seen it too. Your letter confirmed it this morning and Miss May seeing I suppose that I looked rather faint, sent me on a pretended errand for her note books to escape the breakfast table. My darling Mother how thankful you must be. One feels as if one could do nothing but thank Him.[43]

Clearly she felt closer to Him than to him. It seems incredible that Tom did not think, either before or after the event, to communicate directly with his eldest child, who was constrained to scour the gossip columns to discover what the future held for her.

Physical and psychic separation were two wounds Mary Arnold took from her schooling. And this separation explains much of the adhesiveness with which, as Mrs Humphry Ward, she in later life attached herself to powerful older men. Some of these patriarchal figures—such as Benjamin Jowett, Mandell Creighton, Gladstone, Henry James—were valuable and trustworthy guides. Her publisher, George Smith, was the staunchest ally an author could have, as she frequently told him. Others—like Mark Pattison—were more ambiguous in their influence. Still others—like Lord Cromer who persuaded her to start her Anti-Suffrage League—were old fools. And the old men of General Headquarters who persuaded her to write their war propaganda in 1916 were frankly evil. But good or bad, it was automatic with Mary to defer to any man who could assume a paternal role in his dealings with her. This was the cost she paid for never feeling that she had truly won her father's love. ('you do love me, Papa?')

This deference was—from an artistic point of view—a serious flaw. No writer has less of the *non serviam* in her make-up. Her habit of submitting her work to a pre-publication jury of senatorial males for approval explains the whiff of dutiful staleness that offends modern nostrils. Mary Ward was obedient to her elders; and this is not a quality much admired by posterity. Subservience fatally damages even her masterpiece, *Helbeck of Bannisdale*, a work which articulates her mixed feelings about Tom and the repercussions of his Catholicism on the women in his family. 'My first anxiety was to my father', she tells us. She 'confided' her idea for *Helbeck* to Lord Acton—not normally ranked as one of the great literary critics of his generation. This old friend 'cordially encouraged me to work it out'. Then, like some nervous schoolgirl, the 45-year-old novelist duly sought the assent of 'my Catholic father, without [which] I should never have written the book at all'.[44] Permitted to proceed, she wrote a version of the novel which she subsequently heavily toned down—'for your dear sake',[45] as she told Tom. Before publication, the proofs were sent to him in Ireland, for a *nihil obstat*. *Helbeck* thus finally reached print as the epitome of young daughterliness. She deferred to Tom as dutifully as she was later to defer to 'that very necessary person, the Censor'[46] (another old man) in her wartime writings.

Tom Arnold was such a charming fellow that none of his contemporaries seems to have been able to say a word against him. But he was as malign an influence on his family as any wife-beater, drunkard, adulterer, or gambler. Julia Arnold's life after 1856 was a torment: exile, poverty, social humiliation, sexual exploitation and cancer. Tom took a high-spirited colonial girl and turned her into a shrew. The eight Arnold children displayed a high rate of psychic casualty. William—the cleverest of boys—failed at Oxford and resolutely through life refused to accept the promotions and career opportunities offered him (not least by his eager sister Mary). Arthur was the 'black sheep', who became a cadger, a wastrel, was shunted into the army as a lowly trooper, and died—possibly by suicide masked as bravery—in the Basuto wars. Frank qualified as a doctor; but he was unable to hold any position, speculated disastrously, and needed constant loans from his family to keep going. Theodore gambled and was sent—by family subscription again—to the antipodes, where he lived obscurely and made bad marriages. Ethel, the daughter appointed her invalid mother's companion, was temperamentally restless; denied the vocations she wanted (the stage or authorship) she never married and herself declined early into chronic invalidism.

Another lifelong disability which Mrs Humphry Ward carried from her schooldays was poor education. Probably the best quality of day-to-day teaching Mary had was the earliest, at Miss Clough's little establishment, 1858–60. Anne Clough was—as the *DNB* sternly notes—deficient in organizational ability, and not well trained as a teacher. Her syllabus was eccentric (she had her 7-year-olds read Greek history, for instance, from a book loaned by her brother Arthur). But her school had for its time some unusual pluses. Although Miss Clough was personally a devout High Anglican she laid relatively little stress on the drill of religious instruction. The girls were obliged each to read a verse of the Bible aloud in the morning—nothing more. Clough was, by contrast, punctilious on matters of penmanship, giving class dictation twice a week—'always in pen and ink in exercise books'.[47] Apparently at 7 or 8 her charges could take down accurately passages from *Paradise Lost*. In other subjects, Clough would also give marks for neatness, irrespective of correctness. The extraordinary precocity and formal perfection of Mary's letters and exercises, from her earliest years onward, originated in her two years at Ambleside. (This it was that allowed her from the age of 13 to serve as Tom's secretary, when he would let her.) It also seems likely that Mary benefited from the individual attention which Clough lavished on her pupils. It was her practice to write separate timetables and curricula for each of them. But the greatest asset which Clough brought to the education of a girl like Mary Arnold was her lifelong conviction of the equal importance of education for girls.

In some important respects, Anne Clough's subsequent career in public life parallels her best-known pupil's. Born in 1820, she started her school in Ambleside in 1852. Exhaustion led to her giving it up, together with Lake District seclusion, in 1862. She thereafter became a leading promoter of higher education for women, participating in innumerable committees. In 1871, Millicent Fawcett and Henry Sidgwick persuaded Clough to take over the wardenship of a Cambridge residential house for women auditing lectures at the University. This led in 1875 to the establishment of Newnham Hall, which duly became Newnham College in 1879, with Anne Clough as its first principal. In the same year, Somerville Hall was opened at Oxford, with Mrs Humphry Ward as one of its moving spirits and secretary. Thus, the two careers which had crossed at a village primary school in 1858 recrossed two decades later in the opening up of Oxbridge to the female sex.

By contrast, Rock Terrace was a wholly disastrous school environment for Mary Arnold. Clough was—whatever else—an unusual woman to be running a school. She was not, like Miss Davies, primarily motivated by money (otherwise she would have taken more boarders—the major source of income for girls' schools). The run of ordinary professional women teachers before 1870 was dire. Miss Davies's curriculum was years out of date, even by the currently low standards of girls' education in the 1860s. History was still taught in the form of rote-learned 'facts' from *Mangnall's Questions*—a primer devised by Richmal Mangnall for the education of genteel young ladies in the early nineteenth century. Some scraps of this manual's sterile catechism survive in Mary's 1863 notebook:

> Arras long famous for its tapestry.
> Agincourt battle gained by Henry V.
> Amiens here was a peace in 1802 between England and France.[48]

Miss May, with her thin saint's lips, was kinder personally than the horrific

Miss Davies. But there is no evidence that Mary learned much of academic value at Clifton, beyond a certain correctness of deportment, foreign accent, keyboard fingering, and enunciation.

It was Mary Arnold's unfortunate destiny to make her entrances in life a fraction too early. She was at boarding school a decade before the reforms (inaugurated by her beloved uncle, William Forster, in 1870) ensured a minimally decent standard of education for girls. And she arrived at Oxford a decade-and-a-half before the admission of women students. (Her younger sister Julia benefited from both reforms: she attended Oxford High School for Girls and as a home student earned a first-class degree in English literature from Somerville College, Oxford). Surrounded all her adult life by the best-trained minds of her time, Mary was always bitter at the institutional neglect of her brain, simply because it was a female brain. As she saw it, her nine school years had been useless. Even fifty years later it rankled with Mary that her younger brother William, because of his superior sex, should have had a more systematic mental training than she:

> As far as intellectual training was concerned, my nine years from seven to seventeen were practically wasted. I learned nothing thoroughly or accurately, and the German, French and Latin which I soon discovered after my marriage to be essential to the kind of literary work I wanted to do, had all to be relearned before they could be of any real use to me; nor was it ever possible for me—who married at twenty—to get that firm hold on the structure and literary history of any language, ancient or modern, which my brother William, only fifteen months my junior, got from his six years at Rugby and his training there in Latin and Greek.[49]

But he of course was being *trained*. She was being *prepared* for marriage.

3

Oxford: 1867–1871

IN July 1867 Mary Arnold was finally united with her parents and seven siblings.[1] (Two of them—Willie and Theodore—would go back to boarding school in September). It was her first experience of permanent family residence. Tom had evidently had a subscription mounted for him by friends and relatives. The Arnolds now occupied a large, sprawling house in the Banbury Road which Tom had had built and called Laleham after the Doctor's house where he was born. Tom Arnold had come home. Mary had also, as she felt, come home—not just physically to her parents' mansion but spiritually. If Fox How was one pole of Arnold existence, this great university city was the other. She 'slipped into the Oxford life as a fish into water. I was sixteen, beginning to be conscious of all sorts of rising needs and ambitions, keenly alive to the spell of Oxford and to the good fortune which had brought me to live in her streets.'[2] Since she was now in the full storm of her religious phase, it was of course God rather than good fortune whom the 16-year-old girl thanked in her prayers.

No precise record remains of how she passed her seventeenth year, from summer 1867 to summer 1868. But much of it must have been taken up in helping her mother and learning how to manage a large household—skills which were not taught at Clifton and of which Julia was not in fact a notably qualified teacher. Mary remembered and gave a jaundiced picture of Julia's housekeeping in her description of Mrs Hooper in *Lady Connie* (1916): '[She] was the most wasteful of managers; servants came and went interminably; and while money oozed away, there was neither comfort nor luxury to show for it. As the girls grew up, they learnt to dread the sound of the front doorbell, which so often meant an angry tradesman.'[3] Not that Julia's task as 'manager' of Laleham was easy. There were eight resident family members—four of them juvenile and two still in the nursery— and a corresponding entourage of upper and lower servants. There was also a constant coming and going of Tom's tutorial students, some of whom boarded in the house while cramming for university entrance. As a demure, black-clothed, 16-year-old Mary Arnold was taken to tea with married dons and their wives. But it was not until the summer of 1868 that she in any sense had time to find her niche in the academic community.

Also in 1868 Tom began desultorily to take his daughter's higher education in hand. Mary was enrolled for music lessons with James Taylor, future organist at New College, and was soon expert enough to perform her favourite Beethoven in drawing rooms. Tom himself set her to copy out work for him, and gave her a reading list, culminating with a thorough study of Monstrelet's chronicles—presumably to develop her understanding of historical evidence. A jumble of notes survive from this assignment tailing off with the cryptic memorandum 'an incomprehensible paragraph. Mem. Ask Papa.'[4] No answer was forthcoming.

Tom, meanwhile, was embarked on his edition of Wyclif and was mastering Anglo-Saxon. What with these projects, university politics, and his students he had scant time for his daughter. He spent much time behind the closed door of his study (Julia all the while had an ear cocked for the tell-tale sound of Latin chants). Thirty years later in *Helbeck of Bannisdale* Mary Ward painted a bitter little picture of her father's defective sense of his educational responsibility to his sixteen-year-old daughter:

> As to women and their claims, he was old-fashioned and contemptuous; he would have been much embarrassed by a learned daughter. That she should copy and tidy for him, that she should sit curled up for hours with a book or a piece of work in a corner of his room, that she should bring him his pipe, and break in upon his work at the right moment with her peremptory 'Papa, come out!'—these things were delightful, nay necessary to him. But he had no dreams beyond, and he never thought of her, her education or her character, as a whole. It was not his way. Besides, girls took their chance. With a boy, of course, one plans ahead.[5]

Mary found encouragement for her intellectual aspirations in another quarter, namely the Pattisons. She paints a vivid Edenic vignette of her first encounter with this fascinating Oxford couple:

> It was in 1868 ... that I remember my first sight of a college garden lying cool and shaded between gray college walls, and on the grass a figure that held me fascinated—a lady in green brocade dress, with a belt and chatelaine of Russian silver, who was playing croquet, then a novelty in Oxford, and seemed to me, as I watched her, a perfect model of grace and vivacity. A man nearly thirty years older than herself, whom I knew to be her husband, was standing near her, and a handful of undergraduates made an amused and admiring court round the lady.[6]

The 'elderly' (i.e. 55-year-old), angular, red-haired man with piercing blue eyes was Mark Pattison, Rector of Lincoln and the most formidable scholar in Oxford. His 27-year-old wife was 'Mrs Pat'—Emily Francis Pattison—former art student and connoisseur. They had married in 1861. Between 1868 and 1872 this couple 'mattered more to me perhaps than anybody else',[7] Mary records. She was in and out of their lodgings in Lincoln quad all the time. And it was to them, above all, that she owed her subsequent education in life and letters.

Why they took up 'the shy and shapeless creature' who was 17-year-old Mary Arnold is not immediately clear. It is likely that initially they were amused by her mixture of native sharp intelligence and evangelical priggishness. They enjoyed twitting her with the 'speculative freedom' of their talk (Francis dabbled in Comtism and bohemianism; Mark—despite being for professional purposes a clergyman—was a closet agnostic). Miss Arnold's eager innocence confirmed the Pattisons in a delicious sense of their slightly decadent cosmopolitanism. She was a regular guest at their suppers on Sunday nights—gay, unceremonious meals in which the cigarette-smoking Francis would appear in cunningly adorned loose white or grey gowns. Mary would insist, it being *Sunday*, on wearing a high-necked, black woollen frock, defiantly decent. It was piquant sauce to the Pattisons' free-ranging conversations to have this little puritan hanging on their words, torn between hero-worship and moral revulsion at the godlessness of it all.

There were other motives at work. Unformed minds and innocence had in themselves little appeal to Pattison. He was wholly uninterested in his young male undergraduates. Since 1851, he had—as he said—'loathed'[8] teaching. But he had a notorious weakness for young unmarried women. His interest in them, as Mary later noted, evaporated once they had husbands. Another of Pattison's harem was the novelist Rhoda Broughton. (Twenty-seven years younger than Mark she paid him out by the malicious portrait of the arid tyrant 'Professor Forth' in *Belinda*, 1883). In 1879—as the climax of his sexual career—Pattison scandalized Oxford by forming a liaison with Meta Bradley, forty years his junior.[9] The relationship was condoned by his wife who had informed Pattison three years earlier that his conjugal embraces were distasteful to her.

Mary was no beauty. But she had a fine, intelligent face, if perhaps too 'angular',[10] in Mrs Pat's view. The nose was rather pronounced, giving her through life a slightly horsy look in profile. And her hair was a problem, having to be severely pulled back so as not to sprout wildly around her head. In a group, or in company with other young women, she did not immediately stand out, being short and having a tightly limited repertoire of gestures and facial expressions—principally a rather glazed, far-away look. But in conversation her mobile, dark eyes and habit of getting carried away with what she was saying suggested exciting undercurrents of passion. Young dons, who preferred a more doll-like womanliness, found Mary Arnold rather off-putting. Pattison, who had lived among women all his life (as a boy he had nine younger sisters), enjoyed her vivacity and evidently knew how to draw her out. His interest in her, although superficially paternal enough not to alarm her guardians, was sexual. Physically, Mary was an early developer and even at 16 had the form and physical allure of a grown woman.

Mary Arnold's brain was also developing rapidly at that stage of life when every year sees dramatic advances. By her eighteenth birthday, she was an intellectually sophisticated young woman—unrecognizable from the schoolgirl who had been 'undisputed head' of the little world of Clifton. One of the tantalizingly few surviving diary entries from the period indicates a brash confidence in her mental powers and signals the end of her evangelical phase in summer 1869:

> Theodore [her younger brother] and I went to the Radcliffe [Camera] where I had an hour's tough reading in Professor Huxley's article on the 'Scientific Aspect of Positivism'. I was interested but not much enlightened. One must read [Auguste] Comte to judge of his critics. Mr [T. H.] Huxley's arguments against Comte's 'Law of the three States' seemed to me by no means conclusive .... It is curious how entirely the infidel tone of these papers passes me by without causing me any discomfort or disquieting me in the least.[11]

After her marriage in 1872, Mary was to fall under the spell of other powerful Oxford ideologues: T. H. Green, Benjamin Jowett, J. R. Green and Walter Pater. But at this point, Pattison monopolized her. She leaves a picture of their evening tête-à-têtes in his room, sitting before the fire with only his cat for company.[12] On these occasions he would, as she said, open his mind—something he always found easier to do with women than men. It was a remarkable mind. In fact, there wasn't much else to him. 'I have really no history but a mental history', he asserts in his *Memoirs* which begin not with his birth in Yorkshire, but his matriculation at Oxford.[13] For him, as for Mary, Oxford was not just a place, it was a world. In his postgraduate youth he had been a Newmanite; but religion now held nothing for him and he saw the fifteen years of the Tractarian Movement as a dark age. Nor did the trappings of university power lure him, ever since he had been disappointed in his first bid for the Lincoln rectorship in 1851 by one turncoat Fellow's vote. (This trauma is recorded in grotesquely extended detail in the *Memoirs*.) The setback produced in Pattison 'a blank, dumb, despair'[14] which wholly altered the course of his later life. Since 1851 (the year of Mary's birth) he had 'lived wholly for study'. Over the years he had resolutely promoted a cult of modern, hard, Germanic research. This was the mission to which he dedicated his college against the High Church conventionalities of Christ Church and the vocationalism of Balliol—one of which he despised and the other of which he thought misguided. His views were not mellowed by his eventually winning the longed-for rectorship of his college in 1861.

Pattison did not—like Jowett—see Oxford as the training school for the best of England's young men. Oxford for him would be much the same if some Pied Piper came and took away everyone under 40. Nor did he see Oxford as the fortified citadel of Anglicanism, as did Henry Liddon, the highest of High Church clerical dons. The university was for Pattison one thing only: a centre of pure learning. For him the highest Oxford type was the 'student', whose scholarship was not 'polluted' or 'disfigured' (Pattison's words) by any promiscuous desire for publication. It was its own end—even if it never reached finished form—as happened with Pattison's own *magnum opus*.

The numbing effect of Pattison's fetish of scholarly rigour on those less driven than himself is attested to in anecdotes like the following by a terrorized would-be scholar: 'Pattison suggested that I should edit [John] Selden's *Table Talk*. The preparation was to be, first to get the contents practically by heart, then to read the whole printed literature of Selden's day, and of the generation before him. In twenty years he promised me that I should be prepared for the work. He put the thing before me in so unnattractive a way that I never did it or anything else worth doing.'[15] In the same intellectually fanatic spirit Pattison instructed the 18-year-old Mary: 'Get to the bottom of something .... Choose a subject, and know *everything* about it.'[16] In her dramatization of the scene in *Lady Connie* (1916), she recalled that as he said it 'His voice dropped. All that was slightly grotesque in his outer man, the broad flat head, the red hair, the sharp wedge-like chin, disappeared for [her] in the single impression of his eyes—pale blue, intensely melancholy, and most human.'[17]

In later life, Mary Ward was to doubt the wisdom of Pattison's advice. It would have been wiser for the still adolescent Mary to go for breadth rather than depth, essentially unformed and undisciplined as her early mind was. And Pattison's 'getting to the bottom' of subjects carried the terrible risk that life was too short—as it was to prove too short for his own finally unwritten work on Scaliger. But to Mary (unlike the disgusted Seldenite) Pattison's enthusiasm was irresistible. Moreover, he was practically useful as only the institutionally powerful can be. A curator of Bodley, Pattison prevailed on his friend Henry Coxe, the librarian, to give the girl what we would now call a stack pass, allowing her access to the innermost recesses of the place. It was an extraordinary and symbolic privilege. There were no women students at Oxford—would be none for ten years. And as feminists like Josephine Butler had earlier demonstrated by trying unsuccessfully to crash its barriers, Bodley was as much a male bastion as any college. Women—except on a few visiting days during the vacation—were a proscribed class.

Pattison had not apparently indicated what Mary should get to the bottom of. The 'something' she chose was early Spanish. It was, on the face of it, an odd choice of field. And she does not tell us why she picked on the Spain of El Cid and Alfonso. She had never been to the country; she did not speak Spanish; nor did she apparently know any Hispanists. One can, of course, conjecture. A plausible explanation is that like any canny student, she espied a gap, or vacant lot in scholarship. Spanish was virtually non-existent as a subject at Oxford in 1869. There was only one teacher of the language, appointed in 1858; and the school of modern languages was not established until 1903. Spanish studies as such were a perfect Oxonian blank. The twelfth-century *Poema del Cid* was unfamiliar even by title to most highly educated Victorians. 'It is to be feared', Mary Arnold wrote in her 1869 notebook, 'that the majority of English men and English women have no very clear ideas concerning either Spanish history or Spanish literature.'[18] Fear didn't come into it. Mary was overjoyed to find a field where she was entirely unchallenged by the trained Oxford mind.

One plausible motive for the Spanish subject, then, was academic opportunism. There was another cluster of emotional reasons which could have dictated Mary Arnold's choice. Her mother Julia was of Spanish descent, and physically—with her mane of dark hair, passionate temper and brilliant eyes—Mary was (as Tom had observed) 'the image of her mother'.[19] As a country, Spain was in 1868 the focus of considerable romantic interest. There had been a recent revolution and refugees. The intense feelings generated in the late 1860s by the *idea* of Spain are conveyed in George Eliot's melodramatic poem *The Spanish Gypsy* (1868). There was another—perhaps fanciful—factor connected with the topography of the Bodleian Library. Pattison's passe partout gave Mary the freedom of the lower floors, where was the 'Spanish room', with its shelves of seventeenth- and eighteenth-century volumes in sheepskin and vellum, with their turned-in edges and leathern strings. Here Mary might work 'absolutely alone, save for the visit of an occasional librarian from the upper floor, seeking a book'.[20] Mary chose this as her Woolfian room of her own. And what better to work on in the Spanish room than Spanish?

In fact, Mary's work on Spain's literature and history was not to be very fruitful in terms of scholarly output. It resulted in two pedestrian articles in *Macmillan's Magazine* over 1871–2 and a stillborn book project for Macmillan ten years later. But to be known to be doing research gave Mary an entirely new status in Oxford. It meant that Benjamin Jowett could introduce her at his open days as a 'very clever' young woman. It meant that when Hippolyte Taine came to Oxford in 1871, she could converse with him on intellectual topics rather than the food on their plates and what the weather was doing. The French sage duly noted how impressed he was with the pretty young bluestocking in his diary. Their common interest in things Spanish forged an immediate bond between George Eliot and Mary when they first met at the Pattisons for a Sunday supper in spring 1870. After dinner, the ladies withdrew and as they went upstairs to the drawing room, Eliot said 'The Rector tells me you have been reading a good deal about Spain. Would you care to hear about our Spanish journey?' There followed a twenty minute set piece, which Mary Ward reverently enshrined in her *Recollections*: 'As the low, clear voice flowed on in Mrs Pattison's drawing-room, I *saw* Saragossa, Granada, the Escorial, and that survival of the old Europe in the new, which one must go to Spain to find . . . . When it was done the effect was there—the effect she had meant to produce. I shut my eyes, and it all comes back—the darkened room, the long, pallid face, set in black lace, the evident wish to be kind to a young girl.'[21]

It is the last thing Pattison can have intended; but far from burying her under Bodley dust, Mary Arnold's Spanish studies 'finished' her in the sense of equipping the young lady to play a full adult part in Oxford society. No less than presentation to the monarch it allowed her to 'come out'. More than this, it got her talked about in highly useful ways. By 1871 J. R. Green had heard of her 'great fame for learning'[22] and the following year the distinguished historian E. A. Freeman invited the 21-year-old girl to write a book on Spain. She declined, giving her impending marriage as the reason.

4

Stabs at Fiction: 1867–1871

In her eagerness to be a scholar Mary Arnold had not entirely relinquished that earlier ambition to be a novelist. But fiction was something always associated in her mind with leisure and slack moments. Her first two years at Oxford had been too hectic and full for the luxury of day-dreaming on paper. But in the long vacation of 1869 she found herself again at a loose end. The Pattisons were evidently away. On Sunday 20 June, more in a spirit of doodling than anything, she began a diary (something else that she was normally too busy for). Restless boredom is the keynote. She has been left in charge of her 3-year-old sister Ethel, who mercifully is sleeping on the sofa, as Mary sits writing at the table:

> I feel a strong impulse to write to-night but there seems nothing to write about . . . . All is quiet. The sunset is pouring in through the windows lighting up a broad patch of wall on which if I turn I can see my own head and shoulders sharply shadowed. The sunset is not a hopeful one for tomorrow. There is too great a predominance of yellow light and dingy grey cloud. The radiance it throws over everything though intense is neither warm nor cheering. In the distant woods—grey misty and unreal—one misses the massive purple colouring which on a hot evening makes them so prominent a feature in the landscape, and so satisfying to the eye. It has been a wretched June.[1]

In such listless moods, Mary was a novel waiting to happen. And—setting aside the artificial fine-writing and preening of a passage like the above—one sees in it a descriptive energy bursting for expression.

There were other energies and drives preoccupying her. In the diary as it continues over subsequent days Mary sketchily records what was evidently a serious flirtation with 'Mr Price', apparently a young tutor assisting Tom. It was probably Mary's first serious affair of the heart. On 22 June, however, the relationship was clearly on the rocks. After a day 'frittered away in croquet and idleness' with her increasingly inattentive beau, Mary portentously notes, 'Some end must come to this idle self-indulgent life. I feel too ashamed of myself to go on with croquet at 11 in the morning.' 'Monstrelet and the fifteenth century' are being neglected, she chides, and goes on to belabour herself in grand terms: 'I must set to and get it up again else by the end of the vacation I shall have gone far in my own person towards justifying even a Saturday Reviewer's tirade' (a little pomposity of phrase she surely picked up from the Pattisons).

By 10 July it was over. 'Mr Price and I are tired of each other, that is very evident.' It is, she decides, 'a disappointing end', for which, typically—but not without insight—she blames herself: 'I may please for a day—a week—but never for longer. Perhaps it is because I am so anxious to please and so self-conscious.' The diary is now becoming scrappy, and Mary idly mentions yawning over *Sartor Resartus*, 'but Carlyle gets tiresome after a while'. It is, clearly, a dull life.

At this jaded and listless point, Mary's diary finally peters out, to be followed by the draft of a 10,000 word story, 'A Gay Life'. The tale covers an exciting summer (13 May to 4 October) on the Isle of Wight (which the author seems not yet to have visited). Its narrative uses a new and quite complicated method, being told in the form of a diary written by an unnamed invalid girl, wholly and selflessly devoted to her family and friends. 'To think for them and be useful to them in what little ways I can, this is my work.' The narrator's utility—it emerges—lies principally in uttering such consoling sentiments as: 'True Love, true strong love can it ever be given in vain? It may be slighted, scorned, *rejected* but it is not the less a good and blessing to the heart that offers it.' (Mary was evidently not prepared to write off the Mr Price episode as entirely a dead loss.)

Mary's narrative device in 'A Gay Life' derives from those in the best-selling *John Halifax, Gentleman* and *Guy Livingstone*. And wisely she keeps to a minimum the ponderous homily that had burdened 'Lansdale Manor'. This story opens with promising briskness ('So Edward and Lena are engaged!') and sustains a rattling pace thereafter. Lena Warenne, we learn, is the invalid narrator's particular friend: a beautiful but reckless girl. Edward Woolley, her fiancé, is a decent young squire; the possessor of £5,000 a year and Marsdon Hall. All looks fair. But the engagement is threatened by the arrival from Portsmouth of the handsome Sir Arthur Elmore. This dashing aristocrat has £12,000 a year and Glenthorpe Manor, beside which Edward's Marsdon Hall is a poor thing. During a violent June storm, Edward comes on Sir Arthur and Lena making love. He rushes off, and his lifeless body is found the next morning at the foot of the island's cliffs. The narrator is momentarily (but only momentarily) wordless with shock: 'God forgive her! God help her! Edward Woolley is *dead*!' Lena undergoes the statutory redemption by illness from which she emerges pale but pure. And—within two months—she is recovered sufficiently to marry a decent yeoman, Herbert Wentworth. Sir Arthur meanwhile has returned to the mainland to marry Lady Emily Pagnell, a belle of his own class.

Mary optimistically submitted this story to Charlotte Yonge's magazine, the *Monthly Packet*. It was rejected because—as Yonge rightly said—it was too saturated in sexuality, too 'novelish', for her essentially Sunday School readership: 'I do not go on the principle of no love at all, and letting nobody marry, but I do not think it will do to have it the whole subject and interest of the story.'[2] It is to Yonge's credit that she clearly read the story with some care and more to her credit that she was kind enough to be harsh in her just criticism.

The passionate and melodramatic 'A Gay Life' was followed in the diary notebook by a gloomier study in Scottish realism, called 'Believed Too Late'. It too begins with a blissfully happy engagement (evidently a subject much on Mary's mind in the summer of 1869). Maggie Phemister, an orphan brought up in the 'humble demesnes' of her Aunt Jeanie, has won the heart of a lad of her own class, Hamish Graham. She is 'the happiest girl in all Aberdeen.... There was a freshness, an unworldliness and whole heartedness about their love, rare in these hurrying calculating unromantic times.' This idyll is broken by the arrival of an older, socially sophisticated cousin, Robin Macey. Robin is 25 years old, a 'Byronic' man of the world and 'melancholy' in an Heir of Redclyffe way. Robin duly captivates his innocent cousin. Hamish breaks the engagement and returns Maggie's letters. Robin caddishly elopes with Miss Margaret Ballantyne, an heiress. Maggie pines, and is reunited with Hamish only on her deathbed.

A third novel from this summer 1869 period, 'Ailie', survives only in part.[3] One can reconstruct its main story as being set in Italy. The principal action is again a misunderstanding between two affianced lovers, Colonel Musborough and Ailie. After various trials (including heroics by him in the Indian Mutiny), the couple are finally wed. Writing in 1910, Ward claimed that 'Ailie' ('of no merit whatever') was written 'at the age of seventeen' (the word 'eighteen' is crossed out in her manuscript), i.e. 1868.[4] Since an immature-looking 'second volume' survives in a notebook but no full text it seems likely that 'Ailie' was rewritten on loose leaves in summer 1869. This 1869 version was then submitted as to Smith, Elder in August. It is not hard to deduce why she chose this publisher. In 1868 her Aunt Fan had given her a copy of Gaskell's life of Charlotte Brontë. But 'Ailie' was no *Jane Eyre* Smith thought, and he rejected it in September 1869. The young novelist accepted the rejection with studied fortitude: 'Dear Sirs—I beg to thank you for your courteous letter. "Ailie" is a juvenile production and I am not sorry you decline to publish it. Had it appeared in print I should probably have been ashamed of it by and by.'[5]

Mary Arnold had better fortune with *A Westmoreland Story*. She had been helped write and publish this tale by Felicia Skene, whose churchy influence lies like a lead weight on the narrative. An interesting woman, Skene was 48 in 1869 and the author of the mildly sensational *Hidden Depths* (1866), a piously enraged exposé of prostitution in Oxford. Like all her novels, it was written principally to raise money for her favourite charities. Skene had been strongly influenced by the Tractarians. In 1854, she had helped heroically organising nurses in the Oxford cholera epidemic, some of whom later went out to the Crimea with Florence Nightingale. Skene edited the *Churchman's Companion* from 1862 to 1880, and died, a spinster, in Oxford in 1899. Mary Ward—strangely—does not cite Skene in her gallery of Oxford friends in *A Writer's Recollections*. But in the years before her marriage Skene was a major influence on the young girl who devoured *Hidden Depths* and who—as she later recalled—'used to look at St Thomas's Church from the railway with reverence because *she* went there'.[6]

Mary wrote *A Westmoreland Story* 'when I was seventeen or eighteen'[7] (i.e. around June 1869) and it was published in Skene's magazine in 1870. It is notable principally for its lush descriptions of the countryside around Fox How. The story itself is excessively gloomy and uplifting. The beautiful but flighty heroine, Dorothy Morden, jilts a young farmer, Amyas Sternforth, for a clergyman, Mr Paton. As seems a routine hazard in Mary Arnold's fiction, Amyas follows Percy Lansdale and Edward Woolley by falling over a cliff. Dorothy atones by nursing cholera victims (as Skene herself had earlier done) and dies, consoled during her last hours on earth by Paton. The piece earned its author 'a few pounds'. She never reprinted it.

This batch of summer 1869 fiction is all promising but, in the way of early efforts, highly derivative. Stylistically it is a hotch-potch of Charlotte Yonge, Dinah Mulock, G. A. Lawrence, early Ouida, Elizabeth Sewell, and Felicia Skene. But as apprentice work the 1869 stories represented a very hopeful start in the novel-writing line. And Mary had at last got respectably into print at 19. Nevertheless, in later life she chose to see it as a dead end. According to her older self, writing in 1910, Mary 'came despondently to the conclusion that fiction was not for me'. Her next fourteen years were accordingly devoted 'entirely to history and criticism'.[8]

There was another distraction. Tom and Julia resolved in autumn 1869 that she should be married. Having spent her formative years away from home Mary was not by nature domestic or interested in household management. She had not, would never have, the Victorian matron's reverence for 'home' or aspire to be one of its angels. Her lack of such 'natural' womanly instincts was deeply worrying to her parents. On the eve of her marriage to Humphry Ward it was still preying on Tom's mind, who remarked to his mother that 'Mary will have to look to her housekeeping very closely [and] I have already warned her in the strongest words I could find, how absolutely it is her duty to postpone literature and everything else to the paramount duty of keeping a straight and unindebted household.'[9] Her younger sisters would very soon be available to help Julia manage Laleham, and would do it better. One of them (Ethel as it turned out) would be assigned to life-long spinsterdom as her aged parents' companion. This was clearly not a role Mary would ever fit.

It is also possible that the Arnold parents were alarmed by the influence that the Pattisons were having on Mary, particularly Mrs Pat whose looseness was proverbial. Mary may also have been an object of rather too much interest to the young students in the house. Whatever the motives, she was to go to market. To this end the Arnolds invested in a wardrobe for her. (Hitherto Mary had shared one pair of best shoes and one party frock with her young sister Julia, who was large for her age.) Her new clothes were well chosen, as a number of commentators attest—including Taine, who found her sober style a relief against the garish colours in fashion with young Oxford women. And the Arnolds now made their eldest daughter available at the dances, teas and parties where eligible young dons might view her. It was, as it happened, an auspicious time to be husband-hunting. Liberal reforms were in the air. The celibacy rule which bound Fellows was eroding fast (professors and heads of college like Pattison had always been free to marry—but there were only two score of these to go round and all were well gone in years). There was a whole new generation of college tutors on whom there were no marital restrictions.

Had she been shyer, or had she been in some way unmarriageable, Mary Arnold might well—out of sheer boredom—have continued her novel writing after October 1869. But she was pointed elsewhere. It is—regarded from the strict literary-critical point of view—a pity. In her offhand recollection of this fiction-writing interlude, Ward mentions another venture: 'a later and much more ambitious effort called "Vittoria", a novel of Oxford life, which was never finished'.[10] A scrap of this work survives. It is written in a mature handwriting which suggests a date some months (at least) after 1869 and possibly a full year. And it is a tantalizingly accomplished fragment, indicative of a major step forward in technique and conception. The fragment comprises a scene in which the heroine has arrived at the 'overgrown and neglected' house of her aunt in Streatham, south London. It is November. Some mysterious catastrophe lies behind her at Oxford, in the summer. The passage opens with a vivid vignette:

> She paid the cab, dismissed him, and stood a moment at the garden gate, looking so scared and white and bewildered, that the cabman as he drove off turned to look at her standing there under the dreary sky in the pouring rain, a small shrinking delicate creature . . . . Vittoria looked round her with a curious feeling, remembering the summer shower, the waiting, and that sudden flash of consciousness from heart to heart, the remembrance of which stirred the depths within her now as she stood holding her breath for the sound of footsteps within, the heavy November rain plashing on the steps and dripping from the dark ivy leaves.

An old Scottish servant, Martha, at last opens the door. Her aunt is away. In some desperation, Vittoria pleads to be taken in ('You must put me somewhere.'). The scene ends with her sitting quietly in the kitchen, 'resting in luxurious silence within the warm circle of the kitchen hearth'.[11]

The sober quality of the writing and dramatic scene is strikingly good and improved over the stridencies of the earlier 1869 stories ('God forgive her! God help her! Edward Woolley is *dead*!' etc.). In so far as this small taste gives a sense of the whole, 'Vittoria' promised to be a somewhat gloomy, but eminently publishable and worthwhile piece of fiction. Had Mary Arnold's new life not intervened with the Michaelmas Term of 1869, the world might well have had the author's first novel ten years earlier.

5

Marriage: 1870–1872

HUMPHRY WARD has a walk-on part in literary history as owner of the name that his wife, rather quaintly, took as her own. Even during his life gleeful jokes about his nonentity were current. (The best known goes roughly thus: an old friend meets Humphry in the street after many years in darkest Africa: 'You must come and dine with me—and do bring Mrs HW, if there *is* a Mrs HW.') Humphry is never given credit as the formative influence on his wife's intellect, which in fact he was, particularly during the early years of their marriage when she was still a girl-wife. Had Mary Arnold married one of the other eligible young dons in her circle—say Mandell Creighton or John Wordsworth—her later development would have been quite other, in tune with their very different ideals and ambitions. So too had she remained locked (like Meta Bradley) in Pattison's paralysing orbit. Mary Arnold was by most standards a 'catch'. She had an imposing pedigree, a fine mind, an exciting personality, and physical attraction (only the dowry was missing). There is no question but that she *chose* Humphry; he was not a suitor of last resort. What was it that made this young man appeal to her more than others?

Thomas Humphry Ward, usually called Humphry or sometimes T. Humphry Ward, was born in Hull in 1845. The Wards' background was in shipbuilding, before the steam revolution industrialized it. The family had money but was not ostentatiously wealthy. Humphry's father, Henry Ward, was a clergyman who had taken orders late in life at 30, after some years devoted to the life of a landed gentleman of leisure. (Shooting was to be Humphry's favourite sport.) Henry married Jane Sandwith, the daughter of Humphry Sandwith, a distinguished army physician who had worked in Hull in the 1840s. During Humphry's childhood his father had taken over a somewhat uncongenial working-class parish, St Barnabas, in King's Square, a slum area of London by Goswell Road. A stout, hard-drinking, convivial, rather stupid man, the Revd Ward was a parson of the old school, redeemed, as his brilliant young curate J. R. Green thought, by a good heart and a saintly wife, who had brought money and a Wesleyan high-mindedness to their marriage. Henry's main recorded vice was an occasionally reckless speculation in business. But the Wards were well-off and survived Henry's financial plunges and the expense of a family the size of a Scottish clan. There were seventeen children five of whom died in childhood before Mrs Ward herself died aged only 42. The children were clever, and at least one besides Humphry made a name for herself. Agnes Ward weathered consumption and went on to become principal of Maria Grey College in London.

The third son, Humphry was educated at Merchant Taylors' School from 1860. Having been coached by Green, he showed himself an apt pupil. He proceeded as a scholar to Brasenose College Oxford in 1864. There he took a second class in Mods. in 1866 and a first class in Greats in 1868. Evidently not anticipating an academic career he took the Indian Civil Service examination in 1866, which he passed but never took up. Brasenose was a sporting and rather hearty college, and Humphry was a popular undergraduate who played cricket, rowed, and was a first-class shot. He was tall, slim, had a fine, rather narrow face, a widow's peak hairline, and affected long, silky side-whiskers. His college history mentions him principally as the author of some jolly student verses on Brasenose Ale. He also had a happy knack with Latin and Greek verse, which stood him in good stead.

Although he was no scholarly high-flyer, Humphry won a coveted fellowship with extraordinarily little delay at Brasenose in January 1869, by the resignation of a clergyman who evidently wanted to marry. There were seventeen candidates and Humphry—the inside man—was obliged to pass a probationary year before being unanimously elected an actual Fellow in February 1870. His stipend was around £600, which he could bump up with the odd £50 by giving undergraduates tutorials in mathematics (which he hated) or by university extension lecturing to the general public (which he liked and did well). Brasenose was not in the early 1870s inclined to be liberal on the burning question of whether Fellows might marry; indeed it was among the most resistant of Oxford colleges as the reform gained ground elsewhere. Not that this much worried Humphry. He did not see his long-term future at Oxford. On assuming his fellowship he reminded the principal, Edward Cradock (who liked him), 'I believe you know that my intention is not to make Oxford my permanent home or to look forward to a purely academical career.'[1] What he did look forward to was a career in London higher journalism, where he already had useful contacts on the *Saturday Review*. Like that paper's, his politics were Conservative, but radical. (Mary Arnold's at this stage were her father's, stuffily Liberal.)

Humphry was not ordained—a rarity among Oxford Fellows but indicative of courage and advanced thinking. His idol was T. H. Green at Balliol, who had also declined to take orders. Humphry regarded the epic battles of the 'Oxford Movement' and all the 'Essays and Criticism' furore as so much dead-and-gone 'fuss' (T. H. Green's word). Not that he thought about them much; they were simply 'old formularies'—things of the past; as irrelevant to education as compulsory chapel. Humphry had no 'doctrine' to push. His thinking was secular and pragmatic and the preacher who most appealed to him was

Stopford Brooke, a clergyman friend of his and J. R. Green's, who straddled the line where Anglicanism dissolved into bland theistic Unitarianism. Nor did Humphry aspire to be a scholar in the stern Pattisonian sense. He had no 'field' and in his fifteen years at Oxford never published a line that could be called 'research'. Nor did he read much. His 1871 diary records only two books: Scott's *The Pirate* and Flaubert's *Salammbô* of which he noted saucily that it was 'perhaps not the kind of book to leave about in the Rectory parlour'.[2] Books and library scholarship were not his *métier*. He was primarily a teacher, a university lecturer who could lecture well, whether on Chaucer, Virgil, ancient Greece, or the Italian Renaissance. This was another rarity in Oxford, where public eloquence was normally saved up for virtuoso displays in the pulpit and where dons like John Wordsworth took a positive pride in being thought 'dull' and 'difficult' lecturers by their suffering undergraduate audiences. Students liked to listen to Humphry, as the grandees of the university (eminences such as Stubbs, Jowett, and Müller) liked to walk, talk, and have him take tea with them. Much cleverer dons of his own age, like Mandell Creighton, sought out Humphry's company. He was, in a word, 'lively'.

Humphry was one of the new breed of don interested primarily in reaching the undergraduate population; not just the élite first-class minds that Jowett lovingly nurtured at Balliol, but the average, decent young men who swelled the second-class lists before going out into the real world. Lively lectures were one way of doing this. Another was the 'tutorial', a weekly meeting in which the don (not necessarily a 'tutor') would build up a close, pastoral relationship with his student in an atmosphere which went far beyond traditional cramming or coaching. This teaching method—pioneered at Balliol under Jowett and T. H. Green—was gradually percolating through the university. The 'idea' of Oxford—its massive, crusty burden of romantic tradition—did not overawe Humphry Ward.

All this marked Humphry as one of the new liberal (with a small 'l') practitioners whom Matthew Arnold saluted in *Culture and Anarchy* (1869), formers of 'a frame of mind out of which the schemes of really fruitful reforms with time may grow'. Fruitful educational reforms were in the air and in 1870 Humphry could regard himself as in the wave of the future. Internal and external pressure for Oxford to liberalize itself had been building up for some time. Any fool could see what the basic problem was: the colleges were bloatedly rich and the university was starvingly poor (in Humphry's day the income of the one was around £366,000 and of the other £47,000 annually).[3] And the bulk of the colleges' massive endowed income went not into educating undergraduates or into promoting research but into feathering the college nest; that is supplying a luxurious life-style for the resident and non-resident Fellows.

It was a scandalous situation and the 'Oxford Act' of 1854 had instructed that the university professionalize its faculty. Two awkwardly opposed orders of don had duly emerged. These were the prize Fellow—who continued to enjoy his leisurely existence as of yore—and the college tutor.[4] The prize Fellow, after winning his post, could still spend his term-time sketching in London art galleries or promenading on the Riviera if he liked (and some did like). He was free as air—except that he could not marry and should preferably be a clergyman of the Church of England. The thinking behind the celibacy and ordination requirements was a throw-back to Oxford's primeval monastic origins, but not without a certain practicality. The majority of Fellows in the past had given up their collegiate berths on marrying and had gone on to 'livings' (often supplied by their college) or into the professions. This made for turnover. Tutors, by contrast, were always in residence and carried the grinding burden of undergraduate teaching. They had the great freedom that they might marry without losing their jobs but this was outweighed by the manifest inferiority of their condition. In some quarters, they were known as 'college servants'—teaching hacks. The tutor had no voice in college business and his suggestions might be rejected by the vote of non-resident Fellows, or Fellows taking no part in the work of the college. Not surprisingly, there had been no great rush to join their ranks: in 1870 there were no more than a dozen established college tutors. Humphry Ward—who by nature was certainly more of a tutor than a Fellow—would have been a fool to become the thirteenth. Instead he took the awkward joint appointment of 'Fellow and tutor'.

The systematic degradation of tutors as a class was clearly intolerable. More so with the innovation of the 1870 Education Act, which would in ten years push through a new kind of meritocratic undergraduate. Battle lines on 'Academical Reorganization' were drawn up. In 1871 Gladstone (who had represented Oxford University as his constituency from 1847 until 1865) authorized a new commission of reform under the Duke of Cleveland. Great things were expected of this body. In his after-dinner conversations with Max Müller Humphry Ward confidently looked forward to the wholesale reform of Oxford and even—within the decade—the disestablishment of the Church of England. It was a heady moment to be a progressive don.

Humphry was not by nature an original thinker and his mind had been formed by two strong personalities: Walter Pater and J. R. Green. Six years older than Humphry, Pater was a probationary Fellow at Brasenose in 1864 when the boy came up as a freshman scholar. Homosexual, Pater was notoriously fond of 'feminine looking youths',[5] as Pattison cattily noted. The 22-year-old Humphry fitted the bill, and in his final year Pater took a particular interest in him, taking him off for a month in the summer of 1867 to study *à deux* in the seaside town of Sidmouth, which—as Humphry later remembered—'made us rather intimate'.[6] They remained close, often walking and dining together. Humphry sported Helbronner silk neckties and decked his room with Japanese prints and reproductions of Raphael. After he married, Humphry chose a house in the same road as Pater, who was always his closest ally at Brasenose (when he remembered to vote) against the increasingly aggressive conservative faction, led by John Wordsworth, a young Fellow of stern Anglican principles who had been elected two years before Humphry and who loathed aesthetes. In the late 1860s and early 1870s Pater was at his most charismatic, articulating the doctrines that were to find expression in *The Renaissance* with its famous injunction that Oxford's young men burn 'with a hard, gem-like flame'. From Pater, Humphry imbibed a lifelong passion for art—particularly pictures of the Old Masters.

Humphry's other mentor, John Richard Green, was altogether more dynamic than the reclusive Pater. The prototype of Robert Elsmere (although Mary Ward always fiercely denied it), J. R. Green was a small, vivid man with black eyes that flashed when he was in the pulpit. After taking an undistinguished pass degree at Oxford (where he learned to despise the 'crawling dons'), he was a witness in 1860 to the public encounter between T. H. Huxley and Bishop Wilberforce of Oxford. Complacently confident of his power to 'smash Darwin', Wilberforce made an unholy ass of himself. The spectacle profoundly shook young Richard's confidence in the pillars of Anglican religion. Nevertheless, he was ordained deacon at Christmas 1860 and in the same month took up his first curacy at St Barnabas under Henry Ward—an old buffer whom he good-naturedly tolerated for the sake of Mrs Ward and the children whom he adored.

In 1862, Green determined to write a defiantly non-élitist non-Oxonian 'Short History of the English People'. At the same time, he was beginning to be tormented by 'doubts' about religion. He quietened them by throwing himself—at the hazard of his health—into social work among the teeming chaos of London's East End. In spring 1863 he took over a derelict parish in working-class Hoxton and in November 1865 was appointed perpetual curate at St Philip's, Stepney. The church, deserted when he arrived, was soon packed with a congregation of 800 largely proletarian worshippers. He performed heroically during the cholera epidemic of 1866, even comforting infected prostitutes—the lowest of the low. He hated what he called 'indiscriminate charity', middle-class 'relief', which like medieval doles simply perpetuated East End pauperism and the class divide. He wrote savagely on the subject in the *Saturday Review*, where he was a reckoned a star contributor. To Green's way of thinking, the only solution to England's social problems was for the privileged to *work* and *live* with the underprivileged in a spirit of practical fellowship.

All this time, Green was finding the Anglican creeds harder and harder to swallow. He hated the 'damnatory psalms' which Anglican orders of service required him to recite. His historian's conscience rebelled at biblical miracles.

Finally, in 1869, giving health as his reason, he resigned his curacy. He had contracted tuberculosis—an occupational hazard for East End clergymen. Having been given six months by his doctors he in fact lived fourteen more years. This borrowed time he spent largely abroad in sunny climes with his wife Alice (whom he married in 1877) writing, with undiminished mental energy as his body decayed, a stream of history books all of which—like his Short History—were best-sellers.

Green had been deeply affected by the death of Mrs Henry Ward in July 1862, and kept up a friendly connexion with the family. He was particularly attached to Humphry, the human being he had loved most, as he asserted at the end of his life. While the boy was at Merchant Taylors' School, he continued to coach him. As Humphry recalled, 'he shook down the edifice of Bibliolatry on which I had been brought up [and] the Bible became interesting'.[7] It was Green who convinced Humphry against ordination. And most valuably, Green instilled in his young disciple an impatience with what Gladstone called 'Oxford's Agony', its egocentric habit of projecting its dilemmas as things of cosmic importance. Green loved making 'cracks' against Oxford. Above all, it was the physical and mental *laziness* of Oxford which appalled him: 'Oxford is a most enjoyable place .... A charming place, but oh! so idle! Even I, the indolent one, am kindled to indignation at men beginning work at 10 and ending at 1, taking 6 months' holiday, and imagining they have no need for new reading after Baccalaurs.'[8] Green did not exaggerate. A contemporary manual for Oxford undergraduates advised that an 'early' start to the day should be made— 'work should never begin later than ten o'clock', this after 'the inevitable pipe, or cigar and the perusal and discussion of the daily paper'. This work should comprise 'three solid hours of good hard reading' followed by a frugal luncheon. The afternoon should be given over to physical exercise (ideally rowing) and the evening to dinner at six and 'amusement'. Reading after dinner, it was solemnly warned, 'is not only useless but harmful'.[9] After eight weeks of this toil, a long vacation was in order, with travel to broaden the mind.

Humphry was devoted to Green and wrote telling every little thing that happened to him, receiving sage counsel in return. But having won his fellowship he was determined to enjoy some of Oxford's delicious idleness before the back-breaking labour of his life began. First, he intended to travel and see the European treasure house about which Pater had enthused him. And since he did not intend to stay a Fellow all his life, he was in the market to get himself a wife. Indeed, he had at least one candidate in mind: Louise von Glehn. Twenty years old, she was the youngest daughter of a rich German merchant from the Baltic provinces, who had married a Scotswoman, had twelve children, and settled in Sydenham. Green, acting the part of mutual friend, had introduced Humphry to Louise, who was another of his favourites. She was beautiful, clever, cultivated, and rich. (Mary Arnold by contrast was striking, clever, cultivated, and penniless.) But if Louise was not to be his, Humphry would not despair. There were other fish in the sea, and his heart was very much on his sleeve.

Other Fellows were also on the prowl; his brilliant friend Mandell Creighton had similarly decided to marry. John Wordsworth had already made his choice, the daughter of Bodley's Librarian, Henry Coxe. Brasenose College, unlike Merton, would not countenance a married Fellow. The cautious young Wordsworth held off marrying until his own and his bride's families came through in 1870 with enough money to enable him to continue his research in biblical philology without the distraction of having to teach undergraduates (a total waste of time as Wordsworth thought, who already had his eyes on a future bishopric). He been ordained in 1869 and in 1871 Brasenose appointed him college Chaplain, which also neatly solved his housing problems. (A family house in Oxford in 1870 cost around £2,500.) The Revd Wordsworth's main duty would be to make undergraduates' lives miserable by enforcing daily chapel attendance.

1870 was, as Humphry recorded in his diary, a year of mixed gaiety and sadness. It began with the triumph of his election as Brasenose's newest and youngest Fellow. The absence of a colleague in the Hilary Term meant extra lecturing, which was a chore but pleasantly lucrative. For the first time, Humphry had spare cash in his pocket and he splurged on a three-week trip to Paris at Easter. He found the picture galleries, the Bois, the theatres, the churches, the excursions 'very delightful'.[10] It was his favourite word this spring, and much in use with the loveliest part of Oxford's year coming up: 'Never was such a delightful summer term' he noted in his memorandum of 1870; 'The joys of cricket and drags. Even the milder pleasures of the punt on the Cherwell passed into the riper charms of the lawn and the luncheon pleasant party.'[11] The idleness of it all was enlivened by the presence of some interesting refugees from the Franco-Prussian hostilities: 'there was little to do save to read, and bathe and fish, or chat and play billiards with the poor Vicomte. Oh! but there was Marguerite! She seemed as if she were only half a Bretonne—the south had dashed her with "sunburnt mirth". To see her make love to eight people at once was wonderful.'[12]

The war was alarming. Like many Englishmen, Humphry foresaw the total destruction of La Belle France. In order to savour the country before it was crushed for ever under the Prussian jackboot he went again in early summer. Paris was now '*triste* and full of men in uniforms'.[13] In the solemn silence of Chartres Cathedral he heard the roar of the crowd outside greeting the news of a telegram announcing the massacre of 40,000 Prussians. The telegram was a forgery. He returned to England in June to discover that his uncle Humphry Sandwith had already gone to the Front as a surgeon. It was a generally unhappy summer that Humphry spent at his family's summer home in Wales. He fell sick with a mysterious lowering illness. He feared it might be consumption, which was making dreadful inroads into the Ward family. His favourite sister Agnes had contracted the disease and his brother Edward was dying from it. Poachers had raided the coverts, so he could not distract himself blasting the September skies, as he loved to do. He consoled himself with some desultory flirting with a Miss Dunlop, who was spending the holiday with them.

Michaelmas Term 1870 was wretched, positively the unhappiest period of Humphry's adult life. On 4 October Edward finally died. And on 13 November, there followed the sudden death of the friend whom Humphry mysteriously refers to as 'RR' in his diary. At this period it was common for university men to cultivate intense relationships. At Merton, for instance, Creighton had with three others formed a 'quadrilateral', a 'mystic circle' whose members exchanged rings and pledged lifelong fraternity. Posterity perceives such relationships as potentially homosexual and it is likely that in more cases than we shall ever know they found physical expression. Sodomy was as prominent (if unofficial) an element of the public school curriculum as Latin and few men came up to university without at least a theoretical knowledge. And Pater's cult of Greek love added a gloss of glamour to homo-eroticism for those who followed his doctrines.

Humphry was devastated by 'the awful, unrealisable blow' of RR's death which affected him much more than the loss of his brother. It was the suddenness which was most horrifying. 'Never again will any type like his be revealed to me,'[14] he gloomily asserted. He read over his friend's notebooks incessantly, marvelling at the intelligence, now extinguished. Pattison wrote to condole, using the words that Mary Ward later put into the mouth of Wendover in *Robert Elsmere*: 'what knowledge has perished with him. How vain seems all toil to acquire.'[15] Humphry soothed his grief by composing an inscription for RR's gravestone at Littlemore, which he faithfully visited time and again over the next year.

Over the Christmas 1870 vacation he went abroad again, this time to Italy and with his consumptive sister Agnes. On 3 December they started for the south from Liverpool, 'all sick in body or mind or both'.[16] In fact, things went swimmingly. Italy was recuperative and he bounced back. Agnes also improved (she was later to recover altogether in Switzerland) and gave him good advice on how to proceed with his future wooing. With a woman's practical sense about such things she helped him buy a store of ornamental lace as bait to show off to any ladies who interested him. (This was a period when women of the middle class were avid for small scraps of fine material with which to 'trim' hats and dresses.) Most tonic of all, Humphry met an unnamed young lady on the trip with whom he flirted.[17]

At Oxford in January, everything was 'as usual'. Humphry skated on the flooded meadows and enrolled for single-stick lessons. ('England expects!'[18] he wrote sternly in his diary; there was genuine fear of a possible invasion by the all-conquering Prussians.) He also got in some shooting at Stanton. More to the point, he renewed his acquaintance with the family of Sir Benjamin Brodie, Professor of Chemistry, and Lady Brodie, who had a soft spot for him. With three marriageable daughters in whom Humphry was not interested ('dull', he thought) and a house the size of a small college (it was eventually to become St Hilda's), the Brodies would provide Humphry with the entrée to parties, dances, and functions where the nubile young ladies of Oxford might be found. And he appointed as his confidential adviser in affairs of the heart Reginald Copleston, an old Merchant Taylors' man and currently Fellow and tutor at St John's (later he would be Humphry's best man and still later Bishop of Colombo). These two would have routinely long discussions, far into the night in Humphry's Brasenose rooms, as to what his romantic strategies should be.

Julia Arnold had quickly perceived that young Ward was on the look-out for a wife and was not backward in pushing her Mary forward. After skating on Worcester Pond on Monday 30 January Humphry was invited to tea at Laleham. 'Much Beethoven',[19] he noted dryly in his diary. The same events repeated themselves next day, and he brought his lace for the ladies to admire. They were all too full of levity for Beethoven, so Mary showed her paces with Mozart and Schubert, which—like levity—was more to Humphry's taste. At the end of the week he was showing off his lace to a Miss Thornfield. On Wednesday following he met Mary at the violinist Henry Blagrove's evening concert. She delivered herself of the rather mystifying opinion that the first movement of the Beethoven quartet they had just heard was like 'a country bathed in heat'.[20] An arrangement was made to take her the following afternoon to Ruskin's Slade Lecture. For form's sake, Miss Thornfield was to make three. But Ruskin in the event did not show up, and the trio trooped off instead to the Museum and then on for tea at Mrs Pattison's and an '*Exposition internationelle des Dentelles*'.[21] Whatever Humphry had paid for his lace had certainly been repaid in female attention.

Humphry could at this point have proposed to Mary and had his manly interview with Tom Arnold. But he was not inclined to. It was all going too fast. He was not sure he liked Miss Arnold's Beethoven intensities. That urgent desire to please which she chided in herself in her diary a year before *vis-à-vis* Mr Price was another deterrent. Julia Arnold was too obviously throwing her daughter in his path and anyway there was Louise von Glehn, who suddenly announced that she was coming to Oxford on Thursday, 9 February to stay with the Brodies and hear Ruskin's next lecture. She would stay three weeks or so, which would give Humphry all the time he needed to close with her.

When he met Mary on the Wednesday at the Bonamy Prices' house-warming, Humphry took little notice of her.

At Ruskin's afternoon lecture Mandell Creighton observed Humphry talking to Louise, whom he had never seen before. He was particularly struck by a vivid yellow scarf she was wearing—it was his favourite colour. He rushed over and asked 'who is that girl who has the courage to wear yellow?' Humphry arranged a lunch for the three of them. He then contrived to get Louise to himself for 'an entrancing hour'. They walked down the river and round the meadows, discussing Oxford, admiring the water, and disputing 'whether there be such a colour as brown in natural landscape, and talking of home-folk'.[22] That night Mandell Creighton gave a dinner at Merton, attended by—among others—the three Arnolds, Tom, Julia, and Mary. The following day, it was tea for Humphry again at Laleham, by Julia's peremptory invitation. But he was clearly more interested in Louise, whom he took for a tour of Oxford on Monday the thirteenth.

Eager lover that he was, Humphry had organized a Valentine's Day lunch at Brasenose. Among the fifteen guests, there were no less than six marriageable young ladies in attendance, including Louise von Glehn and Mary Arnold. The gallant host had made up 'a camellia for each lady, and an appropriate Valentine with it'.[23] On the college tour that followed Creighton was in good form, and charmed the young ladies. This despite a considerably less attractive external appearance than Humphry's. 'Max'—as he was called by his intimates—had sandy hair thinning on his head and straggly round his chin, gold-rimmed glasses and was physically awkward. But he was reckoned the cleverest young man in Oxford and was possessed of a self-confidence that verged on arrogance. His interest in Louise was very ominous for Humphry. She had already been to Mandell's sumptuous rooms at Merton and had been charmed by his collections of photographs and blue willow-pattern china.[24] He was edging ahead of his Brasenose rival.

Julia meanwhile had clearly seen the signs at the Valentine lunch and warned Mary that if she did not put herself out she would lose Humphry to her new rival Louise. In her desperate attempt to be vivacious Mary now came over as off-puttingly strident. At the Chambers's dance on 15 February Humphry noted that 'M.A. shows her gayer side, being arrayed (externally) in *black*.'[25] But he ostentiously gave his attention to another unnamed 'delightful maiden'. At the Brodies' dance the following evening (a Thursday) Louise was 'gayer than ever'. Mary Arnold, however, was becoming something of an embarrassment: 'M. A. by reason of being overtired takes up certain serious points and rather shocks one by being shocked. If poor Comte had foreseen how even the elect would shudder at his name—he would have changed it!'[26] Clearly some light remark had been made about the French sociologist's proposition that worship of Christ be substituted by the worship of woman. Mary's sophistication had slipped showing a gauche evangelical streak beneath. Things looked bad for her.

It had been a heavy week, and they were all overtired. Humphry felt his Friday's lecture on Plato a burden. And in the afternoon there was a musical tea at the Bonamy Prices', where Mary hurled herself into performance on the piano, playing a sonata of Mozart's with 'a tremendous "alla Turca" as finale'.[27] In response to the compliments that followed she demanded of the company with her maladroit bluestocking earnestness 'is not Mozart Beethoven's inferior in *form*?' At this point Humphry was more interested in Louise's superior form. He contrived to lunch with her over the next three days; but he could see that he was all the time losing ground to Creighton, and suspected that an opportunistic proposal was in the offing. With 'all's fair' ruthlessness at his Merton dinner the following Monday, Mandell placed Humphry well away from Louise to languish between the motherly Lady Brodie and Pater's queer sister Clara. He was, as he bitterly noted, 'much edified'.[28] When they adjourned to the river afterwards Humphry—in torment—'struggled to be disinterested' watching Mandell's dashing courtship of the woman they both loved. The following week, Lady Brodie confirmed his worst fears. Copleston came to cheer him most nights, 'joining laughter and tears with mine'.[29]

The social life of Oxford was gathering pace as the academic year entered its last phase. It was now Torpids Week, and on the University College barge on 4 March, Humphry could see 'there is no mistake about it'. Mandell had won Louise's heart. He walked with Copleston to RR's grave at Littlemore, wrote long, lachrymose letters to Agnes and Green, and inscribed slightly spiteful pen-portraits of Louise and Mandell in his diary. On Friday 10 March, Louise left for London. Humphry was informed the engagement would soon be announced. He commemorated the end of his hopes in his diary: '*Finis!* That terrible 2.2 [p.m.] train carries off L. to London and home.' He went for a three-hour walk. The weather was fine, but 'it seems somehow as if this day were typical of the visit that is just over; bright, keen, stimulating, genial and yet (as tomorrow will prove) not *spring* after all'.[30]

It was very hard. Who would have thought that Creighton—who had not even met Louise until a month ago—had it in him to sweep the prize out of Humphry's grasp, who had known her ever so long? He consoled himself by writing a long pre-Raphaelitish prose poem, 'Spring's Heralds', which he read as it progressed to the long-suffering Copleston. He alone could understand the inner meaning of such gloomy utterances as 'disappointment is the normal atmosphere of that month of March through which life passes'. But having written it, Humphry thought the thing rather good, so he sent it off to his friend George Grove, editor of *Macmillan's Magazine*, who duly published it in April 1871 (by 'W').

Robustness was one of Humphry's most endearing qualities. Nothing kept him down for long. Spring gave way to early summer. The weather was warm and 'the croquet season begins in earnest!'[31] Mary Arnold—now an accomplished player—came back into the picture with Louise's departure on the awful 2.2. The weather remained unusually fine, which meant picnics by the river. As the Trinity Term began Humphry was seeing Mary one way or another every two or three days. But there was '*pas grande chose*'[32] between them. Laleham was a madhouse. For much of May, W. E. Forster was there—now a great man following the triumph of his Education Act. Oxford fêted the Cabinet Minister.

Forster's wife Jane and their adopted children (William Delafield Arnold's offspring) were also at Laleham and it was Mary's chore to act as childminder. The tacit understanding between her and Humphry was that things would reach their natural climax during the ball season, which began towards the end of the month. Miss Arnold was looking particularly demure as Taine noted when he came to lecture at Oxford on 26 May:' "A very clever girl", said Professor Jowett, as he was taking me towards her. She is about twenty, very nice-looking and dressed with taste (rather a rare thing here: I saw one lady imprisoned in a most curious sort of pink silk sheath) .... All her mornings she spends at the Bodleian Library—a most intellectual lady, but yet a simple, charming girl.'[33]

On Saturday 27 May, Humphry went up to meet J. R. Green at the Revd Stopford Brooke's in London, where he was spending a summer visit. They had a long conversation about 'heart phenomena' and the young man confided his plans. He was not encouraged (the financial side of things did not look hopeful) but he persisted. He could not see his future without Mary Arnold. A couple of days later he wrote to his father. On her part, Mary introduced her beau to the Arnolds' oldest family friend, Dean Stanley, who had a 'good talk' with him—probing him on his religious soundness. Henry Ward arrived in Oxford and dined with the Arnolds on 10 June. But—perversely—Humphry now began to get a sudden attack of cold feet. There were other fascinating ladies in Oxford—'the delightful Miss Elliott' and *la belle Américaine*—and he could not help paying them attention, sometimes when Mary Arnold was present. It was a childish demonstration of independence in the face of the impending marriage yoke; but it hurt her. On Monday 12 June Humphry broke into agonized French in his diary: '*je commence à m'en douter ... que je suis entêté!*' All the while in their midnight consultations Copleston was desperately urging his man to pop the question. But he couldn't. At the Exeter College concert on the thirteenth, Humphry sat next to Mary; 'we neither listen nor do anything. My mentor riles me afterwards.' The next day was Commemoration after which there was a fête in Worcester gardens. Julia and Mary were present and the young couple had a long walk and 'talk of intimate things'.

That night was the great Commemoration Ball. Humphry had understood that Mary Arnold would not be coming, her parents simply not being able to afford the expense. But 'to my great surprise she is there with the Reynoldses. We have more talk than dancing. Yet it is 4 before we leave.'[34] In Mary's mind they were now as good as engaged—having spent the best part of a night together sharing heart-secrets. And the mystery of how she had contrived after all to be at the ball was explained much later in her memoir of Felicia Skene, published in 1902:

> One Commemoration, for a special occasion, this girl [i.e. Mary Arnold] wanted to go to a particular ball. But she was already going to one ball and she knew she could not ask her parents to pay for a second. Miss Skene found this out and, I suppose, other matters too, for the girl confided in her a good deal. At any rate, dear fairy Godmother! the morning before the ball, the ticket arrived from her. The recipient hardly liked to take it but Miss Skene insisted, and took the liveliest interest in the ball dress.[35]

If—after thirty years—Mary remembered correctly, the cunning thing must have known that afternoon in Worcester gardens that she would be at the ball, but did not tell Humphry for tactical reasons, thinking that surprise might jolt him into declaration.

The next day was filthy weather, cancelling all luncheon parties. Everyone was exhausted. That evening at a musical evening at the Taylors', Humphry and Mary listened apathetically to Brahms. He walked her back, alone, through the rain to Laleham, which was quite nearby. They stood at the door; she lingered—*and he said nothing*. That night Copleston raged and Humphry had a 'mockery of a night's rest'.[36] Not even Humphry could now hold back. Friday, 16 June's entry in his diary was a triumphant *consummatum est*: 'Ah Happy day! Need more be said? This morning, in a whirl of excitement, I went up to Laleham on some small excuse, to amend yesterday's farewell. In a moment all was said: Mary was mine, and I was happy .... What a change! What worries broken down! What mists dispelled!'[37]

That same afternoon he saw Edward Cradock, the Principal of Brasenose. The outcome was bleak. Humphry could certainly marry. But he would have to revert to being a tutor and give up his fellowship. In the glow of the morning's events it probably seemed a small price to pay for Mary Arnold. In time it would weigh heavily on both of them. There was only time for a walk with his betrothed in the Parks during which he placed the ring on her finger. The next day Humphry was off examining in Lincolnshire. On the way north he stopped in London, where he had 'an immensely long talk'[38] with J. R. Green and his father drank his health in 1830 port. On his way back to Oxford he stopped off at Rugby, where Mary and Julia were visiting Willie, who was now head of school. 'And the journey home—and the evening ... may be imagined,'[39] Humphry gloated in his diary. Back in Oxford, Mary played Beethoven in Humphry's rooms and the next day he gave her a first Latin lesson—a course of instruction he was to continue on and off for the next fifty years.

On Wednesday 28 June they went to see Creighton, who in his capacity of clergyman blessed them. On the Thursday they were photographed by the eccentric maths don, Charles Dodgson—better known to posterity as Lewis Carroll. The triumphal progression continued with a visit to the Matthew Arnolds at Harrow and a climactic journey to Fox How in the Lake District, where Mary was 'quite *tête montée*'.[40] On 16 July the couple had a four-hour clamber up Loughrigg after dinner (taken at the old-fashioned time of three o'clock). It was, Humphry thought, 'a holy day for us—she is never so delicious as when she is showing me her treasures—and Loughrigg is one of the chiefest of them'.

The lovers had for ten days been writing a jointly composed essay, 'A Morning in the Bodleian', which Humphry took to Windermere for private printing. The pamphlet—a lyrical celebration of the library which had meant so much to her over the past three years—was dedicated to Mary's grandmother, Mary, 'from her eldest grand-daughter and her youngest grandson'.[41] It was printed as by 'Two Fellows'.

After Fox How, it was back again to Oxford, where Humphry had to prepare a course of university extension lectures to be given next term in Tunbridge Wells and Croydon. Mary helped him 'immensely' and the lectures were a success. Humphry meanwhile enquired whether his college would arrange a mortgage for him (it's not clear that they did) and bought a house at 5 Bradmore Road for around £2,000. It was one of the new developments springing up north of the Parks. A large house by modern standards, and Gothic in style, it stood in its own ground with a rather squat double turreted outline. The date for the marriage was fixed for 6 April 1872, two months short of the bride's twenty-first birthday. Dean Stanley officiated, and Mary Arnold became Mrs Humphry Ward.

6

Marriage and Oxford: 1872–1878

THE first question that the newly-wed Wards had to confront was whether to stay on in Oxford. By marrying, Humphry had thrown away his fellowship and was now demoted to the rank of college tutor. Partly this was the consequence of Brasenose intransigence. But it was also an unflattering measure of Humphry's worth in the academic market. His friend Mandell Creighton had found himself in a similar dilemma on becoming engaged at exactly the same period. Mandell wrote to his Louise that he was 'distracted by advice from every side what to do; several of the Fellows beseech me not to leave the college .... Others of my friends advise me at all events to keep my Tutorship, even if I vacate my Fellowship by marriage, and stay at Oxford taking pupils'.[1] He was also strongly tempted by the offer of a remunerative teaching job at Harrow School. He was not tempted by the prospect of being a 'college servant'. Creighton's problem was solved when Merton's governing body voted to allow him to become a married tutorial Fellow at a salary of £720 plus examining fees. A brilliant academic career followed, crowned by the Dixie Professorship at Cambridge and a top bishopric in 1897. Similarly, when T. H. Green married Charlotte Symonds in July 1871, Balliol promptly re-elected him to his fellowship. He was, after all, the most brilliant philosopher in Oxford. And had Humphry Ward been a T. H. Green or Mandell Creighton, doubtless a few stops would have been pulled out for him too. But he was not quite in their class. He was good but not so good that rules had to be bent for him.

There were other ways of harmonizing marriage and a productive university career. Humphry's exact contemporary at Brasenose, the strait-laced and High Church John Wordsworth, also engaged himself to be married in May 1869. But he declined to take the final step until summer 1870, when his bride's father and his own parents agreed to make up the amount he would forfeit by relinquishing his fellowship. Thus secured, the cautious young man carried on with the book on comparative philology that the Delegates of the Clarendon Press had commissioned from him. His career thereafter was charmed. He was Bampton Lecturer in 1881, re-elected Fellow in 1882, a professor by 1883, and a bishop by 1885. But neither was Humphry a John Wordsworth. He had not looked before he leapt into marriage.

A salary of £600 was handsome for a heart-free bachelor. It was a bare subsistence for a married Oxford couple with a mortgage and a growing family. A tutor was so hard-worked that it was understood that he had no time for research. But the only way Humphry could increase his income was by the exhausting treadmill of still more tutoring, more reading parties, more lecturing, and more examining, supplemented by whatever bits and pieces he could scrape up by writing for the magazines. 'Harder worked men than ... Humphry do not exist,'[2] Mary Ward told her father in August 1873.

The sensible thing to do was bid Oxford a fond farewell, and pack off to London where Humphry's quick wits and contacts would assure him an editorship where the real money was to be made in journalism. But Mary, who had so recently come 'home' to Oxford, could not face that. So they did not do the sensible thing. The great hope was that when Gladstone's commission of reform reported Oxford would be shaken up and college tutors would be a more important genus. Till then, Humphry had to put up with being disfranchised from the important voting committees in his college. His position was further weakened in March 1873 when Pater published his *Studies in the History of the Renaissance*. Wordsworth led the college attack on this homage to aestheticism as irreligious, amoral, and a disgrace to Brasenose. Pater (who was bold in intellect but timid by nature) amended the second edition; but the implacable Wordsworth demanded that Pater's examining work in divinity be made over to him. 'The difference of opinion which you must be well aware has for some time existed between us must, I fear, become public and avowed'[3] Wordsworth told the older man. Pater was at bay, and Humphry—his closest friend—was pushed further into the college shade. To compound his woes, the promised Gladstonean reforms were shelved and with them the hope for a progressive career structure for tutors within the Oxford system.[4] Humphry was still on his treadmill and would be for the foreseeable future.

The Wards' domestic life was made somewhat easier by the cult of personal austerity which was fashionable among Oxford's younger generation as a protest against the wine-swilling, gluttonous, dandified ostentation of the old school of don. Abstemiousness was *en vogue*. Creighton lectured his rich fiancée, Louise, on the necessity of 'Entsagung'[5] (denial) and when they married in January 1872 they called their house Middlemarch, in honour of Dorothea's noble self-renunciation at the end of George Eliot's novel.

With an engaging self-mockery, Ward vividly recalls the low-key and donnish style of her early married life in Bradmore Road:

> We had many friends, all pursuing the same kind of life as ourselves, and interested in the same kind of things. Nobody under the rank of a Head of a College, except a very few privileged Professors, possessed as much as a thousand a year. The average income of the new race of married tutors was not much more than half that sum. Yet we all gave dinner-parties and furnished our houses with Morris papers, old chests

and cabinets, and blue pots. The dinner parties were simple and short . . . . Most of us were very anxious to be up-to-date and in the fashion, whether in aesthetics, in housekeeping or in education. But our fashion was not that of Belgravia or Mayfair, which, indeed, we scorned! It was the fashion of the movement which sprang from Morris and Burne-Jones. Liberty stuffs very plain in line, but elaborately 'smocked', were greatly in vogue, and evening dresses, 'cut square', or with 'Watteau pleats', were generally worn, and often in conscious protest against the London 'low dress' which Oxford—young married Oxford—thought both ugly and 'fast'. And when we had donned our Liberty gowns we went out to dinner, the husband walking, the wife in a bath chair, drawn by an ancient member of an ancient and close fraternity—the 'chairmen' of old Oxford.[6]

Substitute the second-hand Volvo for the bath chair, and it is striking how little has changed. And since Humphry and Mary had only three well-spaced offspring (as against Julia Arnold's nine confinements and Jane Ward's seventeen) one assumes that another part of the Wards' conscious up-to-dateness was the practice of birth control.

With the Humphry Wards' generation arrived that modern phenomenon, the nuclear family, with its intenser interpersonal relationships and generally higher level of neurosis. Dorothy arrived in 1874. It was an easy birth. Contractions began around three in the morning and the baby (crowned with a mass of black hair) was born at 12.30. Arnold (the hope of the family) was born in 1876 and Janet (the clever child) in 1879. Mary Ward's child-bearing occupied her only half a decade. It was a momentous liberation compared to the never-ending pregnancy, birth and nursing cycle of her mother's married life. And the smallness of her family allowed her to repossess the functions which traditionally the middle classes had delegated to servants. The 1870s generation of mothers 'wheeled their own perambulators in the Parks ... bathed and dressed and taught their children'.[7] They *knew* their offspring in ways that their forever breeding mothers never had.

It was all very modern. But in one respect, Mary was exaggeratedly old-fashioned. From the day she married until her death, she was—for public purposes—Mrs Humphry Ward. Especially when she became a famous woman (and her husband correspondingly insignificant) her name rendered her a figure of slight absurdity—a woman called Humphry. Her motive in being Mrs Humphry was blind loyalty, the same loyalty that led her mother to hold fast to Tom while he trampled on all she held holy. As Mrs Humphry Ward, Mary proclaimed herself utterly and voluntarily her husband's property; bone of his bone and name of his name—even if he didn't set the Cherwell on fire.

During the early 1870s, Mary Ward apparently became the best croquet player in Oxford, and a secretary of the city club. In the mornings—when housework allowed—she continued to read in the Bodleian. And at night, after the children were put down, the Wards would work. She gives an idyllic vignette of domestic nocturnal industry:

I see, in memory, the small Oxford room, as it was on a winter evening, between nine and midnight, my husband in one corner preparing his college lectures, or writing a 'Saturday' 'middle' [i.e. a centre-piece article for the *Saturday Review*]; my books and I in another; the reading lamp, always to me a symbol of peace and 'recollection'; the Oxford quiet outside.[8]

Mary Ward was still picking up an education on the edges of Oxford. But her tutors were different. With marriage, Pattison's 'interest in me changed', as she wryly noted. He remained friendly, but he was no longer intimate. His arid devotion to pure research had been replaced by a more reasonable ideal and more avuncular masters. One was her husband's patron, J. R. Green. Green was, above all, a practical scholar; one who believed in getting things out. As Mary Ward recalled, he cured her other Pattisonian ideal of getting to the 'bottom' of things and squirrelling all knowledge away for some future great work:

'Anyone can read!' he would say; 'anybody of decent wits can accumulate notes and references; the difficulty is to *write*—to make something!' And later on, when I was deep in Spanish chronicles and thinking vaguely of a History of Spain—early Spain, at any rate—he wrote, almost impatiently: '*Begin*—and begin your *book*. Don't do "studies" and that sort of thing—one's book teaches one everything as one writes it.'[9]

The tubercular Green's urgency was sharpened by the consciousness that he personally was in a race with the undertaker. The skull was always on his desk. The no-nonsense straightforwardness of Green's style; the democratic address of his scholarship (aimed at the general reader, not the specialist); above all his pragmatic devotion to getting ideas into print the shortest way (and getting paid for them) were all useful to Mary Ward at this intellectually uncertain period of her life.

According to her own curriculum vitae, Mary Ward devoted the 1870s to 'history and criticism'.[10] But her academic career—so to call it—actually advanced by a series of unimpressive fits and starts. In late 1871, E. A. Freeman (urged by J. R. Green) had invited her to do a book on Spain.[11] But she declined on the grounds of impending marriage. And it is clear that her research was stalled. The article on the Poem of the Cid (which had been meditated for over a year) came out in *Macmillan's Magazine* in October 1871. Its successor, which was written at the same period, appeared in the June 1872 issue of the magazine. She and Humphry put off children for two years; but Mary Ward produced nothing scholarly in the interval. (Perhaps she was working on the abortive three-volume novel 'Vittoria'.)

In the summer of 1874 she was again actively interested in writing. With all the omniscience of a new mother with six weeks' experience she dashed out a broadsheet— 'Plain Facts About Infants' Food'—for circulation among the slums of Oxford. It was both absurd and a rather endearing attempt to put the world right at one theoretic stroke. Ward's instructions to her fellow-mothers began '1. *Never over-feed a child*, and do not feed it at irregular intervals. In nine cases out of ten the reason why a baby is puny and sickly is because it is fed any time in the day, and all day long'. Otherwise Mrs Ward's eleven rules for nursing mothers seem very sensible. Breast-feeding was recommended but if the bottle must be used, watered down tinned Swiss Milk (Nestlé's condensed milk) was safer than raw cow's milk. Milk of some sort was the only possible nourishment for young babies: 'A little while ago', Mrs Ward solemnly warned, 'a child was starved to death in London, because its parents gave it plenty of bread but nothing else.'

More constructively for her literary career she requested J. R. Green's good offices with Macmillan for 'A Primer of English Poetry' to be written by her. (In her slightly self-serving account, Mary Ward lets it seem that it was he who invited her.) It would be primarily intended for young readers and not more than 120 pages long. The 'Primer' would be Mary Ward's first book. On the face of it, the venture was presumptuous. She had no training in the field of English literature. What she did have was a distinguished don father who had written *A Manual of English Literature*, a don husband who was regularly lecturing on the 'The English Poets', and an uncle called Matthew Arnold. Green was a general editor and adviser to Macmillan, and made his recommendation. Terms were offered and accepted by Mrs Ward on 19 June 1874: £50 after the sale of the first 10,000, thereafter a 2*d*. per copy royalty.[12]

Mary Ward assured Macmillan that she would do her best to have the 'Primer' done by the end of September 1874—a wildly optimistic forecast.[13] When she made the contract with Macmillan, she was eight months pregnant with her first child. Dorothy was born on 22 July 1874. On 29 August, Humphry, Mary, baby, and the new nurse, loaded with 'several tons of luggage' made an arduous trip by train north to Fox How for the month. They had arranged to go to Paris over the subsequent Christmas vacation (Mary's first trip to the Continent, and a great intellectual adventure for her). Just how this would-be author thought she was going to produce a book, a baby, a guide for nursing mothers, and run the house in the interstices of all the upheaval of the busiest four months of her life is not clear. She was not thinking rationally.

Nothing daunted, Mary Ward immediately dashed off twenty pages on *Beowulf* and took the sample to Green in his rooms in Beaumont Street in Oxford where he was staying for the summer. As she remembers, 'He was entirely dissatisfied with it, and as gently and kindly as possible told me it wouldn't do and that I must give it up. Then throwing it aside, he began to walk up and down his room, sketching out how such a general outline might be written and should be written.'[14] Green promptly cancelled her contract and reassigned the project to Stopford Brooke, who had a huge success when his *Primer of English Literature* (as it was renamed) came out in 1876. He was not fool enough to try and do it in four months. In her *Recollections*, Mary Ward excuses the débâcle by noting only that 'I was far too young for such a piece of work' and stresses that it was a valuable lesson for her. But at the time, it was devastating. If she could not give satisfaction on *Beowulf*, there was no hope for her as a literary critic. She probably knew the Old English epic better than any woman in England; for some years she had been helping Tom with his edition of the poem (published in 1876). Green—who by now loved Mary—was not telling her to improve her work, correct it, sharpen it, or whatever. He was bluntly telling her that she was simply not up to the task, and never would be. He 'threw it aside'. The set-back was severe.

It was not all dead ends for her. In the early 1870s, Mary Ward had an intercourse with Balliol people which was intellectually rewarding and formative. The Master, Jowett, was always a rather off-putting personage. (Mary noted it as a signal unbending when—having known her eighteen years—he first called her 'my dear' as she grieved over the dead body of her mother.) In the early days she was much friendlier with Jowett's brilliant young protégé T. H. Green (no relation to 'J.R.G.'). After their marriages the Wards and the 'Greens of Balliol' were the best of friends. Mary Ward confided in the Greens in times of stress (which were frequent enough). She would sometimes allow herself to cry in their company and they would console her. And Green's intellectual positions—particularly his rational theism—insidiously became hers.

Green was an embattled figure in the university (a lineal descendant of Cromwell, as his friends like to remember). He declined to take orders (like Humphry) and tended—if anything—towards Unitarianism. But he never joined this or any other Church. He distrusted all ecclesiastical and religious structures. In 1867, he was instrumental in getting compulsory student attendance at his college chapel abolished. Despite his iconoclasm, he had become the most valued colleague of Benjamin Jowett. In the same year as Jowett's election as Master of Balliol, 1870, Green, aged 34, became a tutorial fellow of the college. As an educator, Green was the driving force behind the newly regenerated tutorial system: a semi-pastoral relationship between teacher and individual student, centred on a weekly meeting and essay. It was a means by which humane 'values' could be instilled with the dry stuff of the syllabus. Balliol became a top training school for 'public officials'—the country's civil servants, high politicians, and career churchmen.

Green had painfully developed his own idiosyncratic Christian humanism conceiving God as 'the possible self which is gradually attaining reality in the experience of mankind'.[15] We find his thinking clearly echoed in a letter of Mary's to her father in January 1874: 'Is it not possible ... that God may have designed man to be the only creature in his universe capable of improvement and gradual perfection by means of effort and struggle? We may even conceive that God's creation would have been incomplete without such a type of goodness, as that of the tried and purified human soul triumphant over sin and suffering.'[16]

This was all very well and it squared Christianity and Darwinism very elegantly. But it was plain heresy. There was nothing in the Bible to warrant this nineteenth-century meliorism. It involved Mary Ward and Green in that most awkward of postures: affirming God while denying the truth of his revelation. This necessary quarrel with Biblical writ was agonizing. As Green told Mary (who was clearly experiencing similar pangs) parting with the Christian mythology was 'the rending asunder of bones and marrow'.[17] But it was Green's remorseful conclusion that 'the miraculous Christian story was untenable'. The easiest step forward from this point was into the agnosticism and tenable science of T. H. Huxley. But Green did not want easy steps. While harbouring his doubts he wished to retain intact the whole fabric of Christian civilization. Thus the Church and its central symbolism were not to be destroyed simply because they were incredible. Green's was dissent without rebellion, reform, schism, or denial.

Green's 'agony' was routine at Oxford, where there were any number of dons privately racked with doubt about the faith to which for professional purposes they subscribed. Green was different and made his doubt the occasion for non-devotional works of 'Christian citizenship', as he called it. Doubts paralysed others, they energized him. He was, Ward recalls, 'an energetic Liberal, a member both of the Oxford Town Council and of various University bodies; a helper in all the great steps taken for the higher education of women at Oxford, and keenly attracted by the project of a High School for the town boys of Oxford—a man, in other words, preoccupied, just as the Master [Jowett] was, and for all his philosophic genius, with the need of leading a "useful life" '.[18]

Inspired by his example, Mary Ward too threw herself into a spiritually therapeutic 'useful life', particularly as it concerned the higher education of women at Oxford. The movement had started very gradually in 1866, when the wives and sisters of Oxford dons had won permission to attend university lectures, and had begun to organize informal classes for their sex. Like other young university wives in the early 1870s, Mary Ward was impressed by Dr Jowett's weekend parties, from whose cultivated (but rather strained) discussions women were not excluded. She, Mrs Max Müller, Mrs T. H. Green, and Mrs Creighton duly took the logical next step of setting themselves up as the first joint secretaries of the 'Lectures for Women' Committee. Mark Pattison generously supported the project, with the proviso that ladies attending should dress with appropriate modesty. Other powerful 'liberal' dons threw in their guarded support as well.

The committee held its first meeting at the Creightons' house in 1873. Mary

Ward was elected one of the 'Hon. Secretaries'. The lectures, which began in spring 1874, were held in rooms lent free of charge at the Old Clarendon Building. The secretaries, aided by Clara Pater and Bertha Johnson, sent out circulars describing the lecture programme. The subjects included literature, mathematics, German, Latin, and arithmetic, but principally covered historical topics. The first lecturer, the Revd Arthur Johnson (Bertha's husband), talked to an overflowing room. He was followed by such university notables as William Stubbs, and Henry Nettleship. In the Michaelmas Term 1875, Humphry Ward gave eight lectures on 'Chaucer and his Time'.

The demonstrable success of the lectures led to the formation of an 'Association for the Education of Women' in 1877 with Mary Ward as secretary. Cambridge University meanwhile forged ahead with Newnham, which by 1879 was a proper college. Oxford—as always—moved more deliberately and in a spirit of honourable compromise. The initiative for an Oxford women's 'Hall' (which would be residential, not collegial) split into two camps. One was strictly Anglican, and led to the establishment in October 1879 of the denominational Lady Margaret Hall, which proclaimed a primary allegiance to the Church of England. Miss Elizabeth Wordsworth (John's sister) was LMH's moving spirit and first Warden. The other, non-denominational, project resulted in Somerville Hall. Mary Ward and Mrs Augustus Vernon Harcourt were the first secretaries of the Somerville Committee which was formed in 1878. And it was Mary who suggested the name of the hall. Mary Somerville (who had died in 1872) was the female astronomer who had taught herself algebra and calculus secretly.

Over 1878–9, Mary Ward was furiously busy with the business of advertising the new establishment, raising funds, examining prospective students, appointing a principal (Miss Shaw-Lefevre), purchasing the lease of Walton Manor, remodelling and furnishing the dilapidated property for communal living, incorporating the Hall as a limited liability company. All this was interspersed with innumerable committee meetings at which tempers frayed. Her fellow secretary was meanwhile ill and little help. Mary was heavily pregnant (Janet was born in November 1879). All in all, it was an extraordinary achievement; the 'Doctor' would have been proud of her. But her friends were worried. This self-destructive dedication to good works was excessive. Louise Creighton wrote a warning letter on 17 September 1878: 'Do not lay on yourself a burden which is more than you can bear; hard mental work, a house full of family, the anxiety of a sick mother, is too much strain for you to bear continuously without break. You ... work too hard.'[19]

Somerville Hall opened in October 1879, neck and neck with LMH, with twelve select young ladies in residence (LMH had only nine). Mary Ward had masterminded all the big strategic moves in founding the institution; but she was just as attentive to small detail and donated to the Hall's living rooms some of the blue pots which as decor she particularly admired. She also established the decorous intellectual ethos, carefully designed not to alarm the male custodians of Oxford by any flagrant display of feminist assertion. Both Somerville and Lady Margaret Hall were small and unambitious compared to what was going on at Cambridge. Ward congratulated herself that Oxford's women's halls were in no sense designed to *liberate* women. As she later claimed:

> hardly any of us were at all on fire for woman suffrage, wherein the Oxford educational movement differed greatly from the Cambridge movement. The majority, certainly, of the group to which I belonged at Oxford were at that time persuaded that the development of women's power in the State—or rather, in such a state as England, with its far-reaching and Imperial obligations, resting ultimately on the sanction of war—should be on lines of its own.'[20]

Mary Ward was appointed to the Somerville Council in 1881 and remained on it until 1898, by which time all university examinations in the faculties of Arts and Music were open to women. She did not send either of her own two daughters there, or to any other place of higher education. And she broke her ties with the college after 1908, when she formed her Anti Suffrage League.

Mary Ward's nine married years in Oxford see her emerging as what we would call an 'empowered' woman, able to use committee machinery and lobbying pressures to change the order of things—even orders as set in their ways as those of Oxford University. Her subsequent career charts in one of its main lines an ever more effective skill in the setting up of committees, associations, leagues, conferences, guilds. With this bureaucratic apparatus she would change England. It was to be a main aspect of her genius.

Somerville was the one unequivocal success of Mary Ward's Oxford decade. Humphry also had his one great success. In April 1879 he signed a contract with Macmillan to edit a four-volume (500 pages per volume) anthology to be called *The English Poets*, on a 5 per cent royalty. £300 per volume was allowed by the publisher for subcontracted introductory matter. The democratic impulse of J. R. Green (and his helping hand with the publisher) lay behind this venture which taxed all Humphry's impressive organizational skill. The first two volumes were expeditiously completed in 1880 and comprised selections of the best of English verse—'touchstones', as Matthew Arnold would say. He in fact was recruited to do the general introduction. Author introductions were farmed out to lesser critics (Mary did six; Tom Arnold did the medieval poets; Willie Arnold also chipped in). In the largest sense *The English Poets* is dedicated to Matthew Arnold's confident theorem that 'the future of poetry is immense'. It answered to a general feeling about the centrality of verse in late Victorian culture. Despite condescending reviews (which wounded Humphry) the series was immensely and perennially popular. It went through twenty editions in Humphry's lifetime, bringing him a regular £100 a year.

Had they drawn up a balance sheet of their Oxford years (as did their canny friend Mandell Creighton before he left in 1875) Somerville and *The English Poets* would have been well up in the Wards' asset column. So would the fact that they loved each other. (Humphry gallantly gave his wife Tennyson's 'Lover's Tale' for her twenty-eighth birthday in 1879.) They had three promising children. There had been no major illness, no infant deaths, no bankruptcy.

Against these were a long list of debits. First was the undeniable fact that Humphry's career was going nowhere. In 1880 he was no further forward than he had been in 1870; and at 35 he was no longer a young man. *The English Poets* was all very well as a money spinner and a popular educator. But it did not compare as scholarship with Mandell Creighton's history of the papacy or John Wordsworth's work on the Vulgate Bible. The terrible agricultural depression of the late 1870s bit into the revenues of the colleges, most of which came from rents; there was little free money at Oxford for the extramural education Humphry excelled in. By the end of the decade he could look forward to no promotion; he would never be a professor or a head of college.

Neither was Mary Ward's career in 'history and criticism' going anywhere. Not that she did not have grand ideas. Their visit to Paris in Christmas 1874 had made an enormous impression on the provincial young girl, who since coming from Hobart had never been farther afield than Glasgow. Armed with letters of introduction from Max Müller and Uncle Matt the Wards had met Madame Mohl, Ernest Renan, Gaston Paris, the Boutmys, the Scherers, the Taines, and Paul Bourget. They had seen Sarah Bernhardt with her '*voix d'or*' perform *Phèdre* at the Théâtre Français. Ten pages are devoted to this wonderful fortnight in Mary Ward's *Recollections*.[21] It was one of the formative experiences of her life. In her Francophile enthusiasm, she offered Macmillan a book on France. The proposal was accepted. It verged on *folie de grandeur*. Mary Ward had no qualifications for such a task; it was strange of Macmillan to encourage her.

The French book came to nothing. Six years later Mary Ward persuaded Humphry to write on 26 April 1880 (on the pretext of his wife's ill health) asking that she be excused this task. The commitment, Humphry wrote with an appeal to the publisher's male gallantry, was 'weighing on her mind'. But, he insisted, Mary still 'wishes above all things to aim at achieving something permanent'. What she now proposed was a study of early Spanish literature, 'a book of papers, something after the model of her uncle M. Arnold's Essays'.[22] This too came to nothing. Mary Ward thus arrived at the end of her first married decade with a ghost bibliography comprising an unwritten 'Primer of English Poetry', an unwritten history of France and its literature, and an unwritten study of early Spanish literature. This last landed her exactly where she had been in 1869, with the difference that she was now a 30-year-old woman, not a precociously brilliant teenager.

Their married life in Oxford had witnessed a gradual but inexorable erosion of the Wards' career prospects and of the liberal faction to whom they owed allegiance. The Arnolds no longer enjoyed a pre-eminent place as shapers of the university's ideology; indeed they looked suspiciously like a decayed dynasty. Matthew Arnold, Professor of Poetry in the mid-1860s, had long ceased to be a force and his son Richard had flunked out of the university ignominiously without even a pass degree to save his father's blushes. Tom Arnold's goings-on at Oxford in the mid-1870s were an even heavier cross for Mary Ward to bear. For ten years, her father worked as a tutor and lecturer serenely revising his *Manual of English Literature*, editing Wyclif, and burying himself in Anglo Saxon. He was, during this period, an Anglican; or at least was publicly perceived to be one. There were, however, ominous signs that yet again he was set for some shattering religious perversion. His children would occasionally catch him mumbling in Latin under his breath during church services at St Philip and St James'. On one (perhaps legendary) occasion Julia came back to find him in conference with a couple of priests, whom she pelted from the house with a barrage of kitchen china.

In February 1876 the inevitable happened. Tom announced his intention of 'resuming the practice of the religion which I formerly professed'. Before taking the step he consulted with Catholic friends in London, whither he was pursued by the relentless Julia, who sought the help of the Forsters in saving her husband. 'Wounding' and 'bitter' words were exchanged. Tom contemplated decamping to Manchester. Mary, apparently, contrived to mediate between her warring parents. Julia, Mary explained, was in a 'state of frenzy from the feeling of loneliness and lovelessness'. 'So my darling father,' she implored, 'let there be peace between you for your children's sake.'[23] It was agreed that Tom would put off any decision until the younger children were grown up. But such promises meant nothing when the religious fever was on him. In October 1876, the crisis flared up again. It could not have been a less apposite moment. Tom was set to get the Rawlinsonian Chair of Anglo-Saxon. It was a final step towards total rehabilitation. Tom was in every sense the right man for the job. And, as a professor, he would be one of the thirty-nine-strong charmed circle and rich (the salary was around £1,000). His lustre and patronage would have lighted on Mary, her husband, and his children. The Rawlinsonian professorship would have been the triumphant climax to the Arnold brothers' long and honourable academic campaign to reform the humanities curriculum. But, on the eve of the election, Tom sent round letters publicly announcing that 'any member of Congregation, who thinks of voting for me at the election to the chair of Anglo-Saxon should know that I intend, as soon as may be, to join, or rather to return to, the communion of the Catholic and Roman Church'.[24]

It is hard to think of a more destructive thing he could have done to his career and his domestic comfort. What remained of his marriage was now hopelessly in ruins. He, Julia, and Mary had a terrible Friday meeting during which Julia 'was not in her right mind' and raved. Mary—who was eight months pregnant—had a 'breakdown',[25] which at least had the benefit of diverting Julia's attention. Tom scurried away to London, where he took mean lodgings and where Humphry was dispatched as emissary to 'make things plain' (i.e. on the impossibility of his ever living with Julia again). Mary, meanwhile, was left to put the pieces together as best she could. On 8 November 1876 she bore the male child she had so longed for, and called him Arnold.

Tom Arnold's Oxford career was finished. Students would never come to read with him now. Laleham was uninhabitable and was sold. He was forced to pick up a living in journalism. It was not until 1882 that Newman could secure another professorship for him in Dublin. After this date he earned between £400 and £600 a year, half of which he remitted to the support of his family. Julia remained at Oxford in a meaner house in Church Walk, taking in girl boarders. As the years passed the Arnold children one by one left home leaving only the unlucky Ethel to look after the mother. Julia was not an easy woman to look after, even at her best. She was now at her worst. Not only was she again publicly humiliated and impoverished, she was extremely sick. Breast cancer was diagnosed in 1877, for which the treatment was radical surgery. She was to live ten years in more or less constant pain, eased by increasingly large doses of morphia, homoeopathy, and cocaine.

As usual, the children paid heavily for Tom's quixotism. Mary Ward was plunged back into all the psychic uncertainty of her early childhood. Until the end of Julia's life they were constantly obliged to slip her sums of money they could not spare. (They still had the mortgage on Bradmore Road, and Humphry had taken out heavy life insurance on himself.) Julia was improvident and on at least two occasions the bailiffs were poised to move in. On other occasions she was besieged in the house by angry tradesmen and might have been driven into the Oxford streets had her eldest daughter not been there to rescue her. Mary was forever writing supplicating letters to Tom at the behest of a hysterical Julia, who could not be trusted to communicate directly on the sensitive question of money. As late as January 1880, Mary warned Tom that a visit to his wife (currently being treated for a recurrence of her cancer) would be 'a dangerous risk'.

But it was money that was the never-ending bone of contention. Tom—living in seedy lodgings—would send amounts that left him embarrassingly short but which were still too little for Julia to run a respectable Oxford household. And his censure of her 'improvidence' was ceaseless. On 21 May 1880, Mary was driven to protest to her father that 'I do indeed think that you are rather too hard on Mamma about money matters.'[26] Looking back (particularly in the context of Mary Ward's later earnings), the sums at issue seem paltry. Julia,

Mary pointed out, had received £140 from Tom for the 1880 Hilary Term, of which £22 had been earmarked for Judy's and Ethel's school fees at Oxford High School. Now Julia was short of cash wherewith to put food on the table. Mary vigorously denied Tom's charges of profligacy: 'with regards to last year I cannot think that £120 for the year as a grocer's bill deserves anything like the blame that you have given to it .... It cannot be argued that any one suffering from cancer is physically as capable as a vigorous and healthy woman of managing everything to the best advantage.'[27] Mary Ward was to have to write innumerable letters of this kind for ten years. Often the sums in dispute were as little as five pounds or less. The cumulatively corrosive effect of all this pettiness was enormous.

The direct impact of Tom's reconversion devastated Willie, the academic hope of the Arnold family. The young man had returned to Oxford with golden expectations after Rugby in October 1871. His career at school had been brilliant and he won an open scholarship to University College. His and Mary's boarding school careers had separated them. Now at Oxford they became close friends again—as they had been years ago as toddlers in Tasmania. Throughout the rest of his life, Willie was to be the only brother she was at all close to. But the foolish young man fell in love in summer of 1872, and although he agreed not to become engaged until December 1873 (i.e. after Mods., the first public examination for undergraduates) his studies suffered. He missed a first, despite working 'somewhere between sixteen and twenty hours a day'.[28] Mary was devastated. But she consoled herself that there had been 'unusual slaughter' in Mods. that year. If Willie worked steadily, she told her father, 'he may get a double first yet, and no one will remember that he got a Second in Mods'.[29] Willie—who had none of his sister's grit—was less confident about his prospects in Greats. He wrote despairingly to his fiancée, Henrietta Wale, on 19 November 1875: 'somehow I feel I shan't see my name in the firsts when the list comes out. So you must make up your mind not to be disappointed. Besides that I am not really the equal of the best men up here who get firsts.'[30]

Willie's abilities were never given a fair test. He took his finals in May 1876. At the time he was living at home at Laleham, among all the chaotic upheaval of his father's latest religious crisis and his mother's attendant mental breakdown. No one could do their best in such circumstances; least of all a quiet, sensitive lad like him who had already been baptized an Anglican at birth, rebaptized a Catholic at 5 and confirmed an Anglican at 12.

Willie duly got a second. All things considered it was a creditable achievement. But for a young man carrying his family's insatiable appetite for supreme academic honour the disappointment was severe. Nevertheless, Willie set up in lodgings as a coach and lecturer (his father by now having deserted Oxford). In June 1877 he married his Henrietta. And in 1879, Humphry introduced him to the newspaper editor C. P. Scott, who was in Oxford looking for new staff members for the *Manchester Guardian*. William was invited to the Manchester office on trial, and by the end of the year was writing leaders and art criticism for the paper, a post he held with distinction until spinal illness disabled him in 1896.

For William, his new career in the north—away from Oxford, London, and Fox How—was a merciful release and ultimately a minor fulfilment. Released from the claustrophobic role of being an academic Arnold, he could develop into a man of general cultivation. Despite tempting offers to return to Oxford or London or to stand for Parliament he always refused. Wisely he found happiness by lowering his expectations from life. But for Mary Ward it was another blow. In the best of worlds, her father would in 1877 have been Professor of Anglo-Saxon, Willie a first-class man, Humphry a Fellow of BNC, and she the author of a learned book.

In the *Recollections* Mary Ward skirts over her departure from Oxford at the end of the 1870s, focusing instead on the depiction of the nine years leading up to it as an Edenic interlude. In fact, the Wards came to the end of their Oxford phase in a condition of growing bitterness and anxiety. It was the collapse of the dream with which they had begun marriage. Tom's disgrace was a lasting wound. Julia Arnold's condition bordered on public scandal; and she was clearly dying—if slowly. William had left, with his tail between his legs. Neither Humphry nor Mary Ward had made a mark in their careers. The awful prospect of being another dull, second-rate Oxford couple loomed ever closer.

7

Fighting Back: 1878–1880

THE WARDS' first eight years at Oxford had ended, if not in defeat, then certainly at a distressingly low ebb. Most couples would now have settled for less from life. A modestly comfortable, if undistinguished, existence was still open to them. One did not have to be first class to inhabit Oxford's middle firmament. Humphry could cut back on his teaching, do a bit more journalism, a bit more shooting and fishing, and generally enjoy life more. Mary Ward could potter at her committees and play at some genteel research project. The children would grow up and they would manage somehow. There were rich relatives (notably the Forsters) who had a soft spot for them. But one of Mary Ward's most admirable—and finally rather terrifying—qualities was her pugnacious refusal ever to knuckle under. Rock Terrace had not broken her; neither would Oxford. She *would* succeed. The past ten years were not the Humphry Wards' destiny, but a temporary set-back on the way forward.

Recovery began inauspiciously. In 1875 Henry Wace had been appointed co-editor of *The Dictionary of Christian Biography*. Shortly after he was in Oxford as Bampton Lecturer and he used the opportunity to recruit contributors for the *Dictionary*.[1] Dr Wace was a former Brasenose man and friend of Humphry's, a chaplain of Lincoln's Inn, and an indefatigable member of *The Times* staff—a cultural power broker, in short. The Christian biography venture had been sponsored by the London publisher John Murray as a commercial project and the first volume had proved surprisingly successful when it came out in 1877. Murray decided to expand the original plan. This meant extra contributors at short notice, and Wace was now at his Alma Mater drumming them up. Mary Ward had been recommended to him as someone who might take on the Spanish lives, particularly those of the West Goths—not easy to assign given the academic neglect of things Spanish.

The significance of the *Dictionary* work for Mary Ward was twofold. First, it paid; and the Wards with three children and Julia to support were feeling the financial pinch. Secondly, it would entail systematic and prolonged intellectual effort. Hitherto Mary had expected instant gratification of her grandiose intellectual aspirations, secretly confident that an innate Arnoldian 'brilliance' would triumph over the mere need to work at what she did. She would write a primer of English poetry in two months. She could not conduct a conversation in French with ease, but on the strength of a two-week holiday she would write a book on France. She would write a book of essays on Spain and Spanish literature of Matthew Arnold calibre—having never been to the country. Without even a decent school education and equipped only with a library pass she would 'get to the bottom' of a vast and complex academic subject in a couple of years (something that Pattison himself could not manage in a dedicated lifetime). It had all been delusion—schoolgirlish fantasy induced by the romantic atmosphere of Oxford. Belatedly Mary Ward had learned the freshman's first lesson: there are no short cuts.

Mary Ward was nervous about Wace's invitation. As she recalled, 'the well-trained woman student of the present day would have felt probably no such qualms. But I had not been well trained; and the Pattison standards of what work should be stood like dragons in the way.' However, she 'took the plunge'. It was one of the wisest decisions she ever made. It entailed two or three years of 'sheer, hard, brain-stretching work'[2]—work in which corners could not be cut and in which there was no place for 'brilliance'. It was 'the only thorough "discipline" I ever had',[3] as she later claimed. These two years 'altered my whole outlook and gave me horizons and sympathies that I have never lost'.[4] She returned to the elements of her subject and did the spadework that she had previously skipped. She learned basic techniques for dealing with primary source material. Her 1878–82 notebooks contain vocabulary lists which indicate both a laudable ambition to improve her grasp of Spanish and the limited competence she must have had when she wrote her 1871–72 articles for *Macmillan's Magazine* (e.g. 'Hacienda ... land, domain', 'acuitarse ... to trouble oneself').[5] With Humphry's help she taught herself enough Latin to work through early chronicles and ecclesiastical documents.

It is clear that Mary Ward was given valuable morale-boosting by friends who were more interested in keeping her spirits up than in being strictly honest. The following anecdote in the *Recollections* tells us more than she intends it to:

> Perhaps I may be allowed, after these forty years, one more recollection, though I am afraid a proper reticence would suppress it! A little later [in 1880] 'Mr Creighton' came to visit us ... and I timidly gave him some lives of West-Gothic Kings and Bishops to read. He read them—they were very long and terribly minute—and put down the proofs, without saying much. Then he walked down to Oxford with my husband and sent me back a message by him: 'Tell M. to go on. There is nobody but Stubbs [the Regius Professor of History, currently the most distinguished historian in the country] doing such work in Oxford now.' The thrill of pride and delight such words gave me may be imagined.[6]

It was, of course, preposterous. The idea that Mary Ward's often modest little entries in the *Dictionary* were a match for William Stubbs's *Constitutional History of England* was not something that the future Dixie Professor of History at Cambridge could have asserted with a straight face; nor could Mandell bring himself actually to *say* it—leaving the message to Humphry to convey. Clearly the two men had conspired on their walk on how best to raise Mary's spirits and concocted their flattering fib. Forty years later she was still hoodwinked by their ruse.

Thus fortified Mary Ward finished her 209 entries by 1882. In her own account she stresses how the *Dictionary* work formed *Robert Elsmere*'s central religious doubt. Her discovery while working with her Goths that early Christian 'testimony' was 'non-sane' made impossible literal belief in the miraculous biblical narrative.[7] History made nonsense of gospel (as science had earlier made nonsense of Genesis). But Mary Ward had made this Elsmerean step into scepticism much earlier via contact with T. H. Green and J. R. Green and through reading books like Seeley's *Ecce Homo*. As early as January 1874 (having just read Seeley) we find her telling her father that 'of dogmatic Christianity I can make nothing. Nothing is clear except the personal character of Christ.'[8] When Bertha Johnson painted Mary's portrait in July 1876, she noted in her diary that while sitting her subject indulged in 'very free criticism of the Bible, entire denial of miracle, our Lord only a great teacher'.[9] At most the 1878–82 work for the *Dictionary* may have clinched Mary Ward's doubt as to such things as the Resurrection. But the doubt was already deeply sown.

The pay-off from the two years' 'incessant arduous work' was less intellectual than psychological. Mary Ward had at last done 'work' of modest but real value. Her achievement was certified in February 1883, when, as she told her mother, 'I *am* to examine for the Taylor [the modern language institute at Oxford]. Mr [G. W.] Kitchin ... tells me that I shall be "making history" as the "first woman examiner of men" in either university.'[10] It was a signal honour; and it was one which had been wholly gained by her own efforts. Mary Ward had earned her spurs. She was by 1880 in a position to undertake the book on early Spain which E. A. Freeman had commissioned in 1872. The offer was still open, and J. R. Green (now fast dying) urged her '*Begin* ... begin your *book*.'[11]

But having experienced what scholarly work entailed, Mary Ward was less biddable than she had been as a girl. She did not want to grow old, working year in year out by herself in the dusty silence of Bodley. The book she decided to write in summer 1880 was something quite different and wholly unscholarly. In 1879 the Wards had spent their summer holidays at Fox How. Seeking to entertain her son and daughter, Mary Ward discovered that 'most children's books seemed to me a great deal too clever and written with one eye at least on the grown-ups'.[12] So she determined to write an uncondescending story for very young children. The resultant *Milly and Olly* had a complex narrative framework which mixed elements of Mary's own childhood in the 1850s with the situation of the Ward children in the Lake District in 1879. Milly and Olly are the 7- and 5-year-old children of Dr and Mrs Norton, an Oxfordshire couple in all physical respects images of Humphry and Mary, as the children are the image of Dorothy and Arnold (Janet was not born until November 1879). The Nortons go to the Lake District to have their summer holiday at 'Ravensnest'—a thinly-disguised Fox How. There they picnic, boat on the lake, visit farms, and take tea with 'Aunt Emma' (i.e. Aunt Fan). Enclosed are three interpolated tales: an Arthurian story, a 'Spanish story of a fanciful queen', and the story of Beowulf (presumably recycled from the unlucky 'Primer' venture). 'My children', Mary told Macmillan, 'know [*Milly and Olly*] by heart but then of course it is their own life more or less.'[13]

Mary Ward, who was always shrewd where matters of publication were concerned, showed a fair copy of *Milly and Olly* to Mrs Alexander Macmillan, who liked it. The author duly communicated the publisher's wife's opinion to the publisher, who submitted to the ladies' ambush and accepted the manuscript on 5 November 1880. Thereafter Mary nagged Macmillan unmercifully as to the best illustrator. She was particularly keen to have Kate Greenaway and, failing this leading light of children's book illustration, the Hon. John Collier, Miss Yonge's collaborator. In the event, she had to make do with her friend Mrs Laura Alma Tadema, wife of the famous artist. Mary Ward accepted the available illustrator with reasonably good grace. But she noted on 6 February 1881 that 'I pine a little after Miss Greenaway'[14] and in later life she stated as her firm belief that Alma Tadema's lifeless illustrations had 'killed' the book.

Macmillan paid £60 for the right to print an edition of 3,000 of *Milly and Olly* with a royalty of £25 per thousand sold thereafter. Despite repeated cajoling (at which Mary Ward was already expert) the firm declined to publish until Christmas 1881. She had hoped originally for Christmas 1880, then Easter. It was disappointing. But Macmillan were very generous about advancing their payment, sending it to the author on 6 December 1880, a full year before publication. She accepted gratefully ('at Christmas what housekeeper could refuse?').[15]

Unfortunately the book did not make all the money its author hoped for, at a period when funds were always chronically short with the Wards. By October 1882, Mary Ward was complaining about not having received a statement from her publisher, and wondering whether royalties (on 4,000 sales and over) were not due her.[16] She was over-optimistic. *Milly and Olly* had not done specially well.[17] Nevertheless, at the age of 30, Mary Ward had at last published her first book—even if it was only a children's book.

In 1880–1 there was little time to worry about *Milly and Olly*. Great changes were in the air for the Ward family. Just as they were starting out for a walk one May afternoon in 1880 the second post brought to 5 Bradmore Road 'a letter which changed our lives'. It contained a suggestion that Humphry should join the staff of *The Times*:

We read it in amazement and walked on to the Port Meadow. It was a fine day. The river was alive with boats; in the distance rose the towers and domes of the beautiful city; and the Oxford magic blew about us in the summer wind. It seemed impossible to leave the dear Oxford life! All the drawbacks and difficulties of the new proposal presented themselves; hardly any of the advantages. As for me, I was convinced we must and should refuse, and I went to sleep in that conviction. But the mind travels far—and mysteriously—in sleep. With the first words that my husband and I exchanged in the morning, we knew that the die was cast and that our Oxford days were over.[18]

It is not the whole truth. Humphry was not being offered a position on *The Times*; he was invited to 'try out'—that is, come for a period under probation. There were no guarantees that he would at the end of his six-months trial get either a permanent or a desirable position. And what impelled Humphry to accept the gamble was less the temptingness of the London offer than his dead end prospects at Oxford. Permanently pegged to his tutor's salary of well under a thousand a year the family finances were tight verging on desperate. His daughters would soon need governesses, music lessons, outfits. His son must go to a good public school (Eton, as it turned out). The Wards' gentility was becoming year by year shabbier in the effort to keep up with their class's lifestyle.

The larger background to Humphry's offer involved another gamble. In 1877 the majestic John Delane had retired as editor of *The Times*. Under his superintendence the paper had known its greatest days and its leaders had become a major voice in the country's political counsels. Delane's successor was in the circumstances an eccentric appointment. Thomas Chenery was 51 years old, in poor health, and previously Professor of Arabic at Oxford (1868–77). With the reins of the paper in his hand, Chenery began drastically to change its character. As the official history puts it (with clear disapproval) 'during his six years as Editor, *The Times* became almost as much a learned periodical as a political newspaper'.[19] This new era of donnish literary dilettantism and provincial Oxonian stuffiness drove out many of the old 'political' leader writers and journalists to be replaced by more 'cultured' men of letters from the dreaming spires. Humphry's name had been put forward (possibly by Wace) as a likely recruit. Even if he gave satisfaction during his trial period, there was the greater uncertainty that Chenery's reforms might not 'take'.

It was all very chancy. But on that momentous morning, Mary and Humphry Ward decided to take the plunge. Their Oxford days were over. Humphry would serve one more term at Brasenose, before going to London immediately after Christmas 1880, to live in cheap lodgings in Bloomsbury. If he were offered a permanent position the family would join him in the capital. If he failed (hideous thought) he would return to Oxford. It would be too risky to sell the Bradmore Road house—their only sizeable asset—until the future was firm. Mary, therefore, would stay on at Oxford.

As soon as the offer came in, Mary began to make arrangements for Julia. With the Wards gone, her mother's position would be precarious. Just at the moment (in early June 1880) her house in Church Lane was besieged by irate tradesmen, dunning her for £30 she had run up. She was hysterical. Mary wrote to her father suggesting that Humphry should arrange a £200 loan on Julia's furniture, which was still technically owned by Tom. The sum would cushion her for a year or so at least, and the repayments could be picked up by the family. Mary also proposed that instead of giving Julia money to pay for her provisions a 'grocer's book' should be started, so that she could buy on credit, to a capped amount. Mary felt that, with ingenuity, Julia could take more boarders—as many as eight girls, she estimated (over-optimistically) with doubling up in bedrooms. This would bring in £600 a year, and Tom might cut back his remittances to £200 per annum.[20] But this step would involve immediate outlay on a better class of home help. Her young sister Judy (like Mary a girl more interested in books than pots and pans) was an 'unreliable' cook. If boarders were not going to be driven away by burned porridge, someone would have to be brought in, at about £60 a year plus board. The strain of all this on a daughter as morbidly over-responsible as Mary Ward was crushing. Her guilt at leaving Julia was compounded by the fact that the cancer had flared up again in January 1880.

Lesser women might have wilted. Mary Ward became noticeably more energetic and coping. She took over all the responsibilities of Tom Arnold to his wife. Her letters to Humphry at the same period reveal a perceptible shift in their relations. Increasingly it was she who was now the powerful partner, the source of the family's resolve and sense of mission. On 20 July 1880, she wrote to her 'dearest husband', who was 'grinding' in a London schoolroom to earn some extra pennies examining. The reviews of *The English Poets* had been generally poor. He was very low. Mary undertook to encourage her 'poor old boy' for the struggle ahead: 'Goodnight my own. Cheer up my best beloved, and keep a brave heart. All will be well with us yet and if God is good to us there are coming years of work indeed, but of less burden and strain. All depends on you and me, and though I know the very thought depresses us sometimes, it ought not to, for we have many good gifts within and without, and a fair field, if not the fairest possible field to use them in.'[21] Admirable as the tone of this is, it is the assurance which conventionally the husband should have been offering his faltering wife.

Humphry Ward returned to Oxford and served out his last Michaelmas Term. Mary busied herself with preparations for going—always with the nagging uncertainty that they might not go after all. At Christmas, before he left for London, the family visited Dublin, where Uncle William Forster was Chief Secretary (i.e. Minister for Ireland). There were a number of reasons for making what was—in the circumstances—a burdensome trip at an inconvenient time. A main motive was money. Uncle William and Aunt Jane Forster had always been the rich relatives to whom Mary turned in need. Now, above all, was a time of need. The Forsters (who seem to have been very warm-hearted) came through with a large amount to tide the Wards through until their house was sold. £200 of the loan remained unpaid at the time of their eventual departure for London, nine months later.[22]

Another motive was to equip Humphry for his new position. As a leader writer, he would need a 'line' and political contacts. Forster was heaven-sent. Ireland had become an all-engrossing national obsession. People it seemed in 1881–82 could think of nothing else. When he joined *The Times* in January 1881, Humphry immediately began assisting with leaders and political articles on the Irish question. His first 'first leader', on 25 August 1881, was inevitably on Ireland and strongly pro-Forster.

In the longer term, this week in Dublin was to have profound political consequences for the young couple. Hitherto Mary Ward had been an old-fashioned Arnoldian Liberal. But first-hand experience of crisis conditions in Ireland jolted her as indeed it had also jolted Forster. The Land League had just been formed; Parnell was beginning to organize political resistance to English Rule; a more terrifying resistance was gathering in the streets and countryside. Civil War, no less, was on the cards. It was, Mary Ward recalled, 'A visit I shall never forget! .... Boycotting, murder, and outrage filled the news of every day .... Threatening letters were flowing in upon both [my uncle] and my godmother.' The brilliant social show which the British government put on in Dublin was 'like some pageant seen under a thunder-cloud'.[23]

Administering an Ireland on the brink of revolt was a hideous strain for Forster, with whom Mary felt a 'hot sympathy'.[24] Despite his Quaker background, his Liberal convictions, and his 'sensitive, affectionate spirit', he became convinced that the only way to deal with the problem was by force, coercion, the iron hand of military discipline. This was quite contrary to what his leader Gladstone had in mind. Finally Forster was obliged to resign in May 1882, and four years later Gladstone's whole administration was wrecked on the Home Rule issue. Mary Ward was convinced that the two years her uncle had spent in Ireland exhausted him and precipitated his early death in 1886. She was equally convinced that his strong arm policies were correct, and that Gladstone's appeasement of Irish nationalism was folly. Her conviction was strengthened by the Phoenix Park murders of 6 May 1882. 'This then is Ireland's answer to Gladstone's great move!'[25] she wrote with scathing indignation to her father the day after the outrage. And she was convinced that her newly resigned uncle had been the assassin's intended target. For years after, she would leap to Forster's defence against all critics. Her plucky campaign on behalf of her uncle in fact did her no harm. John Morley, who attacked the Chief Secretary's Irish record in the *Pall Mall Gazette* in late 1882, took Mary Ward's consequent letter of fiery censure in such good part that he promptly (as editor of *Macmillan's Magazine*) became her staunchest patron during her early years in London.[26]

The effect of her Irish experience on Mary Ward was deep and long-lasting. It fuelled the quarrel with Gladstonean Christian optimism which is at the heart of *Robert Elsmere*. It began a drift towards the Conservative Party, one of whose most powerful women spokesmen she was eventually to become; and this in turn led to her becoming an anti-suffragist. Militant feminists and militant Fenians were equally abhorrent to her. And, for Humphry Ward, the consequences were no less profound. That he too was wholly in sympathy with Forster's coercive policies and remained so is clear from the partisan entry on his uncle-in-law which he wrote in the *DNB* (1889). In the short term, Humphry's connection gave him a valuable boost at *The Times* while Forster was still in the Cabinet. But after 1882, when the tide of public opinion and the paper's line turned against Forster, Humphry's position was weakened by the association.

All this was well in the future and in the New Year of 1881 Humphry and Mary Ward had more immediately pressing concerns. He had gone off to his dingy rooms in London immediately after Christmas and scrawled a letter to Mary during his first night's work at Printing House Square on 29 December with the great presses thundering in his ears: 'Here I am in what may be in the coming years all that my college rooms have been since I crossed the threshold of manhood! Will it be so? and should it be so? I have been much perplexed in spirit today as I walked through the noisy London streets in the thick dark atmosphere wondering whether it was good to come or not.'[27] In the following weeks, Humphry Ward slaved by night and slept by day. Mary returned to Oxford, where the house must be sold and everything packed up.

It was to be a stressful few months. For a year, Julia had been having various futile treatments for the recurrence of her breast cancer. In early 1881 further surgery was unavoidable. The operation was successful in gaining another five-year remission; but the wound never properly cleared up, and was to torment Julia horribly until the disease eventually killed her in 1888. Mary Ward's mood in February was morbid and her feelings about departure mingled with an apocalyptic sense that the lights were going out in Oxford and in England generally. She wrote in melancholy strain to Alexander Macmillan on 6 February that in all probability 'this is the last spring we shall see in Oxford and one cannot help feeling rather mournful over it .... George Eliot [d. 1880], Carlyle [d. 1881] and I begin ominously to feel soon Gladstone. And then how poor one will feel!'[28]

In fact, Mary Ward was near breaking point in spring 1881 and she finally exploded in March in a tantrum of rage against the university enemies, who she now saw as driving her and Humphry into the wilderness. The occasion of her outburst was the annual Bampton Lectures. This prestigious series was traditionally given by one of Oxford's luminous theologians. In 1881, the Bampton Lecturer was John Wordsworth. It was for Mary Ward intolerably provocative. Wordsworth and Humphry had in 1870 been neck and neck. Born within two years of each other they were Brasenose Fellows of exactly the same vintage. They had both given up their fellowships to marry. But while over the next ten years Humphry marked time, Wordsworth had forged ahead. The other tutor had in fact played his cards consummately. Believing passionately that university should be 'a nursery of the Christian life' he had created a doctrinal power base within Brasenose as college chaplain. He became a university proctor in 1874. A protégé of H. P. Liddon and later of Professor J. B. Mozley, Wordsworth's career was strongly promoted by the High Anglican party at Oxford. In 1880—as a signal honour—the Heads of Houses had invited him to deliver the Bampton Lectures. The subject was to be 'The One Religion', a demonstration of the new methods of 'Comparative Religion'. It was clearly a stepping-stone to great things, all of which followed very quickly. In 1882, Wordsworth was to have his fellowship restored by Brasenose. In 1883 he was appointed to the Oriel Professorship. And in 1885, aged only 42, he was made Bishop of Salisbury. In spring 1881, John Wordsworth had clearly taken off; Humphry Ward, meanwhile, was pigging it in lodgings in London and—if things went wrong—might find himself positively destitute by the end of the year.

At the lecture on 6 March 1881, Wordsworth's subject was 'the present unsettlement in religion'. Its polemical theme was that current disputes were attributable not to intellectual questioning (which he uncompromisingly aligned with 'pride' and 'avarice') but to 'moral causes' and old-fashioned 'sin'. Pater's cult of homosexuality was alluded to in a veiled way. Mary Ward—evidently not a guest of honour despite being a Brasenose tutor's wife—seethed. In her hypersensitive state, the lecture seemed a direct hit at her and hers:

> I sat in the darkness under the gallery. The preacher's fine ascetic face was plainly visible in the middle light of the church; and while the confident priestly voice flowed on, I seemed to see, grouped around the speaker, the forms of those, his colleagues and contemporaries, the patient scholars and thinkers of the Liberal host, Stanley, Jowett, Green of Balliol, Lewis Nettleship, Henry Sidgwick, my uncle [Matthew], whom he in truth—though perhaps not consciously—was attacking. My heart was hot within me. How could one show England what was really going on in her midst.[29]

The best way to alert the nation was, she decided, an anonymous pamphlet. Entitled 'Unbelief and Sin: A Protest addressed to those who attended the Bampton Lecture of Sunday, March 6th', this counterblast was printed and put up in the shop window of Slatter and Rose, in the High Street. (It rather pleased Mary that the same establishment had published Shelley's notorious atheistical pamphlet.) 'Unbelief and Sin' opens with an attack on Wordsworth.

It then—by way of illustration—contrasts the careers of two hypothetical students, called 'A' and 'C'. 'A' is a believer, but has read history and Christian sources 'critically'. 'C' is blindly obedient to whatever the Church and holy writ tell him; he has no critical faculty. Unsurprisingly, 'A' cuts a rather nobler intellectual figure than his opponent.

There is no evidence that England was particularly roused by the abstruse dialectic of Mary Ward's defence of the Liberal host. And indeed, the country would have had to be fast off the mark. After 'Unbelief and Sin' had been selling a few hours Dr Edmund Foulkes—a crotchety ecclesiastic of Wordsworth's party—drew the attention of the bookseller to the fact that the brochure bore no printer's inscription and was therefore illegal. Proceedings were threatened, and the bookseller nervously withdrew the offending pamphlet, returning a bundle of unsold copies to Bradmore Road.

As was to be her practice in later life Mary Ward had sent pre-publication copies of the anonymous pamphlet to her various friends in high places. They duly felicitated her. As she reports, Green of Balliol came up to her at a college party on the evening of publication day, his kind brown eyes smiling upon her as he said a hearty 'thank-you', adding, 'a capital piece of work'.[30] Mark Pattison wrote a rather frigid letter saying that he had not at first guessed her authorship of the piece: 'It was whispered to me in the street, and I fancy was no secret within the first week of publication. I admire your courage in attacking one of their strong places.'[31]

Unbelief and Sin was Mary Ward's first publication of any intellectual pretension and—as she claimed—was the 'germ' or 'first sketch' from which *Robert Elsmere* sprang five years later. This, however, is to overrate the piece, which remains an arid and parochial item of Oxford controversialism. The significance of 'Unbelief and Sin' is that with it Mary defiantly burnt the Wards' bridges behind them (most of all with the introductory harsh comments on Wordsworth). It would now be wholly impossible for Humphry to go back to Brasenose. Its publication—which she must have known would be tracked back to her—put her in the position of cocking a snook at Oxford and all that it now stood for.

It was not until July 1881 that Humphry was informed by John Walter, proprietor of *The Times*, that he had indeed given satisfaction and was offered a permanent position on the paper's political staff. The house at Bradmore Road was promptly taken by the Henry Nettleships, congenial 'Liberal' friends. They paid £2,500 in two instalments. This quick sale relieved some of the financial pressure but the family would still have an unsettled few months. The Nettleships naturally wished to move in over the long vacation, and Humphry had had no chance to find anything suitable in London. Mary and the children were therefore obliged in August to decamp to cheap lodgings at the seaside in Eastbourne while Humphry scouted the West End property market.

Mary Ward was tortured by various vague ailments over this nerve-racking summer. In early June, she reported a 'violent toothache', which she numbed with large doses of linseed oil, laudanum, and sal volatile. Her toothache was to last two months, driving her to 'hysterics of pain' in which she wished that she could 'beat herself',[32] but instead could find relief only in drugs. Nevertheless, she continued to write letters of extraordinary courage and warmth to Humphry who was (even after his July interview) chronically low. He hated the five days of 'nightwork' a week which writing for the leader page involved. Habituated to Oxford's heavy dinners and late morning lie-ins he simply could not adapt his sleeping pattern to that of the London journalist. Indeed, he never would. Sleeplessness led to chronic depression, which he poured out in letters to his wife. Insidiously taking on more and more of the emotional authority in their marriage, she manfully boosted his flagging spirits. On the occasion of the tenth anniversary of their engagement in June, she summarized their first decade glowingly:

> Ten years of happiness in spite of all minor drawbacks and tiny worries, that is what you have given me darling, ten years of real and great happiness and constant tenderness and care and cherishing. May every girl of 20 who meets her lover today be as happy with him ten years hence as I am with mine . . . . Darling, the next ten years of our life, if this present year goes well, will I hope be even happier than the last. The load of money worries will be lifted off, our children will be more and more interesting, you will like your new work better than the old, we shall have more of a part in the world, and we shall play it strait worthily and so as to help other people as well as ourselves.[33]

A letter like this would have raised the spirits of Job.

There were, however, causes for melancholy which depressed even Mary. In mid-July Dean Stanley, who had married them, died. His death was the occasion for a meditation on religion which indicated that her views were now set in the theistic mould that was to last the rest of her life. 'I have been sitting in the Parks tonight', she told Humphry on 19 July, 'thinking of *historical* religion, and of the basis history affords for faith in God and a spiritual life.' Doctrine could no longer stand. Conscience, she determined, was one key; intellectual endeavour the other: 'Conscience within and knowledge without. God's inner and outer laws, these are the guides of the future.'[34] It was an extraordinary letter for an Oxford housewife to have written to her husband. As she sat in the Parks on that long summer evening, Mary Ward had clearly felt an annunciation. The future would hold great things for her.

Mary Ward left at the end of July for Eastbourne with her three children. Humphry had found them lodgings near Beachy Head, 'out of the crowd round the Pier'.[35] Dots and Arnold ('a regular little fish') looked sweet in their paddling gear, she told Julia, and baby Janet was well. Humphry managed to be with them at the weekends, looking 'browner and ruddier than he has for a long time'.[36] On the whole, despite the toothache, it was a good time for them. On 25 August, Mary returned for a last farewell to Bradmore Road. The sunflowers, asters, and geraniums were still making a gay show. It was sad to leave it, but she could not help noticing how *small* the house was and she shook the dust of Oxford from her shoes without too much anguish.

And anyway, Humphry had found them much grander accommodation in London in late July. It was at 61 Russell Square—by the British Museum and within walking distance of Printing House Square. They would move in during November. As she told Willie, 'It is a delightful rambling old house dated 1745, quite unlike a London house, with some pretty panelled rooms, a most taking front staircase only going up to the drawing room and quite a nice bit of garden.'[37] The house (which was pulled down at the turn of the century) had pretty stucco ceilings and cornices and good wall decorations up the staircase.

The only problem was the sanitation, which was original. (There was a large lead cistern on the front wall, marked '1745'.) The drains were revealed to be in such a dilapidated state that the sale looked likely to fall through. But the pipes were patched up, at least sufficiently to allow the parties to come to terms. Mary had serious doubts about the kitchen (she was already thinking about entertaining), but on the whole felt that they had 'rather a bargain'.[38] The rent was £100 a year for ten years, which was reasonable. The 'premium' (key money) was £900, and constituted more of a problem. They would have to borrow £600 of this, and repay in six years.[39] It would leave them very short, despite Humphry's doubled income. Mary was reduced to cutting back her remittances to Julia to an occasional couple of pounds. But the poverty no longer ground her down, as it had these last years in Oxford. A new world was opening up for them; and these were only temporary inconveniences. The future, Mary Ward knew, held great things.

8

London: 1881–1886

MARY had hoped that the Ward family might be united in London as early as 29 September 1881. But, in the event, reunion was delayed until November by necessary work on the new house. It was frustrating, but exciting as well. Since the Wards were going up in the world and intended to entertain grandly (at least by their Oxford standards) Russell Square was fitted out as expensively as Humphry's new salary and credit would stand. It was an altogether grander establishment than Bradmore Road and Mary took a personal interest in the interior decoration. The nurseries (two of them) should have blue paint and linoleum. The governess's room should have terracotta paint and red paper, as should the servants' room (the two of them would double up). The Wards' master bedroom should have yellow paper and paint; the drawing-room Japanese paper and pomegranate or blue marigold carpeting. The dining-room should have green paint—'the same colour as the Paters',[1] Mary decreed with a small gesture towards their past.

She and Humphry and even the children were revelling in remodelling their London home, she told Willie in an unusually chatty letter in September: 'We are getting a good deal of amusement out of doing up the new house. We have all the paint and paper and every important grate to choose . . . have you seen Norman Shaw's concrete chimney pieces? We are to have them in the drawing room. They are quite lovely and very cheap.'[2] To Mary Ward's satisfaction, Russell Square afforded the luxury of a study of her own. It was, in fact, only a converted 'powder closet' and tiny, but it looked out on to the big plane trees in a quiet corner at the back of the house. In later life she romanticized it as the place where in 1885 she outlined the 'germ' of *Robert Elsmere* to her sister Julia and where three years later she wrote the last words of the interminably revised narrative.[3]

To cap what had been a momentous year, Humphry and Mary Ward spent Christmas 1881 in Algiers. The visit was to see his dying aunt. Accustomed to Lake District gloom, Mary found African scenery terrifyingly sensual. As she told Willie, 'We have had days of astounding beauty . . . sometimes from sunrise to sunset the sky is absolutely cloudless and one sits out or strolls all day basking in the most luminous sunshine I ever saw. All round us are orange gardens and the woods are beginning to be filled with narcissus and purple iris. It is a heavenly climate but how anybody manages to do any work here I cannot imagine.'[4] Worker that she was she wrote her Algerian experiences up for the *Pall Mall Gazette* and frugally stored them away for the last chapters of *Robert Elsmere*.

1882 was an eventful year but not all the events were pleasant. Disappointingly, Humphry came to grief at *The Times*. The paper went to press very late, and leader writers (who had to absorb the night's parliamentary business) rarely got home before four and sometimes not until five in the morning. The hours of work were altogether too much for Humphry. And his star fell with the resignation of Forster in May 1882. He no longer had a Cabinet Minister in his pocket. Exhausted, Humphry took a holiday by himself in Wales in early October. He was in very low spirits (depressed further by news of J. R. Green's terminal condition). Mary confessed to an 'absurd nightmare lately of your being hurt in shooting'.[5] She was apparently anxious about his set-backs at work and may have suspected that her husband was suicidal.

The situation was resolved shortly after Humphry returned to London. In November, Mary told her father the good and bad news that 'Humphry is sleeping better and therefore getting through his work easily. It is now settled that he is to do the art criticism for the paper, which I think he will like and which will at any rate let him off night-work at certain times in the year.'[6] It was not a disaster, but it was a distinct come-down. As a political-leader writer, Humphry could hope one day to be editor or second-in-command of the paper. As art critic he would be a respected member of the staff, but never more than that. His prospects were further reduced when Chenery died of a heart attack in 1884, and *The Times*'s brief flirtation with Oxonian dilettantism came to an end. Humphry was to write occasional second and third leaders (especially in the summer months, when colleagues were on holiday) and some important obituaries and reviews over the next forty years. But in all important respects he had found his level and it was not exalted. Nor did it become more so. As the years and decades passed, the art critic of *The Times* was to become a standing joke in England. Humphry never could come to terms with 'modern' art and loathed anything more radical in style than what Pater had championed in the 1870s.

In the immediate disappointment of Humphry's 1882 demotion Mary Ward thought seriously about his trying for another fellowship. But he was no longer competitive at Oxford. In 1884 there was some thought that he might go into Parliament as a Conservative. (Mary Ward—who was nothing if not flexible—was at the same period urging Willie to enter as a Liberal.) But Humphry's parliamentary aspirations also came to nothing; so too did a later scheme to buy the *Saturday Review* for him to edit. In July 1887, Mary excitedly informed Willie that 'Humphry is to do the authorised life of John Bright—but this is *between ourselves*.'[7] He didn't. Mary Ward finally reconciled herself to the fact that she was married to *The Times* art critic. She did not mope about it but very rationally diverted her ambition on to her own and (in the not too distant future) on to her son Arnold's career. Whatever else, Humphry was secure and good for a thousand to two thousand a year until retirement. She could now concentrate on making her own way. Humphry had had his big chance in life and fluffed it; she had not yet had hers.

It was one of Mary Ward's principal strengths that she was able to think and act strategically. Oxford required one strategy; London another. Her ensuing metropolitan campaign was conducted on three fronts: domestic, authorial, social. The first—though important—was least important. She would have no more children (although biologically another twelve to fifteen child-bearing years remained to her). The larger family income meant that direct care of the children's educational and physical needs could be delegated to 'Fräulein' (the German governess), to the servants, and to nurse.

The larger ménage of Russell Square meant a burdensome management role for the 'lady of the house' and (with whatever inward sighs) any normal Victorian wife would have embraced it as a God-given duty. Not Mary Ward. She recruited a 'companion' to handle the day-to-day running of 61 Russell Square. This was Gertrude Ward, Humphry's younger sister, who joined the household in autumn 1882. She was 23 and had just finished her education at Somerville. Gertrude was an intensely devout, High Church young woman, who was to remain at Russell Square for eight years. As they evolved, her duties were the kind for which you needed the proverbial ten pairs of hands; she was part Mary Ward's confidante, part lady's maid, part secretary, part housekeeper. In the later years of her wealth and fame Mary Ward would have an entourage of female attendants to do these things for her. Until she left in 1891 to live her own life as a nurse, Gertrude was a factotum.

By the stifling standards of the 1880s, Mary Ward was less than a model wife and mother. A model wife would have bred and bustled in the kitchen and children's rooms. Mary protected herself against any sense of matrimonial guilt by various defence mechanisms of which the principal was invalidism. Until the end of her life she was to be forever afflicted by an array of mysterious ailments which totally disabled her from running the house but which—even more mysteriously—did not prevent her from being one of the most energetically productive women in the history of English literature.

Excruciating symptoms of Mary Ward's most persistent affliction, crippling cramp in her right arm and side, began in November 1882, at a period when Humphry was in despair about *The Times* and when Julia was at a particularly low ebb in finances and in health. Mary wrote to Tom on 25 November 1882 to inform him that 'I am just sending mamma another £5. She seems quite without money .... I am afraid the disease [cancer] has made progress.'[8] On his part, Tom continued to shield himself from confronting his wife's extremity by stoking his vexation at her 'over-expenditure'. If she simply looked to her accounts, all would be well. Nine months later, in August 1883, Mary wrote again in desperation to ask her father whether he could do *anything* 'to make Mamma less miserable and harassed .... She cannot get away from home as she has not a shilling wherewith to pay servants' wages and necessary things [and] she is feeling ill again, the lumps in her neck are giving trouble'.[9] Despite the operation in early 1881, the cancer had evidently metastasized to Julia's lymph glands. By March 1882, pain had returned, and she glumly informed her daughter that she foresaw for herself 'a hard ending to a hard life'.[10]

At this fraught period, Mary Ward began to develop symptoms strangely similar to her mother's. In January 1883 she declared herself crippled down her right side: 'There are some symptoms like writer's cramp and others quite unlike.'[11] In February, she confidently reported the disorder to be the result of 'an inflamed gland in the axilla'.[12] In April, she wrote telling her mother that she had just seen Doctor Poor, who bluntly informed her that she need not expect to be well for months: 'according to him there is no gland worry only a muscle slightly thickened under the arm and a general congestion of the nerve centres'.[13]

Whatever the diagnosis, Mary Ward was so disabled that she was occasionally obliged to write with her left hand, or dictate to Gertrude. She was in chronic pain. For much of the later part of 1884, she had her right arm in a sling. Earlier in the year, she experimented with a typewriter, which she found too cumbersome. In August, she was trying 'some strong arsenic tonic'. In December 1884, she consulted a new German doctor, Julius Wolff, 'who is said to be wonderful in the case of nerve ailments'.[14] Wolff (as Janet recalled) transformed Mary Ward's method of writing by making her sit much higher than before, rest the whole forearm on the table, and use an altogether different set of muscles. She thus wrote much of her early fiction perched on 'a miscellaneous assortment of music books'.[15] Wolff seems to have been more a quack than a physiotherapist—and was the first of many such Ward consulted. During her long and protean illnesses, Mary would discuss the virtues of powerful drugs like morphine and cocaine ('it acts like magic')[16] with her mother. It is likely that already she was experimenting with various kinds of self-administered pain-killers.

Invalidism was one way Mary reorganized the Russell Square and subsequent Ward households around herself. Self-confessed incompetence was the other. Anecdotes about mother's being unable to look after herself were legendary in the Ward household. A couple are lovingly recalled by her daughter Janet:

Mrs Ward could never be trusted to keep her small possessions, unaided, for very long, for being entirely without pockets she was reduced to the inevitable 'little bag', which naturally spent much of its time down cracks of chairs and in other occult places. When her advancing years made spectacles necessary for reading and writing, these added another complication to life, but fortunately there was always some willing slave at hand to aid in recovering the lost [object]—or rather her family would half unconsciously arrange their days so that there should be some one. Once she declared with pride to a friend that she had travelled home *alone* from Paris to London without mishap, but on inquiry it was found that 'alone' included the faithful Lizzie [Smith, her personal maid from 1881 to 1920] and only meant that, for once, neither husband nor daughter had accompanied her.[17]

Look at her business correspondence with Smith, Elder or at the minute books of the management committees of the Passmore Edwards Settlement and a wholly different Mrs Humphry Ward emerges: a creature of will and razor-sharp competence. Nevertheless, Mary Ward contrived a pampered domestic existence for herself. As early as July 1886 she told Julia that 'I always now have breakfast in bed at 8, as it seems to suit my work best.'[18] Few healthy mothers of three nursery-aged children began their weekdays in such leisurely style.

Nor over the years was this pampering resented by others in the Ward family. It was an accepted thing by 1886 that Mary was the *de facto* head of their household. Her actions had the authority of a matriarch's. It was she, for instance, who negotiated an offer of secretary to the Clarendon Press for Willie in 1886. (Despite the £1,000 salary, the free house, and his sister's urging he declined; Oxford had already nearly destroyed him once and that was enough for Willie Arnold.) It was Mary in 1886 who got together with Uncle Matt and Humphry to propose that Tom be elected to the Athenaeum ('it is one of the few honours open to literary men in this country, and I should like it so much.').[19] One suspects that it was she who prevailed on her uncle to get Humphry into the club at the same time. Membership was to be the major consolation of his later life. It was Mary Ward who made all the essential decisions about what should be done for the dying Julia.

This growing authority of Mary's was unmanning and obscurely shameful for Humphry Ward. Their relationship since 1871 had evolved through three phases. In the first heady days of their engagement, they had made a pact-like declaration of equality with 'A Morning in the Bodleian'. Written by 'Two Fellows', this charming pamphlet had symbolized the communism of their relationship, based as it was on the camaraderie of high scholarship. Within the marriage, Mary was initially subordinate, as the mores of the age required. A dutiful girl-wife of 21, she helped Humphry (27) prepare his lectures and did little research tasks for him. She bore his children. He on his part taught her Latin in the early mornings—a comely daily submission on her part. She called herself by the chattel name, Mrs Humphry Ward. But over the years, as his career see-sawed down and hers up, the balance of power within their marriage altered.

In the early years in London, Humphry and Mary Ward were on quite different tracks, and behaving suspiciously like rivals. As she threw herself into writing a vast number of articles, he was groaning in his diary about the vast number of columns *The Times* required of him. Her articles were better received than his columns. Now, when they collaborated, it was a sign of his inability to come up to the mark. Thus, in 1881, he was assigned to review John Morley's *Cobden* for the paper at short notice. As Mary recalled, it was too short for Humphry: 'we divided the sheets of the book, and we just finished in time to let my husband rush off to Printing House Square and correct the proofs as they went through the press for the morning's issue'.[20] One infers that Mary stepped in to save Humphry Ward's bacon.

As Mary Ward was writing *Miss Bretherton* (1884), hemiplegic with writer's cramp, and *Robert Elsmere* (1888), 'with my heart's blood',[21] Humphry was slaving on *his* books—*Men of the Time* (1885) and *The Reign of Queen Victoria* (1887). But with the success of his wife as a novelist in 1888, Humphry's career as a writer (or compiler) of books came to an abrupt end. He could clearly no longer compete with a woman who pulled in £10,000 for a single work and who was regarded by prime ministers as the voice of the age. Mary outshone him, outwrote him, and outearned him. As what seems an act of psychic compensation, Humphry began to speculate in pictures. It was a form of gambling which first becomes noticeable in his diaries of 1886–7. Large sums were involved at times when the Wards clearly could not afford large sums. In August 1887, for instance, Humphry gave a London dealer £800 for a Velazquez.[22] (Thirty years later, it was revealed to be a fake.) At the time, a few hundred pounds spent on the Russell Square drains (which regularly collapsed and drove the family out of the house) would have been a sounder investment. In 1892, all the profits on the house that the Wards had built at Grayswood (around £6,000) went on a Cuyp which Humphry decided he *must* have.

Some of Humphry's paintings were displayed at home. In the years of their greatest wealth, the Wards' walls at their country house Stocks resembled a small National Gallery. (The prize items in his personal collections were his Burne-Joneses). But collecting and hanging were lesser pleasures; buying and selling provided Humphry's real thrill. His strategy was to buy low (usually direct from dealers or sometimes junk-shops) backing his expertise to pick undervalued items (typically old masters) which he would later sell high through Christie's or back to West End galleries. He had the occasional stroke of luck (on one glorious occasion he found a dusty Rembrandt in a junk-shop for a few shillings). But altogether Humphry's speculations in the art market were a constant drain on the family finances, and on several occasions he burned his fingers very badly indeed. In letters that survive, we find Mary—with great tact—imploring Humphry Ward to be careful in his buying which typically he indulged most recklessly when she was not by his side. 'You tantalising boy,' she chided him in August 1888, 'to make £52 10*s*. in solid money and then to spend it on a picture which *may* be as doubtful property as the head you brought home last week or the Claude! I can't help groaning. If only you would keep what you make.'[23] If only.

Although the precise financial details of Humphry's dealings were destroyed (most of them by Dorothy Ward, after his death) it seems clear that he was an inveterate gambler and pictures were the form his gambling took. Perhaps the vice was genetic. His father before him had speculated unwisely in business. His son Arnold could not—despite all his gallant resolutions—stop himself playing 'hated bridge'[24] (as his mother called it) and billiards for high stakes. Mary was extraordinarily indulgent of this weakness in both Humphry and Arnold Ward, although it eventually became the heaviest cross she had to bear. In her husband's case, she put it down to 'nerve pressure'; in Arnold's to high spirits. One wonders whether either of the male Wards would have become gamblers had this female been less spectacularly successful. Was this their attempt to catch up with her? Or was it a way of pulling her down to their level again?

Authorially, Mary Ward's London strategy was based (at least initially in 1882–5) on her becoming a metropolitan 'man of letters' like Edmund Gosse or Andrew Lang, both of whom she was now able to observe at first hand. This meant writing with fluency and omniscience on a diversity of topics. At J. R. Green's suggestion she wrote to Willie at the *Manchester Guardian* in September 1881 bluntly proposing a series of weekly papers of 'Foreign Table Talk'[25] concocted out of the French, German, Spanish, and Italian papers. £4 a week she thought would be reasonable reward for this work. The paper's editor was taken with the idea of Mrs Ward's freely ranging over four national literatures and in December 1881 she got the commission on three months' trial.

The *Guardian* work would have been enough for most tyros feeling their way into the literary world. But Mary Ward was voracious. She used Humphry's connection at *The Times* to drum up further assignments from them. Some (such as her leader on Anthony Trollope's death in December 1882) were plums. She also did theological and literary reviews for the paper (e.g. on 'Spanish Novels' in December 1882), which had to be completed to short order. All this while, she was writing a weekly 'London Letter' on politics (particularly Irish politics) for the *Oxford Chronicle*. There must be scores of these incidental writings between 1880 and 1882 which have never found their way into any bibliography of her work. It is fair to say that Mary Ward accommodated to journalistic pressure much more easily than her journalist husband. She was twice the hack he was. She off-handedly complains in a letter of October 1882 that 'three or four volumes of these books a week is about all that I can do and that seems to go no way'.[26]

Through J. R. Green Mary Ward had for some time had an entré to the *Pall Mall Gazette*. Its editor was John Morley, whose *On Compromise* (1874) was a formative text in her intellectual development.[27] She first met him at Jowett's table at Balliol in the late 1870s, and as she walked home later with Humphry they agreed they had been in the presence of 'a singular personal power'.[28] In London, Mary Ward came to know Morley well after opening a correspondence with him in December 1882 on the matter of his criticisms of W. E. Forster's uncompromising Irish policy. The good-natured Morley (who declined to be affronted by Mary Ward's partisanship) and the *Pall Mall* furnished another string to her bow. She also wrote to Alexander Macmillan in October 1882, asking if he would solicit the then editor of *Macmillan's Magazine*, George Grove, to take a couple of articles from her on modern Spanish writers. The magazine in fact took only one. But so impressed was Macmillan with Mary Ward's dash and energy that he seems seriously to have considered her as editor for his magazine when Grove indicated his intention of going to the newly founded Royal College of Music.[29]

In fact the editorship of *Macmillan's* went to John Morley in May 1883, who held it three years until Parliament absorbed all his energies. But this disappointment had its consolations. In March, even before taking up his post at the magazine, Morley wrote to ask Mary Ward to write a monthly '*compte rendu* of some new books, English or French. It is highly desirable that the subject should be as lively and readable as possible—not erudite and academic, but literary, or socio-literary, as Ste. Beuve was.'[30] Whatever else, Mary Ward had sloughed off her Oxford bluestocking image. Morley (who became Chief Secretary for Ireland in 1886 and later Gladstone's biographer) clearly rated her as a writer of flatteringly cosmopolitan range and character.

Mary Ward wrote some fifteen *causeries* for Morley between June 1883 and September 1885, ranging over such topics as 'Renan's Autobiography', 'The Literature of Introspection', and 'French Views on English Writers'. She was meanwhile still turning out the odd piece for the *Saturday Review*, the *Athenaeum* and the *Quarterly*. Much of her 1880s journalism is anonymous and will probably never be known about. But it would seem that between summer 1881 and autumn 1885 she wrote around fifty long and 200 short articles—some million words. It was an astonishing release of writing energy. More so as she was still doing an occasional entry for the *Dictionary of Christian Biography*, examining Spanish at Oxford,[31] writing her first novel, and was all the while 'crippled' with writer's cramp—a disability which seems in the circumstances highly ironical. But all this effort was, as she insisted, so much preparation for the 'serious' book she really wanted to write. What that book actually was, or even what kind of book it might be, remained uncertain.

The Wards' social activity in one sector overlapped with Mary's writing activities. She was very much the Macmillans' author in the period 1881–5, and met a number of star writers through the connection. (She also helped make a star out of an obscure and eccentric Birmingham writer, J. H. Shorthouse. It was Mary Ward who persuaded Alexander Macmillan to accept *John Inglesant* in 1881.)[32] The Macmillan connection entailed hobnobbing with people like the publisher and his wife, the Groves, the Gosses, John Morley, and above all her uncle Matthew Arnold—with whom Mary was at last socially intimate. The Arnolds dined at Russell Square as often as three times a month. Humphry and Mary Ward would also go to Uncle Matt's at Pains' Hill Cottage, Cobham. But these excursions were not to Mary Ward's already rather priggish sense of *comme il faut*. According to Janet, her mother would usually return from Cobham 'full of blasphemies about [Uncle Matt's] precious dogs, who had diverted their master's attention all through the walk and prevented the flow of his wit and wisdom'.[33]

In her early London days Mary Ward was still nervous about meeting purely literary (as opposed to academic) grandees. On 29 November 1882 she wrote to her father with some excitement that she was to be introduced to Henry James and F. Anstey at the Andrew Langs' the following evening. James had recently serialized *The Portrait of a Lady* in *Macmillan's*, and was a real catch. (Anstey—a *Punch* man who had just written the best-seller *Vice Versa*—was less of a target; Mary Ward had no intention of being a comic writer.) This November dinner party was her first encounter with James and she followed it a week later with a forthright note which declared 'I trust we may have other opportunities of seeing more of you and making friends. I am nearly always at home late in the afternoon and it would give me great pleasure if when you are able you would call on me.'[34] In fact James was off to America the next day—his father just having died—and he could not obey Mrs Ward's summons. But on his return the required visit was made, and a friendship which was to last thirty-five years was formed.

Friendship is perhaps the wrong word. Mary Ward was hungry for personal discipleship. The charismatic American novelist (only 39 and still known as 'Henry James, Junior') filled a main vacancy left by the dying ranks of her old father figures ('masters', as she would frankly call them). J. R. Green died in 1883; T. H. Green had gone in 1882. Pattison was lingering with a fatal cancer which would kill him in 1884. (Mary visited the ashen-faced and bedridden Pattison in London; from his deathbed he laid upon her one last Sisyphean task—the translation of Amiel's journal.) James, with his Gallic ideal of the artist's pure mission, diverted Mary Ward's aspiration into new courses.

Like many great men exposed to the oven heat of Mary Ward's worship, James was always somewhat baffled by her. As he wrote to Gosse in August 1895, 'She is incorrigibly wise and good, and has a moral nature as Patti has a voice or Tessa a *chevelure*; but, somehow, I don't, especially when talking art and letters, *communicate* with her worth a damn. All the same, she's a dear.'[35]

Partly it was that James did not understand the needs he supplied for her. Mary Ward wove round the American novelist a fantasy in which she played the part of the loved daughter she always yearned to be ('You do love me, Papa?'). The nature of their relationship, as she saw it, is captured in her recollection of an early visit by James to Russell Square in summer 1885:

It was a very hot day; the western sun was beating on the drawing-room windows, though the room within was comparatively dark and cool. The children were languid with the heat, and the youngest, Janet, then five, stole into the drawing-room and stood looking at Mr James. He put out a half-conscious hand to her; she came nearer, while we talked on. Presently she climbed on his knee. I suppose I made a maternal protest. He took no notice, and folded his arm round her. We talked on; and presently the abnormal stillness of Janet recalled her to me and made me look closely through the dark of the room. She was fast asleep, her pale little face on the young man's shoulder, her long hair streaming over his arm.[36]

It took a few years for the Wards' social life in London to find its distinctive pattern. In general, they conformed to the seasonal rhythm, with social engagements reaching a crescendo in July (when they would be engaged every night) followed by a summer in the country after Parliament recessed for its long annual break. At this period they could not afford to keep a country establishment year-round. From 1883 to 1889, every summer they rented half a dozen front rooms at Borough Farm, near Hindhead in Surrey. It stood in an unspoiled wilderness of common and wood, approached by a sandy track leading from the main Portsmouth Road. Visitors could be entertained (though not overnight) and Humphry could come down at weekends from his work in London. The Frederic Harrisons were neighbours at Elstead. At Borough Farm Mary would read, study classics (with the aim of keeping abreast of little Arnold), English literature, and theology (on which she would write long thoughtful letters to Tom). During these summers she was at her most invalid-like and would self-consciously recuperate from the ardours of London, recumbent on her *chaise longue* in the sunny garden, writing tablet in hand. She would lie long in bed in the mornings, and have the children read to her, or Arnold would proudly instruct her in Greek. During the day, she would sometimes sketch or paint (at which she was surprisingly expert).

In town the Wards were by any standards a gregarious couple and threw themselves into what Mary Ward called 'the whirlwind of London life'.[37] By 1885, they were routinely engaged every night in the busiest summer season and having dinner parties at Russell Square up to three times a week. In the mid-1880s, Mary Ward began reserving Thursday afternoons as an open day for any interesting person who wished to call. Her 'Thursdays' became particularly popular with the literati after the success of *Robert Elsmere*, when her fame acted as a magnet. At their height, as many as forty guests would cram into Russell Square to take tea and make conversation. Mary Ward kept her weekly salon going until the outbreak of war in 1914.

In the early 1880s, the Wards occasionally went to the theatre or concert—but dining in and out was their principal evening activity. Their guest lists mixed old friends with new, distinguished with humdrum. For instance on one memorable (but by no means unique) night, Saturday, 6 March 1886, they entertained to dinner Willie and Henrietta Arnold, the Fyffes, the Leckys, the Frederic Harrisons, Arthur Balfour and three Burne-Joneses;[38] that is, a leader writer on the *Manchester Guardian*, two of the country's leading historians, the country's leading Positivist philosopher, a future Prime Minister and England's most famous living painter. Gertrude who had recently recovered from a month's illness, made up the party.

The Wards' company on such occasions was drawn from a number of sources. There were new literary friends and old Oxford friends. Among the former was Robert Browning. Among the latter, the Paters, Charlotte Green, the Rawlinsons, the Thursfields. Max Müller and the Creightons—when in London—were frequent guests. Two of Mary's sisters, Lucy and Julia, had made interesting marriages, the first to Carus Selwyn and the other to Leonard Huxley. This drew into the Wards' net a contingent set of intellectuals. The Wards and the Sandwiths (Humphry's maternal relatives) had distinguished family members, who duly popped up in the guest lists which Humphry kept religiously in his otherwise supremely uninformative diaries. Visiting littérateurs like Edmond Scherer stayed at Russell Square. Humphry's attendance at the private views supplied access to the art world—the du Mauriers, the Alma-Tademas, Frank Holl, and Hubert von Herkomer were dinner guests from time to time.[39] There was an assortment of political notables such as Morley (wearing his other hat), Arthur Balfour, George Goschen (who was at Russell Square two nights before the historic 1886 Division on Home Rule), Sydney Buxton, Arthur Russell, the Edward Greys (from whom later the Wards bought their country house, Stocks).

The Wards also invited to their dinner table a sprinkling of guests who were merely interesting, like the super-sophisticated (and probably homosexual) novelist Hamilton Aidé, Dr Elizabeth Garrett Anderson (the feminist), visiting Americans like James Russell Lowell (whom Humphry had earlier tried to recruit for the *English Poets*), and the Oscar Wildes, who came to supper twice in 1886. Oscar was evidently *ennuyé* and later claimed that Mrs Ward's religiosity 'reminded him of the sort of conversation that goes on at a meat tea in the house of a serious Nonconformist family'.[40] Mary Ward had her revenge for this insolence in 1896 when she was among the 'well known somdomite's'[41] most implacable persecutors.

It was not enough for Mary Ward simply to be involved in the give-and-take of London social life—even if it did mean having the great and the good at her table several nights a week. As at Oxford, the milieu interested her principally in so far as she could use it, and in so far as it assisted her in fashioning herself into what she wanted to be. Fifteen years before, the first glimpse of Mrs Pat, all in green and silver, playing croquet in Lincoln garden with the sardonic Rector by her side had been Edenic: an allegory of the power and beauty that was Oxford. Writing to her mother in April 1885, she records a similar epiphanic moment. The previous evening, they had dined at the Pearses. Mary Ward sat next to John Morley at table. In front of them was a mass of the loveliest azaleas she had ever seen, while they talked about the *tout Londres*. At ten thirty, the party adjourned to a dance at the Charles Tennants' at 40 Grosvenor Square. Powerful political figures and decorative young men adorned the scene, hanging around the large drawing room, its polished floor stripped for dancing. A. J. Balfour was there (the Wards had recently met him at an Oxford weekend); Alfred Lyttelton –'in the zenith of his magnificence'; Curzon, then a rising Foreign Office star; Henry Cust; James Rennell Rodd (later the British Ambassador of Rome). There were also a company of beautiful women, notably the two young daughters of the house, Laura and Margot. As Mary told her mother, 'the half-lit room, the dresses and the beauty reminded me of a Watteau picture. What a distinct type of life it is. I have been pondering it a good deal this morning.'[42]

The glamorous Laura Tennant (later Laura Lyttelton) and Margot Tennant (later Margot Asquith) were a source of much such pondering to Mary Ward in the early 1880s. She had first set eyes on Laura at a London party a year before. Turning her back on an odious American, Mary Ward had suddenly become aware 'of a figure opposite to me, the figure of a young girl who seemed to me one of the most ravishing creatures I had ever seen. She was very small and exquisitely made. Her beautiful head, with its mass of light-brown hair; the small features and delicate neck; the clear, pale skin, the lovely eyes with rather heavy lids, which gave a slight look of melancholy to the face; the grace and fire of every movement when she talked; the dreamy silence into which she sometimes fell, without a trace of awkwardness or shyness.'[43] Laura was 22 years old in 1884, and had just come out. She was one of the eight children born to Sir Charles Tennant, the illegitimate son of a Glasgow merchant and himself enriched by railway development. Laura and Margot were quickly adopted into membership of the 'Souls'. As Pat Jalland describes them, this group of golden 1880s youth was 'a charmed circle of personages distinguished for their breeding, beauty, delicacy and discrimination of mind. They consisted of about three dozen members of the Wyndham, Talbot, Charteris, Curzon, Brodrick and Lyttelton families, with George Curzon, Alfred Lyttelton, A. J. Balfour, and St John Brodrick at their centre.'[44]

Mary Ward was very taken with the Souls. Like everyone else, she found them beautiful, accomplished and aristocratic—'an intellectual and social *élite*'.[45] But they also possessed something else: 'the maddest delight in simple things—in open air and physical exercise ... a headlong joy in literature, art, music, acting; a perpetual spring of fun; and a hatred of all the solemn pretenses that too often make English society a weariness'.[46] Mary Ward did not delude herself that she could join this exclusive band: she was too old (already); too earnest; married; already rather worn by hard work. But the Souls enlarged her sense of life, and what she wanted from it. 'I saw them dramatically,' she recalled, 'like a scene in a play, full of fresh implications and suggestions.'[47] These implications and suggestions would later—once they were absorbed into Mary Ward's planning—substantially change the course of her and her family's existence.

Mary Ward successfully attached herself to Laura. The young woman visited Borough Farm on Ascension Thursday in May 1884. She was to have come with Arthur Balfour and Burne-Jones. But in the event she arrived by herself. 'Our principal drawing room', Mary wryly recalled, 'was a sand-pit, shaded by an old ash-tree ... we talked all day, sitting under hawthorns white with bloom, wandering through rushy fields ablaze with marsh marigold and orchis.' On the same evening, from London, Laura wrote to Mary: 'I sit with my eyes resting on the medieval purple of the sweet-breathing orchis you gave me, and my thoughts feasting on the wonderful beauty of the snowy blossom against the blue.'[48]

The Wards were invited to Laura's wedding when she married Alfred Lyttelton on 21 May 1885 at St George's Church, Hanover Square. Gladstone (Lyttelton's uncle) gave the speech at the breakfast. Eleven months later, on 24 April 1886, Laura Lyttelton died from complications following the birth of her first child a week earlier. Mary Ward, who visited frequently during the pregnancy, observed in her friend 'a strangely strong presentiment of death'.[49] Nevertheless, the news (which she received at Borough Farm, exactly two years after Laura's visit there) devastated her. 'I think I was simply in love with her from the first time I ever saw her,' Mary told her mother. Social life henceforth was 'a mockery'.[50] Two years later she made Laura the posthumous dedicatee of *Robert Elsmere*, coupling her name with that of T. H. Green. The young, doomed girl and the grey, patriarchal philosopher expressed in their different ways the tension of Mary Ward's whole life.

Mary Ward was profoundly influenced by this relationship, and the whole milieu of which Laura was a principal ornament. But it was not just the elegance and beauty of dynasties like the Cecils, the Lytteltons, the Tennants (and later the Asquiths) which entranced her. It was their power which fascinated Mary Ward. The lifestyle—its balls, country parties, aristocratic pastimes—was intergrafted with the high politics of England. These were also the families who ran the country from their high places. In Oxford, the people who 'counted' were scholars like Pattison or 'Masters' like Jowett. Their mode of life had its amenities, comforts, privileges, and even luxuries. But it was not beautiful. Mrs Ward, in her Gatsbyish perplexity at the Souls and their set, was absorbing a whole new set of ideals. After she came to full emancipation, with her huge earnings from *Elsmere* and its successors, it was to be her great aim in life that her son Arnold should have full membership of this set. A town house in Grosvenor Place (adjacent to the Tennants), a country house, and Conservative politics were all part of the package that Mary Ward put together in her campaign to *be* someone. It was not mere snobbery or social climbing; rather that her experience in 1880s London had given her a new sense of what the 'top' was.

9

The Right Book: 1883–1884

THE Wards' early years in London were among the most hectic and engrossing of Mary Ward's life. They should also have been fulfilling. She was a successful hostess, a successful journalist, a successful mother and wife. Yet she seems never to have relinquished for a minute her ambition to write a great book. Every other success was secondary. But what kind of great book? Clearly her first effort—*Milly and Olly*—was a false start, even if it had been a hit with the junior reader (which it wasn't). For her first five years in London Mary Ward remained unsure whether destiny called her towards a work of literary criticism, of theology or of fiction.

Matthew Arnold's example still had a powerful lure, more so now that they saw each other socially. On 19 July 1883 she wrote to Alexander Macmillan a serious proposal 'which I should never think of making if I did not feel that you were in every way my friend'. Her promised work on early Spanish literature was to be postponed indefinitely.[1] What she offered Macmillan instead of the Visigoths was a 'good critical account of the Romantic Movement in France'. Such a project, she felt, 'could scarcely fail to attract readers'. She would, she candidly explained, need an advance for such an 'attractive' proposition: 'I cannot take time from the weekly and monthly tasks which bring in money in order to spend it on a book on a large scale which will bring in no immediate return.' She coolly suggested £250 as an appropriate advance. If the request were 'quite inadmissible, dear Mr Macmillan, please tear up my letter without showing it to anybody'.[2] The sum demanded was—on the face of it—exorbitant. Since earning her first £5 in 1870, Mary Ward's highest single payment for a work of literature had been the £60 Macmillan gave for her children's tale.

It is to Alexander Macmillan's credit that he declined to tear up the letter and duly forwarded £250 to his pertinacious (but clearly talented) authoress. He cannot have signed the cheque with an entirely easy mind. The work on French Romanticism was as yet quite notional. In fact it would never be written. But the large sum and the implication that she was a serious (that is, financially viable) writer did wonders for Mary's morale. On 22 July 1883 she wrote gratefully, 'your cheque and your letter together have made such a difference in my view of life!'[3] This £250 'loan' (as she euphemistically termed the straight cash advance) was to remain outstanding until November 1888, when Macmillan finally wrote it off against receipts from their colonial edition of *Robert Elsmere*. Ward in later life recalled, 'It was a great boon to me at the time and since we have been in easier circumstances I have often thought gratefully of the prompt friendliness with which Mr Macmillan answered my request, and of the uses which the money was to us.'[4]

The immediate use was for a holiday. In late summer 1883 the Ward family vacationed in Switzerland. While abroad, Mary Ward read the Swiss philosopher Amiel's *Journal Intime*, as extracted and transcribed in Edmond Scherer's 1883 edition. She had been put on to the work by her old Oxford mentor, Pattison. As she later told Macmillan, 'the Rector of Lincoln drew my attention to it, by saying that in *importance* it seemed to him that nothing of its kind had equalled it since Rousseau's *Confessions*'.[5] Pattison was dying when he made this recommendation and what attracted him as 'important' in Amiel's journal was its *ars moriendi* character, something which is heightened by the artful condensation of the massive text as it was published in 1883. The journal climaxes with a harrowing chronicle of protracted death. The last year of Amiel's life (1881) inspired entries of the most poignant stoicism, as the philosopher contemplated his coming expiration (literally) from bronchial asphyxiation. 'I shall die by choking,' he calmly noted, and turned to his journal for consolation.[6]

Mary Ward gives a somewhat misleading account of how she came to do the translation in her *Recollections*: 'I was soon to know Edmond Scherer more intimately. I imagine that it was he who in 1884 sent me a copy of the *Journal Intime* of Henri Frédéric Amiel, edited by himself. The book laid its spell upon me at once; and I felt a strong wish to translate it. M. Scherer consented and I plunged into it.'[7] Her imagination is faulty. In fact, Mary Ward made arrangements with Macmillan to translate the book immediately on her return from Geneva in September 1883. And she began the project evidently unaware whose literary property it was that she was blithely undertaking to translate. As it turned out, royalty arrangements with the Amiel estate were to be complex.

There were other complexities. Mary Ward proposed her translation to Macmillan on 6 September 1883. It could be done she suggested 'in six or seven weeks of hard work'.[8] As often the case with her, it was a grotesquely optimistic estimate. But she reinforced the proposal with the lure that Uncle Matt might be recruited to write an essay in *Macmillan's Magazine* boosting the translation's sales prospects. It was not on the face of it an overwhelmingly attractive proposition—even with the avuncular back-up. Speculative reminiscences by Swiss free-thinkers have not traditionally commanded a large public, even in Switzerland. Mary Ward had no evident qualifications as a translator from the French. But nevertheless, Macmillan offered her £100—half of which was paid in advance—and she went to work.

The Amiel translation was not the elusive great book that Mary Ward so desperately wanted but it was another formative influence on her intellectual evolution. As she yearned for fame, so she was fascinated by obscurity. Amiel's journal focused sharply her foreboding that she too was doomed to live what Matthew Arnold had termed a 'buried life'; that like Dorothea Brooke she would rest ultimately in an unvisited tomb. Amiel had been born in 1821 in Geneva, of French Huguenot extraction. He was orphaned at 12 and had a 'bare and forlorn' childhood. In young manhood, the strongest intellectual influences on him were Swiss Calvinism and German philosophy. In 1849, he was appointed Professor of Aesthetics and French Literature at the Geneva Academy. It was a brilliant launch to a career which subsequently went nowhere. Ostensibly a genius, Amiel published nothing, contributed nothing to the European debate in his subjects. As Scherer put it, 'he awakened in us one regret; we could not understand how it was a man so richly gifted produced nothing, or only trivialities'.[9] Pattison called him 'a soul petrified by the sentiment of the infinite'.[10] But, while wasting professionally, Amiel kept a massive journal, replete with 'the suffering which besets the sterility of genius' (Ernest Renan's phrase).[11]

Pattison could only too easily identify with a mind paralysed by its Germanic 'tyranny of ideal conceptions'. Mary Ward could similarly identify with 'a man who had signally disappointed the hopes and expectations of his friends'. What Amiel represented to her was an increasingly likely future without public success, consoled only by private hypersensitivity. Coming to terms with Amiel involved a wrestle with the paralysis which afflicted Mary through life—physically in the form of 'writer's cramp', artistically in the form of the crippling self-doubt that led her to write and rewrite her novels until the manuscripts looked like battlefields.

By 3 February 1884, Mary Ward had the first volume of Amiel roughly translated (so much for 'six weeks'). And she optimistically informed Macmillan that 'interest in the book is certainly growing'.[12] Always prone to busybody her publisher, she demanded to know how big the first edition was to be, suggested a publication date (entirely impractical) in May; first decided against notes then—after the work was set up in type—changed her mind; had '*Intime*' removed from the running title (again in proof); fussed about payments to Amiel's estate; worried about how best Matthew Arnold's and *The Times* reviews could be used. Macmillan finally received Mary Ward's complete translation in April 1885, and she received the balance of the agreed £100. It had, she admitted, 'been a long and arduous piece of work'.[13] She continued to forward advice as to how the publishers should now go about their business. Could the book not, she asked, be hurried out by 10 October, so as to 'get a good start of the political hubbub of the elections?'[14]

It couldn't. In fact, Amiel's *Journal* was not put out by Macmillan until Christmas 1885. Early copies were sent to Gladstone, the Lytteltons, and

Edward Talbot, the Warden of Keble College. 'I am most anxious the book should sell,'[15] Ward wrote Macmillan on the eve of publication. In fact, it did—moderately well (although a second and revised edition was not called for until March 1888). But the most vexatious thing was that the review by Uncle Matt was so long coming. To do the book good, it needed to coincide with publication in the new year of 1886. In fact, it was not until September 1887 that the piece finally appeared in *Macmillan's Magazine*.

Twenty-two months delay was bad enough. But the review itself was a bombshell. Uncle Matt, it emerged, disliked Amiel's 'Buddhist' philosophy intensely. He found the Germanic posturing absurd. Generally Arnold thought Amiel spineless as a thinker. He made some tepidly polite comments on Mrs Humphry Ward's translation and introduction but silently and quite extensively corrected her renderings from the French.[16] All in all, it was a strict review and coming from that particular quarter it was devastating.

Mary Ward pointedly does not mention Uncle Matt's review in her *Recollections*, reproducing instead the polite eulogies of Jowett and Pater on her book. But she felt his criticisms keenly. Perhaps he meant her to feel them and was cruel only in the interest of a greater kindness to his niece. The moribundities of Amiel's *Journal*—as Arnold perceived—represented a dead end for any young aspiring author. It was necessary that she too should be made to see this. Matthew Arnold's strictures on her début as a philosopher had the same usefully deterrent effect on Mary Ward as J. R. Green's rejection of her 'Primer of English Poetry'. These were not her *métiers*—or at least not unless she did years more work than she was prepared to do.

But it was not self-evident that fiction should be Mary Ward's *métier* although she had by 1885 completed and published her 'first serious attempt at a novel'.[17] (Matthew Arnold as it happened had been none too polite about that effort either, noting caustically 'No Arnold can write a novel; if they could, I should have done it.') *Miss Bretherton* arises directly from Mary Ward's involvement with 'beloved Henry James' after first meeting him in winter 1882. *The Portrait of a Lady* had recently finished its run in *Macmillan's Magazine* in November 1881. That novel and the conversation of its author vastly extended Mary Ward's sense of what fiction could be. There survives in a notebook (which also contains the opening chapters of *Miss Bretherton*) a 'Sketch of an Article' outlining her newly sophisticated theory of the novel in 1883–4. The notes hail the 'rise of a new school of novel' in America. Henry James is confidently proclaimed 'the head of the school' and W. D. Howells its 'most characteristic representative'. American fiction, as headed by James, offered (as she jotted down) a far preferable alternative to the soulless French commodity: 'How different from French realism which is like a last despairing effort to feel, to get sensations at any cost—the art of the decadence one feels with no promise in it. Whereas this American art has the promise of the morning in it.'[18] *Miss Bretherton* was duly conceived as a stylistic homage to James and the new vitalities of American literature.

Mary Ward's connection with James set up the general conditions for her to write a higher kind of fiction than she had hitherto aspired to. But the direct inspiration for *Miss Bretherton*'s theatrical plot was twofold. In May 1883, Mary's youngest sister Ethel had announced her intention of going on the stage. Mary wrote to an alarmed Julia (dependent on her younger daughter as nurse and housekeeper) offering her advice and Humphry's on the matter:

> Our feeling is that it will perhaps be better to let Ethel have her way in amateur acting, that so she may come to think less of the stage as a necessary outlet for her gifts. As to the stage all I hear convinces me more and more that she ought not to think of it, and I have made a good many enquiries of late . . . . Mr Piggott the licenser of plays who has had years' experience says that no young girl of gentle birth and bringing up should go on the stage unless she is driven to it.[19]

All this was in line with Mary Ward's ingrained puritanism, streaks of which could still be found under her metropolitan chic. It was also—like much of her advice to Julia—commonsensical. Ethel's ambition to be an actress did indeed pass away, as did her ambition in 1886 to write novels. And, in her earnest way, Mary Ward evidently had gone into the question of acting as a career for young women with some care. Her interest was further sharpened by the sensation caused by the 23-year-old American actress, Mary Anderson, on her English début at the Lyceum Theatre in September 1883. Anderson was stunningly beautiful but, as all the reviewers lamented, a very limited actress. Her opening performance in the historical drama *Ingomar* was virtually inaudible and physically clumsy. Nevertheless, the theatre was packed by audiences happy merely to look at the American girl with the beautiful face.

Mary, Humphry, and Henry James had entertained Anderson at 61 Russell Square on 30 January 1884 and found her personally charming. Afterwards, at the actress's invitation, they had gone to see her perform in the opening night of her second stage appearance in London, as female lead in W. S. Gilbert's *Comedy and Tragedy*, again put on at the Lyceum. Anderson was still an incompetent artiste, they found. In her diary, Gertrude Ward recorded the party's general disappointment: 'we all agreed that her part was grievously overdone: too excited, too loud, too restless, no self-control, no dignity, no self-expression'. And James, who was also in the box at the Lyceum provided by the luckless actress, raved over what he called Miss Anderson's 'hysterics'.[20] A few days afterwards, in Tottenham Court Road, Ward informed her sister-in-law that the occasion had furnished her with the *donnée* for a novel. At this embryonic stage, she obviously intended to publish it as a serial in *Macmillan's Magazine*.

Having nurtured the plot in her mind for six months, Ward wrote *Miss Bretherton* in a creative rush in August and early September 1884 during the family's summer vacation in Surrey. According to Gertrude, composition lasted 'about six weeks. She used to lie or sit out of doors at Borough Farm, with a notebook and pencil, and scrawl down what she could with her left hand; then she would come in about twelve and dictate to me at a great rate for an hour or more. In the afternoon and evening she would look over and correct what was done, and I copied out the whole. The scene of Marie and Kendal [at the end of chapter 7] was dictated in her bedroom; she lay on her bed, and I sat by the window behind a screen.'[21]

Gertrude's fair copy of *Miss Bretherton* was submitted to John Morley for consideration as a serial in *Macmillan's Magazine*. He returned the manuscript on 8 October, observing that 'there is not story enough'.[22] It would not suit the magazine. But, while declining it for serialization, Morley thought the novel might do well in volume form. His colleague G. L. Craik duly wrote accepting *Miss Bretherton* on 12 October 1884 on half profits, adding: 'whether there is profit or not we shall pay you £50'.[23] The work, it was decided, should be published in one volume of 310 pages.

The novel was speedily put into print. On 4 November 1884 Mary Ward approved the binding and sent back the finally corrected proofs. There followed some minor panic with the cover. On 21 November 1884, Frederick Macmillan wrote that 'I have just heard on very good authority what I did not know before viz. that Mudie has a very strong objection to books lettered in black and as he is a rather crotchety person he is just as likely as not to buy as few as possible of *Miss Bretherton* if we stick to the black letters.'[24] The letters were duly changed to humour the leviathan of the London circulating libraries. As always, Mary Ward was eager to hurry her publishers and she meaningfully informed Alexander Macmillan that 'between ourselves—a review in *The Times* would be much more probable this month than next'.[25] The publisher reassured her that 'not a moment of needless delay will take place'.[26] He was as good as his word and the novel was published at the end of November. 'I wish we could have got the book ready sooner,' Craik wrote, 'but we tried our best.'[27] They gave *Miss Bretherton* top billing in their Christmas advertisements.

As Morley objected, *Miss Bretherton* has too little story and is too much of the Jamesian *étude*. It opens, effectively enough, at the Royal Academy private view of May 1883 (a setting Mary knew well from accompanying Humphry in his new job). The cosmopolitan littérateur Eustace Kendal is struck by the beauty of a young woman in one of the crowded galleries. Forbes, an older artist friend, tells him that she is an actress, newly arrived from the West Indies, Isabel Bretherton, the current toast of the London stage. Kendal subsequently meets the entrancing Isabel at an afternoon party given by the American hostess Mrs Stuart. He later joins the Stuarts in a box at the Calliope Theatre, where Isabel is playing the lead in 'The White Lady'. It is a sad let-down. As an actress, Miss Bretherton has 'physical charm' but no technique, nor any sense of dramatic tradition. Her performance grates horribly on Kendal, steeped as he is in the sophistication of classical French theatre. Isabel, however, has as yet no sense of her inadequacy and offers herself for the main part in *Elvira* which Kendal's American friend Edward Wallace (apparently based on Henry James) has just written. Kendal undertakes to reject her offer at Oxford, where she has visited him for the day. Isabel, who whatever else is not lacking in sensitivity off stage, sees through Kendal's delicate insinuations and forces him to utter his honest opinion of her talent: 'You want a true sense of what has been done and what can be done with your art, and you want an insight into the world of ideas lying round it and about it.'[28]

At the end of the season, the now unvalued applause of English philistines ringing in her ears, Isabel leaves for the Continent. There, she is taken up by Kendal's sister, Madame Marie de Châteauvieux. Under this good woman's tutelage, Isabel learns French and laboriously trains herself in that culture's advanced acting techniques. (This section of the novel is narrated awkwardly, by letters.) After six months, she returns to London and triumph in Wallace's play, for which she is now genuinely equipped. On this occasion, it is not merely the multitude but connoisseurs who applaud:

> She had passed the barrier which once existed between her and the world which knows and thinks, and had been drawn within that circle of individualities which, however undefined, is still the vital circle of any time or society, for it is the circle which represents, more or less brilliantly and efficiently, the intellectual life of a generation.[29]

A reconciliation is effected by Madame de Châteauvieux, dying vaguely of a chill, who entrusts her brother with a memento for Isabel. At their meeting, all obstructions melt away and the novel ends with the lovers' embrace.

Miss Bretherton survives in two versions.[30] The first is the notebook draft in pencil that Gertrude Ward describes Mary as writing out of doors at Borough Farm. This represents the author's first thoughts. The second version is the fair copy which became, with only a few changes, the published text. This incorporates various revisions and corrections made to Gertrude as amanuensis. The first (notebook) version is divided into 'parts' rather than the eventual chapters. These parts presumably correspond to what Mary Ward foresaw as serial divisions. The first part of this early *Miss Bretherton* is generally identical with the opening chapter of the published novel (which remains the best thing in either version). But, originally, Isabel was American, like Anderson. The manuscript second part, which covers the visit to the theatre, is substantially different from its printed version, recording as it does more frankly the Ward–James party's disgust at Anderson's performance in *Comedy and Tragedy*.

Mary Ward evidently changed the story to spare Anderson's feelings, as she changed the actress's American origins to West Indian. In the novel as published, Isabel shows latent genius as yet undeveloped. Her performance is by no means a disgrace, merely technically immature. The other changes (such as the addition of the whole Oxford interlude) seem principally intended to bulk out the narrative. The expansion and the toning down of the awfulness of Isabel's acting style make for a more subtle study of character. But, as enlarged, the work remains awkwardly poised between inflated short story and abbreviated novel and drags painfully in its later sections. How to edit her own fiction was to be a major problem for the rest of Mary Ward's writing career.

Miss Bretherton has other faults, some of which Mary Ward candidly admitted in her 1909 preface. The idea of the work shows too obviously the 'first effect of London on academic inexperience'.[31] It is, in a word, *gauche*. The novelist was girlishly dazzled by the new metropolitan world of private views, literary suppers and first nights. But the major fault, which Ward does not acknowledge, is the botching of the American dimension to the work. In origin *Miss Bretherton* was an exercise in American fiction, based on the spectacle of an American actress in London. By making Isabel of Scottish colonial extraction, Mary Ward dodged what could have been her version of the Jamesian international theme.

Reviews of *Miss Bretherton* were mixed. The *Pall Mall* on 6 December was exceptionally hostile. *The Times* loyally declared itself 'impressed throughout by the refinement and the evidence of culture which underlies the book' and the scarcely less friendly *Guardian* noted that 'into the seductive form of the single-volume novel Mrs Humphry Ward has cleverly thrown a great deal of excellent dramatic criticism'.[32] But indifference and neglect were the principal responses. The unkindest, but in some ways shrewdest, review was the *Athenaeum*'s (20 December 1884), which curtly observed that 'The main interest of Mrs Ward's novel lies in the fact that it is all about Miss Mary Anderson.'

The slur that *Miss Bretherton* was no more than London gossip novelized irritated Mary Ward intensely. She wrote an angry denial to the *Athenaeum* on 27 December asserting that 'Miss Bretherton is not a portrait of any living person whatever. She is an attempt to handle an artistic problem.' For the next quarter of a century, whenever the novel was mentioned, its author obstinately insisted 'that Isabel Bretherton was in no sense a portrait of Miss Anderson'.[33] Given the known facts of the work's genesis, her protests were disingenuous. In conversation with her family, it was an accepted thing that Bretherton was Anderson. Thus Julia, writing to her daughter Mary on 14 December, wondered 'what Mary Anderson feels about it now that she has I suppose read the book. With the exception of the unfavourable mention of Miss Bretherton's uncle and aunt—which I feel sure she will think is meant for her mother and stepfather—there is nothing in the book which ought to annoy her at all.'[34]

In addition to tepid reviews, *Miss Bretherton*'s sales were sluggish. Friends like Henry James, Benjamin Jowett and Mandell Creighton wrote friendly (and in James's case, useful) things about the novel; but not even they marked it a clear success. How then was Mary Ward to proceed? As she reviewed her position in 1885 she had an unwritten and over-ambitious study of French Romanticism for which she had received £250. She had received £100 for her edition of Amiel's *Journal*. But Matthew Arnold had given a firm thumbs down on that venture. And she had written a Jamesian novel to which James had given a tentative thumbs up. But it had earned only a meagre £50. What *was* the way forward?

10

The *Elsmere* Ordeal: 1884–1888

DESPITE tepid reviews and sluggish early sales, Mary Ward was stubbornly optimistic about *Miss Bretherton*'s long-term prospects. Macmillan was less sanguine. On 31 January 1885–just two months after the novel's publication—they advertised a 'second edition' which was manifestly a ruse to clear the unsold bulk of the first printing. Thereafter, the publisher ceased to advertise the work, judging it for trade purposes wholly dead. The author continued to hope. On 5 February 1885, she asked Craik if they were going to reprint soon and if anything were yet known about the American sale: 'I am very anxious the book should make some impression there.' In the same letter, she outlined her plans for a next work of fiction—*Robert Elsmere* as it was to be: 'I hope I may be able to send you a story next October which will be better worth people's interest than this one. I have it all planned, and I shall take, all being well, five quiet months in the country to write it. It will be in two volumes.'[1] (Two predictions in the 'unsinkable Titanic' class, as it turned out.) Like most tyros, Mary Ward had unrealistic hopes for her first novel and was inclined to blame *Miss Bretherton*'s failure to set London alight on the publisher for not pushing hard enough. More significant, perhaps, is the fact that by early 1885 she had determined that fiction—specifically 'popular' fiction—was to be her main way forward.

Mary Ward was keen to get on with her new project. On 26 February 1885 she enlarged on it to Macmillan's. The story would be 'altogether a longer and more substantial piece of work than *Miss Bretherton* and its subject will be one of more general interest'. She was prepared to sell the novel, when completed, 'say for £250, subject to the resumption of my right in it after 2000 copies have been sold'. In the same letter, Ward confessed that the £250 which Macmillan had already advanced in 1883 for a study of French literature weighed on her mind. But the weakness of her writing hand made 'all minute critical or historical work so difficult to me . . . that I have been indefinitely thrown back'.[2] The prospect of her finishing the study was now 'a remote one'. Where this left the £250 advance was not entirely clear. Presumably, it was to be covered by profits in the as-yet-unwritten new novel for which her right arm was quite serviceable.

Macmillan saw little likelihood of any such profit. Craik wrote back a brutally ledger-like letter on 27 February 1885, explaining 'exactly' how the account on *Miss Bretherton* stood: 'We printed 2500. We gave away 71. We have sold 1150. We have sent to America 750. We have on hand 521. The book sells at 6*s*. We gave you £50. We are out of pocket £22.' Nevertheless, Craik offered £100 for 750 copies of the new two-volume novel, with £40 for every 250 printed over 750. He finished his letter with some bracing words of encouragement: 'I sincerely hope that you will write a book that will be largely and permanently successful that will repay you in every way. I hope our offer will induce you to go on with heart. You should always remember that your first book [he meant *Miss Bretherton*] has been unusually noticed and spoken about, but it perhaps did not deal with matters that were of universal interest. If you write a good work with a generally popular subject the results ought to be better.'[3]

It is painful for starting authors to confront the possibility of their ordinariness. Craik was telling Mary Ward—none too gently—that she was not a great writer. She was not yet even a competently profit-earning writer: nor ever would be unless she studied her craft. It was very hard to take, and Mary's first reaction was petulantly to offer her new novel to another more sympathetic taker. She selected George Smith, of Smith, Elder: Uncle Matt's favourite publisher. Once called the prince of publishers (a title which he made no effort to suppress) and the founder of the legendary *Cornhill Magazine*, Smith was now at 60 a Grand Old Man of the trade. He replied on 2 March with 'a few courteous words'.[4] Having read *Miss Bretherton*, Smith, Elder and Co. had 'no hesitation in accepting your offer of your new novel on the terms mentioned in your letter'[5] (i.e £250, for a two-volume work when finished in manuscript, new arrangements to be made after the book reached a sale of 1,500).

Mary Ward was not easy in her mind about deserting Macmillan. Their offer of £100 was extremely open-handed. She already owed them £250 for a book which she knew in her heart she would not now write. Morally, she also owed them £22 for their loss on *Miss Bretherton*. And Macmillan had stood by her and Humphry for years. The moment she received Smith's acceptance she again wrote to Craik with a clearly throbbing conscience: 'I was quite dismayed at the results of *Miss Bretherton*. How do books pay at all if what I suppose is a rather more successful first book than usual ends with a deficit of £22? I must still hope that the sale will go on long enough to recoup you. Of course, I cannot expect you to risk much on another book.' But would she submit to the logic of the account book and accept Craik's £100? No, she would not. Her letter continued: 'At the same time I am sure you will understand the importance of the money side of the matter to me and will not think ill of me for having carried my proposal to another publisher, also a friend, who accepted it at once.'[6] She concluded tartly: 'I can only hope that I shall not always be destined to be a loss to somebody.'

Craik took Mary Ward's defection from the Macmillan list stoically. George

Smith, he replied, 'is a capital publisher'.[7] But Macmillan and Craik must have been chagrined when *Robert Elsmere*—written to their prescription ('a generally popular subject')—went on to become the best-seller of the century. Smith had taken Macmillan's honey to his hive. It was Macmillan who had put the hard work in, coaxing, encouraging and taking risks on the unfledged Mrs Ward. Their house historian, Charles Morgan, writes with some bitterness about the canard which circulated in later years (and which still surfaces) that Macmillan had seen early manuscript chapters of *Robert Elsmere* and been fool enough to reject the novel.[8] They made a perfectly generous offer which, on her part, Mary Ward was insufficiently generous-minded to accept.

In later life, this and other acts of authorial hard dealing (particularly when the grown-up Arnold was gambling away her fortune) earned Mrs Humphry Ward a reputation for graspingness. But her desertion from Macmillan should properly be seen as evidence of a last-ditch determination to succeed. She was prepared to pay any price for success. No period of her life was altogether easy. But—as she later ruefully told Frances Cobbe—the three-years-long composition of *Robert Elsmere* destroyed her health permanently.[9] Against this Faustian expenditure any merely moral unease about disloyalty to a publisher was of minor account.

In her initial optimism, Mary Ward thought that she could finish her new novel in five months. In fact the composition of *Robert Elsmere* dragged out from March 1885 to February 1888. The thirty-five months can be broken down into three phases, each marked by their own kind of pain: (1) 'thinking about the novel', March 1885 to November 1885 (2) 'writing the novel', November 1885 to March 1887 (3) 'revising the novel', March 1887 to January 1888.

Serious thinking about *Robert Elsmere* began with the publisher's acceptance on 2 March 1885. It was a loose arrangement, which encouraged Mary Ward to take her time. Smith, Elder did not draw up any specific contract terms for the new novel until November 1887. There was simply a gentleman's agreement that she would produce a two-volume work of fiction, with a more 'general interest' than *Miss Bretherton*, at some future point—probably by October 1885 (the date she earlier forecast to Craik). Mary Ward in the event did not even start writing her novel until then. The interminable tinkering to Amiel's *Journal* was one impediment. The 'whirlwind' of London social life in early summer 1885 was another. But the main obstruction to Mary's getting down to her novel was Julia's condition.

Her mother was clearly losing the battle against her cancer. The wound had begun to bleed again. Much of Mary's time in late 1885 was taken up with arranging the sale of her mother's Oxford house. Julia could no longer take in boarders; she and her single daughter Ethel moved to a rented house in Bradmore Road—where Mary and Humphry Ward had lived as newly-weds.

Judy (i.e. Julia) had by now married Leonard Huxley in April 1885 which left Ethel shouldering the spinster daughter's burdens with increasing difficulty and—from time to time—a worrying emotional instability.

Among all this distraction Mary Ward continued to meditate on her new story and write it in her head (it was not—of course—'well planned' at this stage, as she had claimed to Craik). At some very early point she outlined the plan of the novel to her sister Julia in the little study at 61 Russell Square.[10] According to Gertrude Ward writing in her diary in May 1885 *Elsmere* had already been stirring in Mary Ward's mind for many months, and seems to have taken root as a firm project around the time of Pattison's death, in August 1884:

> The new story is to be written this summer [1885]. The first time she told me the scheme of it was one evening last November [1884]: she was sitting by the drawing room fire and sketched to me the career of the hero and the development of his opinions. I remember saying I hoped I should not have to write it [to Mary's dictation] as it was so serious and so sad. She has been working it out in her mind ever since, and many a time has been burning to write the first chapter, but she determined to finish up other things first, especially Amiel. Not a line of the new story is written yet. Next week we are going to the Lakes where the opening scene is to be laid.[11]

May 1885 had been a hectic month. It began on the first with the private view at the Royal Academy Summer Exhibition, which ushered in a frantically busy time for Humphry.[12] The Tennant–Lyttelton wedding was on 21 May. As Gertrude notes, the Wards went off to the Lake District a week later, on 29 May, to visit Mary's relatives, the Croppers. Four days after that, Mary visited Longsleddale, near Kendal, lush description of which dominates the first chapter of *Robert Elsmere*. At this point, she began keeping a notebook for ideas for the novel.[13] On her return from the north, the family stayed a few days in London, where they entertained the Matthew Arnolds and the novelist Hamilton Aidé. Then Mary and the children shipped off as usual to Borough Farm for the long summer break, leaving Humphry in London.

But, despite the opportunity offered by rustication, Mary Ward seems not to have begun writing *Robert Elsmere* as planned. Perhaps her arm was still too sore. Perhaps even in Surrey the Wards' social calendar was too busy for concentrated thought. The Burne-Joneses came to visit on 21 June (Humphry was evidently in the process of buying a painting from the artist).[14] On the 27th, Mary Ward returned briefly to town to dine with Henry James. Somerville council meetings were still preoccupying her and requiring visits to Oxford. In August, the family went off for a holiday in Scotland, returning to Borough Farm, where they remained until 30 September before coming back to London for the season.[15]

It was not until autumn that Mary Ward finally got going on *Robert Elsmere*. Even at this late date there were distractions in the form of the general election in November, which the Conservatives won narrowly. Gladstone's 'supposed Home Rule scheme'[16] (as Humphry disdainfully called it) provoked 'much fuss'. During all this hubbub Mary Ward doubtless continued to find odd moments to think about her work in progress. But what, one may wonder, was the plan of *Elsmere* over all these gestatory months, and how did it differ—if at all—from the narrative outline of the final novel? It is clear from Gertrude's recollection that Robert's heterodox 'opinions' and their tragic consequence were one starting point. The religious doubts, charismatic personality, and premature death of J. R. Green (in March 1883) were evidently strongly in Mary Ward's mind and supplied the frame of Elsmere's career. She was an early reader of Pattison's posthumous *Memoirs* in 1885 and presumably knew that his agnostic 'testament' (intended as an appendix) had been suppressed. This agnosticism—in the form of Squire Wendover's bleak rationalism—was made to play a major role in Elsmere's spiritual progress, thus fusing the Green and Pattison elements. And from the notebook begun in the Lake District in June 1885 it is evident that the character of Catherine was the other main building block for the narrative. Mary Ward loved teasing games about who were the 'originals' of her fictions and generally neither her identifications nor denials should be given absolute credence. But she told Benjamin Jowett, who passed it on to Margot Asquith, that Catherine was meant for Laura Lyttelton.[17] This is plausible. Mary was in 1885 besotted by Laura—never more so than in the period immediately around her wedding in May. The first section of *Robert Elsmere* centres on the psychic strain of a highly-strung girl giving herself in marriage; with the obvious transpositions, this was clearly the dilemma Mary Ward had observed in her young friend.

By November 1885, with Amiel finally out of the way, Mary Ward began writing in earnest. By mid-December, she had a first section ('very tame and domestic')[18] to show her brother Willie, who was to be a trusted counsellor throughout the long composition of the novel. It was good progress. Mary Ward's mood was, however, grim and getting grimmer all the time. Julia was sinking and was increasingly cantankerous and depressed. Her mood infected her daughter. 'This is a weary world,' Mary told her mother at the turn of the year on 30 December, adding without much conviction, 'but there is good behind it.'[19] Julia could see nothing behind it all but her own 'hard end' and that not far off. In Russell Square, meanwhile, the twelve days of Christmas were lost to writing, swallowed up in the dressing of the tree, family dinners, children's parties, and domestic theatricals. But on 2 January 1886 Mary shipped the children off to Oxford to stay with their grandmother and five days later Humphry—as he always did—went off to Wales to shoot. It was an excellent season, if very cold; Humphry bagged 200 pheasant on 9 January—a fact he recorded with as much pride as if he had been made Lord Mayor.[20] In London, the New Year weather was 'changeable, cold and disgusting'. Snow fell, thawed, fell again, and lay in the streets until March. On the twelfth of that month, as Humphry noted, there were still 'hundreds of persons skating on the Long Water in Kensington Gardens'.[21] Death continued to cast a pall over the Ward household. At the end of January, Humphry's old friend Edward H. Cradock, the Principal of Brasenose, died.

The domestic arrangements at Russell Square were far from ideal for a novelist at the best of times. Just now in the worst winter for years they were impossible. Mary Ward reported the house to be 'a perfect hospital' in early February: 'all the children and servants ill one after another'.[22] Gertrude Ward decamped to her parents' house for a month, to recover from the epidemic colds and influenza. Nevertheless, the social round went on. On Thursday 4 February the Wards entertained sixteen to dinner; on the next day eight ('a small and pleasant party',[23] Humphry called it), on the Saturday they dined at the Arthur Russells, and on Sunday at the George Rawlinsons.

This pace of social life kept up until 25 February, when relief came in an unwelcome form. The never-very-reliable Russell Square drains had been strained intolerably by the thaw-freeze cycle of the hard winter. Now they burst and the whole family (Mary with a violent cold) was turned out of the stinking house at an hour's notice. Humphry remained behind at his club, while everyone else, including the servants and Fräulein, retreated to Borough Farm until 5 March. While they were away, Humphry consoled himself with the purchase of a Sir Joshua Reynolds, which 'came home',[24] as he recorded, on the same Friday as Mary, Gertrude, and the children. On Saturday, they had fifteen to dinner. On the following Thursday, they entertained no less than twenty-one. Nevertheless, by the end of March, a draft of the first volume of *Robert Elsmere* was complete.

It was by now clear that three volumes (at least) would be needed for *Robert Elsmere*. Smith agreed—in conversation—to add £50 to the author's payment (so bringing it to £300), should sales exceed 1,000. At the same time he made over £200 to Mrs Ward as an advance. One of her first expenditures was a new dressing gown for Julia from Harvey Nichols.[25] By April 1886, Ethel could no longer manage to care for her mother by herself, and a proper nurse had to be found and paid for by Mary at £30 a month. Two weeks later Ethel calmly announced *her* resolution to be a novelist, and revealed that she had sent a manuscript and received a cheque from the publisher George Bentley. Mary dispatched to Oxford what must have been a somewhat awkward letter in the circumstances, arguing that her younger sister was in no position to get involved with *real* novel writing such as she herself was currently engaged on: 'I am delighted with Ethel's news and with the account of Bentley's letter. But both Humphry and I are inclined rather to dissuade her from attempting a serial yet. Why not try a few more short stories first at any rate? I don't think she has any idea of the enormous difference there is between a short story and a long one;—the experience of life, the resources that a long story wants if she is to produce anything which will do her credit.'[26]

Mary Ward evidently felt that she herself knew only too much about such things. She was, meanwhile, staying at Borough Farm for the whole of April. It was a departure from the usual Ward family pattern, but the country retreat proved a godsend this difficult year. She was by herself (i.e. with only Gertrude and Lizzie her maid) and writing at full speed. (The physiotherapist Wolff's exercises seem to have dramatically relieved her writer's cramp at this period.) The bulk of *Robert Elsmere's* second volume was written during April—'with my heart's blood',[27] as she melodramatically told her mother. Her push came to an abrupt halt at the end of April following two shocking blows. On 5 April came news of W. E. Forster's death and on 24 April news of Laura Lyttelton's. Mary's friend died from complications following childbirth, her uncle from exhaustion induced by the Irish troubles. She was devastated. Life—as she told her mother—seemed 'a mockery'.[28] None the less, life had to go on. She returned to London where, on 29 April, the Edmond Scherers arrived for a short stay. (Amiel's *Journal* was doing surprisingly well—Macmillan's edition of 1,250 had all but sold out.)

All through May, Mary Ward (now in London) was incapable of anything but corrections. The second volume was—from an intellectual point of view—the trickiest in the novel and at this stage largely consisted of extended dialogues on faith and reason between Wendover and Robert (exchanges that had their distant origin in Mary Arnold's fireside discussions at Lincoln College with Mark Pattison in 1870). Early summer 1886 was a hectically social end of the season as always, but one in which Mary was increasingly with literary people. On 1 May, the Gosses and Walter Pater came to dinner; on 26 May Robert Browning; on 5 June Henry James; on 29 June at Mrs Susan Jeune's she met a whole gaggle of novelists: Thomas Hardy, Mrs Campbell Praed, Justin McCarthy ('to our disgust'—he was a Parnellite), and Mrs Molesworth—'the child's story teller'—who endeared herself by calling *Miss Bretherton* a 'gem'.[29]

Mary Ward's birthday on 11 June 1886 was celebrated with what was now a standard family ritual, which she described for her mother:

> Judy has sent me a basket full of most delicious roses in honour of the day, and I have had the most fascinating presents from the home party. Humphry has given me a lovely sixteenth-century pendant, rock crystal with enamel mountings. Gertrude a lace scarf like Ethel's in which I shall swagger greatly. Fräulein, a delightful cushion to sit upon when I write instead of the miscellaneous assortment of music books on which I generally perch myself. Arnold a portfolio of his own drawings neatly mounted. Dorothy a pair of bedroom slippers of her own making and Janet a handkerchief of her own hemming and marking.[30]

But hovering over this cosy festivity was a black cloud. As she told her mother in the same letter, 'I must stick to my book like grim death, if I am to keep my promise to George Smith.'[31] The promise was to get the whole manuscript to the publisher by October (again). It was increasingly unrealistic. Despite the helpful cushion, Fräulein left suddenly (there was evidently a row) and finding a replacement ate up a week. Cold winds aggravated Mary Ward's neuralgia in July. Then it was 'blazing', and town was intolerably hot and sticky.

Finally Mary and the children got away to Borough Farm again. During the first week of August, she was still unpicking and rewriting her manuscript. 'Alas! Alas!' she wrote to Julia on 4 August, 'I shan't have got the second vol. done in spite of all my labours.'[32] She had in fact thrown the whole of her draft of the volume 'into the waste-paper basket'[33] and started again. At the same time, she slashed into the first volume, cutting it down to what is the first book (*Westmoreland*) in the printed text, obliging herself to rewrite another sizeable new section that radically changed her conception of Catherine. At this stage—confronted with a novel which would not hold still in her mind—Mary Ward suffered what was a serious physical breakdown and was packed off by Humphry to the Isle of Wight to recuperate. Children and society were now too much for her, and once returned she promptly left town again spending September and part of October at Fox Ghyll near Ambleside (the Forsters' Lake District home—melancholy enough in the funereal circumstances of her uncle's recent death). Here it was in early October that she read the latter part of the second volume of *Robert Elsmere* aloud to the classicist Lewis Campbell and his wife. 'Their praise was very warm,'[34] as she told her father. The most vexatious section of her story—the drama of Robert's spiritual vacillation—was at last falling into place.

The separated condition of her parents as Julia approached death preyed heavily on Mary Ward's mind. Writing to her father on 3 October 1886 she broke off a discussion of Ernest Renan and the existence of God 'for something quite different. I have had an idea in my head for a while past which I have just been discussing with Uncle Matt and Humphry. It is that you ought to be elected a member of the Athenaeum by the Committee. It is due to you, it is one of the few honours open to literary men in this country, and I should like it so much. Uncle Matt thinks it could be arranged so does Humphry [who had just been elected himself]. What do you think dearest?'[35] Mary's strategy was two-pronged. If Tom were elected to the Athenaeum he would, of course, come to London to live, or at least to visit regularly. Meanwhile, it was Mary's other plan to bring Julia to London as well, so that she might die comfortably in a flat near her eldest daughter. What easier, in these circumstances, than to arrange the parental reconciliation that she yearned for?

It was a beautiful dream which inevitably came to nothing. Mary Ward subsequently devoted herself more realistically to squeezing more money (£300 p.a.) from the grudging pockets of Tom (whose salary had recently been raised to £700). Later—as Julia entered the last round of her illness—it was Mary who took on the responsibility of whipping round the family to raise a nursing fund of £150 so that her mother might die without interruption from the Oxford duns. This worry was always on Mary's mind and was exacerbated by the guilty sense that a *really* devoted daughter would at this crisis drop everything (particularly a novel) to attend to her dying parent.

In October 1886, Mary Ward discovered to her horror that George Smith, who had evidently taken her promise seriously, was advertising *Robert Elsmere* as in press. It terrified her; she had not yet written a word of the third volume. But, as she admitted, 'it is my own fault'.[36] (Smith gallantly took the blame on himself: the notice had been sent to the newspapers 'during my temporary absence from London',[37] he claimed.) The episode was highly unsettling, but led to some usefully straight talking. Admitting that the last third of her novel remained to be written, Mary Ward told Smith on 20 October, 'I hope at least to let you have the whole book by the end of December [1886] for publication say in February.' But, she warned, 'I am determined not to let the book go till it satisfies me.'[38] Moreover, she wanted the first two volumes set up in proof—not so that she could correct them, but so that she could revise and (where necessary) rewrite her narrative.

Smith was an experienced enough publisher to be sceptical about her forecast. It had taken Mary Ward twenty months to produce two draft volumes. The idea of her writing a third volume and correcting the whole in ten weeks was far-fetched. And publishers always hate the expensive luxury of authors altering proofs. He 'could see no advantage in it',[39] he told her. But she declined to yield on the point and the argument was postponed until there was enough copy to set up.

Mary Ward was now writing fast again (up to fifty printed pages a week, as she estimated). By 23 November 1886 she had reached the fourth chapter of the final volume—well within sight of the end. In December she and Gertrude Ward went down to Borough Farm yet again, to work on the final chapters by themselves. Mary was feeling increasingly confident. As she told her mother, she would have all but six chapters complete by Christmas, and the whole should be with Smith by the second week in January 1887: 'then a fortnight or three weeks for what revision still remains and I shall have finished the longest and—I hope—the best piece of work I have ever done'.[40] By now, she foresaw quite clearly the ending of her narrative. On 27 December she told her father that 'I have been deep in [T. H.] Huxley's Lay Sermons, Edward Denison's letters and books about the East End for the purposes of *Robert Elsmere*. Only six chapters to write now.'[41] T. H. Huxley was the agnostic scientist whose son Leonard had married Julia Arnold in 1885. (Professor and Mrs Huxley came to dinner with the Wards on 31 December 1886.)[42] Huxley's lay sermons had been delivered to working men in the East End of London and supplied a model for Elsmere's lectures to the New Brotherhood. Edward Denison—a founder of the settlement movement who died tragically young—was a constituent in Robert Elsmere's conception. Another East End pioneer, J. R. Green (who had caught his fatal 'clergyman's throat' in Stepney), supplied the hero's death abroad from tuberculosis.

Mary Ward continued to churn out new and corrected narrative through the early months of 1887. Again it was wretchedly cold. ('At least to June', Humphry noted in his summary of the year, 'it will be known as the coldest, latest, most distressing season within living memory').[43] At the end of January, Mary and Humphry had taken a short break by the sea at Bognor, which she found 'the deadest alive place I *ever* saw in my life. It is so one-horse and shabby and out-at-elbows that we find a sort of amusement in the shabbiness of it.'[44] In early February, she had had the loan of Lady Emily Bowen's house at Hayward's Heath for a week. By 16 February, she was down to the last two chapters. 'It cannot take more than three weeks now,' she told Willie, adding mysteriously, 'it is not a *novel* at all.'[45]

The question was, 'would the novelist would last three weeks?' For some time, Mary Ward had clearly been dosing herself with powerful drugs. Her brother Frank (one of the slightly black sheep of the Arnold family) had just qualified as a doctor in January, and by the summer of 1887 was prescribing sleeping draughts for Mary without which she could not rest at night. On 7 March she confided in her mother 'What a blessing that opium is. What did people do before it and morphia were invented?'[46] Mary was in fact suffering a distressing new battery of symptoms: 'muscle weakness setting in the right hip and leg . . . constant discomfort and pain'.[47] In May, a Mr Hamer was regularly calling at Russell Square, 'to apply electricity to the head'. Probably the drugs helped more.

In other ways, the atmosphere in Russell Square was tense. Mary was not the only temperamental author in the house. For eighteen months, Humphry Ward had been labouring on *his* 'big book' (as Mary deferentially called it). A two-volume conspectus of the United Kingdom in the year of the Jubilee, by various hands, it was also commissioned by George Smith and called *The Reign of Queen Victoria*. Humphry's *magnum opus* was due for delivery at exactly the same date as *Robert Elsmere*, spring 1887.

On 29 November 1886, Mary had told Julia that 'Humphry is dreadfully hardworked just now. What a holiday we will have in April when both the books are done.'[48] But there was much to be done before this release. On 17 January 1887, Humphry resumed the night-work at Printing House Square which was a crucifixion for him. During the 1880s, he was in charge of editing the *Men of the Time* volume—a kind of Top Peoples' *Who's Who*. The proofs for the 1886 volume were due in mid-January 1887. Smith was owed his *Reign of Queen Victoria* three months later. Humphry reacted to this strain in his usual way—by buying works of art with what looks like manic recklessness. His 1887 diary is a catalogue of unaffordable purchases. On 7 March, Mary told her mother ruefully that Russell Square was 'chock full' of pictures.[49]

Mary Ward soldiered on through it all and finally, on 9 March 1887, she informed Tom Arnold that 'this morning I wrote the last words of the last chapter of *Robert Elsmere*'.[50] As she recalled in her autobiography, she emerged from her tiny study 'shaken with tears, and wondering as I sat alone on the floor, by the fire, in the front room, what life would be like now that the book was done!'[51] *Robert Elsmere* had been (to use Mary Ward's own phrase) 'a three-volume baby'.[52] And, as with other Victorian childbirths, delivering the novel had brought her to death's door.

Mary Ward was exhausted but satisfied. There would, she thought, be a fortnight's minor revising to do. But the work was otherwise as she wanted it and she thought the last volume dealing with Robert's defection from holy orders and his moral crusade with the New Brotherhood in London 'the best work in literature I have ever done'.[53] The book would certainly be out by the end of May 1887, she told Julia.[54]

While keeping up her mother's spirits with a stream of chatty letters[55] and occasional visits to Oxford, Mary was working behind the scenes to prepare for the death that could not be far off. In March she wrote to Willie that 'something I am afraid *must* be done directly to make Mamma's and Ethel's life easier to them in the way of money affairs'.[56] She and the Selwyns had been 'laying our heads together' and had decided that a sum of around £175 must be put together 'if what may be Mamma's last year is to be freed from the grinding worry of money'.[57] £25 each from all the close members of the family should do it. Tom was now contributing £70 a quarter; Mary and Humphry £33, plus regular little extras.

On 7 April, Mary Ward went down to Borough Farm. The 'Arctic' spring had given way to a brief spell of 'quite exquisite sunshine'.[58] With the novel out of the way, she could give some overdue attention to her other main object of ambition—10-year-old Arnold. He was, as his fond mother could clearly see, a prodigy. Every morning, at half past seven he woke up before the house was stirring to learn his thirty lines of Milton, before taking out his Macaulay from under his pillow for some judicious study of English history.[59] (One suspects that, even at this early age, Arnie could run rings round Mary.) After a ritual visit to his mother's bedroom for recitation, he would subject himself to a day's cramming in Latin and Greek from his private classics teacher, Eugénie Sellers. The young fellow had already been put down for Harrow.

While Arnold worked on his parsing and the girls learned their humbler lessons, Mary and Gertrude Ward got down to their fortnight's polishing up of *Robert Elsmere*. Mainly, it involved Gertrude's typewriting manuscript sections into a form legible by printers. Mary took a good deal of pleasure in this work.

But, even now, she was plagued by filial guilt. Ethel wrote from Oxford with the harrowing news that Julia was seriously weakened by dysentery and that she (Ethel) was crippled by glandular infections. It was a clear call for Mary to return to her mother's bedside and stay there. It was an agonizing dilemma. 'Smith and Elder are now demanding the book', she told Ethel, 'and I am torn between my desire of being more at Oxford and the necessity of finishing it.'[60] But—if it came to hard choices—*Robert Elsmere* must take precedence even over a dying parent. She remained at Borough Farm working on her novel.

When she eventually returned to London at the end of the month, Mary Ward began dispatching the copy in packets to Smith, Elder. Already, her joy at finishing the work was somewhat clouded by a sense that the typescript did look a trifle *large*. As she informed George Smith, rather warily, on 29 April, 'It will be a very long book. I reckon that the first two volumes will contain about 280 pages each at 30 lines a page.'[61] Perhaps, she added with desperate ingenuity a week later, the book might be printed 'on not very thick paper. It is certainly a long novel'.[62] Mary Ward hoped that the publisher—like a clever corsetière—could hide what increasingly she saw as an ominous corpulence in her brain-child.

Smith wrote back on 13 May with a chilling letter which made clear that such miracles were well beyond the printer's art. They had cast off the first volume and found that it would come to 412 pages—not 280. He added: 'our experience is that great length in a novel militates against its chances of success with the novel-reading public'.[63] They must either reduce the novel savagely, or use a 'solid page' (i.e. compact print), which the novel-reading public also disliked. Three days later, he sent an estimate of what the whole thing would comprise when printed. It came to a whopping 1,358 pages—around three-quarters of a million words. 250,000 comprised a long three-decker. Mary Ward had produced a monster. Smith was nonplussed: 'We are really quite at a loss to know what to do with a novel which so much exceeds the length of an ordinary three volume novel.'[64]

Mary Ward was devastated (her handwriting deteriorates markedly during this exchange of letters). 'I am overwhelmed by the printer's estimate', she told Smith 'and cannot *conceive* how it is that I have so underrated the amount of type-writing.' Would Smith 'let me have a third vol. of 450 pages'? She added, pathetically, 'Mr Walter Pater has already arranged to write about it' (in the *Guardian*).[65] Smith remained unhelpful and Mary Ward asked finally for the typescript to be returned. But on the huge parcel's arrival at Russell Square, like some unwanted ugly orphan, she collapsed utterly. She and Humphry had planned to go abroad in June; both of them (as they fondly expected) with their writing cares behind them. May had been a nightmare month for Humphry as well. It was the exhibition season and he had written no less than twenty columns of art criticism. On 31 May he had his last interview with

George Smith and handed over the preface to *The Reign of Queen Victoria*, thus concluding eighteen months of labour. Humphry had arranged with *The Times* to take the whole of June off, to recuperate with Mary under the Mediterranean sun. Now that plan was dashed by the *Robert Elsmere* disaster and by increasingly bad news of Julia.

Humphry noted in his diary on 27 May that 'M. goes to her mother for three days. By this [time] she has definitely given up the plan of immediately publishing *Robert Elsmere*. Her health is not equal to the necessary work of "cutting down". Alas! Mrs Arnold's state decides us not to go abroad.'[66] Instead, on her return from Oxford, Mary Ward was taken off to Cornwall with the sole aim of recovering her shattered health. In addition to her arm and leg disabilities, she was now chronically debilitated by anaemia and insomnia. She and Humphry left on 4 June, consigning the children to Gertrude and the servants to look after. In Cornwall, they stayed at Hooper's Lodgings, near the Lizard –'a little bleak town built on a plateau'. Mary—always responsive to landscape—found 'the cliffs, the sea, the coves—marvellous'. By 10 June Humphry noted that after writing letters after breakfast she was well enough to spend whole mornings on the beach and 'can gently drive or walk'.[67] Luckily it was exquisitely mild weather, and Mary took particular pleasure in little excursions with Humphry in the hired carriage. By Sunday 18 June, Mary was recovered sufficiently to walk seven miles, and could contemplate the six weeks (as she fondly thought) of revision that lay ahead of her. They returned to London by sleeper on 23 June.

No less than half the novel would have to be carved away. Unfortunately, a clear run at the operation was obstructed by Julia's desperate condition. Mary discovered that the £150 nurse fund was exhausted, and her first task was to whip round the family for yet more money. In the second week of July she went down to Oxford to be near Julia. She stayed at Lady Margaret Hall (uncongenial enough for the founder of Somerville) with her three boisterous children and their governess over the whole of a melancholy long vac.

Julia was if anything slightly better and able to be out a good deal in her chair in the sun. She suffered very little acute pain since she was permanently drugged with morphia. But it was clear to her daughters that the cancerous 'poison' had spread throughout her body. These stupefied afternoons in the sun were a final merciful remission. Mary Ward, meanwhile, was not in the best of health of herself. She could not as she complained get well 'in the head which is provoking and sometimes depressing'.[68] She was also suffering from vertigo—a side effect of her anaemia and general exhaustion. Nevertheless, during the weary months of July and August she contrived to make some useful progress on revising *Robert Elsmere*.

More important, she came to an agreement with Smith as to *how* she would make her changes. It was imperative, she declared, that she should have a constant succession of clean proofs to revise on. On 29 July she came to an agreement that £50 of her £250 should be withheld, to cover the extraordinary expense of setting and resetting. She would, moreover, pay anything that should be further due 'for any considerable alteration'.[69] Why one wonders did she demand this unusual and wasteful arrangement? One reason, clearly, was that she did not want the manuscript to run away with her again. Having the pages set up in print would give her a clear sense of the book's current dimensions. Another motive was her wholly neurotic need like Penelope constantly to weave and unravel her own creations.

Humphry was indulging his besetting neurosis in London, buying and selling pictures furiously. On 6 August he cleared a Terlinck for £200 and bought another Velazquez for £800. On 12 August, his 'Rembrandt' arrived home. He was meanwhile writing up to five leaders a week during August ('many political'), covering for more senior colleagues absent on holiday. The night work prostrated him. On 17 August he bought a brougham at public auction in Clapham. It was not something which the Wards needed. On the same day, the family returned from Oxford and preparations were made for them all to spend three weeks at Castleman House, Eastbourne, which Humphry had rented for £27, 'including plate, linen and houseservant's services'.[70]

Mary Ward did not—evidently—get much *Elsmere* work done at Eastbourne. But she had some constructive thoughts about narrative matters, particularly on the novel's organization into independent 'books' within the overall three-volume architecture. It was now her aim to have the whole revised proofs-cum-manuscript with Smith by 10 October, 'so that it may come out early in November [1887]'.[71] When the family returned on 7 September, the children were promptly shunted off to Borough Farm and Mary settled down with Gertrude Ward (just back from her summer holiday in the north) to tackle the great business of final revision.

There followed a furious shuttling of clean and hugely corrected proofs between Mary Ward, her 'old friends' (Jowett, Grant Duff, Bernard Bosanquet, Lord Arthur Russell, Willie Arnold), and Smith Elder's immensely long-suffering printers. She also suffered—more so as the 10 October deadline loomed. She could not keep up with the printers, she confessed to Smith on 7 October: 'I have had a most hindering attack of headache and bad sleeplessness.' But, she gamely assured him, 'unless I knock up altogether, which is not the least probable' the work would be done by the end of October. 'I am giving up *everything* to it,'[72] she added.

Among the 'everything' was a long-planned holiday in Scotland. Humphry it seemed would get no grouse this autumn. A final desperate push failed. The third volume defeated her. On 5 November she wrote to George Smith, 'I am afraid it is no use—I cannot get this book ready for a November publication. I cannot sleep, and therefore I cannot work at high pressure. The doctors say go to the Riviera, and I am afraid I must.' January 1888 was now the earliest that she could see the book getting out. 'I wish to relinquish *entirely*', she told Smith, 'all claim I may have to any balance of money in your possession'[73] (i.e. the £50 held back to cover extraordinary typesetting expenses). Smith—confronted with this clearly distraught author—wrote back a 'very consoling' letter by return of post.

Humphry Ward had some leave remaining, and the couple left England on 9 November and returned two weeks later. Their children were meanwhile looked after by servants—something that was happening frequently just now. It was apparently the Wards' first visit to the South of France and not what everyone would have considered a restful itinerary: Paris, Cannes, Monte Carlo, San Remo, Genoa, Spezia, Florence, Pisa—then back by sleeper, all in a fortnight. But presumably it was what she needed: and it was cheap. The whole bill was (as Humphry scrupulously noted in his diary) £30 3*s*. for tickets and £88 spending money and hotel expenses.

Bad news about Julia forced Humphry and Mary Ward back sooner than they originally intended. But back in London, she was able once again to turn to the now mangled manuscript and proofs of *Robert Elsmere*. On 17 December, just before the Christmas holidays, she promised everything to Smith by the end of the month. It was the last of her over-optimistic deadlines. It was not, in fact, until 27 January 1888 that she returned the last corrected sheet of the novel for press. The accounts so far could not be entirely pleasing to the author. Printing, correcting and proof alterations, and advertising (including £10 wasted in October 1886) came to £278—some £104 more than for a normal 1,500 run of a three-decker. Smith generously proposed splitting this surplus expense, and keeping the withheld £50. They would still give her an extra £50 (in addition to the £200 she had received in spring of 1886) if *Robert Elsmere* sold 1,000 of its copies and broke even. Copyright would revert to Mary Ward after the first three-volume edition of 1,500 was exhausted. The publisher did not, apparently, have any very great hopes for the novel. Mary was not so sure. She remained 'very curious' as to the novel's effect.

The long boiling down of *Robert Elsmere* rendered the final work more lumpy than was aesthetically desirable. It reads like a saga forcibly condensed into single novel length. Structurally, as William Peterson points out,[74] the narrative is tripartite—conforming to the divisions of the Victorian three-decker. Each part (which could well furnish matter for a whole novel) centres on a particular location. The first, Long Whindale, generates an essentially Wordsworthian atmosphere. Against this Lake District setting is played out the romantic drama of Catherine Leyburn's dilemma: whether to accept the love of Robert Elsmere, or to obey duty and continue looking after her widowed mother, whose support she is. Catherine's final surrender to her suitor would—in a conventional novel—be the climax. Here it merely marks off a section and looks forward to Catherine's later dilemmas as devout wife of a heretic. This first movement of *Robert Elsmere* has emotional power and a fine regional backdrop. But it is deficient in other ways. Robert is insufficiently built up as a character, and comes over to the reader as weak and somewhat epicene. His childhood is a blank; his years at Oxford hurried over; his crucial relationship with his dominating mother sketchy. These are parts of the novel which Mary Ward presumably stripped away in her economizing revisions.[75]

In the second section of the novel Robert—now Rector of Murewell—assumes centre stage from Catherine. The setting is now home counties' Surrey (identifiably the location around Borough Farm). As Peterson points out, the central locus is Squire Wendover's library and its stock of doubt-inducing books. Catherine is relegated to the background, as mother and mute incarnation of the Thirty-nine Articles. The main action revolves around the Faustian duel between Robert and his agnostic Mephistopheles, Roger Wendover. Also involved as contestants in the struggle for Robert's soul are the Oxford ideologues Edward Langham and Mr Grey. Langham—based on Pater—is a spectator of life, unable to act on his love for Rose Leyburn, or even leave an Oxford which he has come to despise. Grey—based on T. H. Green—is a don who has forged a workable, but wholly idiosyncratic, compromise between Christian faith, rational scepticism, and social activism. But lowering over the whole of the second section of *Robert Elsmere* is the character of Wendover. Mary Ward wrote nothing better in her career than Chapter 30 of the novel, describing the squire's night terrors and the crushing spiritual vacuum which finally destroys his magnificent mind. Evidently Mary Ward herself felt that Wendover was something that had to be left intact. Her drastic excisions in the middle of the novel were mainly from Robert's speeches in defence of his Anglicanism. As Gladstone was later to object, this rendered the hero intellectually spineless. Ideally, this section of *Robert Elsmere* should, like *Marius the Epicurean*, have been a full-length novel of ideas.

The third section of *Robert Elsmere* changes mode yet again. It is utopian in style, following Robert to the slums of London, where he sets up his new religious brotherhood. This conclusion is marred by thinness of texture. Robert's creed is insubstantially expounded; not for ten years more would Mary Ward have her ideas on the 'New Reformation' worked out. And her first-hand knowledge of working-class London was inadequate to the needs of the novel. The sub-plot in which Mme de Netteville tempts the inflexibly virtuous Robert is melodramatic and perfunctory. The novel ends in vague tatters, with Robert dead and the reader unsure whether his crusade has been a victory or quixotic futility.

In the interval between the novel's completion and its publication, Mary Ward could devote herself to her family. Attention was in order. Humphry was on a buying spree; his diary notes purchases and sales of a Turner landscape, a

Rembrandt, and two Ruysdaels. In January, Dorothy fell ill with scarlet fever, and was hustled off to quarantine in Hastings until the end of February. Arnold followed after a few weeks. Mary Ward, meanwhile, made numerous short trips to Oxford, while keeping up some sort of social life in London. On one such trip she did the final revises of *Robert Elsmere* 'between whiles' by her mother's bedside.

On 5 February 1888, Julia wrote to Mary in London a short, distraught note 'while the power remains to me'. For the last sixteen hours, she had felt 'that I must be dying'. She could not, she thought, live more than a fortnight. She concluded, the suffering 'is more than I can bear'. She was looking forward to seeing her eldest daughter the day after next: 'I feel *very very* doubtful whether I shall live out the fortnight that you are to be at Hastings [i.e. with her sick daughter Dorothy].'[76] Mary estimated a longer lease of life for her mother. On 6 February she told Tom Arnold that 'I myself do not think it possible that [Julia] should live more than two or three months at the outside. The emaciation and exhaustion are increasing very fast.'[77] There were other worries. Ethel was clearly under strain. The wayward Frank (newly qualified as a doctor) was planning to marry a Miss Valentine—not, Mary drily noted, 'a matter one can be very enthusiastic about'.[78] Money was a never-ending problem; Julia was again absolutely impoverished. Mary sent £20 for immediate household needs. But she did not put off her Hastings trip.

Before making the necessary, final trip to her dying mother's house, she and Humphry spent a few snowy days in mid-February at the Charles Bullens' at Warren Wood, Hatfield. They were, in fact, house-hunting again—this time for a country place. As Mary told her mother in a calculatedly 'newsy' letter: 'I should like to find a house here uncommonly. It is only twenty-five minutes from King's Cross which is close to us and yet it is complete country.'[79] It is unlikely that Julia was in any condition to respond to this news, or to read any of the three-volume *Robert Elsmere* which Mary sent down on publication day, 24 February 1888. It was, in every sense, an anxious period: waiting for death, waiting for notices. Death was certain; good reviews far from certain.

The suspense was 'quite depressing'. And it continued interminably; no big-gun London reviews appeared until a month—and in some cases longer—after publication day. Smith wrote, meaningfully, on 15 March: 'What I am dearly looking forward to is a really good review in *The Times* or some other influential journal.'[80] In fact—despite all the Wards' pull—that valuable review did not appear until well into April. And despite all Willie's influence at the *Guardian*, their review by Pater did not appear until 28 March. While she passed her days in the sombre and hushed house of her dying mother, Mary Ward contemplated the awful possibility that (like most novels) *Robert Elsmere* would sink without trace; it was beneath notice.

It was almost as if the reviewers were waiting for Julia to die. She continued however to fight—lasting much longer than anyone (including herself) thought possible. Mary Ward was consequently able to leave Oxford for her Aunt Forster's house at Burley-in-Wharfedale in the third week in March. She eased her conscience by instructing Ethel to buy (at Mary's expense) a spring mattress for Julia from Jack's (a department store): 'I would like her to have anything she fancies *at once*,'[81] she told her father. London, meanwhile, still was ignoring *Robert Elsmere*.

At the end of March, Tom, Mary, Ethel, Judy, Willie, Lucy, and Frank gathered around Julia's deathbed. Only Arthur (dead) and Theodore (in Tasmania) were missing. It was thus that Mary Ward spent her sixteenth wedding anniversary—ironically in Bradmore Road, three doors from where she had spent her first. Humphry remained in London. His month of intensive art reviewing began on 1 April. On 3 April, Mary Ward wrote to her eldest daughter, Dorothy, that 'Grannie is very very ill, and we must not wish her to live, for her life is only suffering.'[82] But they could not, she reported, persuade Julia to take to her bed. Two days later, she told Humphry that 'My poor mother's life is very near its end.' She was now surviving on small draughts of brandy and morphia, taken hourly, for which she had to be woken up. Before she sank entirely, Mary had unfinished business with her mother. She still wondered about those early years in which she had been exiled from home. Had she been loved, or not? Was she a wanted child? On 6 April, as she told Humphry, Mary roused the dying woman when she was alone in the bedroom with her and asked her to say her (Mary's) name: 'she opened her eyes upon me suddenly and said "Mary, who was my firstborn"'.[83] Willie, meanwhile, had left on the 11.40 train for Manchester; Lucy left for Godalming at noon.

By now Julia's breath had begun to fail. It hurt her to swallow, and she declined any more of the teaspoons of brandy that were her sole nourishment. In a period of extended consciousness, she summoned everyone to say goodbye and asked now to be left alone with her husband, Tom, for five minutes. The day before, she had given him her wedding ring, declaring 'Oh! I might have been a much better wife!' As Mary noted, with perhaps a little wonder, 'It is extraordinary how her feeling for him which in spite of everything has always been the most absorbingly fundamental thing in her has come out during these days.'[84]

On the morning following (i.e. 7 April) Julia died. Mary Ward described the scene in a long and harrowing letter to Willie:

> It was the most profoundly piteous thing. I feel as if I should carry the pity of it in my heart all my life. In the afternoon yesterday [6 April] she was much the same as when you left her: conscious at times but generally sleeping. When Ethel, Julia and I came in after dinner she looked at us with recognition. I bent down and said 'Say Mary once, dear' and she said at once with a sweet look of the eyes 'Mary, *darling*'—then 'Ethel', 'Julia' in the same way. Then she spoke of you and Frank [who had been called away

to the Infirmary], just your names. In the afternoon Miss Skene came. She knew her perfectly. She said to Judy, 'Tell her I love her dearly' and again as Miss Skene was standing beside her, 'I see you' .... When Frank said goodnight to her about 8 she said emphatically 'God bless you'. At ten we all went to bed, or rather lay down, as she seemed peaceful, and were called at 2. Then we found her very restless, talking of great pain and moaning. But she was certainly not quite conscious, though I think she was *partially* so when we first came in for Papa was not with us, and she said twice 'Papa!' and when he came he thought she knew him. She mentioned your name and I am sure she thought you were there too with us, and felt no gap. At 4.30 Judy and Ethel were both so white and worn that I persuaded them to go and sleep and Papa went down soon afterwards. There seemed no change. But after they left she hardly spoke except once or twice the cry 'Oh God! Oh God'—but she swallowed two mouthfuls of brandy which she had not done for hours and nurse gave her morphia twice. At 6, I came to sit beside her and nurse drew up the curtains. I thought she looked very deathly but never having watched the last change before I thought it was only the effect of the morning light after the candlelight. But at 6.30 the nurse looked at the eyes and startled me by telling me to call the others. They came at once. We knelt around her. We asked her if she knew us, but she was past speech or recognition. And at 6.55 after a few gently laboured breaths, with a meekness and piteousness of expression quite indescribable she breathed her last.[85]

In *The Times* that morning, there appeared the long-awaited review of *Robert Elsmere*.

Mary Ward dispatched the necessary telegrams and instructed Lizzie in London to buy her a black hat 'a small one, not very high, with a brim about two inches wide ... and she must trim it with silk and crape'.[86] She was to see Gladstone at Keble College on 8 April to discuss *Robert Elsmere*, a book which had become an obsession with him. The following morning (9 April) she returned to continue the discussion after breakfast, just before going to her mother's 'partial'[87] funeral service at St Giles. The body was finally laid to rest at Ambleside churchyard on 12 April. The day was mild and springlike. Carus Selwyn read the service and the grand old hymn 'Oh God our help' was sung. Even as she grieved, it was now clear to Mary Ward from George Smith's reports of sales that *Robert Elsmere* was going to be a remarkable hit; editions were selling as fast as they could be printed. Fate, however, seemed determined to poison any pleasure she might take in her fame. She returned to London from burying her mother on 14 April. On the next day, Uncle Matt died of heart disease. He fell down dead in the street in Liverpool as he was coming to meet his married daughter, Lucy Whitridge, as she returned with her American husband from New York. It was, as Mary Ward told her father, 'sorrows on sorrow'.[88]

11

Elsmere Mania: 1888

THE critical reception of *Robert Elsmere* falls into a number of phases. First, and most disconcerting, was silence. Newspaper and weekly reviews came extraordinarily late. *Robert Elsmere* was published on 24 February 1888 with the usual trade releases and advertisement. At this period of British journalism same-day reviews were routine for important novels. But the first significant notice of *Robert Elsmere* (in the far-off *Scotsman*) did not appear until 5 March. There was nothing substantial from the English Press until the *Guardian* published a review column on 21 March. And *The Times* did not weigh in until 7 April—six weeks after *Robert Elsmere* was published. Nor was it until late March and April that opinion-forming weeklies like the *Saturday Review*, the *Pall Mall Gazette*, and the *Athenaeum* let the world know what they should think about the new novel.

These first-phase judgements were not unanimous and even reviewers who were impressed with *Robert Elsmere* were loath to commend it wholly. *The Times*'s 'a clever attack upon revealed religion'[1] was a typically mixed response. Other reviewers accompanied their praise with condescending criticism of Mrs Humphry Ward's primitive-seeming (for 1888) fictional technique and her quaint preoccupation with the theological controversies of thirty years ago.

What is most striking in retrospect is the lack of early fanfare. If one had nothing but the popular prints of spring 1888 to go on, *Robert Elsmere* would seem to have been an eminently forgettable production. But, while the reviewers dragged their feet, the novel was making extraordinary headway among library subscribers. George Smith had divided his first run of 1,500 three-volume copies into three specious 'editions', so as to create a momentum to sales. By 16 April, the first two of these 'editions' (i.e. 1,000 copies) had been cleared, and the third was about to be released. Mary Ward got the agreed £50 supplementary payment and negotiated new terms by which every 500 copies sold in excess of the original 1,500 would earn £175 for her. Smith was not at this stage sure whether many more of the three-decker *Robert Elsmere* would in fact be needed.[2] It could be that the library sale was done.

But, surprisingly, demand for the novel from Mudie and his rivals held up and even intensified. Every fortnight, it seemed, another 500-copy edition was called for. By 8 May, the fifth was going like hot cakes and Mary Ward began, as she told Smith, 'to have a fevered curiosity to know when the sale is going to stop'. It was 'wonderful'. But, as she confessed, 'not being used yet to being a popular author—I find it just a little bewildering'.[3] Acclaim on this runaway scale overwhelmed her. She and Humphry had been invited to the Grant Duffs at Twickenham, where 'people want to talk all day and the tables and chairs don't fit my infirmities'.[4] Among the drugs and sleeping draughts, the doctors prescribed another trip abroad, which she and the family took in early June. On her return (18 June) from Switzerland she found *Robert Elsmere* still going strong.

Altogether, seven three-volume editions of *Robert Elsmere* were published between 24 February and the first week of June. This represented 3,500 copies. These were, of course, exclusively books for the library purchaser. No sane Englishman or woman was going to pay 31*s* 6*d*. for a novel—even one as apparently irresistible as *Robert Elsmere*. What this huge library sale meant was that in the large cities of Britain Mrs Humphry Ward had become a cult author. And this despite the coolness and delay of the reviews.

At this stage, *Robert Elsmere*'s popularity was principally a word of mouth phenomenon. Mary Ward records an anecdote testifying vividly to the excitement the novel provoked in its first readers. She was travelling down to Borough Farm from Waterloo in mid-April when, just as the train was starting, a lady rushed along the platform waving a book aloft and shouting to a friend waiting to see her off 'I've got it—I've got it.' She was bundled into Mary's carriage and leaned out of the window to inform her friend, 'They told me no chance for weeks—not the slightest! Then—just as I was standing at the counter [of Mudie's], who should come up but somebody bringing back the first volume. Of course it was promised to somebody else; but as I was *there*, I laid hands on it, and here it is.'[5] There were just the two of them in the ladies-only carriage, and, as the train sped into Surrey, Mary Ward watched her companion devour the pages of the familiar green volume.

The success of this first tidal wave of *Robert Elsmere*'s sales inspired the second phase of reviews. These came from the big guns—the monthlies, quarterlies, and church magazines—and were 'uniformly negative'.[6] They set out to correct what they saw as a dangerously excited populace. An epidemic was abroad, and must be rigorously put down. The most famous of the big-gun reviews came from the Grand Old Man himself, William Ewart Gladstone, in May 1888.

Mary Ward had, in a sense, fixed the Gladstone review—if not as efficiently as she had arranged Walter Pater's glowing article on *Robert Elsmere* in the *Guardian* (28 March). She knew Gladstone of course as a public figure from her Oxford years and from the Laura Tennant connection. He did not know her; except as an anonymous face among the massed Arnolds (with whom he had considerable intellectual sympathy). Mary was insistent, nevertheless, that Gladstone be sent a pre-publication copy of Amiel's *Journal*. It was her hope that he might 'befriend' the book. He did nothing more than politely acknowledge receipt of Amiel's *Journal*, but Mary continued to have hopes for *Robert Elsmere*. In the resounding silence of the novel's early appearance she asked Smith to send a copy to J. T. Knowles, editor of the Liberal organ the *Nineteenth Century*, and a close Gladstone aide. Knowles, she observed, 'seems to be enthusiastic and is handing on his copy to Mr Gladstone'.[7] Knowles also handed on some candid assessments of the author of *Robert Elsmere* who—as he rightly predicted—would be 'made' by a Gladstone review. Mrs Ward, he reported, 'strikes me as very *prim* and rather like a married old-maid in some ways'.[8]

Gladstone had been in the political wilderness since July 1886 and the collapse of his Home Rule measure. He now had time to meditate on the intellectual basis of his political creed and why exactly the English people had rejected it. Elsmere's repudiation of Anglicanism and Oxford (twin essences of Gladstonean Liberalism) he saw as ominously symptomatic. Gladstone put *Robert Elsmere* in the same class as Tennyson's *Locksley Hall Sixty Years After* (1886)—a work which he had excoriated in *Nineteenth Century* (January 1887). Tennyson's poem and *Robert Elsmere* were—in the GOM's view—articulations of the corrosive new thinking that was destroying the fabric of Liberal England. They were pernicious and dangerous.

Gladstone was obsessed with *Robert Elsmere* in spring 1888. Friends reported that he 'talked of it incessantly'.[9] He discussed it at huge length by letter with Lord Acton. One could no more stop reading *Robert Elsmere* than Thucydides, Gladstone unironically declared. His copy was strenuously marked with marginal objections to Mrs Ward's heterodoxy. In early April 1888, Gladstone found himself in Oxford at the same time as Mary Ward. A meeting was arranged by his summons in the drawing room of Edward Talbot (who had married Gladstone's niece) at Keble College—very much enemy ground for the author of *Robert Elsmere*.

Mary arrived at Keble at 7.10 in the evening: just thirty-six hours after her mother's death.[10] She had come from the grisly business of laying out, arranging the funeral, and watching by the corpse. The meeting began with a gloomy interchange, in which Gladstone—intending kindness—recalled that 'though he had seen many deaths, he had never seen any really peaceful'. They then moved to the main business—Oxford's 'long agony', Mark Pattison, the 'state of the country during the last half-century', the rash of marital breakdown (Pattison again), the 'modern girl', the possibility of 'a new construction of Christianity' (much protest from Gladstone at this).

By this route they reached, finally, the nub of their difference: whether or not rational Christians of the nineteenth century could accept the supernatural basis of their faith. With the commanding eloquence of the recent leader of his country, Gladstone rose to a peroration: 'if you sweep away miracles, you sweep away *the Resurrection*!' At this point the dinner bell rang. But so excited was the GOM with his arguments that he proposed a meeting after breakfast the next morning.

Mary Ward duly reported again to Keble College, where there ensued 'a battle royal over [*Robert Elsmere*] and Christian evidences'. By now, her spirits were also high and the heretical novelist gave as good as she got. As she told Humphry, Gladstone 'looked stern and angry and white to a degree, so that I wondered sometimes how I had the courage to go on—the drawn brows were so formidable!'[11] It was a memorable encounter, and one which Mary was careful to immortalize with a written record. She left 'the wonderful old man', posing as it were for posterity's statues: 'standing . . . one hand leaning on the table beside him, his lined, pallid face and eagle eyes framed in his noble white hair, shining amid the dusk of the room'.[12]

Her sticking to her guns did what sycophancy couldn't. Gladstone continued their fight into a third—and this time public—round. His 10,000-word review-cum-essay '*Robert Elsmere* and the Battle of Belief' duly appeared in May's *Nineteenth Century*. It was a coup to make any publisher's mouth water—Prime-Ministerial attention, no less. Not that the attention was entirely flattering. With all deference to Mrs Ward's artistry and intelligence Gladstone branded her novel as brilliant but pernicious—a surrender of all the intellectual high ground to agnosticism and theism. Why did Elsmere not *fight*—do battle—for his belief? Like other crusty patriarchs, Gladstone was quick to discern a fatal lack of backbone in the younger generation.

Gladstone sparked off a higher-journalistic debate on *Robert Elsmere*'s significance. 'What is its "portent"?' was the question everywhere asked. The main discussants were the *Contemporary*, the *Quarterly*, and *Nineteenth Century*—in which Mary Ward published her reply to Gladstone ('The New Reformation') in March 1889. But throughout 1888 and well into the next year there was a spate of ponderous articles and innumerable sermons delivered on the subject of *Robert Elsmere*. Most were defensive in tone; the Anglican establishment girded itself to smash Elsmere as Bishop Wilberforce thirty years before had (as he fondly thought) smashed Darwin.

There was a third and most amazing phase of *Robert Elsmere*'s career still to come. In early July the postman delivered a large parcel of clippings. Humphry saw the foreign stamps and correctly prophesied 'there's America beginning'.[13] Excited by Gladstone's intervention, the novel had become a rage in the United States following publication in mid-summer. It was not—of course—that the average New Yorker or Bostonian was much interested in Anglican scepticism and fideism. It was the intoxicating sense of violent ideological tearing apart which initially attracted them to *Robert Elsmere*. The novel—with Robert's apostasy at its core—suggested that Britain was about to separate Church and State: to become, that is, more like America (something that Americans always see as healthy progress in other countries). Friendly commentators like Oliver Wendell Holmes wrote of *Robert Elsmere* as a beneficial laxative: 'a medicated novel, which will do much to improve the secretions and clear the obstructed channels of the decrepit theological system'.[14] Clearly Britain needed this purge more than the US. But *Robert Elsmere* would also serve to cleanse America of its lingering theocratic tendencies. The great danger, of course, was the romantic attraction of fiction which might—in the paradoxical way of the genre—glamorize what it was anathematizing. (Look, for instance, at what the novel had done with adultery.) This was the sagacious conclusion of the American rationalist W. D. Childs: 'I regret the popularity of *Robert Elsmere* in this country. Our western people are like sheep in such matters. They will not see that the book was written for a people with a State Church on its hands, that a gross exaggeration of the importance of religion was necessary. It will revive interest in theology and retard the progress of rationalism.'[15]

The book trade likes people to behave like sheep and American publishers churned out copies of *Robert Elsmere* by the ten thousand. In the absence of international copyright (still three years off) transatlantic pirates had a field day with Mrs Ward's novel, competing with each other as to who could bring out the cheapest (and nastiest) edition; a dutch auction, which led to copies of the novel being reduced by stores (as loss leaders) to four cents and—final indignity—being given away free with every cake of Maine's 'Balsam Fir Soap', a newly launched product which wished to impress on consumers the literal truth of cleanliness being next to godliness.

Mary Ward actually had a rather higher estimation of the American public than most Arnolds, for whom the country represented the bottom-most pit of barbarism. But even Mary could not but be horrified by the 'vulgarity, the lack of all sensitiveness and delicacy which enters into [American] modes of publication'.[16] Indelicate though its appetites and sensibilities might be, the American public dwarfed the English. To tap into such a large readership and not to be able to market one's literary property was torture. By November 1888, it was estimated that 100,000 copies of *Robert Elsmere* had been sold in America—three times as many as in England. The *ex gratia* £100 which John W. Lovell sent Mary proved him to be an unusually scrupulous American publisher. But it was absurdly incommensurate with the numbers of the book being put into circulation. 'I don't feel called on to be very grateful,'[17] Mary told Tom.

Three options were open to Mary: (1) to grin and bear it—as had generations of British authors before her; (2) to publish first in America—or at least to make her primary contracts with an American publisher who could then lease the copyright to a British co-publisher (piracy was much less a cisatlantic practice). The 'moral pirate' Lovell had already offered Mary £4,000 outright for her next novel. Another American publisher, McClure, offered £1,000 for a 'biblical romance'; (3) to wait for an equitable international copyright agreement which would enable her to collect American royalties. Negotiations were, in fact, already under way and proposals were before the American Congress. As Smith wryly asked in December 1888, 'Does it not seem strange to you that the question of whether a lady living in Russell Square should or should not be a millionairess, is depending on "lobbying" at Washington?'[18] In the event, Mary decided on the third of these courses and waited, fingers crossed, for a successful outcome to the lobbying. She would not, she resolved, publish another novel until the passage of an international copyright agreement.

Why did *Robert Elsmere* provoke this buying and reading mania in Britain, America and even Europe? The overseas popularity is probably easier to account for. There was a familiar coat-tails factor ('Read the novel that took England by storm!') and in America a kind of reverse *Uncle Tom's Cabin* effect. That novel had been an all-time best-seller in Britain not because enslavement of blacks was a British problem, but because it was *not* a British problem. One could relish Stowe's exciting melodrama cushioned by a consoling sense that England had solved Uncle Tom's problem with the Emancipation Bill of 1833. In the same way, America might feel that her forefathers had solved the problems that tormented Robert Elsmere with the constitutional separation of Church and State.

Why *Robert Elsmere* should initially have been a hit in England is harder to answer. It is not to modern eyes a very readable novel—no page-turner, as we would say. The first 'book', with its interminable moral niggling about whether Catherine should accept Robert's proposal and its sub-plots about ruined peasant girls and 'bogies' in the Cumberland hills has the consistency of narrative porridge and—I suspect—has lost *Robert Elsmere* many modern readers. But clearly—as the unknown lady in the railway carriage showed—even the first volume of *Robert Elsmere* had the power to ignite Mary's contemporaries and to drive them to acts of genteel recklessness.

The fact remains that *Robert Elsmere* was a huge success and set up its author for a career of high earning. The receipts from 1888 to September 1889 were beyond the wildest dreams of a novelist whose highest payment to date had been £60. The sale of 3,500 three-volume copies yielded Mary Ward £950. Between July 1888 and September 1889 Smith issued seventeen editions of the one-volume 6*s*. form of the novel, amounting to 38,000 copies in all. At £125 per 2,000 (i.e. 1*s*. 3*d*. per copy) this yielded Mary £2,375. 1,000 copies of a two volume Library Edition were issued in early 1889, which at 3*s*. royalty a copy yielded £150. Macmillan gave £75 for the colonial edition, to which they added £25 for the right to export copies of this edition to the United

States. Following the huge American success of the the novel, they paid over in November 1888 an extra £250 conscience money. European publishers paid her around £240. Lovell, the moral pirate, chipped in his pittance. All this added up to some £4,165. And regular further instalments could be expected from the 2*s*. 6*d*. 'new edition' which Smith released in January 1890. Half England was waiting, apparently. Some 20,000 of the half-crown edition were sold between January and December 1890, yielding £500. And beyond even these depths of the reading public, there were the 6*d*. and 7*d*. editions which would sell till kingdom come, earning a few score pounds every year.

It was all very gratifying. But Mary Ward (who was rapidly becoming an astute business woman) could see that, rich as Britain might be, it was America that should be her richest market. She could also see that Mudie exercised a stranglehold on her all-important first readers. He only took 200 of the first three-volume edition of *Robert Elsmere*. And it is clear that he was under-ordering right up to mid-April (otherwise the lady on the Godalming train would not have had to wait 'weeks' for her novel). Mary Ward determined to do everything she could to overset Mr Mudie and his pernicious three-volume system. Five years later, she succeeded.

12

The Fiction Machine: 1890–1900

THE runaway success of *Robert Elsmere* inaugurated thirteen years of comparative stability and contentment. This was the 'broad middle plateau of life' as Mary Ward called it.[1] She was over these years an established, highly paid author and an esteemed public person—at last someone 'who counted'. Her image—for all the world like the figure-head of HMS *Late Victorian*—was photographed and revered by her contemporaries. She became in the 1890s an accomplished and sought-after public speaker. Causes enlisted her to their committees. Her health—if never robust—at least held up over these years when it most mattered.

Ideologically, she was now her own woman. The Arnoldian do-goodism she had imbibed at her father's knee hardened into brash new Liberal Unionism and a practical philanthropy which found expression in works not ideals. In religion she adopted an undogmatic Unitarianism that was, to all intents and purposes, Christianity of convenience. Her dreams took material form in the shape of her beautiful country house, Stocks; in the Passmore Edwards Centre, her London settlement; in the promising future of her brilliant young son, Arnold; in her corpus of novels, which friendly critics assured her were masterpieces of English literature. The sharp struggle with its life-and-death risks gave way to routines of hard work and dedication. Her life hardened into life-style.

The stability of Mrs Humphry Ward's life-style in the decade and a half after *Robert Elsmere* depended ultimately on money—lots of money. As she reckoned to her brother Willie in 1892, 'we spend now between £3,000 and £4,000 a year with about £500 to £600 a year out of it for giving [i.e. charity—mainly to her family]'. She added, 'owing to bad trade and general scarcity of money Humphry has not been able to make much beside his *Times* income for some time'.[2] Humphry had made a foray into business in October 1888, becoming director of a New Zealand mining company. This had come to nothing. And his speculations in fine art remained a chronic drain on the family resources. His *Times* income hovered around £1,500 a year, depending on the number of columns he wrote.

All this meant that, in the early 1890s, Mary had to provide something over £2,000 a year, just to keep the Ward family afloat. And as the years passed the household needs increased. Growing children needed ever more expensive education (foreign finishing for the girls; Eton and Oxford for Arnold). Gertrude Ward left in 1891, to be replaced in 1896 by a salaried secretary, Bessie Churcher. The Wards' housing became ever grander—culminating in 1896 with the purchase of a country estate in Hertfordshire. Entertaining at Stocks was on a grand scale: shooting parties in winter, golfing, cricketing and tennis parties in summer, weekend parties year round. The servant entourage swelled to around eight. A classics teacher—Eugénie Sellers—was employed to coach Arnold. (Miss Sellers was an impressive woman; later, as Mrs Arthur Strong, she established herself as a leading classical archaeologist.) The girls had dancing, riding, and music lessons. No new doctor with his new nostrum but was not hired by Mrs Ward for her array of ailments. Meanwhile, after 1891, the Wards kept a house in town at Grosvenor Place more fashionably West End (and expensive) than 61 Russell Square. They were enthusiastic purchasers of newfangled gadgets like telephones and motor cars—which meant employing a chauffeur–gardener and providing a house for him. It was a habit formed during the 1890s for the Wards to travel several times a year to Switzerland, the South of France, the Riviera, and—after 1890—Italy. Wardrobes were correspondingly larger and more costly.

The bulk of these recurrent expenses could be met from one source only: novels. Mrs Humphry Ward became a money-generating fiction machine. Novels—especially in her maturity—were many things for her. Not least, they were a craft in which she conscientiously laboured to improve. *Helbeck of Bannisdale* (1898) is, technically, many notches higher than *Robert Elsmere* (1888). Novels were also a laboratory, where Mary Ward worked out her ever-evolving ideas on politics, religion, ethics. But above all novels were instruments for raising large sums of money, without which the Ward family would have gone under the hammer in a year. By 1896 they were no longer *choosing* to live above their means, they were irreversibly committed to it.

Mary Ward undertook the writing of *Robert Elsmere*'s successor at a period of increasing financial pressure. Humphry bought the land for their new house in Surrey in August 1888 and by July 1889 finishing touches to the construction and furnishing were 'swallowing up' £500 a month. The new novel's main purpose—its sole purpose—was to pay for the new house at Haslemere; 'all my book must go to that henceforward', Mary declared in a letter of November 1888.[3]

The subject of the new book was a departure The late 1880s had been troubled years for England. Mrs Humphry Ward's new novel would hark back to the 'social problem' fiction of the similarly disturbed 1840s. It would have a working-class, self-improved hero originally called David Mason.[4] The 'impulse' for *David Grieve* (as it was finally called) came at Borough Farm in July 1888 on an evening walk 'drenched with sunset'.[5] She dashed off a few incoherent pages, centred on the incandescent landscape. It latched on to some ideas formed by Alexander Somerville's *The Autobiography of a Working Man* (1848), which Willie had earlier given her to read. Her growing fondness for Willie—who had been such a tower of strength in the writing of *Robert Elsmere*—also predisposed her to a Manchester novel. On 27 September 1888, she wrote to a book dealer, asking him to procure for her 'half a dozen books or so of a class I am just now curious to collect, viz. working class biographies or autobiographies of the present century. I want anything which will throw light on those processes [and] incidents of self-culture by which men like Thomas Cooper, or William Lovett or Daniel Macmillan rose to knowledge and success in their different ways.' She added, 'Manchester life in particular I should like some record of—something which would enable me to grasp the development of a boy of great gifts, sprung from the the Derbyshire moors, and living the life say of a bookseller's assistant in Manchester.'[6]

This germ of *David Grieve* seems to have had contradictory origins. One main element was Mary Ward's revulsion against the wealth and fame that was pouring in on her after *Robert Elsmere*. She did not want to be co-opted into the aristocratic élite of England as a flunkey—*their* novelist. She wrote in this jaundiced vein to Tom on 5 October 1888, after a visit to the A. J. Balfours at Whittinghame: 'You ask whether our aristocrats were interesting. Not very, I think, except so far as their looks—some of their women were pretty—and their jewels and their possessions were concerned.'[7] In the same anchoritic vein, she informed her father later in the month that 'Society to me is the last chore!'[8]

She was—as always—uncomfortable with comfort. She wanted to *serve* and felt obscurely that service should direct her to the lower, not the upper classes of England. The same self-flagellating idealism that had driven her father to New Zealand would eventually inspire her to set up her play centres, vacation schools, and settlements for the poor and crippled of London. At this stage, in 1888, that idealism took the form of an irritability with the wealth, privilege, and aristocratic ease whose doors had been opened to her by *Robert Elsmere*.

In this mood, Mary Ward found the decent working classes more congenial. She did field work for *David Grieve* in the industrial north and wrote a long enthusiastic letter to Arnold (now at Uppingham) on her return about how worthy the Lancashire factory hands were:

> I had a very interesting time in the North and have just come back full of ideas for David [Grieve]. One day I spent at Oldham where I saw cotton mills and cooperative stores, and almost everything there was to see, and another day Auntie Henrietta and I went out to Bacup a manufacturing town on the Irwell and spent the afternoon and had tea with some mill people—some of the best and nicest people I ever met.[9]

Her political mind at this early stage is more radical and Owenite than at any stage of her career.

More preparatory thought went into *David Grieve* than any other of her novels. The outline of the plot was clear in her mind as early as autumn 1888; in addition to the working-class, mid-Victorian, northern setting *David Grieve* would be a *Bildungsroman*, following the growth (or self-making) of character from childhood to late adulthood. But serious work on the novel did not begin until well into the next year. How to publish was a principal problem. In early March 1889, Smith tentatively floated the idea that she might like to serialize the new novel in the style of George Eliot's eight-part *Middlemarch*. But Mrs Ward distrusted serialization, with its crushing timetable pressures. She already had enough pressures to contend with. *David Grieve* should be another three-decker and it was with this bulky structure in mind that she began to compose the narrative.

On 26 March 1889, Mary sent Dorothy Ward with three chapters to hand-deliver to Mr Aitchison, Smith's manager, to be set up in print before she rewrote them (which she did, several times). It was now an understood thing with Smith, Elder that Mrs Ward should compose on the proofs from the very beginning. Conditions of absolute secrecy were imposed. No one but the compositor was to see the manuscript copy or proofs—not even George Smith himself. Mrs Ward was adamant on this point. The publisher was thus prohibited until October 1891 from reading the novel that was being set up under his nose in his printing house at his expense.

There ensued the usual cycle of over-optimism and disappointment with completion dates. The author first scheduled November 1889, then January 1890, then May. Delivery was impeded by a number of complications other than the usual agony of creation. 1889 was a fearsomely active year—even by Mrs Humphry Ward's standards. In April, she had made her first trip to Rome. Early summer had been preoccupied by the interminable business of getting Arnold in as a scholar at Eton. In July the household had been obliged to move out of Russell Square to a rented country house at Great Missenden. (She felt like the Swiss Family Robinson, Mary Ward told George Smith.) Here it was that she wrote the 'Childhood' section of *David Grieve*. The previous owner of their land at Haslemere, Mr Hudson, was meanwhile making innumerable objections to their building plans (which were proving alarmingly costly). Humphry was displaying symptoms of 'nerve pressure'. And in October there came the bombshell that—without so much as a word of warning—Tom Arnold was going to remarry in two months time. At this moment, with about 200 printed pages of *David Grieve* done, Mary's hand seized up entirely. But by December despite influenza (of which she resolved to 'take no notice')[10] she had about half the novel (i.e. through 'Youth') written and rewritten.

In March 1890 she had the inevitable health crisis which George Smith (who knew about such things) diagnosed as a 'nervous breakdown'.[11] Trips abroad were prescribed, one of which, to Paris, helped her with the tricky French episodes in the 'Storm and Stress' sections of *David Grieve*, which she found most difficult of all to write. By September, she thought that 'two volumes of *David* are all but done'.[12] That winter, the opening of University Hall took up all her time. Nevertheless, she had three-quarters of the novel done by April 1891 and was well into 'Maturity'. With a macabre complacency she told Smith that she now felt the book 'could be printed as it stands even if I were not to live to complete it'.[13] She in fact lived and in the last days of September 1891—drugged by night with sleeping draughts and by day with pain-killers—she wrote *David Grieve*'s final lines. It was a close run thing. As she told Willie on 11 October, 'I was nearly three days recovering energy enough to go on after the chapters describing Lucy's death.'[14] The chapters she refers to (8 and 9 in 'Maturity') offer a painfully close recollection of Julia Arnold's cancer, her operations, and her death.

George Smith was finally allowed to read the corrected proofs in October 1891, and loyally delivered his opinion that it was a 'grand book'.[15] But he had some shrewd objections as well—unalterable as the typeset narrative now was. He thought interest dragged after David's return to Manchester from Paris and the hero's *union libre* with Elise. And he wondered if in describing David's sexual liaison in Paris, Mrs Ward had quite realized 'that there is a *Bourgeoisie* in England, as well as in France'.[16] Early copies of the new novel came on 24 October 1891. Mary Ward—exhausted as any Victorian mother after a difficult birth—made the obvious analogy: 'I am childishly delighted with my three-volume baby; how nice and thin it is!'[17] Not fat, that is, like *Robert Elsmere*. As she wrote this, she was lying prostrated in Italy, where she remained until mid-December, too weak even for sightseeing.

Mary Ward was proud of her achievement in *David Grieve*, and particularly proud of the depth of emotion that informed it: 'only in *Helbeck of Bannisdale*', she claimed, 'did I ever draw so deep again on the sources of personal feeling'.[18] *David Grieve* is certainly one of her handful of great works. But like its predecessor *Robert Elsmere* it suffers from being too many novels rolled into one. A full-length narrative could be spun from *David Grieve*'s first section, which covers Louie and David's upbringing on Needham's farm; their romantically tragic background; and their puritanical (but avaricious) aunt Hannah's withholding from them the knowledge that they have an inheritance. Amidst all this, David gains an education.

This first section of the novel with its self-help theme and densely regional Derbyshire setting recalls George MacDonald's *Robert Falconer* (1868). The next phase of young David Grieve's evolution, dealing with his apprenticeship in a Manchester bookshop and his introduction to working-class radicalism, evokes Charles Kingsley's *Alton Locke* (1850). David's love complications with his employer's daughter Lucy Purcell and with Dora Lomax seem to have been influenced by Mrs G. Linnaeus Banks's *The Manchester Man* (1876). The 'Storm and Stress' episodes in Paris seem, by contrast, to be wholly original (although a faint aroma of *La Dame aux Camélias* permeates them). More surprising, perhaps, is Mrs Ward's daringness. David's *union libre* with Elise Delaunay, and Louie's (admittedly off-stage) affair with the sculptor Montjoie are the frankest sexual episodes in all her fiction. But for all her frankness, the author cannot suppress a persistent *frisson* of disapproval at French morals and that nation's strange propensity, as Robert Elsmere puts it, 'to revel in mud'. Mrs Ward, in short, does not admire the French; particularly the French in love.

George Smith's objection was well-taken and for most readers prophetic. Interest in *David Grieve* flags after David's rescue in Paris by Ancrum and his return to Manchester, his redemption, and—after Lucy's convenient death—his brother-and-sister union with Dora as surrogate marriage. 'Maturity' is a limp anti-climax. The strength of *David Grieve* lies in its gloomy middle sections, particularly the suicide chapter in 'Storm and Stress'. And, major achievement that the novel is, it was damaged in conception by a number of wrong authorial decisions. It was an initial error to select rural Derbyshire: a region which Mary Ward did not know intimately. It is also unfortunate that she chose to render its unlovely dialect with such phonetic exactness. To a lesser extent, Manchester was another *terra incognita* for her. But the main error in her devising of the novel was the selection of a working-class male hero. As Mary Ward's next novel, *Marcella*, demonstrated, middle-class heroines inspired her to a much more powerful kind of fiction.

David Grieve was held back until 22 January 1892 to allow simultaneous American publication. Money arrangements between Mrs Ward and Smith had been very free and easy. 'I shall always be delighted to act as your banker,'[19] he told her, apprehending that the Wards' purchase of new houses was causing her financial difficulties. No request for an advance on *David Grieve* was turned down by him; although he insisted on seeing them as loans, secured by the property of the book.

The all-important question was getting a fair return from America. Mary Ward was determined not to be plundered by those Yankee pirates again. Never again should a novel of hers be given away with bars of soap or sold for four cents. In February 1889, while *David Grieve* was still only the merest outline in her mind, Mary Ward and Smith had weighed up pressing offers from a whole regiment of American publishers. These transatlantic houses—all pirates at heart—banked on international copyright not coming about in the near future. Consequently they made their tempting bids for early sheets of the new novel for serial publication in magazines. To receive their dollars the author would have to send regular instalments of *David Grieve* by steamer across the Atlantic. The dollars were tempting but Mary Ward—still bruised from her *Robert Elsmere* experience—could not trust herself to feed a conveyor belt in this way. Nor in her new vanity as an author was she happy about the sums offered. If Mrs Hodgson Burnett—author of the preposterous *Little Lord Fauntleroy* (1886)—could obtain £3,700 for American rights to a serial why should she, Mrs Humphry Ward, have only £1,400?[20]

George Smith gave some typically sound advice on 9 December 1890. Delivery of the book had already been delayed by Mary's collapses and incorrigible rewriting. Wait, he counselled, to see what the copyright legislation in America 'really means, and when it is to come into operation'.[21] There ensued some nerve-racking cat-and-mouse negotiations with American houses and with Macmillan. Finally, in April 1891, the American Copyright Bill passed into law. Mary Ward (who had taken other advice from her publisher cousin Edward Arnold) told Smith that in the light of this legislation 'my own ideas rise as high as £7,000'.[22] This was clearly higher than his had risen but, nothing daunted, he took the plunge. Her instinct proved correct. Smith wrote in triumph and some astonishment to Mrs Ward on 13 June 1891 that 'I met Frederick Macmillan in the Park [i.e. Hyde Park, where they both rode] this morning. It flashed on my mind that I could sell him the American copyright of your book, and after a long talk (which made me late for breakfast) I promised him that if he made me a firm offer of seven thousand pounds for the [American] copyright, including Canada, before one o'clock today, I would accept it on your behalf.'[23] Macmillan replied accepting.

It was an astonishing *coup*. Moreso as the novel was only three-quarters written and Mary Ward refused to let anyone read any part of it. Macmillan—as they realized soon after with some uneasiness—had bought a pig in a poke. But they had their reasons. As Smith surmized the £7,000 must have been ventured as a publicity stunt.[24] Whatever Macmillan's motives, it meant that Mrs Humphry Ward would receive one of the highest advances for a novel ever paid in England. Before the first library copy of *David Grieve* was sold to Mudie in January 1892, she had in hand £1,750 from Smith, £7,000 for American rights, and guarantees of £675 for French, German, Dutch and Empire rights: some £9,425 in all, with much more to come.

Publication day brought its terrors. Unlike *Robert Elsmere*, *David Grieve* would get instant spotlight attention from the newspapers who were clamouring for their review copies. These were couriered out between 10 o'clock and 12 noon on 21 January. Evening papers had to wait until four in the afternoon to forestall any jumping of the gun. 'The notion of that book being read and reviewed in three or four hours is abhorrent to me,'[25] Mary Ward declared.

A lot was riding on these early reviews. Smith had printed 3,000 three-volume copies, divided, as usual, into two 'editions'. On publication day, only 1,054 sets had been subscribed. This was bad. The reviews were worse. They ranged from the suborned friendliness of the *Guardian* and *The Times* to a veritable hatchet job in the *Pall Mall Gazette* where Mary Ward inconveniently did not have a close relative working. There was the usual balm from 'old friends' like Jowett, who sent in their eulogies by private letter at the end of the month: 'I feel as though the book were entering safe water,' the author told Smith, 'above all as if it were finding the people it was meant to find.'[26]

There was nonetheless still some choppiness ahead. The monthlies and quarterlies—her old enemies from *Robert Elsmere* days—were universally cruel about *David Grieve*. Mrs Oliphant writing in *Blackwood's* was particularly brutal. Both the *Quarterly* and the *Edinburgh* 'bombarded' her. Mary Ward was hurt and resolved to write a fighting preface to the cheap edition of *David Grieve*. But, from a commercial point of view, she was not put out by this critical roasting. It was her intuition (shrewdly correct, as it turned out) that sales of the cheap edition would confound the quarterlies. These 'bogies of my childhood',[27] as she called them, had had their day. Let them worry about *their* sales, literary dinosaurs that they were.

George Smith was, in fact, very keen to get on to the 6*s*. one-volume reprint of *David Grieve*. The three-decker editions had gone rather sluggishly. Normally, of course, Mudie expected a decent delay before the one-volume reprint; otherwise his investment in the super expensive three deckers would be undercut. Time was (in the 1850s) when the delay was two years. Insidiously it had got shorter. In 1892, six months was the decent minimum. Now—in defiance of all trade etiquette—Smith proposed bringing out the 6s. reprint of *David Grieve* in April or May—a mere three months after the three-volume edition had been published. Mudie was furious at the prospect of a spring 1892 reprint of *David Grieve*.

Mudie's rancour was only one of George Smith's vexations. No sooner had the 6*s*. edition of *David Grieve* been in print for a month than Mrs Ward asked for a £600 advance on its sales. Long-sufferingly, Smith dispatched a cheque on 14 June. Then, on the day she received this payment, she wrote asking for another £1,000 advance for her *next* novel—a work which she had scarcely begun to think about. Smith was angry. It was 'a question of delicacy', but he felt he had to be 'quite open' with his pertinacious author. First, his meagre £135 profit so far on the publication of *David Grieve* did not, he pointed out, 'warrant the payment of £1,000 on the commencement of another novel, to be published two or three years afterwards'. This was not how publishers did business. Secondly, if the next novel required the same 'extraordinary expenditure in the way of alterations and corrections'[28] as had *David Grieve*, he could not justify the same generous terms.

George Smith's straight talking chastened Mary Ward. She withdrew her request for an advance (permanently) and nobly offered to take lower royalties in future—an offer which Smith as nobly declined. But she still insisted on having her proofs to work on: 'I am afraid I cannot give up the system of printing as I go along. It has become a literary habit and I must somehow maintain it.'[29] Peace was made between them. And it was hard to quarrel when—after a somewhat slow start and some beastly reviews—*David Grieve* turned out to be a success of Elsmerean proportions. Mary had been right. The British reading public did not give a fig for the opinions of the quarterly reviews. By August 1892 *David Grieve* had sold some 3,000 in the three-volume form and some 17,000 in the one-volume 6*s*. form.

Mary Ward could not but be pleased with the money her two novels had earned. Between them, *Robert Elsmere* and *David Grieve* brought in something of the order of £16,000 between 1888 and 1893. This kept the wolf from the door very comfortably. But what *was* worrying was that these two novels had taken her seven years to write. Three to four years composition was a luxury which the novelist could no longer afford—needing as she did £3,000 a year to keep the Ward enterprise afloat. From now on, she would dance to a faster tune.

The idea for Mary Ward's next novel, *Marcella*, originated in the family move to Stocks in summer 1892. Leaving the 'villadom of Surrey'[30] involved more than grander living quarters. Stocks's Hertfordshire was only half an hour or so from Euston, but socially it was worlds away from bourgeois Haslemere. The Wards were translated from suburban to rural life and (buoyed by Smith, Elder and Co.'s thousands) they assumed a new role as members of England's landed gentry. Mary Ward later claimed that she had been feeling the need to 'come close to the traditional life of field and farm'[31] for some time. But there were social consequences to the move. In London, the Wards were, if not classless, relatively unfixed on the social scale. In the country, like her next heroine, Marcella, Mrs Ward found herself in the awkward role of Lady Bountiful to a population of local peasants wretchedly pauperized and dangerously criminalized by the agricultural depressions of the late 1880s.

The Wards first saw Stocks in January 1892. On 12 December 1891, as Mary records, 'two keepers in the employ of a neighbouring landowner were murdered by poachers, on the skirts of the big wood [Ashridge, actually Stocks ground whose shooting rights had been let out] which from all time has crowned the hill above the house. When we made our first visit, two men were in prison awaiting trial [actually there were three, one of whom was eventually transported for twenty years], and they were condemned and executed before we entered upon our tenancy in the following April. A strong effort was made after the verdict to get a reprieve, but the Home Secretary stood firm, and the law took its course.'[32] This was Mary Ward's *donnée*. *Marcella* would reconstruct 'the psychological origins and the local circumstance of the murder'.[33] An added complexity was given in February 1892, when Mary met the Home Secretary, Henry Matthews, at a country house party. He was currently under pressure to commute the death sentence, which he refused to do. There was, as Mary Ward recalled, 'a good deal of rather hot discussion of the game laws, and of English landlordism in general'.[34]

According to Mary Ward's own recollection, the novel's preliminary outline (up to the execution of the poacher Hurd, presumably) was put down on half a sheet of paper, soon after the initial visit to Stocks.[35] A version of this outline survives in a letter she wrote to Macmillan on 3 May. 'I shall begin work this month ["morning" crossed out] and shall aim at a shorter book done in much shorter time. The subject will be entirely English the questions involved social ["and political" erased] rather than religious, the heroine of our own class ... The book will be a distinct love story and it will end well.'[36]

As the narrative opens, 21-year-old Marcella Boyce is whisked from a bohemian existence in London to the country house, Mellor, where she is expected to act like a lady. She has a father who as a young man disgraced himself in the eyes of his fellow men by commercial misconduct, and her parents, although they cohabit, live separate lives. Despite this and her fierce Fabianism, Marcella's beauty and spirit captivate Aldous Raeburn—the future Lord Maxwell and a true blue Tory. She scorns his class and is sarcastic about the game laws. 'I come fresh into your country life,' she tells one young aristocrat, 'and the first thing that strikes me is that the whole machinery of law and order seems to exist for nothing in the world but to protect your pheasants.'[37] Imbued by vague Ruskinian nostalgias, Marcella revives the almost lost art of straw plaiting among the village women (at wholly uneconomic rates). She befriends a horribly destitute family, the Hurds. Jim Hurd is deformed and dwarfish, and his beaten-down wife ekes out a living on 11*s*. a week when he is in work. Their child, Willie, is dying of consumption and will not survive another winter's unemployment. Hurd poaches to feed his family and in one bloody encounter murders a Maxwell keeper. Marcella sees Hurd's crime as socially caused and she throws herself body and soul into his defence. At the height of her indignation, she even denies God. The trial and the inevitable sentence drive a wedge between the lovers. The wedding is postponed, and finally cancelled. Aldous's place in Marcella's heart is taken by Harry Wharton—also rich and handsome but a socialist. Wharton owns a newspaper devoted to the 'cause of the people'; he runs a farm on co-operative principles; he fights for the downtrodden; he is an apologist for Irish terrorism.

Marcella's love dilemma enacts a general political dilemma of the 1890s. It was also a personal dilemma for *Marcella*'s author. She was Mrs Humphry Ward of Stocks. But her London social work had brought her into sympathetic contact with politicized artisans and labourers, from whom, as she later said 'I learned much .... Those were the days of the early Fabians, and the Fabian Essays. Collectivist ideas were making way in the educated middle class, and all who watched the world with any intelligence had their eyes fixed on the rise and progress of Social Democracy in Germany. Arnold Toynbee's beautiful life had not long closed; the doctrines of Henry George and the single-taxers had been sweeping through the working-class like a tidal wave.'[38] As a political thinker Mary Ward sought some conciliation between the interests of the landed classes and the dispossessed of England. It results in *Marcella*'s extraordinary theorem that the estates of the English aristocracy are property held in trust for the ultimate use of the People. It is ingenious but intellectually dishonest. At root, Mary Ward felt guilty buying up a large parcel of rural England like some *nouveau riche* biscuit manufacturer. The idea that she was merely the stewardess of Stocks soothed her proprietor's guilt.

Mary Ward wrote *Marcella* under financial duress, which intensified as the novel progressed. Moving house and buying houses added considerably to the £3,000-a-year routine expenses of the Ward household. Negotiation with Smith began on 1 June 1892, when she informed him that she wanted half an hour's talk about her next book.[39] Writing began soon after. The novel's extensive literary remains indicate that it was the most thoroughly planned of her works hitherto. She began with 'motives' and such emblematic scenes as that in the first paragraph of Marcella's throwing open her window on to Mellor at dawn. Capsule biographies were prepared for each of the main characters and chapters were thoroughly outlined in advance of composition.

The practical arrangements for the writing of *Marcella* were the same as for *David Grieve*. The author dispatched her manuscript in small packets to Smith's printer to be set up in page proofs. These 'revises' were then scored over and a fair copy (largely autograph, but also using her new secretary and cannibalized proof fragments) was returned to the printer. As before, George Smith was allowed to see the work only when composition was complete. Mary Ward—in her new power as a best-selling author—was fanatic on this question of authorial privacy and the nearest she and Smith came to an all-out quarrel was when she discovered that his son-in-law Reginald was 'dishonourably' sneaking a look at her 'first proofs'.

Health collapses were by now a routine part of of Mary Ward's writing ordeal, but usually they came late in the day when she was worn down. Unusually, the health crisis came right at the beginning with *Marcella*. The novel's germination coincided in August 1892 with a physical prostration more extreme than anything she had experienced since childhood. On 7 August she told Smith that 'I have been on the sofa ever since you left us, often in much pain, and pitying myself a good deal.'[40] Specialists were called in; a nurse was hired; absolute bed rest was sternly prescribed. But—with the bills for Stocks streaming in—she had to write. In the same letter, she informs Smith that 'this afternoon I have sketched out the first two chapters of *Marcella* .... They will have a good deal of my own youth in them, which was very miscellaneous and out of elbows.'

George Smith's wife called on the invalid and reported that she 'wanted a good novel'.[41] Smith gave the matter some thought. Gissing's *New Grub Street* would, he thought, 'make you too miserable, it is not a book for an invalid, though it has a certain strength to it'.[42] He sent instead Gissing's socialist novel, *Demos*. Mary Ward devoured it, and wrote on 19 August, '*Demos* has interested me very much .... But what a curious gospel it is the man preaches!'[43] Curious as it was, Gissing helped form *Marcella*'s political plot as it develops around Wharton—a character who owes something to *Demos*'s demagogic and dishonest Richard Mutimer.

Mary Ward's health gradually mended over the autumn. On 19 October, she reported herself 'nearly well, in fact, thanks to that wonderful drug phenacetine'.[44] At the same time, she reported herself getting on fast ('for me') with *Marcella*. Eighty pages were written and ready for the printer. She promised to finish the first volume by Christmas 1892, and noted that this novel was flowing from the pen more easily than had *David Grieve*: 'something intervened,—a tranced, absorbed state, in which the action of certain normal faculties seemed suspended in order that others might work with exceptional ease'.[45]

The first of *Marcella*'s four 'books' was roughly completed on 22 January 1893. Mary Ward stuck to her resolution not to ask George Smith for an advance. In January 1893, she had another bout of 'obscure neuralgia'. But raw beef sandwiches brought about a quick recovery. She still felt that despite illness *Marcella* 'is more of a pleasure to write than anything has been for a long time'.[46] The experience of autumn at Stocks was invaluable for landscape description and the nearby village of Aldbury supplied a gallery of peasant originals for the novel. In March she thought the novel would be finished in October. She was three months out of her original schedule—which for her was a bagatelle.

By midsummer 1893, it was time to negotiate with Macmillan for American rights. On 21 July, Mary Ward sent Smith a telegram: 'My own idea. Trois mille down 20 per cent royalty. Can you advise me.' Smith thought that '£3,000 would not, in my opinion, be bad terms for you.'[47] It was a come-down from *David Grieve* but as Mary Ward admitted in a follow up letter (the third communication to George Smith in one day), 'of course I know that the Macmillans will not give me £7,000 again .... They have told me so.'[48] But she confidently expected an American sale of 70,000 at a dollar. Smith took the matter in hand. On 24 July he telegrammed, 'arranged royalty zwanzig payment zwei tausend zwei hundert und funfzig.' He had screwed Macmillan's George Brett up from an opening bid of only £1,500. Mrs Ward declared herself 'abundantly satisfied'.

Less money meant faster delivery. On 24 July, Mary Ward told Smith that 'there is no time to be lost I feel if I am to have [*Marcella*] finally ready for press by the middle of January [1894]'.[49] On 8 September she reported herself 'buried in Marcella', despite a hand that was again 'very lame'.[50] But Mary Ward felt confident: 'there will be comparatively little rewriting to do on the [remaining] proofs .... About 520 pages are now in print. There will not be more than 950 pages at the outside.'[51] (The full novel in fact swelled up to 1,028 pages.)

By this stage in September, Mary Ward had written the first two volumes, which brings the action to the execution of Hurd (written at white heat) and Marcella's repudiation of Aldous. On 21 November 1893, she announced that she was coming up to London for a month, where she embarked on writing the crucial third volume, which finds Marcella working as a district nurse in the slums around Bloomsbury. (Mary Ward drew on Gertrude Ward's experiences for this section.) Marcella's lovers, Wharton and Aldous, have meanwhile become opponent Members of Parliament. On 28 November, Mary visited Parliament herself as the guest of her cousin Hugh Oakeley Arnold-Forster, now a Unionist MP. The visit made a profound impression, and in her notebook she jotted extensive observations on the layout of the House. In the same notebook under the heading 'motives to be carried out' Ward sketched a skeletal outline for the action of her last volume, as it centres round the perfidy of Wharton. The heroic socialist of the first two books is revealed as a seducer of working-class women, a gambler, and a traitor to the Damesley steelworkers, whose strike his socialist *Clarion* ostensibly supports. His faults are summed up as 'lack of fighting spirit in him'. Wharton is eventually bribed by the steel magnates, and ends married to the Tory beauty Lady Selina Farrell. With Wharton's apostasy, all hopes for an organized Labour Party in England dissolve.

From references in the text it is clear that Wharton was directly inspired by Charles Stewart Parnell. Parnell had taken the chaotic forces of Irish dissidence and welded them into a party. He survived the libel (propagated largely by forged letters given credence by *The Times*) that he had condoned the Phoenix Park murders. But his lecheries and secret immoralities had betrayed him. His being cited as a co-respondent in the O'Shea divorce case in November 1890 cost him his political career. He died a broken man in October 1891. It was for Mary Ward a parable. The abrupt third-volume degradation of Wharton links to the main theme of the novel: Marcella's rejection of 'abstract' socialism for direct and unostentatious acts of philanthropy which finally—despite all her earlier doubts—enable her to marry Aldous and become Lady Maxwell, mistress of Maxwell Court. The same turn is charted in the death of Aldous's saintly (but socialist) comrade Edward Hallin—a character based on Arnold Toynbee. As he weakens from terminal heart disease, Hallin insists on addressing rallies of workers, although his increasingly Christian mysticism has alienated their support. Addressing a meeting on Land Reform at Bethnal

Green, he is brutally jeered at and breaks down into incoherent pieties on stage. Hallin returns to die at Maxwell Court in high Victorian style. In his last hours he reads Jowett and pronounces to those around his bed that 'there is only one goodness—the surrendered will'.[52]

By mid-January 1894, Mary Ward had nearly finished a penultimate draft of her work, despite 'Arctic weather' and a hand which 'alas! is almost breaking down'.[53] There remained only a hundred of the thousand pages to write. She must, she felt, hold out for another five weeks, to get the work done by 12 February, Macmillan's deadline. But on 18 January the novel hit a rock. The source of the problem went back to 1880, when George Smith had given the influential *Pall Mall Gazette* his son-in-law of two years, Henry Yates Thompson, as a belated wedding present. Thompson promptly changed the paper's politics from George Smith's jingoistic imperialist line to Gladstonean Liberalism. In 1892—while Ward was beginning to write *Marcella*—Thompson had sold the paper without warning to an unknown purchaser (later it turned out to be William Waldorf Astor), who overnight changed the paper's political alliance yet again to Toryism, so stabbing Gladstone in the back, as Liberals thought. The episode provoked considerable criticism of Thompson and it was alleged that he had allowed himself to be bought by his political adversaries. It was an allegation that still rankled with Smith and his son-in-law in 1894, as the final chapters of *Marcella* were being written.

In the novel as it was originally conceived, Mary Ward had her villain, Wharton, rat on the labour movement by secretly selling the *Labour Clarion* to the steel masters. Overnight, the paper's crucial support for the strike at Damesley is thus bought off. Given her embargo, George Smith would not in the normal course of events have read this episode until the eve of publication. But his possible reaction made Mary Ward increasingly nervous. Finally on 18 January 1894 she sent the proofs of the crucial passage with a fearful covering letter:

> Some time ago I made up my mind that the sale of a newspaper to its political enemies at a critical time would make a good peg on which to hang Part III of *Marcella*. But when I was pretty well through it, it occurred to me that last spring at dinner at the Priestleys I had said something to you about the sale of the *Pall Mall* which might make you think I had intended some personal reference by the incident. This idea has tormented me a good deal.[54]

On 21 January, a thunderstruck George Smith wrote back, in extreme distress: 'It seems to me the sale of the "Clarion" will be associated with the sale of the "Pall Mall Gazette" in the mind of every reader who has heard of the latter transaction; that the reviewers will point out the relation; and that it will be supposed that you have made a personal attack on Henry Thompson.'[55] 'Your letter has really thrown me into despair,' Mary Ward wrote back by return. 'The real fact is that I knew *nothing* about the sale of the *Pall Mall* except what the newspapers told me.' But she conceded that 'it is quite possible that unconsciously I may have come nearer the truth in some detail than I knew'.[56] He could not imagine how unhappy she was, she would do nothing to give him pain. Would he come down to Stocks the next day? Smith agreed and there ensued a flurry of telegrams dealing with train times and travel arrangements. On 23 January she sent further chapters about Wharton and the *Clarion* episode for him to read in the train. She was at her wits' end, and would be guided entirely by what Smith and Thompson decided.[57]

At Stocks, Smith and Mrs Ward agreed on how the novel should be changed. In *Marcella*, as published, Wharton does not betray the Damesley strikers by selling the *Clarion* to the employers. He is bribed instead by being offered shares in a syndicate whose value will rocket to £20,000 once the strike ends. Wharton accepts his pieces of silver, changes the paper's line, and is eventually exposed by the investigative journalism of his own more honestly socialist employee Louis Craven. On 25 January 1894, back in London again, Smith reported that his daughter Dolly was content with the change, and 'as to myself I slept last night, didn't I sleep!'[58] In fact the alteration fudges the climax of *Marcella* damagingly. More interesting, however, is the latent aggression which made Mary include it—'unconsciously'—in the first place. Was the episode a warning that she retained the power to bite the hand that fed her?

Despite the mutual relief it gave, the revised *Clarion* episode is a smudgy blot in *Marcella*'s design. But the novel remains one of Mary Ward's best. One of its strong points is that the narrative is more integrated than its predecessors. Like *Robert Elsmere* and *David Grieve*, *Marcella* has a wide social sweep from the rural distress of 'Brookshire' to slum London; from the mansions of Tory peers to trade union committee rooms. *Marcella* also has the tight chronological compass which she normally favoured (few of Mrs Ward's novels extend over more than five years in narrative time). But whereas *Robert Elsmere* and *David Grieve* sprawl and buckle under the pressure of their content, *Marcella* is neatly confined within the evolving mental, spiritual, emotional, social, and moral conflicts of the heroine. And Marcella's painful drift from radicalism to pragmatic Toryism honestly reflects Mary Ward's own emerging sense of herself as helper of London's poor and mistress of Stocks.

With the *Clarion* difficulty out of the way, the completion of the novel was rapid. On 1 February 1894 Mary Ward recorded that 'last night I wrote the last lines of *Marcella*'. Some 'two or three weeks of hard revision and re-writing of Book III' remained. But, as the book was now so near completion, 'I trust the firm will not think that I am breaking a promise I made in a letter of last year, if I ask them now to make me an advance on the sum to be realised by the three-volume edition.' £1,000, she said, 'would make me quite happy'.[59] Smith replied with a cheque. But as usual, he accompanied his payment with a lecture. 'Our profit from the sale of the first 2,000 copies of *Marcella*, after payment of your royalty of 7*s*. 6*d*. per copy will be about £70.'[60] Mrs Ward replied with some adroitness to Smith's moans. It seemed wrong to her 'that the profit of a firm which does so much for me, on my work, should be no more than you say'. She consequently proposed that her royalty on the first 2,000 sets of *Marcella* be reduced to 6*s*. and 7*s*. on any three-volume copy sold thereafter. Smith was dumbfounded. 'I hardly know what to say to your letter .... It is not often that a publisher receives such a letter from an author.'[61] He accepted gratefully.

Publication was fixed in Britain for 3 April and final proofs went off to America on 28 February. From New York, Macmillan's manager Brett wrote on 13 March to say that he had read *Marcella* with great pleasure 'and look for it to have a great success not only on the other side but here as well'.[62] It did. By 23 March he reported that American advance orders were 18,000—somewhat greater than *David Grieve* had commanded. And Tauchnitz (who had sold some 4,500 copies of his three-volume *David Grieve*) offered £250 for *Marcella*'s European rights, which made a pleasant addition to Mary's receipts.

The English picture was less satisfactory—particularly the library order. Smith's manager, Aitchison, reported that 843 of the three-volume set had been subscribed for town and country. George Smith and Mrs Ward pondered this statistic deeply: 18,000 in the US, and less than 1,000 in the UK. It was more irritating since the actual launch of the novel in April was a triumph of publicity: 'Piccadilly', Mary's sister Julia told her, was 'placarded with *Marcella*.'[63]

The reviews of *Marcella* were goodish, and—as usual—neither helped nor hindered sales. Smith printed 2,000 three-deckers divided as usual into two 'editions'. By 13 April all 2,000 were sold and a third edition was in hand. By June 3,000 three-volume sets (six 'editions') had been cleared. The general feeling at Smith, Elder's office in Waterloo Place was that Marcella at birth had considerably more vitality than her 'elder brother' David.

Inevitably, Mary Ward was encouraged by the good news to ask for some cash reward. 'Might I ask you to advance me £500 more as against the sale of *Marcella*?'[64] she wrote on 24 May. Smith obliged. In the same letter, she nudged Smith about the one-volume reprint: 'I suppose too that the cheap edition of *Marcella* will hardly be delayed longer than July?' This innocent supposition was explosive in its consequences for the British book trade. For forty years the libraries (notably Mudie and W. H. Smith) had upheld the massively expensive three-decker and used their power as bulk buyers to prohibit early reprinting of novels at affordable price. The consequence—as all commentators realized—was that the British were principally borrowers of new fiction; the Americans, who had no Mudie, were purchasers. Mary Ward indicated her preference for the other country's system in a letter to an American admirer of *Marcella* on 29 May 1894: 'Your American system of buying books at once compared to our most unsatisfactory library system makes the progress of a book in the States extraordinarily interesting to the writer of it. Its wider immediate diffusion stirs up a strength and variety of public opinion as distinct from press opinion, which is I am sure of very great service to such a book as *Marcella*.'[65]

Mary Ward had a personal grudge against Mudie, whom she saw as a monopolist starving the English reading public of books they manifestly wanted (i.e. her books). In a letter of the evening of 24 May (there had been three exchanges of letters that day), she told George Smith, 'Sir Henry Cunningham [the eminent barrister] told us last night that he had made a tremendous protest to Mudie's against their behaviour in the matter of *Marcella*—which he seems to have told them he regards as a fraud on the public, or rather their subscribers, whom they were *bound* to supply with the new books!'[66] The reprint of *Marcella* was the torpedo that sunk the three-decker and by so doing stripped Mudie of his dictatorial powers. George Smith had a 6*s*. copy out by early July—just three months after first publication. Mudie lamented that this left hundreds of worthless three-volume copies of the novel on his hands.[67] Smith, however, had bitten the bullet. 'I am entirely with you in your wish to be independent of the libraries,'[68] he told Mrs Ward.

It was all very tense. Mary Ward watched the emergence of the 'popular' reprint of *Marcella* in July with nervous excitement and fussed at George Smith about advertisements. She noted too that Heinemann—with his new star author—had upstaged them: 'Here is Mr Hall Caine—to a large extent I believe through skilful advertisement of all kinds—starting with a first edition of 20,000 copies at 6s. I don't want to imitate him—Heaven forbid—still I feel that one can only compete nowadays by a free use of legitimate advertisement.'[69] 'Everything is being done by us that can be done by us,' Smith reassured her on 13 July. 'We have sent show cards freely to the London and Country booksellers. W. H. Smith and Son alone have taken 1,000—and we have sent a circular to nearly every country bookseller.'[70]

The 'popular' *Marcella* was in fact slower taking off than its predecessor. On the day of publication, Smith had cleared only 6,548, against 9,774 copies of *David Grieve*. Mary Ward wondered whether she would not have to cobble together a volume of her religious and critical essays to make up the shortfall in her earnings. But by August sales were up to 12,500, which at 1*s*. 3*d*. royalty yielded £781 5*s*. And by December, the 6*s*. *Marcella* broke the 20,000 barrier. The stopgap volume of essays was gratefully shelved. And things had gone even better in America. On 4 April Frederick Macmillan sent £1,990 5*s*. for the American and colonial editions. By 1 May, George Brett reported that 30,000 of the $2 edition were now in print. By October the figure was doubled.

Marcella was the fastest novel Mary had yet produced: only two years separated its publication from that of *David Grieve*. And it had earned well:

around £6,500 by the end of 1894. But 'well' was inadequate to the Wards' needs. Three thousand odd a year no longer got them by. In the same week that Mary Ward wrote the last lines of *Marcella*, she and Humphry resolved to buy Stocks. To make ends meet, she would have to earn yet more money yet faster for the household.

Now that the American rights bonanza had cooled down (that £7,000 clearly *was* a fluke) Mary Ward had to find other means of maximizing income from her fiction. Dramatization was one option which she would look into when she had time. In the meanwhile, serialization—particularly American serialization—was the obvious next step. As it happened, a particularly tempting (and timely) offer had just come in July 1894 from the American *Century* magazine who wanted a new novel to be serialized in twelve monthly parts. *Century* was a top drawer journal and as they were communicated through the magazine's English agent, Edmund Gosse, the terms were mouth-watering as to money but deterring in some other ways. Serialists had to keep in bounds. *Robert Elsmere* had blown up to 450,000 words; *Marcella* to 300,000 words. The *Century* wanted no more than 150,000 words so as not to squeeze their limited space. And they insisted on seeing the whole work in advance so as to be sure that it was suitable for an American public. Not for them the rigmarole by which George Smith was excluded from examining his property until it was too late to change it.

Finally, after interminable wrangling, a contract was hammered out over the summer months of 1895. The arrangements were unusual, and designed to protect Mrs Ward's anemone-like sensitivities while satisfying her now enormous appetite for money. Smith—as primary purchaser—would buy the copyright of the new novel, *Sir George Tressady*, outright for £10,000, payable when the work was set up in type. (It was, incidentally, the highest lump sum Smith had ever given for a novel and equalled Longman's legendary payment for Disraeli's *Endymion* in 1880.) Smith would then make his own deal with the *Century*. After much haggling the *Century* agreed on £3,600 for the serial rights. A compromise was reached on prior examination of the material. She would—via Smith—let the *Century* editors see the first half of the book, at which point they would have the option of withdrawing. Publication of the new novel would begin in the November 1895 issue of the magazine and would run an exact year. Book publication of *Sir George Tressady* (by Smith, Elder in Britain and Macmillan in America) would be held back until 25 September 1896.

It was all very disciplinarian and wholly against Mary Ward's writing instincts. She would never have gone along with all these restrictive clauses if Stocks did not have to be paid for. That transaction—which was not completed until March 1896—loomed over all phases of *Sir George Tressady*, from first thoughts to first edition. Unusually, first thoughts for *Sir George Tressady* came quickly. As early as 16 August 1894, Smith congratulated Mary on 'having found a promising idea', noting 'You are going to quite a new field in adventuring on the mining districts.'[71] As it happened, Mary Ward had been nurturing a mining melodrama since September 1888, when on a visit to the Dugdales at Merevale she had—as she reported to her father—heard a most interesting family tale:

> just beyond the park, visible from all the windows, lies the coal-pit where [Mrs Dugdale's] husband lost his life six years ago. It is a splendid story. He went down to save ten men who were cut off from the shaft by a fire in the pit. There was an explosion, he was stunned and lost his companions. By a miracle he was found after a couple of hours and brought up, but so burnt that even his wife who was waiting at the pit's mouth hardly knew him, and he lived six days, in terrible pain, borne without a murmur and died. An ordinary English gentleman who would have hated to be called a 'hero' or 'martyr' and yet what a story, what an end.[72]

This 'splendid story' was reserved for the last chapters of her new novel, in which the mine-owning hero, Sir George Tressady, gives up his life to save the union strikers who have defied him. Onto this idea Mary Ward grafted the life drama of Sir George, a pococurante aristocrat Tory MP who makes a disastrous marriage to the light-headed Letty Sewell and who is saved from moral ruin by the incandescent goodness of Marcella—now Lady Marcella Maxwell of Maxwell Court and good fairy to the sweated women workers of the East End of London.

By 19 November 1894, the author had a sketch of the whole narrative on paper. There was one rather risky novelty to it. As a framing device, *Sir George Tressady* would be a sequel to *Marcella* with that heroine carried over as presiding angel. Mary Ward decided on this step in the face of George Smith's publishing wisdom that sequels always sell less. She persisted, one suspects, because she simply did not have months to invent new main characters and themes. Time—particularly undistracted time—was now in almost as short supply with her as money. In addition to the never-ending Stocks business, the Passmore Edwards Settlement (the successor to University Hall) was in its most critical phase: planning for the new structure began in February 1895 and building was not finished until October 1897. The laying of every brick was a matter of importance to Mrs Ward.

By 6 January 1895, she had nine of *Sir George Tressady*'s twelve numbers written. She was ill, of course, and needed an operation for an unspecified internal problem. But she would postpone this and had no doubts that 'under any circumstances the book will be finished by the end of February [1895]'.[73] But she did have some doubts about the *tone* of the story. The *Century*'s tighter wordage made her unload her style—cutting out habitual descriptive flights and essayisms. What remained looked awkwardly rushed, depending stylistically on short paragraphs, dashes, and exclamations. She sensed that the narrative was somehow jerky and unleavened. With a rare insight she noted to Smith, 'If only I had more humour... It seems to me I appreciate it as much as other people. But I cannot produce it.'[74] In later years, she marked *Sir George Tressady* down as something of a failure for its style and manner. It lacked, she declared in 1911, 'freshness of attack'. There was no other novel 'that I would so gladly revise'.[75]

There were other more material problems. Mary Ward now realized with some alarm that she had a novel virtually complete but the bulk of her payment for it would be withheld for eighteen months until September 1896, when the American serialization came to an end. She needed to generate money to fill the gap—ideally without too much labour. So on 14 January 1895 she tempted Smith with the sketch of 'a village story based on a tragic incident that occurred in our neighbourhood last summer, which rather pleases me'.[76] As a rule, Smith did not like Mrs Ward writing short fiction, which 'cheapened' her value as a full length novelist. But he gave the go-ahead on her village tragedy, which she wrote up in two weeks as *The Story of Bessie Costrell*.

The story came in an effortless surge of creativity that surprised its author and was, she was sure, 'one of the best things I have ever done'.[77] It tells the story of John Bolderfield, a farm labourer, who on retiring entrusts the box containing his life's savings—£71—to his young niece, Bessie, and her older husband, Isaac Costrell, an austere chapel-goer. Bessie is light headed. She eventually succumbs to temptation and steals some of the money to spend on drink. John Bolderfield returns from a journey and the truth is forced out of Bessie by an unforgiving Isaac. She throws herself down a well. If hardly the best thing Mrs Humphry Ward had done in fiction, *Bessie Costrell* is told with remarkable economy and lack of condescension to its humble personages.

Smith agreed to publish *Bessie Costrell* in three instalments in *Cornhill Magazine*, May–July 1895. This would yield some £560 with the expectation of as much again from the volume issue of the novella in Britain and (through Macmillan) in America. *Scribner's Magazine* serialized the work in America and the *Revue des deux mondes* in France. Around 15,000 copies of *Bessie Costrell* were eventually sold in the first year in England and 20,000 in America. Altogether, it was a profitable fortnight's work. Not least because reviewers adored it, the *Critic* (13 July 1895) going so far as to call Mrs Humphry Ward 'the greatest woman novelist of her day'. The only fly in the ointment was that George Smith would not encourage her to repeat the experiment.

The completion of *Sir George Tressady* was smooth and surprisingly pleasant. When they received their first half for inspection in July 1895 the editors of the *Century* were ecstatic, telegraphing 'Congratulations on our greatest novel!' In February 1896, George and Reginald Smith (the son-in-law who was more and more taking over the reins of the firm) were permitted to read the whole novel bar the last chapter, which duly came on 23 March. For this privilege they might send the author a second £1,000 on account. She had done the book, earned the payment, and would now have her 'little operation'. And, after infinite heart-searching, Mary Ward told Smith that she and Humphry had taken the plunge at last and engaged themselves to buy Stocks. The contract would be concluded at Michaelmas 1896.

Sir George Tressady, as Smith predicted, is, like most sequels, something of a comedown from *Marcella*. It nevertheless has points of interest. It was by far the most political novel that Mary Ward had hitherto attempted. The intricate business of the political in-fighting around Aldous Maxwell's new Factory Act and Marcella's organizing of sweated workers in the East End are new ground in the fiction of Mrs Humphry Ward. As in *Marcella*, the drama of political conversion is excitingly handled, climaxing in Sir George's voting against his party in the final debate on Aldous's bill (something, however, that Gladstone thought politically improbable). There are numerous incidental good things in the novel, such as the hero's incurably feather-brained mother.

The weakness of *Sir George Tressady* principally resides in Marcella and her role in the narrative as incarnation of moral goodness. There is never any expectation when Sir George falls in love with her that she might fall—or even flirt with her admirer. There seems every possibility, by contrast, that Letty Tressady will succumb to *her* unscrupulous male admirer; something that makes this sub-plot one of the more interesting features of the novel. Improbably, Marcella contrives to mend the crumbling Tressady marriage, simply by displaying her angelic goodness to the errant wife. There is both a logical completeness and at the same time a surrender of all plausibility when, at the end of the novel, the vision of Marcella (not his pregnant wife) appears as pure divinity to Tressady as he lies dying in the mine shaft:

> As he lifted his sightless eyes he saw the dark roadway of the mine expand, and a woman, stepping with an exquisite lightness and freedom, came towards him. Neither shrank nor hesitated. She hurried to him, knelt by him, and took his hands. He saw the sweetness in her dark eyes. *'Is it so bad, my friend? Have courage—the end is near'* *'Care for her—and keep me, too, in your heart,'* he cried to her, piteously. She smiled. Then light—blinding, featureless light—poured over the vision, and George Tressady had ceased to live.[78]

Sir George Tressady received 'respectful' reviews in England from the usual friendly places. First-day English sales on 25 September 1896 were 12,500, which was hopeful. But, by January 1897, only 23,700 had been sold. 'Rather slow'[79] was Smith's verdict. But the American returns were much more worrying. By March 1897, Macmillan had sold only 13,000 of *Sir George Tressady*, against 30,000 of *Marcella* in its first six months. The *Century*, with its 170,000 circulation, had simply killed the book. Macmillan made it clear that they could not do business with Mrs Ward again if she insisted on American serialization. All this posed awkward problems as Mary Ward worked out her strategy to maximize income for the next novel—*Helbeck of Bannisdale* as it was to be.

George Smith asked discreetly around and discovered that nothing like *Century*'s payments could be expected for American serial rights, and then only from magazines below the 'top plateau'. And even if they arranged a concurrent serialization for the new work in an English monthly (say Knowles's *Nineteenth Century*) Mary would probably jeopardize home sales of her novel in book form. There was nothing for it but to give up serialization, at least for the while. Smith, Elder were prepared, nevertheless, to advance £5,000 on account of a 20 per cent royalty for the English 6*s*. edition of *Alan Helbeck's Passion* (as the new book was provisionally called). Mary would additionally get a 3*d*. royalty on colonial sales (these rights were also bought by Smith). Smith, Elder's £5,000 advance was not, however, what modern authors understand by the term, but a 'loan', extended at 4 per cent annual interest. The publishers took as their collateral all receipts for the novel, including the American. They gave Mary a first instalment of £1,000 in June 1897.

£5,000 was by most standards a princely payment. For Mrs Humphry Ward it was short commons. In 1897, butchers' bills alone for the Wards ran to almost £200 a year, and total rents, food, servant wages, and new furnishings to around £2,000. The recent purchase of Stocks was, as she told her father in May 1896, 'a great pull on our resources'.[80] A month later she had to pay the first £1,000 towards down payment. In December 1896, Mary discovered to her dismay that her brother Frank Arnold had incurred debts of £3,000, for which she was joint guarantor. It was increasingly clear that her other brother, Willie, was never going to work again, and would have to be supported as a long-term invalid—primarily by his rich novel-writing sister, it went without saying. Young Arnold Ward was now in residence at Oxford, and nothing must be stinted in that department. Dorothy and Janet Ward were regularly travelling to France and Germany for their languages. Janet was (in 1897) attending her first balls, and needed a new wardrobe. In addition Mary Ward undertook early in 1897 to give a matching £1,000 towards setting up the Passmore Edwards Settlement. To cap it all, Humphry resolved to buy a Gainsborough.

It was in these pressing circumstances that Mary Ward wrote what most readers consider her finest novel, *Helbeck of Bannisdale*. The idea evidently came to her in the last days of October 1896, when she told Willie that 'I have got a really good subject in my head for the next book—romantic, and haunting and original.'[81] The subject had come to her during a trip to the Lake District—always a powerful stimulus. While driving on the outskirts of Kendal with her cousin Mary Cropper, she had been told the story of Sizergh Castle. It had a great ruined thirteenth-century tower, and a melancholy history to do with a decayed Catholic family. The story and its setting preyed on Mary's mind next day, during the seven-hour railway journey to London. But by the time her train had reached Euston, 'the plot of *Helbeck of Bannisdale* was more or less clear to me'.[82]

Given Tom Arnold's tergiversations, it was a tricky subject and Mary Ward first consulted the senior Catholic statesman Lord Acton, who 'cordially encouraged me to work it out'.[83] On 15 November 1896, she timidly asked her father 'Would you mind my dearest, if I chose a certain *Catholic* background for my next story?'[84] 'Background' was an understatement. What Mary had worked out on the train was the story of a free-thinking fatherless young girl, Laura Fountain ('not myself', as she firmly claimed), who is transplanted from Cambridge, where she has been brought up among godless intellectuals, to a decayed Lake District mansion, Bannisdale Hall. There she lives with her stepmother's brother, Alan Helbeck. Despite a sixteen-year difference in age they fall in love and become engaged. But Alan is an austere Catholic. At first, they plan a mixed marriage. Then Laura decides to convert. But her father's spirit comes to her as she sits by the bedside of her dying stepmother. He tells her, 'Laura, you cannot do it—*you cannot do it!*'[85] Thus torn between her Catholic lover and her atheist father Laura drowns herself. Alan survives to become a Jesuit priest.

The religious theme was sensitive; more so as Mary Ward included a gallery of gullible, ignorant, and sinister Catholics in her cast. Even more sensitive was the love relationship at the core of the novel; the generation gap and the strong hint of incest were suggestive of a father–daughter passion. It was not something to make a Catholic father with a famous free-thinking daughter easy—especially as that father had recently married a younger woman. Nevertheless, Mary always maintained that she had adequately and well in advance consulted 'my Catholic father, without whose assent I should never have written the book at all'.[86] None of Tom's letters of paternal assent seem to have survived. He evidently raised no insuperable difficulty. But he had an enigmatic way about him that Mary found hard to read at the best of times; and it is hard to believe that he actively approved of *Helbeck*.

The clinching inspiration for *Helbeck* came from Levens Hall, at Milnthorpe, close to Sizergh Castle. The Wards rented this rambling Elizabethan house for three months from March 1897. It was another financial burden—but one which repaid itself quickly in terms of raw material for the new story. Levens merged with Sizergh to create Mary Ward's mental image of Bannisdale Hall and got the book off to an unusual flying start. Mary Ward arrived at Levens 'alone' (i.e. with her daughter Jan, her secretary Bessie Churcher, her maid Lizzie and a troop of servants) on 6 March 1897. She described her first impressions enthusiastically to Dorothy (currently in Paris). It was, as they first saw it, a 'wonderful grey house rising above the river in the evening light'. In the twilight, they rushed with lamps through the buildings 'with the delight of a pack of children' excitedly exploring the 'intricacies of the upper passages and turrets', the great peat fireplaces; the strange stucco birds running round the ceiling of the great hall; the kitchen, gleaming in the 'scanty lamplight' with rows of copper pans and moulds.[87] Levens in short was an adventure and food to the novelist's imagination. The disgusted servants whom Mary Ward had brought with her saw Levens differently. It was a great, draughty, run-down, inconvenient, freezing barn of a place. (The spring weather of 1897 was—even by Westmorland standards—'appalling'.) They all protested bitterly and en masse went down like ninepins with rheumatism.

Mary Ward persisted in her love for Levens, despite its hideous inconvenience and exorbitant cost. The Hall, she was delighted to find, had a ghost—'the Levens Lady', as the locals called her—who 'must certainly [be] put into the book',[88] she told Humphry. There was much else to put in. A crate of her books arrived on 11 March and she threw herself into reading up on religious history, eight hours a day. She was, after a week of this, 'suffocated by Catholicism'.[89] (Among the first books she read at Levens were William Ullathorne's *Autobiography*, *The Life of Abbé Blanchard* and Bishop James Grant's *Life*.) Mary Cropper, who lived not far away in Ellergreen, took her to see a yeoman family in the Fells, who gave her the models for the Masons in *Helbeck*. She was increasingly swept up in ideas for the novel, and on 24 March she told Willie that 'I believe the subject might be as effective as the subject of *Robert Elsmere* if only I could sink deep enough into it, but that is the whole point. Oh!, how strange the Catholic world is!'[90]

She had by now (26 March) written the first chapter, 'with a glow of pleasure'.[91] The de rigueur opening landscape picture faithfully re-creates what Mary Ward saw from her window at Levens—'a desolate scene, on this wild March day; yet full of a sort of beauty'.[92] By April the weather somewhat improved and domestic routines were sufficiently settled for visitors. Henry James came and kissed Mary's hand 'affectionately' when he left. But, after the effortless first chapter, she was now bogged down in *Helbeck*. Clearly as she could see it in her mind, the story would not come. Only four chapters were written in the three months at Levens, and an overload of 'research' was threatening to clog the original narrative conception with thesis as had happened in the early composition of *Robert Elsmere*. But now she could not afford to take three years over a single novel.

By the end of May, when they returned to London, Mary Ward was suffering the familiar trouble sleeping and was taking injudiciously large doses of trional. Her pulse was racing. Her skin was mysteriously inflamed. She was stiff down all the right-hand side of her body. On the trip back from Kendal, Humphry was obliged to book a whole first-class compartment, so that his wife could lie at full length behind drawn blinds. The 'train shaking' was nevertheless agony to her. Her London doctor, Henry Huxley, immediately dosed her heavily with quinine.

On 9 June Mary Ward was 'indescribably glad'[93] to return to Stocks. But it was bitterly cold still. And she was so worried about money that she 'hardly knew which way to turn'.[94] As always, the anxiety destroyed her sleep, even when drugged. The novel continued to drag. By the end of August, only six of the projected twenty chapters were written. Her arm was now so neuritic that she was reduced to dictating—something that she was convinced damaged a writer's flow. 1897 was in other ways a hectic and exhausting summer. The Jubilee in July and the opening of the new Passmore Edwards Settlement in October both cut into her writing time. Christmas was even more of a distraction with balls at the Rothschilds' Tring Park mansion and elaborate family theatricals for the benefit of the local Aldbury children. But Mary Ward soldiered on, and by January 1898 had a first draft of *Helbeck* in readable form.

Tom Arnold came over in the New Year and the ordeal of the paternal imprimatur for *Helbeck* was gone through. Mary found that she would have to 'rewrite it almost entirely' for her father's 'dear sake'.[95] She withdrew from her family altogether for this last, exhausting pull. On the morning of Friday, 25 March 1898 at Stocks, she finally wrote the last page of her revised draft. By this point, she was unable to walk unaided.

Between January and March George Smith was at last allowed to read the entire novel in proof. 'I cannot say that I am in love with Laura,'[96] he reported. And like all English and American publishers of the period he hated unhappy endings, which he was convinced lost an author readers. But he recognized *Helbeck* as a powerful performance. Reginald Smith was more enthusiastic and 'effervesced' about the work. They happily paid over the £4,000 balance of her advance and Mary Ward left immediately on 1 April for the Continent for the sun and rest she so urgently needed.

The proofs of *Helbeck* were corrected as the family travelled in France and Italy, settling finally at their favourite villa at Cadenabbia, near Lake Como. They were the most complicated proofs Mary Ward had ever had. Neurotic lest she make any factual errors, she had entrusted a set to another reader, the Catholic theologian and Oxford professor William Addis. On the whole, Mary Ward had made few blunders. But Addis had an objection that went to the heart of *Helbeck*. One of the many complex motives in Mary's writing the novel was to 'explain' Tom Arnold's desertion of Julia and his children in terms of the peremptory demands of Catholic faith. She pointedly referred in the text to the sixteenth-century story of Saint Francis Borgia. When his wife was mortally sick, the legend went, he prayed for her and God told him that he could have her cured if he so desired, 'but it is not expedient for thee'. Francis dutifully stopped praying, and the poor woman died. The widower duly disposed of his eight children (a meaningful detail), took the vows, became Vicar General of the Jesuits, and was eventually canonized. This was the 'expedient' destiny God had in mind for him. None of which would have been possible if the wife had inconveniently recovered. For Mary Ward, it was a perfect exemplum which made sense of Tom's walking out on his family in 1876, leaving for ever a wife with cancer and *his* eight children. But Addis firmly insisted that Catholicism was not and never had been inimical to the family as an institution in this way. 'St Francis Borgia', Addis told her, 'had no right to take up, as he did, *new* work which cut him off from his children.'[97]

It tormented Mary Ward, although she obstinately kept the Borgia material in her narrative. Her torment was heightened by the fact that Tom (who had a third set of proofs) did not send his comments on the last sections of the book. Mary interpreted his silence as meaning that the erotic-religious climax was 'too painful' for him and sent a stream of increasingly urgent letters, to which he did not reply. George Smith, meanwhile, was sending another stream of urgent letters, demanding his corrected proofs. 'Perhaps you hardly realise how anxiously I have been waiting for what you have to say,'[98] Mary wrote to Tom on 20 April. She clearly hoped that it was indifference that accounted for his silence (something that she had always found difficult to read in her father). As clearly, he was intimating that she should not publish. But she had to publish. Returning £5,000 to Smith, Elder was not within the realm of possibility unless she wanted to move out of Stocks into the poor-house. The last revisions (unapproved by Tom) were reluctantly sent off by registered post from Florence at the end of April. She was exhausted and apprehensive about the novel. 'Sometimes I have dreadful nightmares and depression about the book,'[99] she told Humphry on 3 May.

It was in other ways a ghastly summer. When they returned to England she learned that Willie's mysterious spinal complaint was locomotor ataxia—a terminal complication of syphilis. At the end of the same month, May, Gladstone died. At his great age, it was hardly an unexpected shock. But the lying in state at Westminster Hall was profoundly depressing to Mary Ward. In defiance of her doctors she had risen from her sick-bed (bronchitis was currently the main health worry) to pay her last respects to the remains of the Grand Old Man. An appropriately dignified ceremony would have been some consolation. But the bare, undraped coffin was 'sadly lacking in poetry and solemnity'.[100]

There were problems with the publication of *Helbeck*. The Boer conflict was looming and wars are notoriously injurious to the prospects of new novels. If it was to be a 'short war', should they wait? asked Reginald Smith.[101] Like Mrs Ward, he was a staunch imperialist and could not think that a crew of rag-tag Afrikaner farmers could hold out against the might of the British Army. But she was firmly for going ahead, and decreed publication on 10 June 1898—the eve of her forty-seventh birthday. Smith, Elder acceded to her 'strong wish' in the matter.

The early reviews were good. '*Helbeck* came out this morning,' Dorothy wrote in her diary. 'We sent for all the papers. Thank Heaven most of them are delightful, some interesting too, and very precious. Mother and we are all happy so far.'[102] Gratifyingly, intellectual Catholics did not generally condemn *Helbeck*, as Mary had feared they might. Even the *Tablet* was favourable. Inevitably, the more fanatic Catholics, particularly the Irish, resented *Helbeck*. And there was at least one novel written in refutation, *One Poor Scruple* (1899), by Josephine Ward ('Mrs Wilfrid Ward', as Longman mischievously entitled her; she was no relative). Tom Arnold's direct references to *Helbeck* were notably cool. But his anger (if there was anger) seems to have discharged itself in disagreements with Mary about the Boer War. Tom still had radical and anti-imperial sentiments, as did his Nationalist wife. These quarrels about the war soured the father and daughter's relationship over the last two years of his life, and were a lasting wound to her after his death.

Gently as the Catholics handled her, Mary Ward came in for some buffeting from Anglican dogmatists, who thought (as did she in her heart) that the novel was too favourable to the opposition. Her old opponent John Wordsworth, now a bishop and stuffier than ever, wrote a furious rebuke on the day of publication. He had not read *Helbeck*. (What? *He* read a *novel*!) But his eye had been caught by the review in *The Times*: 'I would venture to ask you,' he pontificated, 'in all good will and in recognition of your power, if you have done well as a teacher responsible for great gifts to make [the narrative] end as you have done.'[103] Mary Ward was indeed nervous about Laura's suicide and her presentation copy to Leslie Stephen was bound up less the last chapter.

Mudie's subscription for *Helbeck* had been a disappointing 2,704. But the novel sold a healthy 18,817 in its first six weeks in England. American sales over the same six weeks were again disappointing. Macmillan only sold 11,400 of the $2 *Helbeck*. But George Smith remained hopeful. 'I think the public *must* find it out,'[104] he told Mary. He was right. *Helbeck* proved to be the novel of Mrs Humphry Ward's with the most long-term stamina. By November, American sales had picked up to 1,175 a month. Her American public was still loyal and she rewarded them with an American heroine in her next novel. Eighteen months later, in March 1900, *Helbeck* had covered its £5,000 advance, and was earning afresh for Mary.

Most critics who allow Mrs Humphry Ward any literary achievement at all assume that it is confined between the years 1888 to 1898 and that she falls off as a novelist after *Helbeck*. The decline can plausibly be given a number of explanations. After 1900—and particularly after 1905—her inexorable downward slide as a popular writer meant that she had to write faster for less—two pressures that ate away at her always precarious artistic confidence. She had other pressing concerns and often downright anxieties in her last decades—her social work, Arnold's gambling, Humphry's health, anti-suffrage business. But the principal cause of the decline after *Helbeck* was—arguably—something else. She had no one left to write for. Ever since she first put pen to paper and dedicated 'A Tale of the Moors' to her grandmother she had seen herself catering for a charmed circle of readers—older men (mainly) and women who really mattered to her.

One by one this Sanhedrin were dying off. Mandell Creighton and George Smith both died in the first year of the new century, Gladstone in the penultimate year of the old. Jowett was long gone (he died in 1893). Most important was her father, who died in November 1900. Significantly, Mary concludes her *Writer's Recollections* in this year. Part of her was henceforth forever frozen over. And possibly it was the harder frozen in knowledge of the fact that he had not after all approved of his daughter's Catholic novel.

In the ten years between *Robert Elsmere* and *Helbeck of Bannisdale*, Mrs Humphry Ward earned getting on for £45,000 from her fiction. She was probably the highest-paid woman in England. But as 1900 approached she was still not 'comfortable'. Like Scott before her, she had discovered that a novelist in Britain—however much he or she earns—cannot emulate the truly rich classes. Novels supplied wealth sufficient for considerable luxury but not enough to live like a member of the landed gentry. But the Wards would come nearer than most scribblers.

13

Families—the Arnolds: 1890–1900

MRS HUMPHRY WARD'S career as author and public woman was ostentatiously motivated by self-sacrifice. Like the mother pelican, she gave her life's blood for others. What others? In the largest sense, her ideals were philanthropic. Mary Ward's mission was to serve her fellow man. But in practice—like the pelican again—her most wrenching sacrifices of self were on behalf of her family. She was always ready when her nearest and dearest were short—often enough they were. But her idealism soared above mere hand-outs. She wanted her family to be a power in the land; a radiant source of sweetness and light. And here Mary Ward's aims were sadly frustrated. The Arnold dynasty was a wholly spent force with the death of Matthew in 1888. The Ward dynasty—if there was to be one—was hobbled by the ordinariness of Humphry. The single hope of the two family branches (as Mary saw it) was her extraordinary son Arnold. Mary Ward's strategy as family woman in the years of her power thus followed two tracks: first to support the many lame ducks who carried one of her surnames; secondly to promote the glorious career of Arnold Ward, who proudly carried both of her surnames.

The death of Julia shattered the bonds of physical community in Tom Arnold's family. What connected the Tom Arnolds in subsequent years was Mary—as banker and indefatigable problem solver. After 1888, her relations with her parent and siblings was one long series of chores, expenses, and—all too often—disappointments. The boys were the principal disappointment. By the laws of genetic averages, Tom Arnold's four male offspring should have boasted a crop of distinguished Victorians. Instead his sons were all failures and mediocrities.

Of Mary's many siblings the most shadowy is Arthur Arnold (Tom's third son), who was born in 1856 at Fox How, in the upheaval of the family's return from Tasmania. Arthur's was a restless and doomed life ever after. He and Mary never seem to have been close, even after she came 'home' to Oxford. Alone of the Arnold children, Arthur reverted to Catholicism with his father in 1876. Alone of the Arnold boys, he did not go to Oxford University (or any other). He was shiftless and dishonest. In 1876, Arthur was packed off to Tasmania. But transportation did not work. Arthur returned in March 1877, leaving a trail of debts behind him. 'That wretched boy', Tom declared to Julia on 27 March, 'seems born to heap trouble on our heads and shame on his own.'[1] He would cadge money and was always getting himself in scrapes. Tom—who was in desperate straits himself in 1877—tried to find his wayward son various lowly positions in trade. Nothing suited. Finally, he managed in January 1878 to place Arthur as a trooper (the lowest serving rank) in the cavalry. After training, the young man was shipped out with his unit to South Africa. The callousness of some of his letters, recording the cheerful slaughter of 'kaffirs', appalled his parents. He died in the Basuto War, in September 1878, aged only 21, shot through the lungs. On learning of his death, Tom told Newman bleakly that the boy had been 'incurably loose about money matters'.[2] But he was still an Arnold, with an Arnold's pride. His last message had been 'God's blessing on his parents and family'.

Theodore (Percy)—Tom's second son—was born in 1855 in Tasmania. Like Willie (his elder by three years) he was subjected to the destructive yo-yoing of Tom's conversions between 1855 and 1876. At Oxford, his sister Mary evidently had charge of him during his school holidays from Rugby. Theodore was not, like her and Willie, a clever child. He matriculated as a non-collegiate student at Oxford in October 1872 (aged 17) but did not graduate until 1878. Unable to settle down in England he returned to Tasmania, with £300 subscribed by his family to give him a 'good start' farming in the colony. Over the subsequent years regular sums were remitted to keep him started. Mary was the main donor and collector. In January 1889 she was worried by Theodore's lack of progress, feeling that after all this time 'he ought to have made some steps towards a fresh settlement'.[3] By 1890, she was sending him a regular £100 a year. He was remiss about acknowledging the cheques, which made the family justifiably suspicious as to where the money was going. In 1891 he informed Mary of a 'Kaurie gum plan', which she 'did not much like'. With good cause: two months later Theodore was again penniless. Mary's cheques for £50 and £75 continued to be dispatched once or twice a year. In December 1893 there were new grounds for anxiety. Theodore returned to England for a six month 'furlough' and indicated an urgent need to see the Wards at Stocks over Christmas: 'Ethel and I both imagine that it may mean that he has lost money at cards on board ship, and one wonders how he is going to finance his six months at home',[4] Mary Ward sighed. Of course, she knew. Theodore was going to finance his furlough by sponging off his rich sister. She gave freely enough, but was affronted by her colonial brother's 'sulky and ungracious manners'.[5] He did not show to advantage in her new country house. The following October (he had overstayed three months) Mary gave Theodore £45 for his passage back to the antipodes. This time Hokianga, New Zealand, was his destination—Tasmania had evidently not worked out for him.

Mary saw Theodore off at Fenchurch Street Station in the custody of a young Yorkshireman called Hodgson, whom she generously tipped 'to look after' her brother and stop him gambling his stake away again. 'Fortunately,' Tom noted,

'[Theodore] has never taken to drinking,'[6] so there was still hope he might make good. Eighteen months later Hodgson had got himself into an obscure 'Maori scrape'. Theodore, meanwhile, acquired a farm and a wife. Neither were lucky for him. For the next three years Mary committed herself to sending him a £20 dole.[7] The marriage went bad, and finally broke up in 1899. 'We must help him,' Mary told her father, adding 'if only he doesn't fall into [a marriage] equally improvident as soon as he is free! Poor, poor old boy.'[8] He married again, and had children. Nothing is known of his first wife, Mary Seymour-Kane, nor of his second, Grace Parkin-Moore. In 1914, his sister Mary was still remitting him £100 a year.

Frank (Francis Sorell), the fourth and youngest of Tom's sons, was a larger drain. Born in 1860 in Dublin, Frank was emotionally vulnerable in 1876 when Tom's second conversion to Catholicism occurred. He entered Christ Church in 1879, won a minor scholarship, took his BA in 1883, with a second in natural science. He then studied medicine and after some unspecified difficulties qualified B.Med. in 1887. Like other young Arnold males, Frank had great trouble settling down and earned a reputation in the family for 'weakness in judgment'. In March 1888, he justified it by declaring his intention to marry his Miss Valentine while Julia was on her deathbed. The engagement fell through. The following January Frank quarrelled with his superiors at the Oxford Infirmary and was obliged to leave. He evidently knocked about for a year or two, taking locums where he could. 'Poor old Frank!' Mary exclaimed to Tom in June 1890. 'I wish his prospects were better.'[9] She gave him £75 in October 1891, to set him up in Bradford. Three months later, in January 1892, she organized a family subscription of a further £100.

In the same year Frank married Annie Reed Wilkinson. Annie 'quite took our hearts,'[10] Mary told Willie on first meeting the young lady in February. More to the point, Annie had a wealthy father, connected with the well-known Wilkinson's Sword steel firm. Marriage temporarily steadied the young doctor. In 1895 the couple settled in Lancashire. Mary must have felt with some relief that her part in Frank's career was played out. But in December 1896, her brother revealed that he had lost heavily in business speculations. His father-in-law called on Humphry, who had guaranteed Frank's borrowings from the bank. Frank's total debts, Wilkinson told an astonished Humphry Ward, amounted to £3,000. He, Wilkinson, could not cover this sum as *his* losses were five times as great. Mary (via Humphry) undertook to pay £250 a year as their share until the debt was cleared. Thereafter Frank settled down as a modest general practitioner.

Of her brothers, Mary was closest to William (Thomas) Arnold, Tom's oldest son. Born in 1852, he was her nearest in age and the only sibling she had played with as a child. As she later confessed, she felt some jealousy for Willie during the years that he was his father's darling and winning prizes at

Rugby. But their comradeship was re-established when he returned to Oxford as an undergraduate, Mary helped him over the disappointment of his second-class degree and his imprudently early marriage in 1877 to Henrietta Wale. Thereafter, she took his career in hand. It was Mary who got Willie Arnold a position on the *Manchester Guardian* in 1879. And she chivvied him in 1891 to do what she had made Humphry do in 1882—demand a rise to £1,250 from his editor, C. P. Scott.[11] In April 1895 she urged him again to fight for £1,500. On other occasions Mary Ward successfully lobbied to get Willie offered the secretaryship of the Clarendon Press, and encouraged him to stand as a Liberal Member of Parliament. Willie politely declined all his sister's attempts to make something of him. He infuriated her by his blank refusal to be ambitious. He was—as she could see—wasting his talents, burying himself in his provincial paper, for whom all his writing was anonymous. He was the paper's principal leader writer and a chief architect of its enduring journalistic character. But when he left Manchester after seventeen years, few in the city—even his regular readers—knew his name.

For all her irritation, Mary Ward rather envied the peace which 'Willie and Het' enjoyed; childless, self-consciously unachieving, and provincial as they were. He loved his Roman history, Keats, his wife, his dogs—above all his freedom from having to be an Arnold. His sister perceived, with something like envy, that he was *happy*. In her mind's eye, she always remembered him as he was in 1890, 'standing before the fire in the drawing-room of the pleasant house in Nelson Street [Manchester],—alert and vigorous, his broad shoulders somewhat overweighted by the strong, intellectual head, his dark eyes, full of fun and affection, beaming on the guest who had just arrived, perhaps, from the south'.[12]

Willie had fifteen good years in Manchester. But in spring 1896 he suffered what was first diagnosed as a nervous breakdown. Then 'bladder mischief' originating in the kidneys was identified as the culprit. He came down on long leave to London in summer 1896 to recuperate. Mary used her contacts to get him odd jobs for the magazines. Willie did not, however, mend. The pain intensified and his embarrassing incontinence persisted. 'Obscure nerve trouble' was added to the list of diagnoses. Over the next year he took the waters at Bath and was seen by numerous doctors. By February 1897 it was clear that his working days were over. Still he wanted to return to Manchester. C. P. Scott wrote to Mary Ward urging her to persuade Willie against any such 'madness'.[13] Henrietta, meanwhile, had collapsed into a condition of permanent hysterics and was—as Mary despairingly told her husband—'absolutely devoid of common sense'.[14]

Mary Ward took charge of the last ten years of her brother's life. Her first act in 1897 was to 'make right' his money problems with an immediate £100 to allow him to go to Menton with Het. Perhaps the Riviera sun might work its healing wonders on him, as it always did with her. If it did not, Henry Huxley (Mary's own London doctor) gave Willie a store of morphia to deaden the pain in his lower body. At this stage his problem was still suspected by some of his physicians to be stones in the kidneys. In Italy he unwisely submitted himself to an operation for this supposed ailment. His sister Mary was 'doubtful'[15] (her own recent experience of surgery had made her very wary of the knife). Inevitably the operation was useless. The pain got worse.

In April 1897, Frank advanced the theory that Willie was suffering from echinococcus—a parasitic disease caught from his beloved dogs licking him on the face.[16] (Frank was not, on what evidence one has, a very brilliant fellow.) 1897 was a savagely dark and cold summer and Willie continued to decline. He and Het stayed in Grosvenor Place, Stocks, or the lodgings Mary had procured for them. He was able to attend the opening of the Passmore Edwards Settlement in October; but Humphry had to find a hansom with rubber tyres and even so Willie had a fit during the journey. All shaking, however slight, was now agony to him. In May 1898 locomotor ataxia was finally diagnosed by the specialist, Dr Gower, whom Mary had hired. It was incurable and fatal, he informed her. He would also have told her that the disorder was certainly syphilitic in origin and probably contracted a decade earlier from a prostitute. Willie was not told what was wrong with him.

Willie's would be a protracted death as his central nervous system collapsed under the microbial invasion. By now he was addicted to the morphia which had been too generously prescribed by Huxley and Frank. His new doctors determined to wean their patient off it so as to allow other drugs to work. On her part, Mary urged her brother to contemplate the 'moral degeneracy'[17] that dependence produced. The task proved hopeless. By summer 1899 Willie's 'morphia habit' was worse than ever and his pain so great that only a sadist would have tried to take the narcotic away from him. C. P. Scott, who was clearly as fond of Willie as everyone who knew him, kept on his *Guardian* salary until January 1899, when the arrangement was severed by a final payment of £700. From now on Mary Ward alone supported Willie and Het. Luckily they had no children, but their medical bills were incessant. In May 1899, Mary set the couple up in a flat in Carlyle Square, Chelsea. Every six months or so, she chipped in £100 to tide them over.

There was one consolation in all this dreary business and expense. Willie's childlike dependence on her enabled Mary Ward to come close to him in ways that she never could with other members of her Arnold family. It was almost as if they had a second try at their lost childhood. She explained it to him in a tender letter of 19 June 1897:

> Yes dearest, it was a real loss that we did not know more of each other in our first youth, but it was at least as much my fault as yours—more—I think. In point of character we both developed late—I am sometimes amazed to remember how childish and

unformed I was for years after my marriage—we were both self-conscious and clever and conceited! (I was anyway) and then worst of all, our home life had done little or nothing to bring us together. How little we *saw* of one another as children—I often think of it .... I think few sisters could love a brother more than I do you, and for many years now you have been one of my best friends and counsellors—second only to Humphry.[18]

In 1894, Mary Ward reckoned that hand-outs to her family and 'Humphry's people' were running at £500 a year. By 1900, her three surviving brothers alone were costing her that amount and more. Her sisters were somewhat less expensive relations and on the whole more prepossessing than the males. The three of them—Lucy, Ethel, and Julia—had been close among themselves in Oxford, while Mary was farmed out at her boarding schools. The pre-pubescent Arnold girls were a strikingly pretty trio and caught the eye of Charles Dodgson (Lewis Carroll) in 1870, as he walked in the city streets.[19] The little Arnolds—especially Julia—soon became the eccentric mathematician's favourite photographic models. He invented a word game, 'Doublets', in their honour, took them to the theatre, and was generally a fairy godfather. Dodgson kept on a relationship with the unmarried daughter, Ethel, well into her young womanhood, consoling her on the death of her mother (he was a clergyman as well as don, photographer and author). At the end of her life Ethel—now a keen amateur photographer herself—recalled her Oxford days with Dodgson as sunny intervals in an otherwise 'grey and melancholy childhood'.[20]

Of her sisters, Mary was initially closest to the eldest, Lucy (Ada), who had been born in 1858. Lucy made a good marriage in 1884 to the Reverend (Edward) Carus Selwyn. Six years older than his bride, Carus was a former Fellow of King's College, Cambridge. When he became engaged to Lucy, he was Principal of Liverpool College. Lucy and Carus would not have met in the normal course of events. But in 1876, after Tom Arnold's second conversion to Catholicism, the 18-year-old Lucy had been adopted by a rich quaker uncle, John Cropper of Liverpool, who had married Tom's sister Susanna.[21] It was through the Croppers that Lucy met her husband-to-be in the north as an eligible young lady. Three years after the wedding, in 1887, Carus was appointed to the headship of Uppingham. As headmaster's wife, Lucy was 'mother' to thirty boarders. She had additionally, seven children of her own—all under 10 in 1893. It was to the inexhaustible Lucy and Carus that Mary entrusted Arnold in 1889, for his year's cramming before going on to Eton.

The Selwyn connexion was regarded as a feather in the Arnold cap (even if Cropper money had helped put it there). The Uppingham school hall during the Christmas holiday of 1890 saw the four sisters together for the last time. Mary wore a new frock in blue and black; her sister Julia wore black and heliotrope; Ethel made a fainter impression in pale pink. Lucy (pregnant as always) bustled between them as hostess. A dozen or so children were eagerly at the tables and tree. Thomas Arnold sent a celebratory telegram, which was read out, Dorothy played the piano and Mary 'had a polka with Leonard [Huxley, her brother-in-law] and danced so vigorously that when it was over I sat down somewhat abashed and thought of Vauvenargues' maxim "Il ne faut pas sortir de son caractère".'[22]

Mary Ward last saw Lucy at the school prize-giving in July 1894. Two months later, she received the thunderbolt news that her gentle sister was dead. A blood-clot had lodged in her brain. She fainted and died in fifteen minutes, before the doctor even arrived. Mary remained convinced that the last painful childbirth was the real cause. Seven pregnancies in nine years had killed her sister: 'to see the hands on which only thirty hours before so many tasks had pressed folded—last Monday—in the stillness of death, made for me one of the saddest moments of my life.'[23] Lucy was buried, 'in a dream of beauty, in the brightest October sun',[24] at Ambleside beside her mother. Carus could scarcely be dragged from the graveside.

There were complications for the survivors. Carus was, as Mary observed to Humphry Ward, 'the last man to manage such a household [as Uppingham] by himself'.[25] Boyish in manner, he had depended utterly on his superbly competent wife. Maud Dunn, a cousin of Mary's and Lucy's on the Sorell side, had been living with the Selwyns and helping in the school. Mary Ward liked Maud. But Carus's proposal that she should continue to live with him alone in the house 'without another and older woman'[26] shocked Mary profoundly. She had an interview with Maud, 'feeling a wretch all the time'[27] but plainly pointing out the dangers of the arrangement. Maud stuck to her guns, and refused to decamp. She and Carus remained together. The school grew to 400 pupils, and Maud became the second Mrs Selwyn.

After Lucy's death, Mary Ward grew closer to her younger sister Julia ('Judy' Frances). Born in 1862, Julia was 'a tall raven-haired girl with a face which managed to be thin without being angular'.[28] According to her oldest son Julian she had a tremendous sense of humour. Fully as clever as her oldest sister, Julia was much better educated. After attending Oxford High School she was one of the first home students at Somerville (on a Clothworkers' scholarship). In 1882, she took a first-class degree in English. At Oxford, Julia Arnold met her future husband, Leonard Huxley. A son of the famous Darwinian, Professor T. H. Huxley, Leonard was strongly advised by Benjamin Jowett not to marry Julia—at least not yet. He had been a brilliant undergraduate. Should he marry, he could never go into law or politics with their long delays. He would have to take up schoolmastering with at best the prospect of a modestly paid headship at the end of it (and, with his notoriously agnostic father, most trustees would be reluctant to promote him). Nevertheless, Leonard decided that Julia now was worth the sacrifice of future fame. They married in April

1885, and went to live at Godalming, where he had taken up a teaching post at Charterhouse the year before.

Of all the Arnold children, Julia most resembled Mary in her drive and fierce family pride. In defiance of Charterhouse sensibilities, the Huxleys' house at Godalming was called Laleham in tribute to Arnold of Rugby. Julia did not make Lucy's mistake of shattering her body with multiple pregnancies. She had just five well-spaced children between 1887 and 1899 (four survived). As a tribute to Mary (whom Julia saw as *the* living Arnold) Aldous, born in 1894, was named after the hero of *Marcella* (1894). The two sisters were increasingly close as the century came to an end. Leonard's brother Henry—a young, fashionable, outgoing, West End doctor—became Mary's favourite physician. Mary took a keen interest in her brilliant Huxley nephews—Julian (who adored her), Aldous (her godchild), and Trevenen. They all spent holidays at Stocks and it was in the grounds there that the 13-year-old Julian saw the green woodpecker that set him on a lifetime's serious bird-watching.

The Huxleys had no need of Mary's money; and their pedigree was as distinguished intellectually as that of the Arnolds. But there were other ways in which the author of *Robert Elsmere* could serve them. As Jowett predicted, Leonard's career stalled after fifteen years. There were no higher rungs for him at Charterhouse; he was trapped in a job unworthy of his abilities. Mary shifted about and landed 'Our Len' (as she called him) a position with Smith, Elder in 1901 which led eventually to the editorship of the *Cornhill Magazine*. This elevation was entirely Mary Ward's doing. The family left Laleham, and set up in another house nearby. Gratefully released from her matron's duties, Julia took out a bank loan and established her own experimental girls' school, Prior's Field, which opened in January 1902. Within two years it had fifty pupils, and eventually enlarged to 200. It was 'what I've always wanted to do',[29] she said. Prior's Field was a success, as was Leonard Huxley in literary London.

Ethel, the youngest daughter, presented a more intractable problem and one which eventually proved even beyond Mary Ward's power to fix. Born in 1866, Ethel (Margaret) Arnold was four years younger than Julia. She too had gone to the Oxford High School, but—unable to win a scholarship—missed out on a university education. There was some hope in 1884 that she might squeeze in at Newnham, Cambridge, but this too fell through. Ethel had also missed out on a husband. She evidently came close. On 22 March 1884, Julia wrote to Tom Arnold that 'I was not without hope that Ethel might not be very long before following Judy's example, but a nasty horse accident has put a stop to my hopes for her.'[30]

Her three sisters' marriages left Ethel by 1885 in sole charge of her slowly dying mother and a scrimped household. Clearly one daughter had to stay behind in these circumstances, and Ethel was that daughter. That parents should not have to die alone was a main rationale of the large Victorian family. Some daughters—Dorothy Ward, for example—accepted the maternal companion's role with joy. Ethel did not. She had wanted more from life, and her ambitions had been one by one ruthlessly stifled. Having joined the Philothespian Society (later OUDS) in 1883, Ethel had declared her intention to become a professional actress. Mary advised strongly against it.[31] In 1886, Ethel (who was a fluent writer) enterprisingly got herself an invitation to write a novel for Richard Bentley and Co.'s *Temple Bar*. Mary wrote (via their mother) a chilling letter: 'both Humphry and I are inclined rather to dissuade her from attempting a serial yet'.[32] Ethel lacked 'the experience of life, the resources that a long story wants'. Intimate knowledge of the workings of cancer did not, apparently, fit the bill. Clearly Mary Ward had been told about Ethel's authorial ambitions so that she—with her contacts in the London literary world—might *do* something for her sister, as so many powerful people had done things for Mary.

After Julia's death in April 1888, Ethel Arnold suffered a health collapse. The family hypochondria took early hold on her, and she was already periodically crippled by rheumatism, lumbago, and assorted nerve complaints. Mary allowed her sister to recuperate at Borough Farm in the summer. In early August (with some financial assistance) Tom—still himself racked with guilt about Julia's death—took his daughter abroad. 'He really is not fit to have charge of Ethel,'[33] Mary observed wearily to Humphry—but then who was fit to have charge of her? What did one do with a mother's companion after the mother was dead? Ideally, there should have been some Victorian form of suttee to dispose of the unnecessary no-longer-quite-young woman. Tom Arnold did not much help. He wanted more domestic and spiritual comforts than his daughter could provide. He was—even in his late sixties—a handsome and sexually attractive man with a fine-drawn melancholy face and a mane of silver hair. As soon as was decent (and without consulting with any of his children) he married in January 1890 a Catholic woman, eight years younger than himself. Good-naturedly, he indicated that Ethel might, if she liked, live with the newly-weds in Dublin.

Mary Ward was put in a dilemma. She certainly wanted her sister Ethel provided for. But even more than this she wanted appearances preserved. She had been singed by public scandal too often in her girlhood, and fear of what people might say was now an overriding consideration with her. Ethel could not, she finally determined, live with a Catholic stepmother: 'It would hardly be natural or possible.'[34] Reluctantly she again accepted the provider's role. She found lodgings for Ethel in London and gave her free run of a small spare bedroom at Russell Square. Mary and Willie used their contacts to put journalistic assignments their sister's way in *The Times* and the *Manchester Guardian*. In Dorothy's 1890 diary her aunt Ethel makes a peripheral appearance as part of the unconsidered furniture of the famous novelist's life. Ethel breaks her arm in January. A couple of months later she is bedridden in her lodgings with rheumatism—no one visits her except Dorothy. When well, Ethel serves as an all-purpose minder, taking the children down to Oxford or chaperoning the girls on their expeditions into the London streets. Mary finds her a summer home in Bedfordshire—sufficient for 'one servant and a friend'.[35] A year later, in 1891, she helps install Ethel in a small London flat.

Ethel Arnold, meanwhile, lived with a series of companions—genteel unwanted spinsters like herself. She was always 'difficult', and fell out with Leonard Huxley. She dabbled in writing, never with much success. In 1894, she published a sensitive novel, *Platonics*, which made no impression. She did a translation of a book about Turgenev. She did a small handbook on the British Constitution for use in schools. But there was no 'living' to be had in this dilettante authorship, and Ethel clearly lacked the obstinate power of sticking to a literary task which had sustained her sister through the lean years. In 1895, Mary Ward proposed that Ethel might work for Nelson's, the Edinburgh publisher, as a literary adviser. It came to nothing. In September 1896 Mary asked her friend Frederick Macmillan if he could give Ethel a permanent reviewing engagement 'something like what I used to have'.[36] Even if he could, Ethel Arnold was not—like the 1882 Mary Ward—a reliable contributor. And her views were—as Mary thought—decidedly 'odd'. Ethel was, for instance, pro-Boer and pro-suffrage. Explosions were inevitable.

There was a crisis in Mary Ward's relations with her sister Ethel in spring 1899. Mary was staying in some style at the Villa Barberini near Rome, writing the opening chapters of *Eleanor*—something which turned out to be rather difficult. Ethel announced her intention of coming to stay. It was entirely against her sister's wishes. 'I ought not to have any extra weight just now,' she complained to Humphry. 'I am under legal contracts and the strain of the book is already serious.'[37] Nevertheless, ignoring Mary's prohibition, Ethel came. She was never an easy guest and just now was impossible. It was suspected she might have consumption. She arrived by train, looking 'like a little ghost'.[38] Every night she coughed so as to keep everyone awake and could herself only sleep with the aid of twenty drops of chlorodyne ('blessed stuff',[39] Mary noted). Dorothy was assigned to sleep in her aunt's room, lest she have a crisis in the night. Her presence in the villa gave Mary 'uncomfortable dreams'[40] and bad-tempered days. In reaction to all this disturbance, she put Ethel in the novel as Manisty's mad and hypochondriac sister Alice.

What seems principally to have saved Ethel from Alice Manisty's self-destruction was her passion for photography. She seems to have taken it up seriously in 1898, enrolling for a course at the Regent Street Polytechnic. The hobby provided an outlet for her frustrated creativity. Over the years she lived with a succession of women friends. In the early twentieth century, she undertook lecturing tours in America trading on the fact that she was the sister of the famous Mrs Humphry Ward. Around 1905 Ethel Arnold took up residence in one of the Wards' cottages at Stocks. 'Auntie Ethel' figures faintly on the edge of Dorothy Ward's accounts of family get-togethers over the years, always looking frailer; although eventually she outlived Mary by a full decade, dying in 1930.

There was also, of course, Tom Arnold for Mary Ward to worry about. He collapsed immediately after his wife's death in April 1888 and needed a period of recuperative travel before being able to return to the Royal University of Ireland. After things settled down, Mary's long term plan was to have her father looked after by Ethel, who now lacked an occupation with Julia's death. In order to firm up the arrangement, Mary took her sister over to Dublin in July 1889. They found Tom's rooms to be a disgrace: 'never swept and never dusted. The dirt of everything is oppressive—and this with a payment of £2.18.0 a week!'.[41] Mary promptly found her father better lodgings and installed Ethel as housekeeper.

This tidy arrangement did not, however, suit Tom. Ethel Arnold was prickly, non-Catholic and neurotic. Besides which he had other, less orthodox plans for his future, namely marriage. His bride, Josephine Benison, was a Catholic woman from County Cavan. Her father had been a landowner, a magistrate and a Protestant. Her mother had been a Catholic and Josephine—partly under the influence of Irish nationalism—had converted in 1853. After Tom Arnold's return to England in the early 1860s she continued to correspond with him. Since his return to Ireland in 1882, their relationship had been close.

Tom did not trouble to discuss the matter of his marriage in advance with his daughter Mary, who replied to the news on 23 October 1889 with rather pained felicitation:

> I was not altogether unprepared for your announcement and certainly dearest I have nothing but sympathy and affection to give you with regard to it. To all grown-up children I suppose a father's second marriage must always carry with it something infinitely sad and moving. But at the same time I recognise that you have owing to our family circumstances many needs your children cannot supply, however much they might wish; and I rejoice that you will find in Miss Benison that fundamental sympathy which is and ought to be precious to all who hold their faith strongly.[42]

It was hardly a warm letter. Nevertheless, by the following February Mary had come round sufficiently to send the newly-weds some furnishings for their new home. Tom was in the scholar's invariable state of penury and never too proud to take from his rich daughter. In June 1890 she offered to settle his Oxford debts—adding cautiously, 'if they are not more than £100'.[43] He accepted. In 1891, she offered to stand the couple a holiday in Italy: '*David* [i.e. *David Grieve*] shall bear all the expense!'.[44] This offer was, however, declined until 1899, when Mr and Mrs Tom Arnold came as Mary Ward's guests to her villa at Castel Gandolfo near Rome. As Mary recalled, 'Never before, throughout all his ardent Catholic life, had it been possible for him to tread the streets of Rome, or kneel in St. Peter's.'[45] There was a certain irony in his heretical daughter crowning his life with this gift in his seventy-seventh year.

He was an enviably robust 76-year-old. Freed from the burden of a recriminating wife he was cheerful and genial in ways that Mary had never before observed. He celebrated his new vigour in summer 1890 with a walking tour in the Wicklow mountains. As a married man he visited England regularly. He began to work more energetically on early English history and Anglo-Saxon literature. (When his 'Notes on *Beowulf*' came out in September 1898, Mary Ward dutifully fixed *The Times* review.) By the mid-1890s, there were some ominous signs of physical decay—occasional giddiness, 'nerve pressure', and weakness of the eyes, especially when marking exam papers. But, if anything, Josephine Arnold was the more chronically sick of the two. There were of course no children. Perhaps because she found nothing sexually threatening in her stepmother, Mary Ward—after the first wary meeting in London in 1891—adopted a good-natured condescension towards the other woman. 'She is really a good sort' Mary wrote to Humphry in 1896: 'if only Nature had made her a little fairer to look upon!'[46]

After a bracing walking tour in Scandinavia ('the Beowulf country') in 1898, Tom decided to write his autobiography. Mary was only too willing to assist. Lest in his modesty he aim too low she sent him a copy of Bismarck's recent autobiography for his November 1898 birthday. Her comments on the manuscript of *Passages in a Wandering Life* were tactfully critical: she suggested the author look into his memory 'more closely on certain interesting points' (points, for instance, like his catastrophic conversions to Catholicism—entirely glossed over by Tom). The narrative could do with 'more Rugby'; she noted with surprise, for instance, that he said nothing about his father's death—of which he had been the principal witness.[47]

Although she could not have expected it in the shock of his remarriage, after 1889 Mary Ward's relationship with her father became more confidential and genuinely compassionate. This note is heard clearly in a letter she wrote him in February 1890, replying to an expression of remorse on his part: 'You must not be depressed about yourself my darling. No one looking back over your life can feel anything but that it has been a life lived not for the world but for the spirit, and what shall it matter to any of us but *this* at the last? There is the real fight.'[48] There were other intimacies between them. In her letters to Tom, Mary admitted what she admitted to no one else—how much she missed Julia. On Boxing Day 1888, she told him that 'many nights lately I have gone to bed crying over some detail of her illness, or the operation which suddenly struck me as I undressed in the room with her two photographs, the one taken just before the first operation and the one taken in the garden'.[49] She felt the loss again keenly while laying out the new garden at Haslemere in April 1890; whenever Arnold won a prize; or when she was travelling abroad. 'I always feel it unjust that I should have had so many more of the kind earth's pleasures than she!'[50] she confessed to Tom shortly before his death.

14

Families—the Wards: 1890–1900

AFTER 1888, Mary was *de facto* head of the Humphry Ward family: the main provider and the main decision maker. Her rule was a judicious mixture of edict (especially where 'appearances' were concerned); joint edict with Humphry (particularly on financial matters); and passive manipulation (she was, for instance, expert in the use of her invalidism to control the domestic environment). On the whole, Mary ran the Wards' affairs fairly and wisely. But as a matriarch she had one incorrigible weakness: any hint of desertion by her loved ones devastated her.

This reaction was spectacularly provoked in early 1895, on one of the very few occasions that Humphry and Dorothy jointly left her side for any length of time. The occasion was not in itself that worrying; Humphry had engaged to give a series of lectures on old masters to audiences on the Eastern seaboard of America. His performances earned him £50 apiece and altogether would just cover the £400 first-class passage expenses for him and Dorothy—very small beer compared to what Mrs Humphry Ward now commanded as a novelist. But Humphry liked travel and he looked forward to the Yankees' 'ferocious hospitality'.[1] He loved dining and gossip and could talk about old masters till the cows came home. Nor were he and Dorothy to be away for long—leaving in January and returning in early April. But for Mary Ward, the episode was a three month nightmare. On 31 January her heart stopped when she opened *The Times* only to see the headline 'Atlantic Liner Sunk—386 Lives Lost'. She was convinced that her loved ones were going to be likewise shipwrecked. 'Oh you shall never go away from me like this again. Such pangs are not worth while,'[2] she told Humphry. Nor did they.

It is easy to understand why Mary Ward was so terrified by separation from Humphry and her children. Her own childhood had been one long banishment. The wounds would never heal. But Mary's neuroticism was selfish and even dangerous for those she had charge over. Her not sending her daughters to university can only be explained as over-fond possessiveness. They would certainly have done well and Janet would probably have excelled. Apart from brief periods on the Continent for their languages or to be 'finished', her girls were entirely educated at home by 'Fräuleins' and tutors. Dorothy never left her mother's domestic orbit. As late as 1900, the two Ward girls were dressed identically—a uniform of family servitude chosen by their mother.

With Arnold Ward, Mary's possessiveness was even more smothering. He did not go to school until he was a lumbering 13, and then only with much heart ache on his mother's part. (Young Arnold bore the separation remarkably well.) At school and university Mary was forever prying and meddling and even when her son was a fourteen-stone moustached swell would tenderly remind him to wear a vest in June. In what time was his own Arnold Ward fell in with a fast set and things went badly for him after Oxford. When in 1903 he determined to make a new start in life in India, Mary Ward talked him out of it with the promise of a parliamentary career. Her dream was that he might become a local MP, set up his base at Stocks, and thus never leave his 'mummy's' side. The punitive gods granted her wish.

Humphry was, on the whole, happy enough as Mrs Ward's husband. It would have been nice to earn large sums of money himself, but Mary's providing it was the next best thing. He had his cigars, his shooting, the Athenaeum, his pictures. He lived for moments like that in January 1893 when for £35 he picked up a couple of Rembrandts worth £1,000. More often—like all gamblers—he lost; but disappointments were easily forgotten in expectation of the next *coup*. He had a certain eminence in the London art world. In 1891 he was made a trustee of the new Tate Gallery. Buffers (as they now were) like Burne-Jones, Alma-Tadema, and Millais got on well with him. More modish artists treated 'old Humphry Ward' with friendly condescension. Like his paper, *The Times*, he was an institution.

Humphry Ward's standing in the art world is testified to by an anecdote which is much repeated in lives of Oscar Wilde as the source of the famous 'I wish I had said that'—'You will Oscar' episode. On one occasion, looking at Whistler's paintings, Humphry dared to tell the dandified artist what was good and what less good in his work. 'My dear fellow,' retorted Whistler, 'You must never say this painting is good or that is bad. Good and bad are not terms to be used by *you*. Say "I like this and I don't like that" and you'll be within your right. And now come and have some whisky. You're sure to like that.'[3]

Mary had one last go at making something of Humphry Ward in October 1895. She confided in her brother Willie the 'wild idea' of buying the *Saturday Review* from Frank Harris, 'who has dragged it into the depths and now wants to sell'. Some rich Liberal Unionist might finance the paper, she thought. 'Then if Humphry put money [whose money?] into it and became editor we might set up a fine family organ.'[4] Mary evidently foresaw herself, Tom, Arnold, and Willie all happily contributing to a morally regenerated organ. But Humphry Ward at 50 did not have the energy for such adventures.

Mary Ward had been a close and attentive mother in her Oxford years. But with the move to London there was little time for children and there was money to employ servants to be close and attentive. Humphry was at work many nights and sleeping during the day. There were elaborate dinner parties at home or away three or four times a week. Mary was either turning out a flood of articles or wrestling with an all-important book. Among all this, child-rearing was often delegated to servants. Mary (particularly during the summer weeks at Borough Farm) would have the children visit her in bed in the morning to read and recite. She in turn would now and again read them some improving book (like Southey's *Life of Nelson*) in the evenings. But her contact with her young children was affectionately intermittent and distant. She loved them dearly but remotely.

Dorothy Ward's earliest surviving diary is that for 1890.[5] It was her sixteenth year, Arnold's fourteenth, Janet's eleventh. Her immature entries give a worm's-eye view of daily life at Russell Square. Mary Ward is typically off-stage: away writing for weeks on end in the country; preoccupied with University Hall business; or at home but 'tired' and secluded in her private room. On Sundays, the children are sent off to Stopford Brooke's Bedford Chapel, where they ingest a suitably Unitarian brand of Christianity. (Mary Ward does not go with them; nor Humphry.) When he is home from Eton, Arnie bullies Jan and on one awful occasion almost breaks her arm. But Jan is a tough little thing; later in the year she is thrown off a 'treacherous' pony on to her head. She bears the pain heroically, but Dorothy suffers agony at the prospect of telling her mother about the accident. Mary Ward will certainly in her panic want to put down every four-legged beast in Surrey.[6]

The girls' education is in the hands of their live-in German governess, Fräulein Zimmer, who is difficult. There are weekday visits by a private governess—the dashingly beautiful and clever Eugénie Sellers. The girls go out for piano lessons and Dorothy—who lives for music—is agonized by her inability to excel ('Oh! when shall I get control over my fingers!').[7] Occasionally they make a day trip to Godalming for some lessons from Uncle Len. Dorothy's red letter days (there are half a dozen in the course of the year) are quiet Sundays when her mother lets her oldest daughter read to her in bed, or when just the two of them go for glorious 'rambles' or drives over Blackdown. A whole series of distantly related Wards and Arnolds drift in and out of the diary's account. There is little or no mention of Father. Dorothy gets a shilling pocket money a week, out of which she must buy gloves, birthday presents, writing materials, and stamps, and pay fines levied for such things as scratching the piano.

Almost from childhood, Dorothy was trained to serve as her mother's aide. As early as August 1888, Mary instructed the 14-year-old girl to 'please think about the housekeeping while I am away'[8] (at the Rothschilds, no less). It is likely that in 1888 Mary told herself that Dorothy's subservience would be a prelude (indeed a useful training) for marriage. But service to mother became Dorothy's permanent destiny with the defection of Gertrude Ward on New Year's Day, 1891. Gertrude, the eleventh of Henry Ward's dozen children, had come to live with her brother Humphry and Mary after graduating (in English and History) from Somerville. For eight years, she had been an amanuensis, confidante, and housekeeper. During this period, a passionate loyalty to her sister-in-law combated a High Church call to do good works. All through 1890, the younger woman had steeled herself to leave. In her diary for 20 March she recorded that 'at night M. and I got on to the subject again—which I had not meant to do as she was very tired. She can't see my point of view in the least and thinks it would be deliberately wrong of me to go away and leave her.'[9] Gertrude finally did go, to become a district nurse in London. Four years later, in November 1895, she sailed as a missionary nurse to Zanzibar. 'It will be a great wrench to us all, and I wish I cordially liked it'[10] was Mary's tart comment to Louise Creighton.

Dorothy threw herself into Gertrude Ward's vacated role with unalloyed joy. Serving mother was everything she wanted from life. Like all of Mary Ward's most trusted female aides, she was—in time—a devout High Anglican. By 1890 and her sixteenth birthday, she was in charge of the household: superintending the servants, arranging for the reception of visitors, writing 'invites', making flower arrangements, going on shopping expeditions for such things as 'mother's silk undergarments' or gloves. In 1892 when Mrs Humphry Ward undertook her first lecture tour through Manchester, Liverpool, and Leeds, Dorothy (still only 18) acted as companion and secretary, indefatigably and invisibly keeping all straight for her mother. 'I don't know what I should have done without her'[11] Mary told her son.

Devoted as she was, Dorothy was inferior to Gertrude Ward in some important respects. She was not university-trained and, unlike Gertrude, Dorothy could not secretarially help her mother write her fiction. To fill this function, Mary in 1895 (a fearfully busy year) temporarily employed a trained stenographer, Miss Bessie Churcher. Bessie—like Gertrude—was High Church. When Mary Ward offered her a permanent post, Miss Churcher experienced agonizing doubts about the propriety of working for the heretical Mrs Ward. As she told Mary on 24 August 1895, 'There is nothing I could wish more for myself than to come back and work for you again if I felt that I as a church woman could do it quite honestly but of this, I am not sure.'[12] Mary Ward contrived somehow to reassure her. Bessie Churcher worked for Mary for the next twenty-five years, making up one leg of the female tripod on which all her later writing and social work was supported. Dorothy Ward was the second. The third was Mary's personal maid, Lizzie Smith, who had joined the Ward household in June 1880 and who remained in their service until senile dementia struck, immediately after her mistress's death in 1920.

At the age of 24, Dorothy Ward had absorbed much of the responsibility for the day-to-day running of the Passmore Edwards Settlement, the play centres, and all the other social services connected with it. Her essential support role is evident in her diary entry for Friday 25 March 1898. Literary history records it as possibly the most important date in Mrs Humphry Ward's career—the day she finished her masterpiece:

> Helbeck of Bannisdale was finished this morning at 11.30 Stocks—M[other] came up by 3.1 [train], tired out, happy, yet sad too! I met her [at] Euston, then PES, boys' tea—45 of them—4 of our helpers for to-night had fallen out, but owing to frightful weather the attendances were smaller than usual, and all went well. I had the singing class alone and gave them a lesson on *tone*. Then home in a cab, dress and all dine Lovelaces, Wentworth Ho[use]. The Tennysons, Lushingtons, Morpeths, etc. Wrote PES letters all morning.[13]

Had Mrs Humphry Ward been obliged to write those letters all morning, give those slum children their tea, or teach the girls' choir about *tone*, the world would have had to wait for *Helbeck*—probably indefinitely. Yet, with Dorothy's genie-like assistance, her decks were always clear.

By 1901, Mary Ward no longer thought of Dorothy as a sexual object. That others might do so came as something of a surprise to her. There had been some raised eyebrows that 'your mummy lets you go back by that late train to Stocks alone',[14] she uneasily jested in a letter to her daughter. Mary continued to have occasional twinges of remorse about monopolizing Dorothy's young womanhood. She would rationalize it by pointing to the girl's willingness to sacrifice herself. But still the guilt would sneak back. 'Dorothy, is as usual "an angel in the house!"' Mary Ward told Louise Creighton in 1905. 'What she gets through for everybody is really wonderful—but it *does* make her happy! Only more and more I think that people with such a marvellous unselfishness don't marry!—and that troubles me.'[15]

Dorothy Ward was 30 and her chances of finding a husband were remote. And now that she was no longer marriageable, she was permitted to be an adult. In May 1903, she opened her first bank account at Barclay's. A year later, Mary gave Dorothy her own cottage, Robin Ghyll, in the Lake District. Over the years, Dorothy Ward's emotional energies were channelled into series of intimately confidential friendships with women her own age like Kate Lyttelton and the American Sara Norton. In later life, she became a tender godmother to her niece Mary (Janet's daughter) and enjoyed a comradely friendship with her PES co-manager, Mrs Ward's secretary, Bessie Churcher. But throughout life, uncritical heroine-worship of her mother was Dorothy's overriding reason for being.

Janet—the baby of the family—was altogether different from her sister. More physical in childhood (Julian Huxley remembered her as a keen archer) she was more intellectual in adulthood and more independent at every stage of life. Janet Ward had a brain and a mind of her own. Mary Ward did her best to cultivate the one and curb the other. Janet could have gone to a good boarding-school and to university. Instead, she was educated at home.

In lieu of college, Mary set her daughter at the age of 17 to a mammoth labour—translating Adolf Jülicher's commentary on the New Testament. The Sisyphean task occupied four years, or about as long as an undergraduate degree.

Janet Ward's translation, after much correction by her mother, was duly published in 635 pages in 1904. It was a formidable achievement. But what, one may wonder, did Mary Ward intend by imposing this ordeal on her daughter? Presumably it was designed to break Janet—to give her a taste of the self-flagellating discipline by which Mary had controlled her own 'wildcat' nature as a schoolgirl. To the same end the young Janet was enrolled as a regular 'volunteer' worker every Saturday at the settlement.

Janet Ward nevertheless went her own way. When eventually she married it was someone whom her mother did not initially quite like. Janet's wholly unceremonious mode of life thereafter deviated strikingly from that of Stocks. By 1908 Janet was openly contemptuous of the stuffed shirts, superannuated grandees, backwoods Tories, and aristocratic boobies with whom Mrs Ward loved to mix. And yet Janet always admired her mother personally to the point of idolatry. She was a saint, but like other saints gullible and an easy victim. The principal victimizer—in Janet's uncompromising analysis—was Arnold. Her brother (as his sister saw him) was a scrounger, a cad, and ultimately little short of a murderer.

Anyone reading Janet Trevelyan's life of Mrs Humphry Ward could be forgiven for not even realizing that the lady had a son. Arnold gets just two terse sentences and no mention in the book's index. But from the day he was born in November 1876 until the day she died in March 1920, Arnold was the most important thing in Mary Ward's life. Whenever she mentions the boy in letters there is an uncontrollable croon of maternal pride. 'Arnie's latest craze is drawing,' she tells her mother in March 1886, 'and he is really beginning to show a considerable turn for it. But the child is so clever that he could almost do anything—up to a certain point of course.'[16]

He was, in fact, a very bright child and one who found it easy to play the part of infant prodigy. In April 1886—aged 9—he delighted Mary Ward by writing a poem on the current Belgian riots. 'Really a most stirring performance,'[17] his mother thought. Two months later, she tells his grandmother that 'I read [Arnold] a paper of Grant Allen's last night on the relation between the humming birds and the swallows which is very Darwinian in theory and it was curious to see the fascinated attention which the child took in the new point of view.'[18]

Nothing was stinted for young Arnold. At Borough Farm that year (1886) he was given a pony of his own and Eugénie Sellers was brought to stay for a fortnight and teach him Thucydides and Homer. 'Her enthusiasm for Arnold is very great,' Mary Ward told her mother, 'and she and I plan out his education together.'[19] The following April (1887) during the family's Easter vacation at Borough Farm Arnold is recorded by his proud mama as waking every morning at half past seven to learn his thirty lines of Milton. Then—if Mary Ward is to be believed—the little fellow would for relaxation take out his beloved Macaulay from under his pillow and immerse himself in the history of England for an hour or two, before the family began to stir downstairs. At the end of the year he began to keep a 'political diary'. It begins, 'There is imminent danger not to say absolute certainty of war between Austria and Russia.'[20] On the back cover, with more candid enthusiasm he scrawled down the England v. Australia test scores.

Humphry and Mary Ward's first intention was that Arnold should go to Harrow. (Rugby—the first choice for an Arnold boy—was perhaps thought too far away from his home.) But despite Miss Sellers's coaching, he was not in a position to compete with boys who had had the advantage of a prep school. And Mary Ward worried about his health, fearing he might be too delicate for the outside world. He indeed had a chronic tendency to colds and a 'nervous' throat, which led him to lose his voice in the presence of strangers. After much heart-searching, Mary Ward resolved to send her son to Uppingham for a year, before trying him for Eton entrance.[21] Carus Selwyn could teach him classics and Lucy would mother him. Mrs Ward took her boy on the train to Rutland in January 1889. 'He went from home very cheerily,' she glumly told Louise Creighton, 'but it was hard to leave him. What an epoch it seems to make in life.'[22] Arnold did well at Uppingham, coming third out of seventeen in his lower sixth class. It seemed that intellectually he was ready for big school. Nor did he after all seem too sensitive in health or character for the company of other boys. In June 1889, Mrs Ward and her girls made an exploratory visit to Eton. She liked what she found. The physical beauty of the school—particularly the chapel—pleased her. The teachers she found to be steeped in a congenially 'Greenian school of thought'.[23] The Revd Henry Bowlby, an assistant master, told her he would give Arnold a place in College. The boy's 'home training in religious matters will be fully respected',[24] she was reassured.

All that remained was for Arnold to win his scholarship. Bowlby had invited Mrs Ward to stay at the school while the boy took his examination in July. A succession of telegrams was dispatched to London, informing Humphry how each paper had gone. At night, after the daily ordeal, Arnold was coddled and made much of and stuffed like a Strasbourg goose for the next day's exam. There were fifty-four boys competing. When the results came out, Arnold Ward was placed fourteenth. It was a good, but slightly awkward result. Arnold was on the reserve list; he would get his place at Eton 'when a vacancy arose'—which might be as late as a year away. It was not quite clear to his mother that this was a triumph: 'Will he be announced in *The Times*?'[25] she asked Humphry, rather nervously.

In fact, Arnold got into the school in January 1890 as one of the select band of seventy or so King's Scholars who lived in the original college founded by Henry VI. By April, a brimmingly proud mother could inform her relatives that 'Arnold has done brilliantly at Eton, coming home with the Trials prize, form prize and pupil-room prize.'[26] Arnold continued to shine. Eton—with its emphasis on classics, debate, and sports—suited him. In 1891 he won the Consort's German Prize. Like his father he had a happy knack with Greek and Latin verse which regularly won him more prizes. Aged 14, he was also writing little bits for the *Field*. He made the College wall side in 1894. He was a good all-rounder on the cricket field. By 1892, Mary Ward had already entered him for Balliol. ('I hope I shall be there to see him and welcome him,'[27] the dying Benjamin Jowett had told her.) In the spring of his upper sixth year, Arnold went through the ritual ordeal of swotting for his university entrance. Mrs Ward wrote Bowlby an alarmed letter, '*begging*' that he would not let her son overwork. Overwork Arnold did, and he duly won a Newcastle scholarship, in March 1895.

It was probably the happiest period of Mary Ward's life. She rewarded Arnold with two weeks' hire of the Eton cricket pro, Matt Wright. In July, her son took the field with the Eton XI at Lord's, to play against Harrow. 12,000 spectators—Mrs Humphry Ward among them—were there for the biggest match of the public school sporting calendar. Arnold scored seventeen in the first innings and took 'two of the best Harrow wickets' with his fast bowling. It was a close finish which had the crowd roaring. Eton lost, but Mary Ward felt that for the first time she 'really understood what the Englishman *is* at his game'.[28]

Arnold's Eton career was thus crowned in glory. But during his five years at the school there had been some disturbing moments. At the time, they seemed nothing more than the usual schoolboy rubs. But with hindsight they were prophetic. He had evidently got 'squiffy' a couple of times, and told Dorothy about it, who told Mary. He had persistently overspent his allowance and Mary had worried interminably on the subject. 'I don't understand why you have needed to spend so much more this summer than last,'[29] she complained in 1891. There was a rather unpleasant episode in August 1894, when Arnold as treasurer of the cricket XI misspent or lost the team fund and had to be bailed out by the ever-ready Dorothy. Mary Ward was continually worried about her son's 'levity'—schoolboys in the 1890s singularly lacked the earnestness Dr Arnold had inculcated into his boys half a century before.

Most disturbing for Mary and Humphry Ward were the 'signs of insubordination and wilfulness'[30] which Arnold began to display in 1891. Over the years, these developed into a running feud with his house and school authorities. The first serious rebellion blew up in November 1891, when the confirmation issue came up and smouldered over the next four years. Mary Ward was by now quiescent herself on the issue of religious observance and regularly counselled Arnold not to antagonize his masters: 'your time for ruling will come'.[31] But Arnold continued to rebel on 'Sunday questions' and the religious instruction given by the school's senior teachers. His provocations reached a climax in early 1894, when Humphry was obliged to go down to the school to plead with Bowlby and the head, Edmond Warre, that Arnold not be sent down. It looked as if the boy might have to go to Oxford a year early, or spend a last year at Winchester. But Warre finally relented. A mightily relieved Mary Ward bound Arnold over to be on his best behaviour: 'Meanwhile do nothing for *the sake* of controversy—that is a plain duty for all of us, and above all for a boy at school. Many differences of opinion—although not all—can be rightly avoided, and put aside, without loss of truth.'[32]

In her overpowering love for Arnold Mary was an impercipient mother. It was for *her* that he rebelled. Since childhood she had drummed into him the shining example of Robert Elsmere's heroic *non serviam*. And now apparently she wanted him to behave like the Vicar of Bray. Even more confusing was Mary Ward's desire that Arnold should excel while at the same time retaining a humble indifference to his excellence. She lectured him on the subject in May 1893: 'This passion for mere "winning" is something quite new in you. You should learn to take all your successes more chivalrously, and I tremble lest by our delight in them at home we have helped to lower your standard.'[33] It was Mary herself who for seventeen years had indoctrinated Arnold with the belief that if he was not pre-eminently successful he was nothing. Now she would have him treat his success as if it was a small thing.

These were small clouds in the summer sky of 1895. The glorious fact was that Arnold Ward was going up to Oxford in triumph. And after Oxford? Mary Ward already had that planned out. Arnold should be a great politician: 'Do you ever wish for public life dear son, and for an honourable part in the great change that the next twenty years must see? If you do it is a worthy ambition and [one] your mother shares for you. It can only be achieved by hard work now and at College .... The changes in social organisation and the development of labour and democracy which the next generation must see fill me with thought and on the whole with deep sympathy. I should like to see my boy in the struggle—when the time comes.'[34]

This future Prime Minister of England went up to his college in October 1895. He immediately took to university life with its 'allowances ... for human nature'.[35] Balliol he found to be the finest hotel in England—if one of the more expensive. (He was soon exceeding his allowance.) Humphry Ward gave him an engraved reproduction of Botticelli's *Primavera* for his rooms; Aunt Fan a large mounted photograph of Dr Arnold; his mother donated Dr Jowett's sermons and Turgenev; Dorothy kept regular hampers and supplies of clean clothes coming.

In his heart of hearts, Arnold Ward had two great ambitions at Oxford. One was to star in the Union. The other was to play cricket for the university. As a cricketer, he had not a bad first season scoring 424 runs for his college at an average of 36 and taking 54 wickets. Over the next three seasons he was a stalwart of the Balliol team and became secretary of its club. But persistent injury dogged him. He developed tendon trouble in his bowling arm in 1895 and his action was never quite right again. (In one awful match in May 1896 he bowled seven consecutive wides.) As a bat Arnold was not quite good enough for the top level. In his very last term at Oxford he finally got a trial for the University XI. Were he chosen it would, he felt, 'retrieve the disasters of three years'.[36] He did in fact play for Oxford against Surrey in June. But he failed to win his blue.

Arnold's achievements in the Union debates were similarly tantalizing. His 'line' (aided by any number of briefings from his mother) was fashionably anti-Liberal. He attacked Home Rule, Cecil Rhodes, women at Oxford ('confessedly a measure for the ignorant, for the mob'),[37] free trade, 'our coward's policy of splendid isolation', and socialists. As a speaker, he had some small triumphs. The newspaper report of the Union debate in November 1896 on whether 'England has passed its Zenith' actually recorded that 'Mr A. S. Ward made the speech of the evening.' But his uncertain throat and tendency to lose his wits under pressure were against him. It was also his bad luck to be a contemporary of debaters of genius like John Buchan and Hilaire Belloc. In the November 1897 debate, for instance, when Arnold got up to speak as Belloc finished, the room emptied. The speech that followed was, he told his mother, a 'dismal failure'.[38]

More seriously, Arnold Ward failed to get elected to the Union in March the following year (1898) 'for reasons composed almost exclusively of bad luck and crass stupidity on my part'. The stupidity was allowing the Balliol vote to be split by not coming to peace terms with a college opponent. He wrote a suicidal letter to his mother: 'so ends my career at the Union ... it has been a career of complete failure, not uncontributed to by misfortunes, but mainly due to great absence of sagacity and forethought, as well as inferior speaking, on my part. Rather ominous for the future, is it not?'[39] With hindsight it was.

Socially, things had gone better. At Balliol in November 1896 Arnold made the most useful friendship of his life, with Raymond Asquith, eldest son of the statesman. Raymond took to Arnold (whom he found less 'ugly' and hearty than other college men)[40] and cultivated him—despite warnings from his tutors that this was not a worthwhile connexion. With Raymond Arnold formed a convivial 'Horace Club'—devoted to conversation, dining and drinking. Raymond introduced Arnold to his father, his stepmother Margot (née Tennant), and his young sister Violet. In November 1897, Arnold could delight Mary with the news that he had entertained Herbert Asquith to dinner at the Gridiron. The two young men went on shooting parties together at Stocks and in Scotland. And it was Raymond who apparently introduced Arnold to the card-play that was to be the bane of his life.

Arnold's studies also went well. After a rather shaky start he took a first in Mods. in 1897. He won the Chancellor's Latin Verse Prize in the same year. In December 1898 he was elected a Craven scholar ('the first university scholarship in the family since Matt's Hertford',[41] Humphry Ward noted). Finally he took a first class in Greats in 1899. 'Thank God,' was Mary Ward's terse reaction when she received the telegram in August. She had admitted to herself, she told Tom Arnold, 'just a possibility of failure'[42] (i.e. a second).

Arnold Ward was a double first. In the normal course of events a fellowship was the logical next step. There was the possibility of one at Magdalen—but it did not materialize. It was a main handicap that all Mary's powerful friends (Jowett, Pattison, Pater) were dead. And there seems to have been a prejudice against Arnold in the university. As Julian Huxley recalled his cousin had a brusque and arrogant manner that put people off. There remained a distant possibility of an All Souls fellowship, but that would not come up for a year or two. To fill in the interim and broaden his mind, Arnold sailed off to Egypt in November 1899 as a 'special correspondent' for *The Times*. Humphry had arranged it. If the fellowship fell through, Mary Ward planned that he should read for the bar in London, where she could remain close.

15

Homes: 1888–1900

MUCH of the Wards' energy was devoted in the 1890s to finding appropriate places to live—accommodations, that is, which were appropriate to their raised and ever-rising status, to the different needs of their children, to the complexly divided lives which the two parents led. More than with most families, the Wards' homes defined them and articulated their ambitions. Humphry Ward was—initially at least—more ambitious than Mary in his desire for grander residences than 61 Russell Square. He never really recovered from the drains débâcle of 1886. On 18 August 1888—with *Robert Elsmere* in its twelfth edition—he bought four-and-a-half acres of land at Grayswood Hill, near Haslemere in Surrey.[1] Two days later the family picnicked on the imposing eminence where they would build their new house. It had 'a pearl of views', right over the 'vast tangle'[2] of the Sussex Weald as far as the South Downs. Mary Ward outlined her plans to Humphry on 30 August in some detail: 'My idea is to build a very complete house on a small scale, except for one or two large rooms to give one a sense of space. I should like for instance to have a small servants' hall and a work room. All these things add so much to the comfort of life.'[3]

Work started in October 1888, Humphry having contracted the distinguished services of Edward Robson as architect. The final stages of construction were supervised from a rented house a stone's throw away. The fact that there *were* houses a stone's throw away was—although they did not fully appreciate it at the time—a fly in the ointment. Nor was the actual construction plain sailing. Their building plans had by legal agreement to be approved by the previous owner, a Mr Hudson—'the most unbearable of men'—who niggled interminably. His restrictions as to size, Mary told her father in spring 1890, had spoiled the 'first *coup d'œuil*'[4] of the house—making it look like just any other Surrey villa.

Mary Ward undertook the writing of *Robert Elsmere*'s successor pressured by the needs of the new house. By July 1889 building work and furnishing were 'swallowing up'[5] £500 a month. It was not pressure that brought out the best in her. In November 1888, Ethel collapsed, having borne the brunt of nursing the dying Julia and the bereaved Tom Arnold. She was at the moment living with her father in Dublin. A week earlier, Mary had totted up her receipts for *Robert Elsmere* as £3,100. Despite this sudden wealth, she nevertheless sent

Ethel £30 on the strict understanding that it was a *loan* to be repaid. 'I wish I could give it to her,' she explained to Tom, 'but having bought our land we are now trying to get together money for our house at Haslemere, and all my book must go to that henceforward, or nearly all.'[6]

Lower Grayswood finally became habitable in early summer 1890. By May, the main rooms were papered (a detail to which Mary Ward gave minute attention). In mid-July, the last of the furniture was arriving. On the fifteenth of the month, the maids slept in the house for the first time. Humphry Ward's pictures were ceremonially hung and the Bechstein ('a superb instrument') was placed in the entrance hall. There was no room for it in the drawing room. On 18 July the whole family slept under the Grayswood roof for the first time. 'Oh! God, bless our lives here!' a humbly jubilant Dorothy inscribed in her diary. 'May they be lived usefully, for the happiness of others.'[7] Her sixteenth birthday, four days later, was, she declared, the happiest of her life.

Mary Ward did not—in the event—adore Grayswood. Her initial desire for it had been the result of a strong but temporary urge to 'settle down' and live the *petit bourgeois* life of a professional London couple made good. The 'villadom of Surrey' offered *rus in urbe* convenience. It was a refuge from town and yet only forty-five minutes by train from Bloomsbury. Nothing could have been more convenient for a *Times* journalist and his wife. But even before she moved in, Mary Ward began to be unhappy. The 'house is *small*', she told her father in March 1890. And, small as they were, the buildings looked 'terribly conspicuous on the hill side'. Perhaps, she added forlornly, 'years and *young* trees will soon temper them'. But over and over her vague discontent focused on the 'newness' of it all—the lack of 'old trees'.[8]

'New', in conjunction with 'rich', was currently a loaded term. The *nouveaux riches* were not a party that Mary Ward wished to be associated with. More so when, to her horror, other families just like them began building *their* dream villas in the neighbourhood of Lower Grayswood. There was an aroma of new paint and bourgeois vulgarity about the place. As Janet later recalled, 'Americans walked in, taking no denial, and once in mid-August, when the youngest child [Janet] tactlessly won a junior race at the Lythe Hill Sports, with all Haslemere looking on, there were paragraphs in the evening papers. It would not do, and I think the house at Haslemere was doomed from that day onwards.'[9]

To cap it all, running the brand new house proved to be neither convenient nor easy. Christmas 1891 was a squalid nightmare. Lizzie and the head manservant, George, fell ill with the influenza which was raging through the country. The cook, Apted—used to the torpid flame of coal ranges—scorched her face lighting the new gas oven and needed medical attention. Dorothy, meanwhile, was away being 'finished' in Paris and Gertrude had left for good. The 12-year-old Janet fell ill as excited 12-year-olds will at Christmas. A flustered Mary

Ward found herself obliged to cook the festive meal and tend the sick. She felt herself to be what in her most insecure moments she feared she really was—the daughter of an unsuccessful Oxford boarding-house keeper. The iron entered her soul. They would leave the new house, scarcely having warmed it. A fortnight later, Arnie told his astonished sister Dots (still in France) that 'On Monday Mother and Father went down to the Thursfields for the night and saw a lovely old house near Tring which belongs to Sir Edward Grey's mother and is probably going to be let.'[10] By the end of February, the arrangement to rent Stocks was signed and sealed.

There were factors other than snobbery and Mary Ward's personal comfort at work. Since her early years at Fox How, she had been sensitized to houses' atmospheres. She loved the resonant feeling that the places she lived in were 'haunted' by previous tenants (preferably literary or otherwise noteworthy tenants). Lower Grayswood had the sterile deadness of a hospital ward. It was clean, new, small, and soulless. Mary Ward's need for a brew of romantic association (not least to trigger her novelist's imagination) had been demonstrated in 1889, when—having found it could be rented for a song—the family spent the autumn at John Hampden's old house, in Great Missenden, Buckinghamshire. This historical monument was lacking in every amenity currently being expensively installed at Lower Grayswood. They had had to bring with them a 'truck load of baths, carpets and saucepans'. The furniture was 'decrepit, scanty and decayed' but, Mary Ward told Bertha Johnson, 'it has breeding and refinement, and is a thousand times preferable to any luxurious modern stuff'.[11] It was draughty and cold, even in September. The servants complained sullenly of rheumatism. It had no gas and no electric lights. But Mary Ward loved it. Hampden had the 'ghosts' (including Queen Elizabeth) she needed. She could write there. And it was large—Arnie and she could scarcely sleep on their first night there for excitement at 'the size at it'.[12] For months afterwards she toyed with the fantasy of taking a long lease on the house.

Stocks was old (120 years). But it was neither as large nor as decrepit as the Hampden house, the upkeep of which would have strained even Mrs Humphry Ward's earning powers in the early 1890s. The house belonged to Sir Edward Grey who had inherited it through his mother. It was also in the close neighbourhood of the Rothschild mansion at Tring Park. After *Robert Elsmere* brought her to their attention, Mary Ward had been on visiting terms with this powerful Jewish banking and political dynasty. Her first visit was in August 1888, when she was still a new literary lion. Mary visited Tring Park—'this house of wealth and pride!'—frequently over the next two years. She found the Rothschild décor 'Monte Cristoish' verging on bizarre. The Baron—an avid naturalist—was assembling at Tring a menagerie of exotics: emus waddled and kangaroos bounded across the alien Hertfordshire countryside. Nor did

Tring Park offer the *society* Mary most admired, 'if by society one means what the French understand by it—good talk and a real exchange of wits'.[13] But, despite herself, Mary Ward could not but be seduced by the special train which transported her from Euston to Tring Station, feeling like a Russian princess. And—however stupid the conversation—this 'society' contained within it the real power of England. Her first glimpse of this world had the epiphanic force of her first sight of Mrs Pat playing croquet at Lincoln, or of the Tennant girls dancing ethereally among the company of 'Souls' at Gloucester Place.

The power élite of the 1890s and Edwardian England was dominated by a country-house society in which the upper-classes met as a supra-party *monde*, cultivating the arts of higher gossip and aristocratic leisure. Mary Ward was exposed to this milieu with the fame of *Robert Elsmere* when the novel attracted the interest of Gladstone and set him talking about her among his friends. She undertook a whole round of country-house visits that autumn. Mrs Ward at first felt somewhat out of place in these grand surroundings: 'it is difficult', she declared, 'for plain literary folk who do not belong to it to get much entertainment out of a circle where everybody is a cousin of everybody else and where the women, at any rate, though pleasant enough, are taken up with "places", jewels and Society with a big S'.[14] Nevertheless, on the whole she was much impressed by the male society she observed in these houses of the great. Life there, she told her old Oxford comrade Bertha Johnson, 'is lived intellectually, on the widest and and freest of all possible planes'.[15]

By the early 1890s, 'Souls' like A. J. Balfour had grown up and serenely put away childish things to take over leadership of their country. Mary kept up her connection with the set through Margot Asquith and their mutual Oxford admirer Benjamin Jowett. Mary Ward was never fool enough to think that she could, like Margot, who married Asquith in 1894, herself attain full membership of the country-house élite, any more than she hoped herself to be an Oxford don in 1870 or a Soul in 1885. But, established at Stocks, and with the right contacts, Mary might do great things for her son Arnold, whom already she saw as a politician in the making.

By 1890, Mary Ward's combined success as a novelist and Gladstone's favour had made her—against all the odds—a 'person who counted'. But she could not invite powerful friends like the Rothschilds, Lord Acton, or—perish the thought, Gladstone—to see and share her 'old gowns and milk puddings'[16] at Lower Grayswood. The Wards, having sold their Haslemere house for £9,000 on 23 May 1892, moved into Stocks on 14 July of the same year. Mary Ward described their new home enthusiastically to her father in Dublin. It was, she assured him, 'not grand in any way' and promptly went on to extol its grandness. It had 'old-walled and yew-hedged gardens, a small bit [i.e. 300 acres] of beautiful park, an avenue of limes like a cathedral aisle'.[17] It was surrounded by the woods of Ashridge, beyond which was the fine

Berkhamsted Common. Initially the Wards took Stocks on a seven-year lease, at £400 per annum. Almost at once, the house paid for itself by inspiring the plot of *Marcella*—something that Lower Grayswood could never have done. The house also brought with it a whole range of social responsibilities. The Wards became—by the mere fact of residence—a family of local consequence. Every January, they would put on theatricals for the village. Mary Ward visited the parish poor, and gave them doles of vegetables and coal. The Wards as a family began attending church regularly, something they had never done in London. They became close friends of Canon Henry Wood, the Rector of Aldbury. Humphry eventually took a seat on the parish council, Dorothy on the village school management committee.

The family had their first ball at the house in Christmas 1892. The first Stocks shooting party was held in December 1894—a modest little affair to celebrate Arnold's various triumphs in his last year at Eton. Two years and a lot of game preserving later, the winter function had grown. In November 1896, the Wards had for the annual massacre of Stocks pheasants the Alfred Lytteltons, the Sydney Buxtons, the Hon. William Peel, George Duckworth, Lord Brownlow, Henry Wheatley, and Sir Donald Mackenzie Wallace. Everyone, of course, stayed a few days as a house guest. In due time the *sportif* Arnold would have his tennis courts, cricket pitch and nets, and even his nine-hole golf links at Stocks. There was not a grandee in England (even the Duke and Duchess of Bedford) whom Mrs Ward could not entertain with equanimity at her country house.

The problem with all this was, of course, that the Wards (unlike their landlords the Greys) did not really *belong* in a home like Stocks. They were—as malicious tongues must have said behind their backs—a pair of middle-class Londoners who had unexpectedly struck it rich. Rich—but not sufficiently so to live with any security among the country house set. The Wards' finances were precarious. A change in the taste of the reading public or a slump in the sale price of fine art could easily wipe them out. Humphry and Mary Ward had virtually nothing in the way of material assets. They had originally intended to rent out Lower Grayswood, keeping the property as a nest-egg. Its value had soared in a couple of years to £9,000. Holding on to it would have been a very wise decision. But Mary was persuaded by Humphry to sell. They made the predicted killing on the sale, and the whole of the profit (around £5,000) was promptly sunk by Humphry in the purchase of a Cuyp. Effectively this represented a wholesale transfer of Mary's money—earned by her pen—to Humphry's art collection.

Meanwhile, the Wards' London address had also changed. In November 1890, they were made an offer on their Russell Square lease. The site was wanted for the future Imperial Hotel. It was a good offer, and the house was looking shabbier than ever. Humphry found something grander at 25

Grosvenor Place, on the corner with Chapel Street. It was 'a great bargain'. Since it had only a nine-year lease, there would be no premium to pay. It had amenities like modern plumbing and electric light—both signally lacking at Russell Square. Mary Ward weighed up all the pros and cons in a letter to her father: 'Of course, Grosvenor Place is terribly "smart" and I dread the extra society it might bring. Still, I must think of what Humphry likes and what will be pleasant for the girls as they grow up. And it would be well perhaps for me to be somewhat farther from the [University] Hall ... for my family life and my literary work ought to come before it.'[18] It's a sagacious and in its way generous-spirited letter. But one cannot help noticing how—unconsciously—Mary Ward assumes that the decision is hers to make. The Wards duly left the old house in Russell Square on 17 January 1891.

The decision to buy rather than rent Stocks was formed after Mary finished *Marcella*, in April 1894. But it was not until March 1895 that she heard confidentially from the Greys' agent, Crawford, that the house was on the market. Humphry was meanwhile away, lecturing in New England. He replied, 'You thrill me by the news that Stocks is to be sold! But oh dear oh dear how can we buy it without making for ourselves a burden too heavy to be borne? One must think and cast about.' Somewhat reluctantly and at his wife's prodding, he admitted that 'Meyer will buy the Cuyp for *six*'[19] (i.e. £6,000). But it was evidently not a sacrifice he much relished. Nevertheless, he would do it. Stocks would fetch—Crawford had reckoned—about £18,000. 'It would be easy to borrow the rest,'[20] Mary optimistically added (easy that is for the author of *Robert Elsmere*, not for the art critic of *The Times*). In the short term George Smith would lend them £10,000 as a bridging loan. In the long term the Greys would leave a sizeable portion of the purchase money of Stocks on mortgage at 3.5 per cent (that being half a per cent more than they would get from a regular deposit account).

In the event finances proved no obstacle. The Wards paid some £7,000 down, which allowed them a few thousands over for current expenses after Humphry finally sold the Cuyp, a Gainsborough, and a Greuze. The sale of Stocks was finally agreed in March 1896. Together with the Grosvenor Place rent the Wards were thus committed by 1897 to £750 a year in standing payments on their two main residences. Mary Ward at once began to feel nervous about these new millstones round her neck. On 19 July 1897 she told Humphry (who was again speculating rather recklessly in pictures) that 'money matters worry me a good deal just now, especially at night'.[21] She lost even more sleep in April 1899, when her husband informed her of his intention of buying a Gainsborough, having just had one of his Constables 'questioned'. 'Dearest,' she wrote with unusual firmness, 'I hope it is good for us, this uncertain earning of large sums! "What shall it profit a man, etc". I do so dread as one gets richer losing the the capacity for poetry, for disinterested enthusiasms for all that really makes life. Don't you? I know you do! I have often seen you fall back on some literature of high ideas and imagination after some exciting business, as though you felt the danger as I do.'[22]

And yet, although Mary Ward could see how money was inexorably coarsening their lives, she was just as much to blame as Humphry in forging the golden shackles. There were any number of points in the 1890s when Mary Ward could have pulled in her horns and lived within her substantial means. She chose not to. And the reason was Stocks. She had resolved to live there in a certain style. Coming down from that level, after she had once tasted its delights, would have been intolerable.

16

Respectable Genius: 1890–1900

THE dominant tone of Mrs Humphry Ward's mind and conduct in the 1890s was one of deafening 'Respectability'. Appearance—or 'what people will think'—was her fetish. Social conformity is usual enough in middle age, as a kind of hardening of the moral arteries. But in Mary Ward's case there were other factors than her advancing fiftieth birthday. The 1890s was a period when society and history demanded that the thinking person take sides. The Oscar Wilde scandal, the emergence of naturalism in fiction, decadence in poetry, modernism in art, the establishment of the Independent Labour Party, the Boer War, Ibsenism—all these forced the English intelligentsia off their habitual fence. As a woman of consequence, Mrs Ward had a position to keep up. It was symptomatic when in February 1894 she told her father that *Marcella* 'ought to please more and offend fewer people than anything I have done yet'.[1] Not offending was now important to her.

She was quick to take moral offence herself. Charles Stewart Parnell, for instance, offended her mightily; less for his Irish nationalism—although that murderous cause was bad enough—than for his flagrant adultery with Mrs O'Shea. (Mary Ward almost certainly used her behind-the-scenes influence to harden *The Times*'s libellous persecution of the Irish patriot.) She had the same reflexively moralistic reaction to the Prince of Wales's involvement in the 'baccarat scandal' of June 1891 and his rumoured misconduct with Lady Brooke. 'Why cannot the man behave himself like a decent English gentleman,'[2] she expostulated to her father, unaware apparently that English gentlemen had been at each others' wives since time immemorial. She heartily approved of Oscar Wilde's prosecution—even if he had been a protégé of Mr Pater's and a guest at Russell Square. She bullied Wilde's publisher, John Lane, into further hounding Aubrey Beardsley, so as to purge English culture of the infection of yellow-bookery and sodomy.[3]

Mary and Humphry Ward saw 'the Danish play of *The Doll's House*' in June 1889 in the company of the Creightons. Although the deferential manager bowed them to a complimentary box as guests of honour, Mrs Ward was not impressed: 'none of us admired it', she told Arnold. It was 'too improbable and extravagant'.[4] New French literature was worse than extravagant. 'Zola's *L'Oeuvre* is disgusting beyond words,'[5] she declared in September 1890. She concurred happily with *The Times*'s hard line on Zola's English publisher,

Henry Vizetelly, and that 69-year-old man's three months' imprisonment in 1889, for daring to translate French filth. When Zola was prosecuted in 1898—at the height of his heroic campaign to clear Dreyfus—Mrs Ward sternly opposed a public statement of support by English writers: 'It is not our business,'[6] she declared.

Mary Ward would sometimes privately yearn to treat scenes such as Letty Tressady's extra-marital temptation 'with French or Tolstoi-ish frankness'.[7] But 'English custom' restrained her. It was not that she did not know the depraved ways of the world; but that children and servants must have such things curtained from them. '*Pas devant*' was Mary Ward's watchword. What others saw as the tyranny of the young novel-reader she held to be a sacred trust. When discussing literature with her own young readers, she never missed the chance to ram home an incidental moral about the wages of sin: 'Tonight I have been reading a number of Burns's songs to Dot and Jan,' she wrote to Arnold in May 1891. 'What a great man was Burns—and just for want of self control he died poor and sad and miserable at 37. It is a comfort that we possess a few long-lived sane *respectable* geniuses like Wordsworth and Tennyson.'[8]

The swamp of modern art Mary left to Humphry Ward, and his *Times* reviews. Both Wards had a 'healthy' English scorn for 'those clever madmen—Monet, Degas, Manet and the rest'.[9] They passed their healthy English aesthetics on to their children. In her diary for 11 January 1911, Dorothy Ward recorded: 'Arnold and I went to the Post-Impressionists this afternoon and had a good laugh.'[10] Between them, Mary and Humphry could take much credit for keeping England two jeering generations behind the rest of Europe in art appreciation.

It took longer for Mary Ward to purge herself of the taint of religious heterodoxy—always the least respectable aspect of her public character. Of course, people worried less about such things as the secular twentieth century approached. But Mary was still dubious enough a figure in the 1890s for her devoutly Anglican secretary, the aptly named Bessie Churcher, to have strong doubts about the propriety of working for the author of *Robert Elsmere*. The fact was, however, that the author of *Robert Elsmere* was increasingly unhappy about the positions she had taken up in that book. Although she rarely mentioned it, throughout the 1890s Mary Ward felt ever more strongly the tug back to the orthodox religion which she had formally repudiated in the 1870s. For the first time in two decades, she heard morning service in Eton Chapel, while visiting her son Arnold in 1891. 'The words that sound so alien to us now' made her profoundly happy she told her son. 'I never felt a keener sense of discipleship to Jesus of Nazareth.'[11] Her motive for attending the College chapel—a calculatedly public act—indicated another pressure nudging her towards orthodoxy. Arnold Ward had antagonized his masters by his recalcitrance over 'Sunday questions'. Mary Ward intended to show him (and Eton) that even as notorious a heretic as she could toe the line.

Mary Ward had for some years been an intermittent attender at Stopford Brooke's eccentric Bedford Chapel in Bloomsbury. Brooke had been in younger years a successful and orthodox clergyman—for a couple of years he was even chaplain to the Queen. His preaching was famous. After 1874, when his wife died, he went a little crazy. Friends subscribed to buy him the lease of Bedford Chapel. Here he quickly attracted huge congregations from the surrounding middle classes and university people. As many as 600 crammed in to hear him on Sunday mornings. His sermons were *tours de force* and notoriously wide-ranging; he was as likely to talk about the beauties of Shakespeare or the latest strike of gutta percha workers as the Gospel. He was known as something of a socialist, which did him no harm in Bloomsbury.

Like his friend J. R. Green, Brooke was a believer in the 'human Christ' and a disbeliever in biblical miracles. In 1880, without any great fuss, he seceded (his word) from the Church of England. He could not any longer believe in the miraculous element in Christianity. Although his doctrine was essentially Unitarian, Brooke dogmatically refused to join that movement, which he saw as a 'sect', not a 'National Church'. Nor could he stand the bleakness of their form of service. 'They cling to ancient uglinesses as if they were sweethearts,'[12] he told Mary Ward. At Bedford Chapel Brooke continued to perform the lovely Anglican forms of service as laid down in the Book of Common Prayer, with the small omission of the Creeds and those parts of the liturgy containing direct reference to miraculous incarnation. He also administered Holy Communion with the same credal deletions. Thus, for the first time in good conscience since the early 1870s, Mary could savour the ineffable joys of the Eucharist.

It did not, however, last. In 1895 the lease for Bedford Chapel expired. Brooke—who had been ailing—resolved to give up the fight. Mary, however, was very keen that he should continue, if not at Bedford Chapel then somewhere else, and made vigorous efforts to find him new premises. She failed and the chapel closed in August 1895. There was nowhere else in England that administered Holy Communion as Brooke had—in Anglican style without the offensively superstitious elements. 'Yes I do not wonder at your regretting the communion service,' she told Brooke in August 1895; 'I feel with you the loss of old association, of connexions with the great past of the whole of the ancient church and its formulary.'[13]

With Brooke gone, Mary Ward's sense of loss was that much sharper. There was nowhere she could now take Communion, no shelter in the lonely wilderness of her theism. Like Brooke, she had always resolutely refused formally to join the Unitarian Church, identical as her views were with its. In an unusually intimate letter to Dorothy eighteen months later, she described her agonizing sense of 'outsideness'. It was, she told her daughter, 'part of the cross that is laid upon *us* at the present moment .... Our Master felt it before us in the face of an organised belief and ritual from which he was breaking away.'[14]

Denied the sacraments, Mary Ward continued to thrash about in spiritual agony. Not being able to take Communion any more made her feel, literally, excommunicated. It went beyond doctrine or liturgy and brought back in her middle age all those agonies of separation which she had felt as a lonely little girl at Rock Terrace. This being shut out from the great Anglican family supper made her want to die, as she had then wanted to die. Her hard-headed Liverpool cousin, Mary Cropper, suggested in Easter 1898 that Mary should simply accompany her to Communion service. She could mumble the responses or keep silent at those parts of the Creed which worried her. It was sensible advice. But Mary Ward's nerve failed her.[15]

In this susceptible condition, Mary Ward was inevitably drawn to the faith of her father—more so as Tom now radiated such content. Could she not join him? The Catholic Church had ritual enough even for her starved appetite. She was deeply affected in September 1895 by an enigmatic remark of Tom's to the effect that 'I think Newman would have gone with you a long way ... farther than you think.'[16] Newman's hand had stretched out across the globe in 1854 to grasp Tom Arnold. Could it not now stretch across the decades to grasp Tom's daughter. The emotional appeal of Catholicism was heightened by the illness which she now accepted as her lifelong lot. 'The Catholic passion for "suffering" ', she told Louise Creighton in 1897, 'has been very suggestive to me.'[17] Catholicism confronted as inevitable that 'one must suffer and one must die'. And it sublimated that suffering through *imitatio Christi* into 'ineffable joy'.

Although Mary Ward masked it as 'research', it is clear that, while writing *Helbeck of Bannisdale*, she was studying Catholicism with a view to possible conversion. As Tom had told her, she could go further with Newman than she thought. What finally stopped her (and Laura) going all the way was confession. This 'uncovering of the inner life and the yielding of personality'[18] constituted a last step she could not take. There was something buried deep in her that no other person must know.

At the end of *Helbeck of Bannisdale* Laura drowns herself, torn as she is between love for Alan and a final inability to accept Catholicism. It is a strikingly pessimistic conclusion. Robert Elsmere agonized on leaving the Church, but there was still a life outside it and even religious fulfilments of a kind in his New Brotherhood. Once Laura accepts her outsideness, there is nothing for her but the river. Melodramatically heightened, Laura's predicament mirrors her author's in 1898. Mary Ward's sense of exclusion had also reached the pitch of desperation. And in her desperation Mary had passed beyond longing to anger. It was sheer 'injustice', she told Mandell Creighton in 1898, 'which excludes those who hold certain historical and critical opinions from full membership in the National Church, above all from participation in the Lord's Supper'.[19] She resolved to take the bull by the horns and wrote an article length letter to *The Times* in early September 1899. The occasion was the so-called 'Crisis in the Church' provoked by the High Church party and their successful demands for more ritual and incense. Mrs Ward entered a collateral demand for those like herself 'whose Christianity no longer depends on miraculous narratives'.[20] If the High Church party could have their neo-Catholic toys, 'liberals' like herself should be able to take Communion, under the provision of a conscience clause which would allow them to be silent at crucial points of the Creed.

Mary Ward's letter was provocative in the highest degree. Legions of indignant clergymen fired off letters. The furore looked on the face of it like *Robert Elsmere* all over again. But there was a difference. In 1888, Mary Ward had asserted the necessity of leaving the Church. Now she was begging to be readmitted. She wanted to be a respectable, communicating churchwoman again. This was what underlay her provocation. She was knocking on the Anglican door, not assaulting its structure.

Unsurprisingly, the Church of England declined to rewrite its liturgy, leaving out the bits that were inconvenient to Mrs Humphry Ward. Nor was it reasonable to expect it to do so. Mary Ward's position on the 'New Reformation' was highly illogical as a number of correspondents were quick to point out. The assent to the virgin birth of the Saviour and to his Ascension were not—as she claimed—removable details but the very core of the Communion of the Eucharist. Take away the miraculous elements, and there was nothing in the service but pretty words and symbols. It was out of the question. But Mary Ward continued to hope over the next decade that the 'Modernist' movement would eventually reform Anglicanism in the direction she desired and so let her back in. It never did.

In the meantime, she quietly made her peace with Anglicanism. Dorothy Ward's diaries mention it as a routine thing that on Sundays in the country her mother attends morning service. The Wards used the 'Stocks pew' at St John's Church in Aldbury. In 1903, it was a cause of 'pain' to Mary Ward that Janet's fiancé, George Trevelyan, should be 'wholly non-Christian' and that he should propose 'a civil marriage'.[21] By now, she was sensitive to any public show of heterodoxy on religious matters.

Although it has to be speculative, it may be that Mary Ward contrived to take Communion, even after Brooke closed his Bedford Chapel. There is no definite evidence on this, but a number of suggestive clues. Writing to Creighton in 1898, she refers to 'those who can now only share in [the] Eucharist on terms of concealment and evasion'.[22] Was she herself doing so on these terms? Eugénie de Pastourelles, a character with whom Mary closely identified herself in *Fenwick's Career* (1906), is an agnostic in town, but hears Mass and takes Communion at a 'little church' in the country.[23] Dorothy mentions in 1907 taking Communion at the little country church at Aldbury,[24] although evidently she was not confirmed until after both her parents' deaths. It would seem that the rector (Mary's good friend the Revd Henry T. Wood) was not strict on this point.

Another possible clue is Mary Cropper's enigmatic comment in a letter of 1920: 'I wonder if any clergyman has refused you the Holy Communion. I cannot believe it.'[25] It would have been an odd thing to wonder had not Mary Ward been in the habit of at least occasionally taking Communion. And finally, at her own wish, Mary Ward was buried not with her mother at Ambleside but in the churchyard at Aldbury. It is at least conceivable that the reason was that the friendly Revd Wood had over the years (possibly privately, certainly discreetly) administered to her the sacraments that meant so much to her.

The turn to respectability in Mary Ward's attitude to the monarchy was quicker and more direct. In her girlhood she had been a staunch little Jacobin and had no time for kings and queens. By the 1890s her views on the monarchy were unimpeachably loyal. On hearing in 1894 that Queen Victoria 'likes Marcella' she returned the compliment. 'I have long felt very warmly towards H.M.', she told her father. 'Of course she is full of limitations and prejudices—stupidities perhaps. But I am not sure that these very prejudices have not done good .... It would never have done to have had too clever a woman in such a place! Anyway—Republican as I used to be—I am pretty clear now that we owe her a great debt.'[26] In May 1895 Mary and Dorothy Ward attended a royal drawing room, suitably decked out and ceremonially photographed in their finery. On the occasion of the Jubilee in June 1897, Mary chartered a 'buss' for the Huxley children, so that they might view everything from the top deck. The celebration was the 'most interesting and thrilling thing I ever saw',[27] she loyally declared.

Mary Ward was never particularly fired on the issue, but she would have seen herself in her youth as something of a sympathizer with utopian socialism ('new brotherhoods'). By the 1890s, her attitudes to the proletariat *en masse* had hardened into paternalism mixed with occasional apprehension. She was furious with the rural working classes for voting in the Liberals in 1892: 'that England should be at the mercy of these totally ignorant agricultural labourers!'[28] she expostulated to her father. London's totally ignorant labourers filled her with a more persistent 'despair';[29] but she had some hopes for the deferential and well-regimented Lancashire workers whom she had met during her fieldwork in the north for *David Grieve*. Regimentation in fact rather appealed to Mary Ward, and she read 'General' William Booth's *In Darkest England* with interest in November 1890.

Although she could not be said to be sympathetic, Mary Ward was curious about the upheavals that were producing the Independent Labour Party. In 1895, she attended a meeting of the Social Democratic Federation in an underground cellar in Lisson Street. She was 'of course incognito'—shrouded in veil and shawls. 'They talk nonsense but most of them are good fellows'[30] she concluded. But—unlike Queen Victoria's—theirs was not a stupidity which she had any use for. During the 1890s, Toynbee Hall had become a hotbed of socialism. Mary Ward was resolute the same thing would not happen to her settlement. Inviting A. J. Balfour to become a trustee of the Passmore Edwards Centre in 1897, she assured the Conservative statesman that 'the settlement as such will be wholly uncommitted to any particular view'.[31] Arnold Ward, more candidly, suggested that she should advertise for 'some honest Tories'[32] to give the institution a backbone.

Mary Ward—true to her 'Souls' belief that sensibility and mind transcended party—did not ally herself publicly to any specific political group. Tories, Liberals, and even socialists like Graham Wallas (who gave her tutorials for the radical sub-plot of *Marcella*), were all of them welcome at her Thursdays and at Stocks weekend parties. But privately her affiliation after 1886 was to the Liberal Unionists. Like many of her generation, Irish violence had made the Gladstonean mainstream of the party uninhabitable—affectionate as she remained to the GOM personally. And like many of the Liberal Unionists (not least their leader, Joseph Chamberlain) she drifted over the years into the fringes of Conservatism. She happily embraced Tariff Reform, and the quirky pro-imperialism that was part of the package. In fact, by the end of the 1890s, her support for the British Empire had become unpleasantly strident and aligned her more with Liberal Imperialists, like Edward Grey.

After Kitchener's victory at Omdurman in 1898, Mrs Ward fairly crowed. 'The pride of the English name it seems to me was never greater—Kipling will have to write a new Recessional,'[33] she told Humphry. When the Boers rose up in 1899, her views were wholly punitive. 'They will have to learn to live peacefully with the people who financially run their country,' she declared. English rule was 'the natural discipline appointed for them by Providence'.[34] A year later, she hailed the British Army's successes as 'signs that at present we *are* fit to rule, and are meant to rule'.[35] To give her credit, she did not gloat at 20,000 Boer women and children starved to death under English rule in concentration camps. And the country's crass jubilation and the 'khaki election' of 1900 frankly appalled her. 'The English are not pretty in victory,'[36] she observed. But the victory was dear to her and confirmed God's approval of England. As Janet Trevelyan notes, 'the end of the [Boer] war found her more staunch an Imperialist, more definite a Conservative'.[37]

It was on the home front, particularly with regard to women's rights, that Mary Ward's growing conservatism was most pronounced and pernicious.

In 1879 she had been *the* moving spirit behind setting up Somerville. With whatever private reservations, she had then thrown herself into the cause of advancing women's interests and achieved something of enduring worth. Ten years later, she had swivelled round to an intractable opposition to her sex's emancipation. (Somerville eventually disowned its founder.)

Although the votes-for-women movement had been on the march for twenty years, it had made no impact in Parliament since John Stuart Mill's day. Mary Ward's active hostility to suffragism was ignited by the news that a private member's bill was coming up imminently and might—given Salisbury's recent pronouncements—have some government approval. At a dinner on 20 January 1889, she found herself opportunely between Lord Justice Bowen and J. T. Knowles, editor of *Nineteenth Century*. Over their meal, the three of them 'concocted a women's manifesto against women's suffrage which we mean to launch when the critical time comes'.[38] All that was required was that Mrs Ward should drum up the necessary women for this spontaneous demonstration. This she did with the help of Knowles's friend, Frederic Harrison. Among the 104 eminent members of the sex conscripted for the 'Appeal against the Extension of the Parliamentary Franchise to Women' were a number—like Louise Creighton, Charlotte Green, and Beatrice Potter (Webb)—all of whom subsequently became stalwart fighters *for* the suffrage. Their signatures on the appeal were to haunt them for years; but in 1889 they merely did what their friend Mary Ward asked them to do.

It was not all plain sailing. Frederic Harrison was England's leading Comtist. True to his positivist faith, he believed that the role of woman was to be a divine object, passively worshipped by man. Mary Ward could not swallow that degree of subordination, and there were a few tense moments when Harrison explained that by organizing—even if only for the purpose of putting their names to a petition—women were betraying their womanhood. But with a bit of give and take she, Harrison, and Knowles contrived to get the 'Appeal' into the June issue of *Nineteenth Century*, where it created the expected terrific stir.

It was followed in July by a 'Reply' from Millicent Fawcett and Mrs Ashton Dilke. These suffragists made some telling points about the fact that the 104 signatories to the 'Appeal' featured 'a very large preponderance of ladies to whom the lines of life have fallen in pleasant places'. It was easy for the privileged among the sex to say they did not need rights. In August, using Louise Creighton as her front-woman, Mrs Ward delivered a hammer-blow 'Rejoinder'. The Appeal had appended to itself a detachable coupon, which sympathizers could fill in and return. The August edition of *Nineteenth Century* thus contained the names and addresses of no less than 1,200 anti-suffrage ladies. The two lone pro-suffrage voices were drowned out in what was made to look like an overwhelming and deafening plebiscite of 'anti' voices. The extension of the franchise, Louise plausibly claimed, was 'distasteful to the great majority of women'. Mary Ward's 'Appeal' was a propaganda triumph, and in histories of British feminism is plausibly credited with helping hold back the cause of votes for women for years. It gave male opponents (whose brain-child it was, and whose cat's-paw Mrs Ward was) the powerful ammunition that women themselves—the cleverest women in England no less—did not want the vote. Why force it on them?

The immediate crisis was averted, and the cause of votes for women went back into the doldrums for a decade. The opponents' tactics over this period were simple and dishonest. MPs would assure their constitutents that they were for the extension of the franchise in principle, secure in the knowledge that whenever it came up in the House as a private member's bill it would be talked out by spending inordinate time on previous items on the order-paper.

So things remained until 1897 when due to an oversight Mr Faithfull Begg's Women's Suffrage Bill made it to a second reading. The pent-up energies of the suffragists were mobilized and a petition bearing a quarter of a million names was delivered to Parliament. On the night of the debate, the House was besieged by supporters. MPs trying to skulk in were shouted at or buttonholed and reminded of their pledges. The Conservatives (who were in power) caved in, and the bill passed by seventy-one votes. It did not, of course, have a hope of passing into law; any number of sabotages and ambushes were available to the bill's opponents at the committee stage. But this 'theoretical victory' marked a new high water mark in the campaign. A governing party was now on record as favouring votes for women.

The parliamentary events of 4 February 1897 provoked an explosion of wrath in Mary Ward. 'Was there ever anything so disgraceful?' she asked her father. It had made the House of Commons 'ridiculous', she declared; 'I don't know when I have felt so angry about anything.' It was not so much the monstrous regiment of women who enraged her so much as the ranks of men who supinely declined to *govern*. 'The tone of the Conservative members in private is the most cynical thing you ever heard,' she complained, 'there are not six men in the whole House who want the change, and yet there they all go and vote to rid themselves of what they are pleased to call their "pledges" to a "pack of women".'[39]

No one, least of all herself, has convincingly explained why Mary Ward was so hostile to the cause of women's rights. It was not as if she was surrounded by 'antis' (as they were called). Her sisters Julia and Ethel were actively for the suffrage. (They were also, like Tom, pro-Boer.) Even her septuagenarian Aunt Fan was 'pro'. Apart from Mary and the ever-faithful Dorothy, the Arnold–Ward females were sympathizers to a woman. Mary's own daughter Janet was—after the age of 19—pro-suffrage. What, then, were Mary Ward's reasons for obduracy? The 'Appeal'—which she of course framed—had used a quasi-Darwinian argument: 'We believe the emancipating process has now reached the limits fixed by the physical constitution of women.' It also alleged that women were biologically 'lacking in sound judgment'. But one cannot think that in her heart Mrs Ward conceived herself an inferior species to those 'totally ignorant agricultural labourers' who swept Gladstone in to power at the election of 1892. A related line of argument which Mary Ward was to reiterate over the next thirty years was women's inability to participate in the physical activities on which industry and empire depend. Votes were equated with male brawn. Women, she claimed, can never share in such labour as 'the working of the Army and Navy, all the heavy, laborious, fundamental industries of the State, such as those of mines, metal and railways, the management of commerce and finance, the service of the merchant fleet on which our food supply depends'. But this is a deeply flawed argument. Mrs Ward was surrounded by puny male servants at Stocks whose main business it was to wait on table or bring Humphry his letters on a silver tray. Why should such as these have the vote, and not Mary or the indefatigable Bessie Churcher, who could rise at five, type Mary's manuscripts, work all day with crippled children, and cycle however many miles home? 'Women are not naturally voters', she declared in 1908 on setting up the Anti Suffrage League. Why not?

Mary Ward's support for the anti-suffragist cause was indefatigable and unwavering even when (as eventually happened) there was no woman of intellectual respectability in England prepared to share a platform with her. Her obstinacy cost her and her MP son Arnold dear and is one main reason why Edwardians were so keen to abuse and posterity to forget Mrs Humphry Ward—traitor to her sex. This intransigence is very strange in a woman otherwise so adroit. There seem to have been three main reasons for her recalcitrant anti-feminism—all essentially emotional and irrational. The first was that men—patriarchal, powerful men—asked her to help them. And she had never been able to resist wanting to please and serve father figures. The second was Mary's ineradicable belief that for women to want the vote was somehow unseemly. One of her favourite dicta on the question of the suffrage was: 'For Heaven's sake, don't let us be the first to make ourselves ridiculous in the eyes of Europe!'[40] This question of how one 'looked' in the eyes of others was terribly important to her. The third reason—and in many ways the most sympathetic to posterity—was her horror of 'militancy', which she instinctively associated with the Irish outrages that had terrified her in the 1880s. 'Constitutional' suffragists she could tolerate, even if she disagreed with them. But suffragettes were hardly better than the Fenians.

For the public in 1888, Mrs Humphry Ward was a famous novelist and thinker. For the public in 1898, she was an institution, an embodiment of middle-class, late-Victorian values. As such, she became a target of a convenient kind for those who did not share those values; specifically, the younger generation. 'The Great Mary', as Ezra Pound called her, was every ambitious young writer's black beast. Spattering Mrs Humphry Ward with smart sarcasm became a rite of passage. Oscar Wilde for instance claimed in 1891 that reading *Robert Elsmere* 'reminded him of the sort of conversation that goes on at a meat tea in the house of a serious Nonconformist family'.[41] Oscar did not dare say this in the first 1889 version of 'The Decay of Lying', published in *Nineteenth Century*—Mrs Ward's home journal. He inserted it maliciously in the reprinted version of the essay, two years later. Few bright young writers dared take on Mrs Humphry Ward face to face until well into the twentieth century.

But, as time went on, others followed Wilde's insolent cue. Scorn for what Mrs Humphry Ward represented was formative in Virginia Woolf's evolution. Reading Mrs Ward, she once said in her smart way, was like catching flu. That his *Eminent Victorians* baited the (dying) Mary Ward was deliciously pleasing to Lytton Strachey. Her fury was 'a triumph', he declared. He had hated her for years. In a letter to his brother in 1909 he described her as 'that shapeless mass of meaningless flesh—all old and sordid and insignificant'.[42] Max Beerbohm and others had a club, in which Mrs Humphry Ward was reviled as 'Ma Hump'[43] and she was the inspiration of one of his most maliciously funny literary cartoons, in 1904. Her great offence to Beerbohm was that he found her 'dull'. When H. G. Wells first made love to Elizabeth von Arnim, it was in the open air, her naked buttocks pressed against a copy of *The Times* featuring one of Mrs Humphry Ward's moral tirades.[44] The old lady's unwitting presence gave zest to the couple's adulterous frolics. (Mary's article in the paper was actually directed against the young Rebecca West, who was also to have her fun with Mary.) Mary Ward figured in the sexual fantasies of other younger writers. The mother-dominated Arnold Bennett, for instance, stimulated himself by picturing her heroines being gang-raped by a whole army of 'brutal and licentious soldiery'.[45]

By 1918, vilification of Mrs Humphry Ward had reached the level of a minor art form. Rebecca West pictured the old lady as writing 'with an umbrella in one hand', poking away irritably at the degeneration of the times. Mrs Ward's career, West jibed, was 'one long specialisation in the *mot injuste*'.[46] Even Aldous Huxley, who claimed to love his aunt, nevertheless cleverly abused her. In Huxley's 1920 satirical fantasy (his first published work of fiction) 'The Farcical History of Richard Greenow' the hero, Dick, reads *Robert Elsmere* at the age of 8. He grows up a stern rationalist and pacifist, but insidiously the Elsmerean bacillus has infected his sensibility. A second personality—the florid female novelist Pearl Bellairs—takes the adult Dick over. In this other identity, he/she becomes a best-seller. Miss Bellairs is a hilarious (and transparent) spoof of Mrs Humphry Ward, who gave up writing fiction of ideas for romance after 1900. Dick—who has kept his split personality unknown to the world—dies deliriously raving anti-German war propaganda (a clear hit against Mrs Humphry Ward's war writing).

Huxley's squib can be taken as an allegory, like Woolf's 'flu'. The English literary system had to be purged of the Great Mary's influence. Modernism needed her cultural space. Respectable genius was now a contradiction in terms. But it does seem hard. As Enid Huws Jones records, a copy of 'Richard Greenow' was found lying at Mary Ward's bedside in 1920 and 'it was probably one of the last books she read'.[47] It cannot have given her any pleasure.

17

Health: 1890–1900

THE refrain running through all Mary Ward's letters and private papers from 1883 until her death is that of ill health. One could write her biography as a sixty-nine-year medical case report or an anthology of the age's female invalidisms. Over the years she was racked by a baffling array of chronic, acute, subacute, psychosomatic, and organic ailments. At first sight it looks like a catalogue of sheer physical bad luck, bad genes, and bad medicine. But reading the solicitous accounts in her daughter Dorothy's diaries, one perceives rhythms and even a strategy in Mary Ward's recurrent health crises. Sickness was by the 1890s installed as a component in a complex overwork–collapse cycle on which her productivity—and possibly her creativity—as a novelist depended.

The first distinction to be made is between illness or incapacity as such and the breakdowns precipitated by the intense but temporary strain of writing. Physically, the Arnolds seem to have been predisposed to persistent lowering complaints of two types—colds and rheumatism. As a baby, Tom Arnold irritatedly noted his daughter's vulnerability to any passing infections—her perpetual cold, hacking cough, and dire teething pains. She was, he told his mother, 'a very sickly child'.[1] At Rock Terrace, the schoolgirl Mary suffered never-ending colds and blinding headaches. Sickness at this stage of her life was (as *Marcella* confirms) a self-punitive rebellion against authority.

But she was also pathologically infection-prone—something that afflicted her from birth to death. From childhood onwards, Mary suffered (especially in winter) from 'septic' throats, from inflamed or 'lumpy' tonsils, and earache. On the evidence of Dorothy's diaries her mother regularly caught at least a dozen heavy colds or influenza infections a year. So did her children. (It was Mary Ward's bequest; Humphry enjoyed indestructibly good health until the onset of his arteriosclerosis in 1912.) So bad was young Arnold Ward's throat that in 1900 he had an operation to remove his tonsils and adenoids—a prophylactic measure that was just becoming fashionable. Nevertheless, throughout his adult life Arnold suffered from even more colds than his mother (although they may have been in later years euphemisms for hangovers). Arnold's 'throat' (like his grandfather's stammer) was partly psychosomatic—he would for instance lose his voice when in the presence of strangers. But it was also an inherited condition that he shared with his sisters. Janet Ward had a 'horrid granulated swelling at the back of her throat'[2] touched with electric cautery in February 1895 by the family doctor Mary trusted most, Harry Huxley. Dorothy Ward, like her mother, was forever dripping and snuffling with colds in the head.

Rheumatism was another epidemic Arnold complaint. It normally struck its victim in mid-adulthood. Dorothy, for instance, records her first serious attack in 1898 (a 'tingling stiffening pain, which I think must be rheumatism').[3] She was 24. Mary Ward records having a vexing 'pain in the back'[4] as early as 1874 (she was also 24). By the 1880s she was a frequent and by 1890 a chronic sufferer. By her early twenties, Mary's youngest sister Ethel Arnold was two or three times a year 'horribly bad with rheumatism and complications'[5] (i.e. lumbago) and bedridden. In the mid-1890s, Mary Ward and her younger sister, Julia Huxley, were sharing their preferred prescriptions for their 'rheumaticky' pains. (They finally settled on nux vomica—a strychnine compound—twice daily as preferable to coal-tar-based medicines.) This rheumatism almost certainly was exacerbated—if not caused—by excessive attention to posture, deportment, and modes of dress and social conduct that entailed women holding unnaturally rigid positions for hours on end.

In the early 1880s, Mary Ward began to experience the first agonizing symptoms of 'writer's cramp'—a condition which may have had rheumatic origins or complications. The first sharp onset was in 1882–4, when the pain was so severe that she was obliged either to write *Miss Bretherton* with her left hand, or to dictate the narrative to her amanuensis Gertrude Ward. Photographs throughout life after this date often show Mrs Ward with her right hand dangling limply like a dead fish on the end of a line. Janet Trevelyan suspected that her mother's ailment was not in fact true writer's cramp, which is a kind of local tendonitis, but a pinched nerve. Certainly the doctors who first treated it were of the opinion that Mary had either a 'thickened muscle under the arm' or 'an inflamed gland in the axilla [armpit]'.[6] The condition continued throughout life and became regularly acute during the most stressful phases of her novel-writing. At such crises, Mary Ward's normally firm calligraphy would disintegrate into an illegible pencilled scrawl, and she would cut letters short, complaining tersely of 'hand' or 'arm'. But the advantage of this cramp was that she never felt obliged to waste her limited writing energies on merely secretarial labours, which she routinely left to the dutiful (and never cramped) Gertrude, Dorothy, and—finally—Bessie Churcher with her professional shorthand and typewriter.

Over the years Mary Ward adopted other strategies for easing the pain of her 'lame hand' or 'neuritic arm' (neuritis was in fact the diagnosis her physicians favoured towards the end of her life). Dorothy became expert at 'stroking'—gently massaging—her mother and so enabling her either to sleep or relax. In 1883, Mary Ward was taking an arsenic tonic, which she seems to have continued for some years. A whole range of anaemia treatments (dietary or diet-supplement) were thrown in—on the grounds presumably that women's disorders were always complicated by poor blood. The physician Wolff prescribed a set of physio-therapeutic exercises ('gymnastics', as Mary quaintly called them), which gave her relief when the cramp was at its most acute in 1884. In the 1890s she tried 'Swedish rubbers' (i.e. masseurs), who seemed to have lifted her spirits for a while. In February 1903, she was consulting a 'new German', Herr Maschik, who administered a drastic 'electric vibration treatment'. After a promising start ('we both liked him and felt he knew what he was about',[7] Dorothy noted) the jolts from Maschik's electrical machine made Mary so 'depressed' and 'itchy' that—on the advice of a third German doctor, Fürth—she discontinued the course. As was often to be the case, Mary Ward's primitive home remedies were the best palliatives. In 1895, she reported discovering 'a clever way of propping the arm so that I can write in fair comfort'.[8] In 1909, she described herself to an American correspondent 'writing as I always do in a corner of a sofa, with a small support for the right arm, and a writing board'.[9] In the last decade of her life, this apparatus also served her to write in bed, propped up on pillows.

If rheumatism was the woman's cross, the Arnold men notoriously suffered from bad hearts. Angina carried off the Doctor aged only 47 and Uncle Matt aged 66. Matthew Arnold's symptoms began as early as his twenty-fifth year. Tom Arnold began reporting in the 1890s ominous symptoms of giddiness and weakness after physical effort. In fact he survived until 77, to die of bronchitis. But—significantly—it was deterioration of the heart which eventually killed his daughter Mary. She first reported 'sharp pains in the heart' in 1907. By 1910, she was regularly feeling 'heart-y' after exercise or excitement, and was taking strychnine drops under the direction of her London doctor, Huxley, and her Stocks doctor, Brown. In her mid-sixties, cardiac degeneration produced tell-tale symptoms of dizziness and debility, which prevented Mary Ward in her last year from walking more than a hundred yards or so. Her diet, which after 1890 was always excessively rich in eggs, rice pudding, yoghurt, and cream (invalid food which was easily masticated and digested), cannot have helped her heart.

Mary Ward's eyes seem to have been remarkably sound organs. Although (unlike her intellectual sister Julia) she took care never to be photographed wearing them, she wore reading spectacles by 1897. In 1910, she had a scare about her sight—triggered evidently by overwork and financial worry—and for a while she found it comforting to write in a darkened room and to avoid light. Her pupils were meanwhile excessively dilated—which may have been an unrecognized side-effect of some medication. But in general her eyes were sharp, even in her sixties. Her observation of landscape, and wild-flowers particularly, remained keen until the end of her life, and looking at the detail of the world was one of her purest pleasures. She also had good hearing, and attended concerts until her last years with no difficulties although she worried about damage over the years to her Eustachian tubes from all the colds she had suffered.

Her teeth were another matter altogether. They had been sadly—not to say criminally—neglected in childhood and early youth. The young Mary Arnold records stoically plunging her head in a bowl of icy water at Rock Terrace to ease the extremity of toothache. All through the momentous year of 1881 Mary Ward was again 'in hysterics' with toothache and dosing herself with sal volatile, linseed oil, and laudanum to kill the pain. The better dental care and hygiene of the late Victorian and Edwardian period saved her children from similar agonies. And Mary Ward herself underwent extensive remedial dental surgery in later life, having four or five bad teeth pulled out under gas in 1905–6. But it seems certain that decayed stumps (probably of her wisdom teeth) were the cause of the facial 'neuralgia' which at times made it impossible for her to go out, if even the slightest wind were blowing to fan her hypersensitive cheeks. Mary Ward had yet another bad tooth taken out in March 1919 ('one in the lower jaw which has plagued her for months', Dorothy noted).[10] So rotten and crumbled was the embedded root that another operation was required to clean out the gum—all of which strained Mary Ward's failing heart and spirits terribly. Some months later her old friend Mary Cropper suggested that she should have all her surviving teeth extracted; 'doctors are oddly shy, or careless, about mentioning it'.[11] As Cropper pointed out, 'pirea' (as she spelled it) had an insidiously poisoning effect on the constitution and could account for Mary's chronic 'seediness'. Twenty years before, this shrewd advice would have been well worth taking. As it was, in 1920 Mary Ward's system was not up to more general anaesthetics than strictly necessary.

Mary Ward's weight was never excessive although she constantly worried about it in the 1890s and early 1900s. She was at this period mildly stout in an acceptably matronly way. Sporadically she dieted and typically she overdid things. In 1905, for instance, she was losing a pound a day by her 'fasting *diners*' and causing Dorothy some anxiety. At this period, the inertia of Mary Ward's invalid mode of life—which entailed whole days of bed rest, whole weeks stretched out on *chaises longues*, and locomotion for months on end only in a bath chair—made it well nigh impossible for her to work off body fat by exercise. In her youth she had been an unusually vigorous girl—especially when in the Lake District. To celebrate their engagement, she and Humphry had taken a four-hour hike to the top of the neighbouring Loughrigg. But, after the age of 35, Mary Ward took no exercise beyond the occasional short walk at strolling pace (two miles and three-quarters was 'too much' for her mother, Dorothy thought in 1908). Even as a young wife, Mary did little or no housework and after Oxford no strenuous gardening. Unlike the middle-aged

Henry James she seems never to have learned to cycle—a form of exercise Dorothy and Janet kept up well into middle life. Croquet was dropped by her in the 1880s. Childbirth in the 1870s marked one great fall-back in her activity; menopause in the mid-1890s a second, which reduced her to genteel paralysis.

It was a great pity that one of her army of doctors did not see that coddling and passive medical 'tonics' (such as strychnine, Sanatogen, and changes of 'air') were producing rather than alleviating symptoms like sleeplessness and the chronic muscular 'flabbiness' which tormented Mary Ward's later years. It is demonstrable that underneath all the medically induced feebleness there remained a tough constitution. These reserves of toughness were brought out by the rigours of war in 1914. To everyone's surprise, Mrs Ward then showed herself quite capable of digging for victory in the vegetable garden, after all the Stocks groundsmen had left for the Front. Even more amazingly, for her three war propaganda books she bustled with the energy of a woman half her age, visiting the fleet in Scotland and the Front in France, spending hours in open cars in winter and scrambling up the sides of warships or trench slopes 'with the agility of a squirrel'.[12] If she could do this in her late sixties with a weak heart, there is no reason to suppose that in her forties she could not have been a vigorous, healthy woman—had the right regimes been advised at the right time.

Lack of exercise also conduced to the other bane of the second half of Mary Ward's life—insomnia and its attendant 'depression'. But the main causes of her 'bad nights' (as Dorothy called them) were overwork (particularly in the critical phases of writing) and—more chronically—money worries. Sleeplessness was always a problem during Mary Ward's adult years. Even in Oxford in the 1870s it was her habit to stay up into the small hours working, after the children were in bed. At this period—in the strength of her young womanhood—she evidently took pride in her stamina. But insomnia became a worry during the London years (when Humphry's 'nightwork' disturbed her sleep) and a never-ending torment during the financial disasters of 1913 and after. From 1888 onward, Mary Ward routinely dealt with insomnia by the use or abuse of sleeping draughts, prescribed by the more easy-going of her physicians. Her brother Frank and her relative by marriage Harry Huxley were notably free in providing her with powerful sedatives. Chlorodyne and trional boosted by aspirin as a pain killer were Mary's potions of choice although *in extremis* (as in 1906) she would use morphia for weeks on end. In later life, her doctors seem to have weaned her somewhat from the stronger soporifics.

Despite all these long-standing constitutional ailments, Mary Ward outlived most of her siblings, and all her sisters bar Ethel, who was never subjected to the wrenchings of Victorian child-bearing. But various accidents over the years combined (together with incompetent doctoring) to keep Mary Ward permanently below par and not infrequently brought her to death's door. In

December 1895, for instance, she informed George Smith that she needed a small operation to rectify 'an ailment, originally due to a slight accident'.[13] Mrs Ward was understandably reluctant to go into physical detail with her publisher about the nature of an intimate problem. But to her father in November 1895 she explained that 'I have developed a slight new ailment, akin to piles, though not piles, which makes all sitting painful.'[14] One assumes that the condition was a fissure. She had surgery in January 1896. As Janet Trevelyan reports, the operation was 'clumsily performed .... After it, she lay for days in such pain as the doctors had neither foreseen nor prophesied, while the nervous shock of the operation itself was aggravated, one night, by the antics of a drunken nurse, who came into her room with a lighted lamp in her hand and deposited it, swaying and lurching, on the floor.'[15] Worse still, the operation did not work. By December 1896, the complaint was as bad as it had ever been. The outcome was an incurable disability which caused Mary Ward low-level pain and private embarrassment for the rest of her life. Her home remedies entailed an overdependence on aperients. Excessive diarrhoea needed then to be controlled by such binders as bismuth. And by her last decade the walls of her stomach had become ulcerated by the excessive intake of bowel-controlling drugs. Anorexia and malnutrition were the further consequences of the cycle.

In what was a continuously precarious state of health small domestic accidents which would have been of no account to a healthy middle-aged woman carried great risk for Mary Ward. In summer 1902, she stumbled while visiting her sister Julia at Godalming and fell heavily on her right side. For six weeks she was disabled during which, as she told Willie Arnold, 'all writing is labour and pain'.[16] It was not pain she could put off. She was concluding *Lady Rose's Daughter* at the time and was harried by deadlines. The effect of such accidents was to make Mary Ward even more nervously valetudinarian and anxious about the dangers of living a normal life with its normal knocks. She wrapped herself in cotton wool.

The most intractable and puzzling of Mary Ward's adult illnesses was what she termed 'side', or with grim humour, 'the old enemy'. Its first symptoms made themselves felt in 1892, with a few weeks of 'obscure neuralgia ... in the right breast and side'.[17] Her London doctor (then Mr Hames) vaguely diagnosed 'rheumatic gout' and 'internal catarrh'. Starvation was prescribed. The diagnosis was later refined to that of 'floating kidney' and 'loose liver'. The pain meanwhile intensified and laid Mary Ward so low at the beginning of writing *Marcella* in August that she seriously anticipated death. Kidneys do not, one is now told, float or come adrift from their internal moorings. This imaginary ailment conveniently covered a range of enigmatic symptoms some (as probably in Mary's case) the result of delayed complications from childbirth.

The 1892 attack of 'side' subsided after a few weeks' bed-rest. But another and even more agonizing attack occurred in May 1896 on the Wards' journey through Switzerland homeward from Cadenabbia. Motion—particularly the shaking of train rides and sea voyages—were horrific things for Mary Ward for a few years after. (It was fear of 'side' that prevented her from capitalizing on her American popularity with a visit to the country in 1895.) But on this occasion in 1896, the pain was worse than anything Mary had previously known. In the Göschenen Tunnel she was convinced she was going to die and in fact fainted. The whole right-hand side of her body, including her leg, was affected. The Swiss doctor she consulted in Lucerne wisely prescribed large doses of morphia and brandy for the immediate distress and less wisely diagnosed internal neuralgia of the ilio-inguinal (i.e. groin-hip) and upper sciatic nerve as the cause of the pain. 'A very rare condition',[18] he somewhat gratuitously added. Kidneys, he determined, had nothing to do with it. He ordered Mary to wear a patent surgical corset designed to lift the weight of the upper part of the body off the affected nerves.

'The old enemy' continued to strike at regular intervals during the rest of her life. But Mary Ward became expert at fending off the worst of its effects with her private store of pain-killers. And despite its visitations she managed to travel to the Continent several times a year and even dared a visit across 'that terrible Atlantic'[19] to America in 1908. During the 'side' attacks, which lasted from a few days to weeks, Mary would run a moderate temperature of about 100 degrees. She would have 'disagreeable diarrhoea' and would feel as if she 'had a skin too few'. This was all bearable enough. What was unbearable was the intermittent spear-like jabs of pain. In the climax of her drugged agony, Mary Ward would be so 'stupid' that—as with Sarah Orne Jewett in August 1892—she could not physically recognize people who came to Stocks to see her.

Despite scores of diagnoses and treatments (which included at different times raw beef sandwiches and all-milk diets) the nature of the complaint—which returned with dreary regularity—remained obscure. As late as August 1918, Dorothy Ward noted in her diary, 'Quite suddenly and unexpectedly Mother has been laid low with a violent attack of "side"—only *how* it sets one wondering, in fresh suspense, what "side" really is?'[20] Dorothy was evidently coming round to thinking it must be linked in some way to her mother's heart disease. A year later—in August 1919—the riddle was finally solved. Gallstones. Frank Arnold told Dorothy that he 'could now feel the lump of them ... quite distinctly'.[21] In Mrs Ward's increasingly feeble condition he was against an operation.

Given the remote nature of this disorder's symptoms—which can afflict points as distant as the sufferer's shoulder—gallstones may conceivably have been the root cause even of her 'writer's cramp' as well as of myriad other malaises over the years. The stones are, of course, worsened rather than dissolved by the high-cream dairy foods and egg custards which formed the staple of Mary's diet. (The most common gallstones are actually formed out of cholesterol.) The Victorians had various effective treatments for gallstones, and, had they been treated early on, the last twenty years of Mary Ward's life would have been blessedly easier. It is surprising that the condition was not at least guessed at before 1919. Edwardian medical authorities reckoned that women made up 75 per cent of sufferers, and cited tight lacing and post-childbirth complications as causative factors of gallstones. On all counts, Mary Ward was a prime candidate.

The other major health problem which afflicted Mary Ward in later life was eczema—a condition which doctors now term exfoliative dermatitis. This irritation of her exposed skin began gradually in the late 1890s and early 1900s. As Victorian doctors noted, eczema often coincided in women with the onset of menopause. Like asthma, it is thought often to have a nervous origin. With Mary Ward, it came on insidiously. In Italy in March 1903 Dorothy recorded in her diary that 'walking or driving against wind for only a few minutes brings on a prickling and makes [Mother] feel as if she were feverish all over'.[22] By 1905, the condition had become the 'plague' of Mary Ward's life, more so even than 'side', which at least offered remission between attacks.

The eczema grew into a never-ending torment. It attacked Mary Ward's arms primarily—covering them with a mass of white blisters which after maddening irritation swelled up, ached horribly, and eventually scarred. By 1908 the condition was capable of erupting anywhere on her body. Mary Ward sadly noted that even her face was scarred by the spread of the irritation. This eczema was complicated at its height (particularly in 1906) by styes and boils. Benzol-based lotions, and zinc ointments seemed if anything to aggravate the condition. Sun, wind, cold, anxiety, tiredness were all irritants which could trigger off a full-blown attack. Mary Ward's dermatitis remained obdurately incurable to the end of her life, inevitably flaring up at moments of stress. And it was a major factor in her disastrous health breakdown of 1906, after which she was never a well woman again.

Over the years Mary Ward became an expert amateur apothecary and travelled everywhere with a portable medicine cabinet, full of pills, lotions and a 'thousand-and-one little drugs'.[23] This private drugstore evidently contained her currently favoured sleeping draughts and such all-purpose pain-killers as the addictive sulphonal (which she often administered to her family). The mainstay was the analgesic and febrifuge salicylate (later trade-named 'aspirin'). This she routinely took 'for any acute pain'. In 1892, she became enamoured of phenacetin ('a wonderful drug'),[24] to which she attributed her rapid recovery from the attack of 'side' that preceded the writing of *Marcella*. Janet Trevelyan dates her mother's passion for pharmacy as beginning with this miracle cure: 'phenacetin and all its kindred "tabloids" came into common use at Stocks from that time onwards, in spite of the mockery of her friends. Mrs Ward developed an extraordinary skill in the use of these "little drugs" and would often baffle her doctors by her theories of their effects.'[25]

There were some effects neither she nor her doctors knew about. Phenacetin (acetophenetidin) is an aniline derivative. It has strong anti-neuralgic properties, but if taken in large doses the drug can cause kidney damage, which is why it was withdrawn from the British market in the 1960s. It is likely that in her enthusiasm for phenacetin Mary Ward took excessive amounts of the wonderful drug at a time when her kidneys (floating or not) were suspected of being disordered. Around 1900, she is also recorded as taking large doses of quinine—sometimes under the direction of Harry Huxley. He may have assumed that like Humphry she had caught 'swamp fever' (i.e. malaria) during one of the the Wards' stays at Venice in the early 1890s and that this was an underlying cause of her chronic low fevers and lassitude.

Morphia (injected or taken orally) was not a tightly restricted drug in the late nineteenth century. Mary Ward knew it well. She had administered it to her dying mother and was profoundly grateful for the opiate's power to ease suffering. Her experience with her terminally ill brother Willie Arnold was less reassuring. Willie—suffering from locomotor ataxia—was first prescribed morphia by Harry Huxley in June 1897, to ease his spinal pain. In Manchester, his brother Frank continued to prescribe ever larger doses (in the context a humane act). By 1898 Willie was wholly addicted. When—in the last stages of his illness—he was moved to London, the specialist whom Mary had hired decreed that the dying man be weaned from his 'morphia habit'. Willie's withdrawal agonies were heart-breaking and ultimately unavailing. By 1903 his morphinism was as bad as ever—despite the 'moral degeneracy'[26] which Mary warned him about.

Mary Ward herself used morphia on many occasions, to deaden the pain of 'side'. But on the whole—having seen its effects on Julia and Willie—she seems to have been wary of the drug's power and regarded it as an emergency measure. She had no such compunctions about cocaine, which like many Victorians she conceived to be a fairly harmless analgesic and pick-me-up. It is evident that she kept supplies in her private pharmacy most of her life. Cocaine was first extracted as a chemical from the coca leaf in the 1850s. But up to 1883, only a few pounds were available in Europe and it was regarded as a purely experimental substance. In 1884, for the first time, several tons were imported from Peru as a commercial venture. Cocaine swamped the middle classes of Europe; young men took it openly at the opera, as a smart refreshment. Mary Ward—never one to lag where new drugs were concerned—is first found mentioning cocaine in correspondence with her mother in October 1887. It 'works like magic', she observed. Ten years later,

Mary Ward gave a nerve-racking public lecture at Glasgow on 'The Peasant in Literature'. The audience was 800-strong and the event was a major ordeal for her. She got through it, as she told Willie, only 'thanks to cocaine, egg and port and other amenities'.[27] A few years later, staying in Paris, Dorothy noted that after the journey 'Mummy felt very bad, however her cocaine lozenges saved her.'[28] On a suffering family friend in April 1905, travelling with them to Paris, Mary Ward's cocaine lozenges were said to have 'the most wonderful effect'.[29] On another occasion, Dorothy casually mentions an attack of 'side' so sharp that even 'Mother's cocaine powder' did not help, which suggests that Mary used the drug routinely whenever this recurrent ailment was severe. The alternations of extreme lassitude and the furious bursts of writing and committee work characteristic of Mary Ward's later life may well have been aided by the intermittent use of cocaine, particularly as a counteractive agent to the sleeping draughts which she took nightly during the crises of novel writing, family worry, or social campaigning.

The sum effect of all Mary Ward's ill health and disease was chronic debility and exhaustion. Dorothy almost every week records her mother as 'tired', 'dreadfully tired', or 'really tired and upset'. On such occasions, Mary would miss family dinner—or on some occasions leave abruptly during the meal—and retreat upstairs with the faithful Lizzie. (It was not until well into her twenties that Dorothy Ward had free entrance to her mother's bedroom suite.) For whole days, her family would not see her except for ritual brief visits and greetings. After 1900, Mary Ward's official activities at the settlement, at play centres, at invalid schools, and at her innumerable committees were routinely delegated to Bessie Churcher (who became much more than a personal secretary) and to Dorothy. In her extended spells of convalescence she would be totally immobile and would spend the warm part of summer days lying on her *chaise longue* on the lawn at Stocks, being pulled around in a 'lugchair' or pushed in a 'bathchair' and physically carried up stairs by a servant—all the time on eggshells lest she be 'shaken'. Every effort—however slight—carried its risk. In July 1898, for instance, Dorothy noted fatalistically, 'Poor M., after being much better, suddenly got worse before lunch and stayed upstairs till dinner. She attributes it to having strained something by shutting a window in her sitting room.'[30] To forestall such accidents, Mary Ward was waited on hand and foot.

Towards the end of writing her novels, Mary Ward's brain would be 'all wool'; her right arm would be locked with cramping pain; her side would throb; her skin would be an envelope of flame; her stomach and digestion would be disordered; sleep would be narcotic torpor; but still she would write. At the end of it all she would collapse entirely with physical exhaustion and depression. Her doctors would insist that she be carried off to recuperate, preferably in the Mediterranean sun or the Tyrolean mountains. She would be carried to the boat train in a mass of cushions, blankets, shawls, furs, and fussing helpers. Over the period of the next few weeks, she would gradually regain the power of walking. Her letters home would become sprightly. Then she would begin to think about the new novel and sketch out ideas. A head of energy would build up. And so the treadmill would begin again.

Illness of Mary Ward's kind was a full-time occupation. Had she been a less resourceful woman she could have done nothing but suffer and convalesce. That she managed to do so prodigiously much else in the literary, social, and family way is a wonder. Partly it was the outcome of using her illness constructively. In the first instance, invalidism was a good reason for not having more than three children. Her mother had borne up to her forty-first year. Mary could argue that she was too frail to fulfil any such biological destiny. When the Wards moved to London, Humphry was landed with long periods of night work at Printing House Square and even after he became art critic was often obliged to spend the night in his office. This—together with the establishment of her little work room in Russell Square—gave his wife considerable space (and time) of her own within the home. When the Wards moved to larger dwellings, she increased her territory to a small suite of rooms. Being an invalid allowed Mary to set up an inviolable zone of privacy to which she could always retire pleading indisposition.

Much of Mary Ward's time at home was thus spent in seclusion—with her family but not among them. Routine women's duties were naturally excused her. No one expected Mrs Ward to interview kitchen maids, discipline 'dr**k' cooks, or make out invitation cards and flower arrangements for dinners or 'Thursdays'. These were Miss Dorothy Ward's chores (Miss Gertrude Ward's before her). Should her mother need new 'silk undergarments' or leave her handbag somewhere in London, Dorothy was dispatched by taxi or underground to fetch them. Should Humphry Ward need a female companion on his travels—Dorothy was always on call.

Mary Ward's frailty ensured that a large part of the family income (most of which she earned anyway) necessarily went on her entourage and comforts: the personal maid, Lizzie Smith; the personal secretary Bessy Churcher; the devoted spinster daughter, Dorothy, who required a standard of living higher than that of a mere female companion. The chauffeurs the Wards employed after 1905 were exclusively for Mary's benefit. Her love of flowers meant extra expense on a skilled class of gardener at Stocks. She had doctors in the country and in town and after 1892 regularly sought second opinions and consultation with specialists. All this cost money. Mrs Ward always travelled first class. (When she was alone, Dorothy travelled third.) The life-threatening nature of her ailments gave her in this way a power of passive domination over her family. They (particularly Humphry) never questioned anything that was 'good' for her. Their annual vacations were timed to fit her authorial needs for recuperation, stimulation, and convalescence. Even the Wards' shuttlings between Stocks and Grosvenor Place were primarily geared to suit the rhythms of her social and literary work. The family's movements revolved around her as ailing materfamilias, always with the unspoken fear that, should they cease to do so, she might die. And then where would they be?

18

The Passmore Edwards Settlement: 1892–1900

NOT one person in a thousand entering what is now called Mary Ward House at 42 Queen Square, Holborn, knows who she was. Not one woman in a hundred thousand who gratefully drops her child off at the morning play-centre knows that this same Mary Ward founded the service which uniquely liberates the modern working mother. Of her more than any other Victorian novelist (excepting Anthony Trollope and his pillar boxes) may one say, if you want monuments look around you: not in the libraries, but in the London streets, any day of the working week.

The Victorian settlement movement—of which Mary Ward House and Toynbee Hall are the most famous relics—took off in the 1860s, when a younger generation of clergymen threw themselves into the 'abyss' of working-class London. Foremost was J. R. Green. In 1863, at the age of 26, Green discovered Carlyle, was inspired, and voluntarily took charge of a derelict parish in Hoxton. He retired on the grounds of ill health before the end of the year, but returned in 1864 as a 'mission' curate to St Philip's, Stepney, where he remained until 1869—nominally a priest but primarily a social worker among the unfortunates of the East End. Meanwhile, 'General' William Booth (for whom Green like most Anglicans had little time) had begun his first meetings in a tent on the Mile End Waste, in 1865. Within a decade his Salvation Army would muster a quarter of a million Christian soldiers. And the issue of the forgotten underclass of the capital was given huge publicity in 1869 by James Greenwood's *The Seven Curses of London*. This tract, with its shock-horror rhetoric, whipped up one of England's periodic hysterias about the socially disadvantaged.

Something must be done. But what exactly? J. R. Green was convinced that what should not be done was simply to give the poor bigger doles of money and provisions. He despised the 'relief' by which the middle classes squared their consciences with annual £5 notes, coal, and soup. Green believed passionately that the middle classes must themselves work and—more importantly—reside among the poor if they wanted to help the poor. His belief was shared by his Oxford friend Edward Denison, who settled in squalid lodgings in the Mile End Road in 1867, so as to learn the facts of social inequality at first hand.

In 1868 Ruskin summoned Denison (now a Radical MP), J. R. Green, and the Revd Brooke Lambert to a meeting at his house in Denmark Hill.[1] The young men put forward, for the first time, their idea for London settlements—colonies of middle-class pioneers who would live voluntarily in slums, with the aim of civilizing their uncouth neighbours. There was no direct outcome of this meeting; largely because of the physical frailty of the principals. Green had contracted tuberculosis and was headed for lifelong convalescence. Denison had also ruined his health living in the East End, and died in 1870—the first martyr of the movement. Lambert had a health collapse in 1870. Ruskin was a visionary, not an organizer.

The task of realizing the settlement ideal passed on to a still younger clergyman, Canon Samuel Barnett. In 1872 (aged 28) Barnett took over St Jude's in Whitechapel—legendary as the most criminal and depraved parish in England. Barnett was physically indestructible and an organizer of genius. He and his young wife Henrietta set up education classes for adults and children, a penny bank, a maternity society, a parish library, annual flower shows, art exhibitions, and concerts. The Barnetts civilized their cockney savages to a remarkable degree. St Jude's became a beacon in darkest London.

Among everything else, Barnett was a born propagandist and did not hide his light under any bushel. During Eights Week in 1875, he made the first of a series of proselytizing visits to Oxford. Initially, he was invited by a school friend of his wife's, Gertrude Toynbee, sister of Arnold. Arnold Toynbee was a brilliant young economist who is famous as the inventor of the concept 'industrial revolution'. He was also a protégé of Benjamin Jowett's and of T. H. Green's, whose ideals of 'active citizenship' he had imbibed. Arnold Toynbee became a passionate convert to what the Barnetts were doing in London and made several visits to St Jude's—although his delicate health could not stand the strain of prolonged stays in the East End.

The settlement idea finally took off in 1883. J. R. Green and Arnold Toynbee both died in March of that year—two more martyrs for the cause. Toynbee's last public statement was a moving apology to the working classes of England: 'We—the middle classes, I mean, not merely the very rich—we have neglected you; instead of justice we have offered you charity, and instead of sympathy we have offered you hard and unreal advice.'[2]

Emotions ran high in Oxford with the deaths of its alumni Green and Toynbee. It was decided to set up a series of memorial 'Toynbee Lectures', to be delivered to the populations of England's large towns. This, however, was a mere gesture. Action was called for. What that action should be was outlined on 17 November 1883, when Barnett read at St John's College his paper 'Settlements of Working Men in London'. Effectively, it called for a new 'Oxford Movement' to regenerate England. And specifically it projected a non-denominational 'college', to be set up by the young men of Oxford in the East End of London. Barnett enlarged on the idea in an address to the Union in December 1883. Toynbee Hall duly opened in the Hilary Term of

1885, on the site of a disused boys' school near St Jude's in Whitechapel. The premises had been bought for £6,250 and were rebuilt in the neo-Elizabethan style favoured by Victorian philanthropy.

Barnett was employed as warden of the new Hall, at a salary of £250 a year. There were thirty rooms accommodating seventeen 'residents'. Middle-class intellectuals, writers, social workers, aspiring clergymen, rabbis, and politicians all made up the shifting resident population. Activities in the centre were organized around university extension lectures, *conversazione*, a 'Denison Club'—devoted to social debate, cricket, chess, even tennis (once a court was laid down), reading parties, a 'Shakespeare Society', and numerous self-help groups. Mrs Barnett, for instance, ran a Children's Country Holiday Fund, which gave thousands of deprived youngsters their first experience of rural England. It was Barnett's ultimate dream that Toynbee Hall would one day become a great East London University.

This then was the background to Mary Ward's own foray into settlement-making. She had been a particular protégée of the two Greens and of Jowett in the 1870s. Mrs Ward had, in fact, offered her own public memorial to 'JRG' and Toynbee in the character of the hero of *Robert Elsmere*. Humphry on his part had been active in university extension lecturing from its start, and had strong childhood connections with London's East End (J. R. Green had once been his father's curate). There was every reason why Mary Ward should have made herself active in settlement work once she became famous in 1888.

But why did she not throw herself into support for one of the two settlements (Toynbee Hall and Oxford House) which Oxford had already set up? Why start her own? An obvious answer is the egotism that consituted so much of her moral and creative energy. However much she might laud self-sacrifice in her novels, Mary Ward was not a follower by nature. And there were other reasons for avoiding the two pioneer settlements. Oxford House—which had been founded by Keble College—carried the kind of doctrinal baggage which Mary Ward disliked. It was too churchy. In principle, Toynbee Hall was more to her taste. But recently, at the end of the 1880s, it too had become a hotbed of a doctrine which Mary Ward also disliked—socialism. Barnett had been openly sympathetic to the dockers in their great 1889 strike for 'the docker's tanner' (i.e. a minimum wage of 6*d*. an hour). Toynbee men took a direct supporting role in the strike of busmen, in June 1891. Organized in the East End by the fiery socialist John Burns, this dispute had led to violence and intimidatory picketing. During the 1890s Toynbee Hall gradually became associated with militant causes like suffragism, Home Rule, and support for the Boers, all of which were anathema to Mary Ward. She never actually broke with Barnett (he helped interview potential wardens for her settlement, for instance).[3] But she could not but see the direction which Toynbee Hall had taken as a betrayal of its Jowett and T. H. Green principles of impartial, apolitical goodness.

The first recorded hint that Mrs Ward might found a settlement on Toynbee lines is found in a letter to her father in May 1888. Henrietta Barnett had forwarded to her a letter about *Robert Elsmere* from 'a genuine working man—though evidently one who has raised himself . . . . It has interested me a great deal,'[4] she added. Mary Ward kept in touch with the Barnetts and visited Toynbee Hall in summer 1889. She was struck by the fact that in the library '*Robert Elsmere* had been read to pieces, and in a workmen's club which had just been started several ideas had been taken from the "New Brotherhood" '.[5] By now, her ideas were taking solid shape. On 11 November 1889, Gertrude Ward recorded in her diary that 'Lord Carlisle came and had a long talk with M. about a proposed Unitarian Toynbee somewhere in South London' and a few days later that 'Mr Stopford Brooke came and had a long talk with her about a "New Brotherhood" they hope to start with Lord Carlisle and a few others to help.'[6] In February 1890, Mary wrote to her father that 'I have been very busy with a project of a sort of small Toynbee Hall in London on Unitarian lines . . . . The basis of it is to be definitely religious—Christian Theism.' The venerable Unitarian James Martineau had given his name 'after much hesitation'.[7]

South London was certainly a possibility. It did not have the romantic squalor of the East End that was so irresistible to novelists. But the slums around the Elephant and Castle and Walworth Road were as bad as anything north of the river. Mrs Ward, however, had other ideas. For some time she been very taken with her friend Stopford Brooke's Bedford Chapel in Bloomsbury. Brooke's 600-plus attendances (which included many self-improving working-class people) suggested that there was an opening for settlement work in West Central London.

There were the added attractions that the Wards lived themselves at this time in Bloomsbury and that there were ideal premises available: University Hall in Gordon Square, by University College. Built in the 1840s the hall had originally been a residence for Unitarian students. University Hall had never been used to capacity and with the establishment at Oxford in 1888 of the wholly Unitarian Manchester College it had fallen into disuse. The property was disposed of by UCL to the neighbouring Dr Williams's Library (who still own the gothic-looking premises). They were prepared to rent this gloomy dormitory and its cavernous common rooms to Mrs Ward and Dr Martineau for the purposes of a residential settlement. The accommodation was not ideal; but University Hall had the advantage of being at the frontier of three segments of West London's population: the professional middle classes (people like the Wards or the Leslie Stephens); the poor, whose St Pancras and King's Cross slums started on three of Bedford Square's edges; and the University of London—the 'godless place in Gower Street'. It made for a potentially fruitful social mix.

Having decided to found her settlement, Mary Ward moved with great speed. A powerful and non-sectarian committee was formed in February 1890. In addition to Brooke (still the darling of intellectual Anglicans), Carlisle (a Unionist peer), and Martineau (the most respected Unitarian in the country), the committee contained three women of very different character: Mary Ward, the feminist Frances Power Cobbe, and the upper-crust Dowager Countess Russell. The axis was perfect. Committee meetings took place at 61 Russell Square, and were dominated by Mary. 'The rest of us were simply admiring and sympathetic spectators of her enterprise and zeal,'[8] one observer later recalled. By March 1890, a circular was issued. The principal aim of University Hall would be to encourage 'an improved popular teaching of the Bible and the history of religion in order to show the adaptability of the faith of the past to the needs of the present'.

Although Mary Ward resolutely declined to ally herself formally with the Unitarians (as some of her committee wanted her to) University Hall appealed to this notably open-handed community. The first subscription raised a covenanted income of £700 for three years and amassed a working capital of £500. By September, Mrs Ward had selected a Unitarian minister, Philip Wicksteed, to be the first warden. A Dante scholar and later a prominent (if theoretical) socialist, Wicksteed had needed some coaxing from Mrs Ward to accept the position and was, in fact, never to be entirely happy in it. At the final interview, she greeted him with the jovial words, 'I want to *wrestle* with you.'[9] By October, the first ten residents were chosen and an impressive curriculum of Sunday lectures set up.

A formal opening ceremony was scheduled for 4 o'clock on Saturday, 29 November, at the Portman Rooms in Baker Street. Mrs Ward agreed to make the main speech. In itself, this was no great ordeal. All that was required was to trot out her well-worn arguments about the reconstruction of Christian belief on rational not superstitious bases, etc., etc. But Mary Ward—40 years old as she was—had never spoken in public before; not to so much as a drawing room full of friends. There would be 450 people present—a veritable horde.

She was, as her daughter Janet remembered, 'terrified at the prospect' and practised neurotically in the empty hall. Dorothy recorded the event itself in her diary: '*The* day of the meeting. Dearest Mother's paper made, I feel sure, a great impression this afternoon. She read it beautifully. It was an awful sensation though, when she got up to begin. I almost felt as if it were myself!'[10] Dorothy Ward was not her mother's sternest critic, but for once she was right. The paper—delivered in her low but curiously penetrating voice—did make a great impression. More so as it followed a rambling and inaudible address by Wicksteed. 'As I sat down', Mary told her father, 'I was conscious that I had got through my task better than I could have expected.'[11]

Mrs Ward basked in congratulation at the party and musical entertainment which followed at University Hall. And the success of that afternoon launched her on a career of public speaking (always faithfully attended by 'Dots' in the wings or front row). By November 1892, she was serenely addressing 4,000 people in Manchester Free Trade Hall—treading in the footsteps of great orators like Bright and Cobden. And when the anti-suffrage forces recruited her in 1908, it was her platform skills rather than her pen that they most wanted.

Over 1891–2, the ostensible activity of University Hall revolved around courses of lectures, principally on biblical criticism. They did not set Bloomsbury alight. And those who came to hear the 85-year-old Dr Martineau quaver about the Gospel of St Luke or Professor Knight expatiate on the intellectual case for theism were overwhelmingly the middle classes rather than the poor of St Pancras or the young counter-jumpers employed in Maple's and the other department stores in Tottenham Court Road. At University Hall the genteel talked to the genteel and took genteel cups of tea afterwards before returning to their genteel Bloomsbury terraces. Wicksteed later confessed to Janet Trevelyan that in these early years 'I was uneasy all the time.'[12]

Things were brought to a head by the residents (sixteen in number by April 1892). Increasingly they balked at the role imposed on them by Mrs Ward's management. They did not want to be 'students' and reconstruct Christianity. They wanted to reconstruct London. Some of them were sullen in their demeanour to Mrs Ward and her distinguished visitors. And soon they were beyond sullenness into rebellion. As early as winter 1890–1, the residents (under the leadership of Alfred Robinson, an idealistic lawyer in his early forties) raised their own funds to take over a small building in the slum neighbourhood behind Tavistock Square. With conscious irony, they called this unimpressive edifice Marchmont Hall. It comprised two small rooms which could hold, at a squeeze, some 150 people. Against some initial resistance from locals, the residents set up a programme of social activities and entertainments, modelled on what Toynbee Hall did.

Marchmont linked itself to the local community by a system of 'associates'—fifty working men and women from the neighbourhood who were made to feel that they had a direct stake in the place. There was no warden; no centralized authority or management. The unofficial hall was secular and highly—not to say embarrassingly—successful. Two things seem particularly to have been valued by the ordinary King's Cross people who used Marchmont and avoided University Hall. There was no preaching. And there was no middle-class atmosphere to make them feel out of place. Working mothers would call in for a moment's relaxation and gossip with their baskets heavy from shopping. 'We had the loveliest time there, damp and all,'[13] one recalled thirty years later. Children were parked there on Saturday mornings and almost spontaneously recreation was devised for them. Boys' and girls' clubs were organized on week nights. An old piano was installed. Over classless tobacco and tea the residents could jaw endlessly about Marxist theory with actual labouring men. At night Marchmont functioned as a jolly pub without beer.

It was all very vexing for the warden of University Hall, 500 yards away. Marchmont was bursting at its grimy seams, while he was having difficulty respectably filling the large lecture-room when Mrs J. R. Green lectured on the development of English towns, or whatever. Mary Ward was equally vexed. She made some unsuccessful attempts to bring Marchmont to heel but the truth was the place rather overawed her. And her touch faltered when it came to dealing with the poor on equal footing, without the ceremonies of organized philanthropy. She came over as too much the *grande dame*. Thus in November 1891, while wintering in Florence, she told Arnold (at Eton) that 'I am going to amuse myself with making a photograph book for Marchmont Hall of Italian Cities. Then at Christmas I can have the boys in detachments [i.e. at University Hall] and talk to them about it.'[14] Actually walking herself across the Gordon and Tavistock Squares to visit Marchmont Hall was psychologically difficult for her. She gave her first address there on the night of 30 May 1892 (a full eighteen months after Marchmont had opened). Again, she seemed to think that they would be fascinated by her luxurious holidays in Italy. She talked for an hour and a half (far too long, she later realized) and showed lantern slides of Florence. 'I think it went well,'[15] she dubiously told Arnold. It is likely that the *de haut en bas* exhibition of her holiday pictures went down rather badly—at least with the residents.

Marchmont Hall was manifestly a rebellion against Mrs Ward's leadership. And she was hampered in putting down this rebellion by the fact that her leadership was so remote. Toynbee Hall had the charismatic Canon Barnett in attendance. He did not spend long weekends at his country house writing fashionable novels and giving shooting parties to the aristocracy. Nor did he winter in Florentine *palazzi* when London turned chilly. Mary Ward's deepest instincts—even from the first—were to hold aloof from the hurly-burly of the settlement; to retreat into a protective, above all *moneyed*, life-style. When the Wards took over their new London apartment in Grosvenor Place in November 1890, Mary had noted as a main attraction of the place that 'it would be well for me to be somewhat further from the Hall ... for my family life and my literary work ought to come before it'.[16] Bluntly—let others settle in the settlement, and 'reside' with the poor as Toynbee, Denison, and J. R. Green had advocated.

Unable either to absorb or abolish Marchmont Hall, Mary Ward decided in 1893 to accept it and redesign the settlement anew around its manifest success. Henceforth, University Hall (under some new name and new dispensation) would dedicate itself to 'practical and social work' of the kind pioneered by the residents. It was a sensible, flexible, and extraordinarily courageous decision on her part. And it released her to do what she did best: run committees and charm money out of rich and powerful people.

Mary Ward's achievements for her settlement over the years 1893–7 were astounding (even more so when one considers what else she was doing). As a first step, an associates' programme was inaugurated on the Marchmont model, to broaden the base and create a buffer between the residents and the organizing committee. There were 160 associates by July 1894—mainly members of the local working-class community, entitled by their small subscription to privileged use of the settlement facilities. Mrs Ward embarked on a fund-raising lecture tour which took her all over the country, particularly the north, where Unitarianism was entrenched. The target at this early stage was £5,000, for new buildings. (Eventually they were to cost three times this sum.) It was not easy work, as there were now twelve settlements in London, all worthily competing for money.

Five- and ten-pound notes from friends and the odd hundred from her publishers all added up. But what Mary Ward most urgently needed was single large benefactors. She had no luck with with the Scottish American Andrew Carnegie; but by sheer charm and persistence she persuaded the veteran philanthropist Passmore Edwards (enriched by newpaper and magazine proprietorship) to back the new settlement. Edwards was an aggressively self-made man who liked endowing secular libraries and reading rooms. The workers' education aspect of the proposed new establishment appealed to him as something that might foster a new generation of Passmore Edwardses. His first offer in May 1894 was £4,000 for the building, so long as some other donor would provide the site.

Mary Ward promptly went to work on the Duke of Bedford, the biggest landholder in Bloomsbury. After several visits to Woburn, the expenditure of vast amounts of flattery and infinite frustrations ('Dukes are delusious,'[17] she wrote wearily to Humphry in August 1894), Bedford eventually made over a large plot of land (1,140 square yards) on the south east corner of Tavistock Square. The site was symbolically nearer to Marchmont than University Hall. The terms—extraordinarily generous for such a prime site—were 999 years leasehold for £5,000.

By early 1895, the socialist politician Graham Wallas and the architect Norman Shaw were helping Mrs Ward draw the specifications for the new settlement. It should be five storeys high but must not look an arrogantly high building. It would have electric light (unlike University Hall's gas). It would have a library dedicated to the memory of T. H. Green and facing the quiet gardens; a workshop; classrooms; a smoking-room (its walls 'adorned with paintings representing outdoor sports'); a large dining-room with folding tables from Heal's which would allow it to double as a lecture room; a great hall on the first floor over the front of the building capable of seating 500 and with acoustics suitable for concerts; a gymnasium in the basement; hotel-standard kitchens; accommodation for eighteen residents and further accommodation for servants for the residents.

Using her now practised 'wrestling' techniques, Mary Ward persuaded Norman Shaw (who was not keen) to judge a public competition for the design of the new building. It was won by two young architects, A. Dunbar-Smith and Cecil Brewer (both former students of Shaw's, as it happened). The result was publicly announced to the cheers of the residents in the large lecture room of University Hall on 8 March 1895, the two young men appearing very white-faced and '*emotionnés*'. Shaw thought their Voyseyish plan had 'the touch of genius'—a judgement which posterity has confirmed. A rather grumpy Passmore Edwards wished 'we had entrusted the whole thing to Norman Shaw himself'. Mary added sardonically (and privately): 'Très bien!—if he had been willing to pay for it.'[18]

She was, in fact, becoming very adept at making Edwards pay for things. His commitment was jacked up to £7,000 then to £10,000. By June 1895 Mary informed her father that they had £19,000 collected or in pledges and needed another £3,000 for furnishings. As the estimates ballooned still further, she threw herself into a furious last round of subscription drives during 1896. In January 1897 Philip Wicksteed (to his immense relief) was replaced by a new warden, R. G. Tatton. A Balliol man (a former Fellow) who had worked for some years in London education, Tatton was particularly interested in university extension work. He was also made of tougher stuff than his predecessor. He bluntly told Mrs Ward that—as the best-paid author in England—it would not be amiss if she donated some cash herself to the new settlement. Rising to the challenge she immediately pledged £1,000 (although such were her outgoings at the time that she could afford it less easily than Tatton thought). On his part, Edwards gave ever more money, amounting to a whopping £14,000 by opening night. He allowed that the new institution should in gratitude bear his name. The settlement's other main patron, the Duke of Bedford, gave £1,500, the Bloomsbury site on a generous lease, and the use of two large adjoining gardens at a notional rent of £10 per annum.

The Passmore Edwards Settlement was opened in two phases. The first opening on the evening of 10 October 1897 was primarily to advertise the new building and its facilities to the local community. The associates (now 250 strong) distributed 8,000 leaflets round the area. The hall was packed on the night, with all its 500 seats taken. Mrs Humphry Ward's entry on stage was preceded by the overture from *Die Meistersinger*. She went on to give the main address of the evening, a long lecture on 'Social Ideals'. Essentially it was an austere restatement of her familiar religious and ethical positions. But she prefaced her remarks with a graceful tribute to the residents sitting in the front row: 'What unites us all here tonight [is] the real success, so far as buildings and funds allowed, of the social work carried on by the old hall in Marchmont Street. That success has mapped out our path for us'.[19] That success, and the failure of her original University Hall idea, she might ruefully have added. They had won and she was large-minded enough to submit to their victory. It was not complete surrender, however. Mrs Ward sternly reminded the Residents that the new hall was not 'committed to a party or a school of economic affairs'. Social work yes; socialism no. As they left the new building, Mary Ward turned and 'for the first time we saw the beautiful windows of the Hall shining out into the London dinginess like a beacon of light'. She loved such symbolism and it was, she told Willie, her 'thrilling moment'.[20]

Mary Ward was by now a master of publicity. There were twelve settlements currently in London and she must get high visibility for PES or it would decline into drab insignificance. A second grand opening was organized for the night of 11 February 1898. It was shrewdly calculated to attract the attention of the national Press and to that end Mary Ward exploited her authorial fame to the full. 'Robert Elsmere's Scheme at Work' was the *Daily News*'s crass but eye-catching headline. The general theme of the reporting was that this was the dying Elsmere's dream made reality. On the night itself, Mrs Ward took a back seat on stage, allowing John Morley to make the main speech under the chairmanship of Lord Peel. She presented herself—effectively—as the silent Madonna of the settlement. It was, as Dorothy Ward noted, 'a great success'.[21]

Ostensibly the Passmore Edwards Settlement's mission was one congenial to Tatton, the warden in residence and hopefully the Canon Barnett of the place. PES would sponsor a mixture of adult education and wholesome recreation for the working classes—free of overt Christian or other ideological doctrine. A programme of concerts, clubs, classes, lectures, debates, and sporting events was set up. As a conscious departure from University Hall, the male collegial ethos was less in evidence at PES. The residents, of course, remained exclusively men and they had exclusively women servants to clean their rooms and wait on them at table. But, socially, PES was more open; it had a constant supply of coffee on the go and it was not unusual for some unfortunate woman reeking of gin to be brought in during the day to be sobered up before going back to her family. Girls from a nearby mineral water factory would use the gym during their lunch breaks and after work, thus keeping themselves out of mischief. Consumptive girls were sent to convalesce at one of the cottages owned by the Wards at Tring.

Successful as these activities were, it was with children—particularly the children of the poor—that the Passmore Edwards Settlement was to make its major impact on London (and world) history. Here again, humble Marchmont showed the path. In 1894 Miss Mary Neal, a local Church social worker, had begun a Saturday morning 'playroom' for children in the cramped Marchmont premises. A born teacher (and a leading authority on morris dancing), Neal had used the old upright piano to organize dances and sing-songs, and had devised an impromptu programme of games and stories to keep her children amused.

Mary Ward had registered Miss Neal's success as something to build on. The first Saturday (16 October) after the new settlement opened, she had Dorothy and Janet don their serge work smocks and troop off to Tavistock Square with Bessie Churcher to set up the inaugural play hour in the big front hall of the settlement building. Janet Ward (who was 18) recorded in her diary the 'perfect pandemonium' which met them: 'there were at least 120 children to deal with. We also had to give each child a pair of list slippers to put on over its own boots, and this was a tremendous business and took over half an hour. Miss Neal made them a little speech before we began the games, and then we all formed rings and played Looby Loo and others of that stamp for nearly an hour more.'[22]

Things rapidly became better organized; principally by breaking the mass down into small sections of ten or so children, run by 'helpers' (and elder sisters) under the general supervision of the Misses Ward and the Misses Neal and Churcher. 'Drills'—quasi-military routines—were devised for keeping order. By the end of the first year, 650 children were coming to each Saturday session. By October 1898, 800. It was the size of any school in England, a fact that had not escaped Mary Ward. She mused to Willie, 'sometimes I wonder whether Grandpapa's 6th. form system could be adapted'.[23] It was a bizarre thought, but indicative of how she saw what she was doing as essentially Arnoldian.

By 1902 no fewer than 1,200 children were being taken care of, the most even the new building could hold in safety.[24] Saturday mornings had been supplemented almost from the first by weekday evening sessions, from 5.30 to 7 o'clock. These were designed to keep children off the streets until their parents got back from work. Mary Ward persuaded schools in the St Pancras and King's Cross district to follow the PES example. Initially teachers worked in play centres after hours aided by voluntary helpers. In 1905—when the pressure to set up more centres had become explosive—the system was brought under the central control of a cadre of monitoring and steering committees.

By 1914 there were 1.5 million attendances recorded in the London area and, finally, the whole apparatus was handed over to the London County Council to administer in 1942. By this date, the servant class had disappeared from English society and the now nannyless bourgeoisie wanted to buy into this child-minding facility. It had another surge in the 1970s, when it became the norm for middle-class mothers to work (as working-class mothers always had) and when single parent families became socially more acceptable. In the 1980s, the service pioneered in a small room in Marchmont Street had become prized as one of the essential amenities of modern urban living in Britain.

At PES the evening and weekend 'recreation school' was supplemented by a 'Vacation School', first opened in 1902. This was a service for the numerous children left alone in the parental house or loosed into the London streets during the interminable weeks of summer holiday. It meant setting up a whole-day curriculum at the centre. A headmaster, Mr Holland from Highgate, was recruited (unpaid). Students from the Maria Grey teacher training college assisted. It helped that the Principal was Agnes Ward, Humphry's sister. As an adjunct to the vacation school a children's country holiday fund was set up to get children out of London smoke and grime into the sunshine.

It was, however, with handicapped children that Mary Ward had her most spectacular success. In 1899, PES inaugurated as a pilot scheme its invalid children's school. Like the vacation school, this was designed to use the settlement building and facilities during the daytime hours of the working week, when it was relatively deserted. But given the nature of the children's problems, the organization of the invalid school had to be less ad hoc.

Working in liaison with the London School Board (particularly its congenial progressive members such as Graham Wallas) a formal application was drawn up. The proposal was to set up a small school in specially equipped ground-floor premises at PES. It would require extra building, which would be financed privately. But local authority subvention was requested for special furniture and equipment (such things as a go-cart for the playground, easy-access toilets, and 'Merlin chairs' with special arm and foot rests). It was further requested that the authorities supply a qualified teacher. PES—in addition to the customized premises—would provide the full-time services of a nurse (something that Mary Ward shrewdly saw as vital to the smooth running of the school) and an 'ambulance' (i.e. school bus) to take the children to and from their homes. This necessary vehicle and horse were donated to PES by the distinguished Bloomsbury physician Sir Thomas Barlow. The names of twenty-five selected children were offered as the school's first intake.

There was the inevitable resistance to the PES proposal on the grounds that it would cost hard-earned ratepayers' money. None the less, with the inside help of Wallas, the objections were overridden. In January 1899 the London School Board approved the PES scheme and agreed to support it financially. On Monday 28 February 1899 the school opened. Twenty-eight children rode in on a specially adapted omnibus (the ambulance was not quite ready). Janet Ward records that 'it was pitiful to see their excitement and delight at the new adventure, their joy in the "ride" and their wonder at the pretty, unfamiliar rooms, each with its open fire, its flowers from Stocks, and its set of Caldecott pictures on the walls, which greeted them at the end of their journey'.[25] They were also met at the end of their journey by Mrs Humphry Ward and various dignitaries. The little invalids were duly put on parade for their first medical inspection by the nurse and by their teacher, Miss Milligan.

Always practical, Mary Ward paid particular attention to the children's diet. As was standard for London schoolchildren, they each brought daily a penny-halfpenny 'dinner money'. Mary Ward perceived that, although this might be sufficient to cater for ordinary children, her 'cripples' (a term she used unpejoratively) had special needs and peevish palates. She supplemented the school's food supplies with eggs, cream, hot meat, fresh vegetables and fruit—supplying a 4*d*. dinner for the statutory penny-halfpenny. Appetites and general health improved. The meals—another innovation—were served by voluntary helpers: 'dinner ladies' as later generations of English schoolchildren would call them.

The PES invalid school was a success and enrolment rose to forty within a year. As with play centres, PES blueprinted future educational development. Four new centres for invalid children were eventually added, the first two at Paddington and Bethnal Green, in September 1901. In 1903 a separate school building was opened at PES, thus taking pressure off the central settlement facilities. By 1906 there were twenty-three 'special' schools for the physically defective in London, caring for nearly 1,800 children. The PES play schools, summer schools, and invalid schools were eventually copied country- and world-wide. They left an indelible mark on the education of young English people in the first half of the twentieth century, not least in their stress that education was not just learning but socialization; 'play' or 'recreation' were as vital as arithmetic. In their way, the PES schools were as formative as Dr Arnold's Rugby and fed as main influences into the establishment of the post-war British Welfare State. And, like the 'Doctor', Mary Ward could take proprietary credit for the achievement. PES was her brain-child. It was only justice that, in course of time, it should be renamed Mary Ward House. Founding ideas are more important things than pounds sterling.

Mary Ward—wisely enough—realized that actually dealing with children was work for younger and less fastidious hands than hers. She took no active part in the pell-mell of the big hall. The notion of her leading a raucous chorus of 'Ten green bottles' from the piano or conducting a 'drill' or wiping dripping noses was as unthinkable as Queen Victoria doing the Lambeth Walk. Mrs Ward did, however, give the occasional sedate 'story-telling' to a group of children specially selected for their dutifulness and intelligence. The first took place on Friday, 21 January 1898. Eighty children were rapt by the dulcet voice reading aloud from Scott. This, however, was symbolic and ceremonial. And it was designed to enhance the 'Fairy Godmother' role Mary Ward had devised for herself, as were the hampers of fresh vegetables and bouquets of flowers which arrived regularly from Stocks. She would sweep down in state to judge boxing and wrestling matches; or would give prizes for the girls' embroidery. She was the invariable guest of honour at PES concerts and theatricals.

But—behind the scenes—Mrs Humphry Ward was much more than a mere figurehead. Although her title was modestly that of 'Honorary Secretary', she controlled the settlement by her chairmanship of a bafflingly complex network of management committees, guilds, associations, clubs, and leagues. The bureaucracy of PES was entirely accountable to her. She took care never to let things get away from her, as they had in 1891 with the establishment of Marchmont Hall. Nor could she be ousted by palace revolution; PES (unlike Marchmont) was monstrously expensive to run. There was only one person who could raise the needed income for the settlement—Mrs Humphry Ward. It was her regular letters of appeal to *The Times*; her lecture tours (which took her as far afield as America in 1908); her readings from her work (as at PES in 1901); above all, her constant petitioning of *her* rich and powerful friends and acquaintances which raised the cash. Without her as its financial engine, the settlement would break down in a month. No one was going to resist Mrs Ward's right to chair meetings. Unless, that is, they wanted to go back to the two poky rooms in Marchmont Street.

All the emphasis on children represented a deviation from the adult services originally conceived as the settlement's (and the settlement movement's) main aim. By 1900, Tatton was observing (good-naturedly enough) that his building was in danger of being overrun by juveniles at play. And there was, in fact, resentment against the children—particularly the handicapped—among the residents (around a dozen at any one time) and the associates (400 in number by 1900) who wanted the premises as their club. The building was made over exclusively to the adults at 7.30 p.m. but none the less their sense of centrality was infringed. It was as if they were after-hours users.

Mary Ward was determined to keep her settlement out of controversy; particularly the kind of agitation that Toynbee was increasingly linked with. It was—apart from anything else—financially prudent. In 1907, the Duke of Bedford threatened to withdraw his support if the residents continued to work for Progressive candidates in the local elections. His 'support' in this instance meant use of the two adjoining gardens—removal of access to which would have closed down all the school services. One of the main attractions for Mary Ward in directing the settlement's work primarily to children was that it was an intrinsically apolitical and non-religious activity. It did not matter if a little spastic were Irish Catholic; or if an urchin who came on Saturday morning was the offspring of a Marxist shop steward.

And children were innocent in other ways. Sex was a never-ending source of difficulties with adults at PES. Encouraging community without promiscuity at dances or parties was a delicate business. Alcohol, of course, was banned. But the atmosphere in the men-only, smoke-filled billiard room—particularly at the weekends—looked suspiciously like that of a public house. All in all, the shift to children's work was a highly workable compromise. There was an insatiable demand, and few social risks.

Between 1888 and 1898, Mary Ward—a chronic invalid who while at home in Stocks spent much of her time in a bath chair—earned an estimated £45,000 for her novels and raised at least the same amount for her settlement. It was a phenomenal achievement—something to rank with those of Victorian heroes like Brunel and Florence Nightingale. Stocks and the Passmore Edwards Settlement testified in bricks and mortar to the scale of her achievement. Nevertheless, Mary Ward could not be entirely happy as the twentieth century dawned. She did not really want to write romances; a 'Life of Christ' was what called to her. But such were her family's needs for money that she would now never get off the novel-writing treadmill to do that great work. Nor was PES what she had originally yearned for in 1889. A school of Bible studies, a new *understanding* of scripture was her vision then. Instead of which, she had the world's best kindergarten. And, to keep it going, she was indentured to a lifelong series of lectures, fund-drives, and pestering of her friends.

Mary Ward nursed one small ewe lamb in memory of that lost settlement vision. Benjamin Jowett—'the Master'—had died in 1893. Two years later, in 1895, she established the Jowett Lectures. These were a series of addresses, on scholarly biblical themes, to be given annually by a visiting speaker and funded with £100 taken from the PES general kitty. One of Mrs Ward's executive committee colleagues, Estlin Carpenter, noted in some perplexity on 29 March 1895: 'Your lecture scheme is exceedingly interesting in itself, but I do not quite understand its connection with the settlement.'[26] There was none—at least not with PES. The connection was with the original settlement of 1890, a dream that Mary Ward conceded was now lost forever.

19

Eleanor: 1900

UNLIKE the older Dickens or George Eliot, Mary Ward did not with years grow in artistic strength. 1900 sees the onset of a gradual decline which becomes precipitate after 1905 and vertical by 1913. There were a number of causes for the decline, most of them painfully obvious to the novelist herself. Her American, colonial, and British readerships remained gratifyingly loyal; but like her they were by 1900 middle-aged and losing cultural and economic power. Mrs Humphry Ward—despite all efforts—failed to recruit younger readers, for whom she represented antediluvian Victorianism. Things were not helped by her implacable crusade to deny women the vote—as offensive to most under-thirties as a campaign to send little boys back up chimneys.

Mary Ward's health, always poor, became progressively poorer after 1900 and collapsed in 1906. Meanwhile, the money needs of her surviving family became ever more insatiable. Arnold's careers in law and politics were expensive. His love of gambling was ruinously expensive. One daughter married in 1904; but the other remained at home and had to be supported like a lady. Humphry, after the onset of his arteriosclerosis in 1912, was semi-retired from *The Times* but continued to indulge his passion for acquiring costly works of art.

Ambitious remodelling was constantly going forward at Stocks. In 1900, for instance, a nine-hole golf course was laid down. In 1907 a wholesale reconstruction of the house was undertaken, which proved unexpectedly costly. All these financial obligations forced Mary Ward to write faster. Whereas at the beginning of her career she was taking three years for a novel, by the end she was producing two a year. By founding the Passmore Edwards Settlement, Mary Ward had committed herself to its upkeep. She had to lecture, write, and lobby for its funds, and sit on interminable committees to keep it smoothly running. Managing PES would have been a full-time occupation for any healthy spinster: Mary Ward was an invalid novelist with a large family to support. Various costs were paid; in health, in peace of mind, and most of all in quality of writing. In the last ten years of her life, she was turning out what charity would call romantic pot-boilers.

Helbeck of Bannisdale had been well received on all sides: by the critics, by Mary's friends and family; by the public. Once the American returns were in, George Smith told Mrs Ward the good news in September 1898 that her next work could again be serialized in the United States and that she could confidently anticipate first-year earnings for it of around £8,000. It was gratifying. But even this amount was inadequate to the Wards' now gargantuan needs. No one could deny that Mary Ward was doing well from new novels and reprints. But well was not good enough. She needed to increase her income from entirely new sources. A number of possibilities came to mind. She could—like Dickens, Thackeray, and Uncle Matt—enrich herself with an American lecture tour. There were offers in plenty. She told her brother Willie with some amusement in October 1896 that 'An American gentleman implores me to go round the States with fifty readings from my books, and asks me to name the lowest terms! I think for £1,000 a night I would oblige him—wouldn't you?—not for a penny less.'[1] Joking aside, lecturing in America required greater physical robustness than Mary Ward felt she now possessed. It had, everyone knew, killed Dickens.

Then again, she might write the short stories and novellas which were all the rage at the turn of the century. *Bessie Costrell*, after all, had earned her some £1,500 and taken only a fortnight to do. But the objection to short stories which George Smith constantly urged on her was that they would surely cheapen the value of her name, making 'Mrs Humphry Ward' a drug on the market. This would undermine the sale of her blockbuster novels. And anyway—unlike younger writers such as H. G. Wells or Arnold Bennett—Mary Ward was not fertile in ideas for fiction. Increasingly, as she got older, she had to borrow or find her plots. She could not spin them out of her own mind.

An easier option was to 'work her copyrights' with a de luxe collective reissue. George Smith had pioneered the practice with his 'biographical' editions of Thackeray's work. These were finely bound, shamelessly expensive volumes which sold as sets, and were handsomely embellished with illustrations and prefatory apparatus. Mrs Ward herself had been recruited to write the introductions to Smith's seven-volume Haworth Edition of the Brontës' works, in 1898. It had been an interesting assignment. She had visited the famous Parsonage in bleak February 1899 and drank in its associations standing 'in the room where Charlotte and Emily died'.[2] A sonnet ensued—'Pale sisters! Children of the craggy scree.'[3]

Mary Ward's Brontë prefaces (published 1899–1900) earned her a pleasant £1,000 (400 guineas of it from America) and were the best sustained critical prose that she ever wrote. And as she worked on the Haworth Edition she could not but wonder why Mrs Humphry Ward should have to wait until after her death for this canonization and the revenue that came with it. Why not a collective reissue of her own work now, when it could do her some good? Her *œuvre* was now seven strong and fully equal to that of the Brontës. She raised the question of her collective reissue with Smith as early as October 1898. He answered cautiously that the idea was 'interesting'. But it was not something to rush into. Co-publication with an American firm was an essential preliminary. Production costs for these editions were dauntingly high. There was also an awkward problem with Mrs Ward's copyrights. Smith owned all the major novels in England bar *Miss Bretherton*, which would be cheap to purchase. But the American rights of her bestsellers were owned by Macmillan. And their American manager, George Brett, was piqued that Mrs Ward intended to take her novels elsewhere after *Helbeck*. (She was piqued on her part because he kept nagging her to make her subjects more 'American'.) Brett indicated that repurchase of his American copyrights and stock would cost a whopping £4,500. He was not prepared to be co-publisher himself (and thus write off the value of his Ward books in print). And his asking price was prohibitive to others. One by one, likely American houses politely turned Smith, Elder's overtures down.

Mary Ward never gave up her dream of a collective reissue of her work until, quite disastrously, she got her wish in 1912. In the meantime, she turned to another potential source of income. A number of novelists in the 1890s—notably Hall Caine and J. M. Barrie—had enjoyed tremendous success with dramatized versions of their fiction. Mary Ward was keen for some of these rich theatrical pickings. There had been American interest in a dramatic version of *Bessie Costrell*. But that story, like all her Victorian fiction, was too prosy. She resolved that her next novel would be written specifically with an eye to future adaptation as a play. There would be a new stress on sharp dialogue, melodramatic crisis, and extended pathos.

But first, this new novel would have to be written as a novel. The early stirrings of *Eleanor* are found in April 1898 when Mary told her brother Willie that she intended taking a villa next spring 'for three or four months [so] that I may write an English story in a Tuscan setting'.[4] By now it was an invariable habit to trigger her creative imagination by immersion in a location. Italy (in the event Rome rather than Tuscany) would serve as well as the Lake District had done for *Helbeck* and would—with any luck—be less draughty. The actual *donnée* for *Eleanor* came to her by a very roundabout route in summer 1898. The Neville Lytteltons were at this period staying in Stocks Cottage. Neville was a colonel in the Army serving in Egypt in 1898, where he performed gallantly at the great victory of Omdurman. Dorothy Ward had for some time enjoyed an intimate friendship with Kate Lyttelton, Neville's wife. Mary was also fond of Kate, and resolved to take her mind off the Colonel's dangers by giving her something useful to do. She consequently encouraged the young woman to work during her grass widowhood on a translated 'selection' of Joubert's *Pensées*. (It was eventually published by Duckworth in 1898, with an introduction by Mary.) Joseph Joubert had been a witness in 1800 to the unhappy love affair of his friend François René de Chateaubriand and Pauline de Beaumont. As well as being his lover, Pauline was Chateaubriand's intellectual helpmeet. But he later abandoned her for a more beautiful woman. Mary Ward, as she informed Mandell Creighton, found the plot for her new novel in the couple's 'work together on the *Génie du christianisme*, his desertion of her for Madame de Custine and her [i.e. Pauline's] pathetic death in Rome'.[5]

The characters of *Eleanor* were based on this trio, but 'Anglicized and modernized' (and, in one case, Americanized). Edward Manisty, the brilliant but perverse politician and author, took the Chateaubriand role. That of Pauline was taken by the exquisitely frail Eleanor Burgoyne, who helps Manisty on the great study of modern Italy which he is writing. Lucy Foster, the puritan New England miss, is the Madame de Custine figure who lures Manisty away from Eleanor, who duly expires in Rome—nursed by her American rival (Mary Ward's added romantic twist).

An outline of *Eleanor* was written up in a couple of days and read to the family at Stocks on 23 November 1898. It was then typed by Bessie Churcher and sent to Smith three days later, who thought it 'admirable'. Good terms were forthcoming. Mrs Ward was offered a £6,000 'advance' (i.e. a loan at the usual 4 per cent, all receipts as collateral) for the English rights, payable when the novel was complete in manuscript. This was in anticipation of her standard 20 per cent royalty on the 6*s*. edition. Even more gratifying were the American arrangements. Smith secured a new publisher, Harper's of New York. They would run the new novel as a twelve-part serial through *Harper's Magazine* from January 1900. It would be illustrated by the distinguished artist Albert E. Sterner, whose work Mary Ward liked intensely—although he never got Manisty quite right for her. Harper's would pay the author £200 an instalment for *Eleanor*. With a £600 advance on their 30c. royalty for the book version of the novel, she was thus assured of £3,000 from America.

To prepare for her new novel, Mary Ward undertook in November 1898 a course of lessons from an Italian girl who came to the London house. The country (which she had been visiting on and off since the mid-1890s) was now her favourite continental resort. France—which she had preferred in the 1880s and early 1890s—was by the end of the century morally offensive to her. Zola, decadence, Dreyfus—all demonstrated a national 'debauchery'. Morally more salubrious, Italy was also in 1900 a paradox of the kind that stimulated Mary Ward's hypersensitive historical imagination. As Manisty puts it, 'sometimes I have felt as though this country were the youngest in Europe; with a future as fresh and teeming as the future of America. And yet one thinks of it at other times as one vast graveyard; so thick it is with the ashes and the bones of men.'[6] Mary Ward had resumed the habit (begun thirty years before in her engagement) of studying the classics for half an hour or so every morning. By dogged persistence, she had made herself a decent scholar, and could respond to the 'stupendous fragmentariness' of Rome. Present day Italy also thrilled her. It was a stage on which momentous problems were being played out with the clarity of a chess game. The see-sawing conflict between the 40-year-old State (younger than her) and the age-old Catholic Church fascinated Mary. Nor was it clear, even to her, where her deepest sympathies lay.

Humphry Ward acquired an *appartamento* in the Villa Barberini, at Castel Gandolfo in the Alban Hills, fifteen miles by train from the capital. It was often used by people from the embassy. The Wards arrived there on 23 March 1899, in a caravan of three *vetture* loaded down with warm bedding, books, stoves, and comfortable items of furniture. They lumbered up the ilex avenue to the house ('like a cool green tunnel', Janet remembered) to be met by 'the delightful little butler Alessandro and his stately sister Vittoria'.[7] In some wonder the family trooped over the suite of a dozen rooms, dominated by two huge central *saloni*. There were a few worn carpets and 'wonderfully ugly wall papers'. But the apartment had magnificent views from balconies east and west over the Alban Lake and the Campagna. To the north, one could glimpse the dome of St Peter's through the notch in a stand of stone pines. On a nearby hillside were the ruins of the villa of the Emperor Domitian.

Humphry deposited his wife, then left her with the girls, Lizzie, and Bessie and returned to London (it was not yet his month's holiday). He prophesied Mary would decamp to a comfortable Roman hotel before the week was out. It was indeed a bitter tail-end of March. Snow fell and the Wards awoke two mornings later to find 'the Alban Lake lying like steel in its snowy ring'.[8] The apartment was meant for *villeggiatura* in the summer; refuge from the city heat. It had only two bedrooms with fires and was impossible to keep warm. The locals clearly thought Mrs Ward was an English madwoman.

But it was more than accommodation she had wanted, and the Villa Barberini was the perfect stimulus to her pen. She wrote the first chapters of *Eleanor* in early April as a mere transcript of what she saw around her. In the opening of the novel it is spring in the 'Villa' (unnamed but recognizable in every detail) and Manisty struggles to write an important book about Italy, aided by his soul-mate, Eleanor. When the weather improved the Ward family (rejoined by Humphry in May) made an excursion to the Lake of Nemi, to view the ruined temple of Diana. Henry James was with them. He was dressed (as Mary recalled) in his immaculate short coat, with a wide-brimmed summer hat shading 'the smooth-shaven, finely cut face'. James enjoyed showing off to the Ward ladies his Italian and took a great interest in an Italian peasant lad whom they met on their expedition. The episode—including some florid description of Nemi—went whole into the novel. In April, Mary Ward (in the company of a seething Bessie Churcher) was at St Peter's for the Pope's Easter appearance. It too was duly conscripted into one of *Eleanor*'s purplest passages.

At the end of April, Mary's younger sister Ethel Arnold turned up at the villa, despite telegrammed instructions for her to stay away. Ethel was very sick (possibly consumptive), emotionally disturbed, 'looked like a little ghost',[9] and could only sleep at night with the aid of heavy doses of chlorodyne. She became Manisty's maniac sister, Alice, who descends similarly unbidden on the villa. Alice's mad scenes and monologues are, in fact, some of the most gripping passages in *Eleanor*. Infuriating as it was to Mary, one cannot but wish she had been bothered by her sister more often. Ethel obstinately stayed on until the whole party returned in June, by which time half the novel was drafted.

The twelve bedrooms were an irresistible draw to others of the family and friends—many of them more welcome than Ethel. Notably, Mary's father Tom Arnold came to stay at Villa Barberini (his physical appearance is grafted on to Father Benecke in *Eleanor*). It was for him a spiritual climacteric. As Mary Ward recalled in her *Recollections*, 'Never before, throughout all his ardent Catholic life, had it been possible for him to tread the streets of Rome or kneel in St Peter's.'[10] Grateful as he was to his rich daughter, who could give him this gift, Tom and Mary quarrelled bitterly about the Boer War a few months later—a final rift between them that haunted her after his death.

The Catholic sub-plot to the novel, concerning Father Benecke's betrayal by his Catholic superiors and his brutal excommunication, was suggested to Mary by the case of Dr Schell, rector of the Catholic University of Würzburg. Schell wrote a 'Liberal' book—*Catholicismus und Fortschritt*—which was put on the Index; he recanted, and later withdrew his recantation. Mary Ward was at a dinner party in Rome, where the English Cardinal Vaughan (a man whom she despised as a superstitious zealot) heard the news of Schell's rebellion. 'Oh, poor fellow!' the Church dignitary muttered, with what she took to be 'easy pity'.[11] This—much heightened and melodramatized—was the kernel of the Benecke plot.

Eleanor is in terms of narrative art one of Mary Ward's most disciplined efforts. But it is marred by a growing self-importance of authorial manner. The 'Dedication' illustrates this tendency clearly enough:

To
ITALY
The beloved and beautiful
Instructress of our Past
Delight of our Present
Comrade of our Future
The Heart of an Englishwoman
Offers this book

This address on equal terms to a country reflects back on the dedicator as pomposity. More so as virtually every actual Italian who appears in the book is either a doltish peasant, a comic servant, or a contemptibly ignorant cleric. Italy is incapable of producing a genius like Manisty, who can *understand* Italy. Even Lucy Foster, the American, has to be 'taken in hand' by the incomparably more sophisticated English women who teach her how to dress her hair and her body to advantage. England is the indisputable master and mistress race. It alone boasts specimens of the calibre of Manisty and Eleanor.

There are other weaknesses in *Eleanor*. The melodrama, for instance, is overdone. To make Eleanor interestingly 'delicate' it was not necessary to have her husband (a drunken brute) scoop up their child and hurl himself over a Swiss cliff. Henry James picked on another conceptual flaw in the novel. Eleanor is made too perfect. When Lucy comes and wins Manisty's heart (too easily won, one might think) it strains credulity that Eleanor would befriend her rival quite as tenderly as she does. Nor that, as she dies of longing for Manisty (tended by an adoring Lucy), she would never utter a word of jealousy or recrimination. The relationship between the women lacks what James called 'anti-thesis'.[12] Mary Ward refused to credit that two gentlewomen could actually fight for a man.

She had the novel complete in manuscript by March 1900, and the balance of Smith's £6,000 was paid over. George Smith was permitted at last to read the corrected proofs in late July and pronounced it 'a grand book'.[13] *Eleanor* was duly published on 1 November, in Britain and America. The new novel clocked up record advance sales: 17,000 were subscribed in Britain; 30,000 were sent out on sale or return in America. 'It is the book of the year here,' a jubilant Harper's informed Mrs Ward. She had a hit. But she was allowed no time to relish the triumphant launch of *Eleanor*. On 7 November, she was called to Dublin, where Tom Arnold was dying. She described his last hours to Willie (himself mortally ill). As always, she was a connoisseur of deathbed detail:

> He sleeps heavily a good deal or talks to himself. Then a cough wakes him and he will say two or three sensible things. I told him just now that you had wired and asked him for a message for you. He said as far as I could make out his 'best best love'—But the words were very indistinct. At any rate I am sure he knew that it was Willie and that he was sending his love. He knew me quite well—when I arrived between six and seven—said 'Mary!' with a look of pleasure, opened his arms and said God bless you. Later on he talked of Leonard's book [i.e. his *Life of T. H. Huxley*] and said it would rank among the 'massive' historical or scientific biographies of the time. Then he asked for news of *Eleanor*, and quite laughed a moment when he heard the news of the sale.[14]

Mary was shooed out of the room soon after, to make way for the priest, Father Darlington, who had given Tom the viaticum the day before. On the tenth, she watched her father 'just fading away .... He lies with his beautiful grey head a little raised, the sunshine coming freely in through a large south window, breathing very fast, 51 to the minute'.[15] Mary asked him if he loved her and forgave her everything. 'He made a motion as though to put up his hand to my cheek, with a sweet faint smile.'[16] He then lapsed into coma and died on 12 November. Congestion of the lungs and heart failure were given as the cause of death. He was 77 and did not, Mary thought, suffer much. Frank, Judy, and Ethel arrived for the end.

Looking through her father's papers (strangely enough, with Tom's wife still in the house) Mary Ward came across 'the most touching and remarkable things—things that are a revelation even to his children'.[17] Doubtless, she chose to destroy most of these things. She described the memorial service on 15 November to Mandell Creighton. It was held in 'Newman's beautiful little University Church, the early mass, the academic costumes, the bright morning light on the procession of friends and clergy through the cypress-lined paths of Glasnevin, the last "requiescat in pace" answered by the amen of the little crowd—all made a fitting close to his gentle and laborious life.'[18]

The period from *Eleanor*'s first conception in early 1898 to its triumphant launch was a season of death. Gladstone had died in May 1898. In October 1899 Mary's beloved aunt and godmother, Jane Forster, had died, blind but 'very peaceful' and in no great pain. In January 1901 Mandell Creighton died, aged only 58. All England, Mary Ward thought, felt the sense of catastrophe, 'as of some great tree fallen'.[19] In January 1901 Queen Victoria died. Mary's historical era was officially ended. And in April 1901, George Smith died.

20

Best-Selling Novelist, Failed Dramatist: 1901–1905

IN the wake of *Eleanor*'s success, Mary Ward was induced to move quickly on to a follow-up novel. Despite Reginald Smith's suspicions that his American partner was not financially sound, Harper's was again chosen for the unwritten new novel. The terms (finalized in February 1901) reflected Harper's warm feelings about *Eleanor*. The author would again get £2,400 for the American serialization, which would run from May 1902 to April 1903. But the advance on American book sales was upped from £600 to £2,400. And Reginald Smith was, like Harper's, inclined to be open-handed. Mrs Ward could have an early advance of £5,000, payable in full when the new novel was complete and in their hands. As usual, she would be charged 4 per cent and the American payments would be collateral for the advance (in fact, the so-called 'advance' was more in the nature of a bridging loan; English royalties would come nowhere near £5,000).

The new novel was provisionally called (in July 1901) 'A Woman of Title'. It was renamed in October 'Lady Henry's Companion', before finally settling as *Lady Rose's Daughter*. Mrs Ward's publishers still did not quite like it. Her previous novels had all had as their title the hero's or heroine's plain name. This new tendency to periphrasis might unsettle the loyal readership, they thought. But finally they allowed the novelist to call her novel what she wanted. And she was proved right. In line with her dramatic aspirations, the new mode of title signalled a move from simple portraiture to plot as the heart of her fiction.

Reginald Smith had the work complete at the end of January 1903 and communicated his approval by letter, from France. He had read the manuscript in a restaurant in Boulogne, he told her, and had been moved to tears. 'When the hardened publisher is touched, surely the far larger public will respond,'[1] he observed hopefully. *Lady Rose's Daughter* was published on 5 March 1903 with the usual domestic ceremonies and votive offerings. Jan and Dots got up early to buy roses and carnations to take their mother in bed. They found her already surrounded by the morning papers—all gutted to get at the review pages. Mary Ward was, Dorothy noted, 'very happy' with most of them, especially the friendly old *Manchester Guardian* which arrived after breakfast. But the 'D. Mail, D. Express and D. News were odious'.[2]

On the day of publication, unable to walk and sustained only by her cocaine lozenges, Mary Ward left for Paris, en route for the Villa Bonaventura, at Cadenabbia. This villa by Lake Como, which Humphry and Mary had originally rented from Alfred Trench, was now her favourite recuperative resort. Recuperation was in order. Her eczema had flared up more severely than ever. It would not be until the end of March that she could walk unaided, or take a drive in any comfort. She was, as Dorothy put it in her diary for this month, 'very seedy'. Meanwhile, *Lady Rose's Daughter* triumphed. The prepublication subscription in Britain was a healthy 16,600—only slightly short of *Eleanor*'s 17,000. Within the month, just on 20,000 had been sold. Of these the abominated Mudie took only 3,380, each of which was reckoned to have six readers thus robbing Smith, Elder of five purchasers. Harper's—who had no Mudie to eat their profits—did predictably better, outselling England by three to one. By December 1903, they had cleared 113,000 copies. It was an extraordinary achievement. One thinks of best-sellerism as a late twentieth-century phenomenon; but hardback sales on this scale would be respected by Harold Robbins.

Mrs Humphry Ward was now, to all intents and purposes, an English writer for the American public. It was something that had an insidiously distorting effect on her art. She was shouting her novels across the Atlantic; serving an alien master. The disproportion of her trans- and cisatlantic publics is evident in a statement of earnings drawn up in October 1904, after *Lady Rose's Daughter* had been out eighteen months:[3]

Serial Publication
  (*Harper's Magazine*) £2,400 0*s*. 0*d*.
Royalties
  American and Canadian £7,413 16*s*. 7*d*.
  English £1,624 1*s*. 8*d*.
  Colonial £131 5*s*. 0*d*.
Continental editions in English
  Tauchnitz £250 0*s*. 0*d*.
Translations
  Norwegian £15 15*s*. 0*d*.
  Italian £15 0*s*. 0*d*.
  Serialization in *Revue des deux mondes* £346 15*s*. 9*d*.
  French book version £21 16*s*. 8*d*.
Cost price copies to *Ladies' Home Journal* £110 0*s*. 0*d*.

Total £12,328 11*s*. 8*d*.

The extraordinary fact, confirmed by this account, is that Mrs Humphry Ward was making six-sevenths of her literary income overseas, the overwhelming bulk of it (five-sixths) in North America.

Lady Rose's Daughter is a weaker performance than *Eleanor*. In it, Mary Ward deserted the previous novel's up-to-date historical setting for a vaguely drawn Victorian England 'a good many years ago'.[4] The plot is a *mélange* of high melodrama and silver forkery. A powerful old socialite—Lady Henry Seathwaite—runs a Wednesday afternoon salon (a romanticized version of Mary's 'Thursdays') in London's West End. Almost blind, she has entrusted the running of her Wednesdays to a young companion, Julie Le Breton, about whose origins there is some romantic mystery. It emerges in the course of the story that Julie is in fact the illegitimate daughter of Lady Rose Delaney, a *femme incomprise*, who deserted her military husband for an atheist artist. Julie is unaware of her shameful parentage.

All this is antecedent to the novel's narrative. Lady Henry's paranoid tyrannies and Julie's machinations form the first movement of the plot. A crisis breaks (in Chapter 9) when Julie presumes to hold a 'Wednesday' against her *patronne*'s express wish. She and her company are surprised by an enraged Lady Henry, risen from her sick bed. A rupture between the ladies follows. The second half of the story centres on the gradual revelation of Julie's birth and its effects on her relations with two lovers: the adventurer Captain Warkworth and Jacob Delafield, heir to the dukedom of Chudleigh and disciple of Tolstoy. Warkworth dies for his country in Africa and Julie finally accepts—though without any great love—Delafield's proposal. She is finally reconciled with an unregenerate Lady Henry.

The main attraction of *Lady Rose's Daughter* is its rendering of the conversation and gossip of the English power élite as they congregate under Lady Henry's and later Julie Le Breton's roof. But *Lady Rose's Daughter* has some glaring faults. As Janet Trevelyan noted, salons—as cultural power brokerages—are a French, not an English institution. The word 'reception' is very colourless and thin in its Anglo-Saxon associations. Mary Ward had, as with *Eleanor*, drawn the idea of her plot straight out of French history and it did not travel as well as the other had. In mid-eighteenth century Paris, the powerful hostess Madame Marie du Deffand (who like Lady Henry was old and nearly blind) had befriended the young Julie de Lespinasse and introduced her into society under a bond of silence as to her (Julie's) origins—she was the natural daughter of the comtesse d'Albon. The two ladies had later fallen out and Julie had gone on to eclipse her old benefactor as a hostess in Parisian high society. But whereas the historical Julie eventually came to a suicidal end, Mary's fictional Julie is rewarded with the romantic novel's conventional happy marriage to Jacob. In her 1911 preface, Mary Ward admits it as a fault, and excuses herself on the grounds that she lacked the 'nervous energy'[5] to make Julie destroy herself. There was probably another motive. Mary Ward had come under fire for Laura's suicide in *Helbeck* both from the Bishop of Salisbury (on religious grounds) and from George Smith (on publishing grounds). Happy endings may have been good morality and good business. In this case they were not good fiction.

Taken in the light of Mary Ward's career development, *Lady Rose's Daughter* has a number of points of interest. Ideologically, it articulates her belief that women's political power is best exercised indirectly by the creation of quasi-domestic environments (here the salon) through which men may be influenced. Another symptomatic innovation in *Lady Rose's Daughter* is its attention to the psychology of old womanhood in the person of Lady Henry (a name with some echo of 'Mrs Humphry'). This novel introduces what Françoise Rives calls the *cycle de la mère*,[6] a series of novels by Mary Ward preoccupied with the ambitions and frustrations of powerful matriarchs.

One can plausibly surmise why Mary Ward should have become so engrossed in the matriarchal theme at this time. In the autumn of 1903 she took over full responsibility for her son Arnold's future. He was 27-years-old, disillusioned with the bar, and intended to throw up England as a bad job and make a new life for himself in India. Mary Ward urged him to give up his 'India plan altogether'. She added, by way of inducement, 'it is *quite* possible that I may be able to help you quite materially'.[7] Help him with a career in politics, she meant. It was, as she must have known, extremely risky. Getting Arnold into Parliament and giving him an adequate income to make his way in political life would add thousands a year to her bills.

In 1904, Mary Ward took out heavy life insurances, which she used as collateral for private loans from Reginald Smith. And she girded herself to write harder. She also became a more grasping author. She increased her demands for advances to exorbitant (as Smith thought) levels. She insisted that her royalty rate be increased to an unprecedented 25 per cent. It was not her nature to do this but a forced response to the extra burden of motherhood which she had assumed. Her graspingness also derived from the sense that there might not be many earning years left to her. In 1905, Humphry Ward was 60. He still looked '*wonderfully* young',[8] as Dorothy loyally insisted. But he was suffering a number of inexplicable falls, and was perceptibly duller than he had been. Suddenly, Mary Ward realized that she was—the word was inescapable—an *old* woman. Nor had chronic illness rendered her young for her age.

Her fragility was borne home heavily by the death of Willie, her younger brother and childhood playmate. After the agonizing and humiliating suffering of his locomotor ataxia, eased only by his equally incurable morphinism, Willie Arnold finally died on 29 May 1904. He and his wife Het had been travelling in the Basque country in March and April. The previous winter had been awful ('I have never known him so bad,'[9] Dorothy noted in September 1903). But there had been an apparent improvement. Then, without warning, a blood vessel burst in his head. He was brought back to England, and his sister Mary was called back from Italy. There followed: 'a week's unconsciousness, and the end. It was very merciful, without any conscious pain, and all the few utterances that came through the cloud to us, were like the sayings of one living in a cheerful dream.'[10] Mary packed the shattered Het off to Switzerland. 'One will never lose the pity of it,' she told Henry James in August. Willie's death 'took me far from my work and it was long before I could settle to it again'.[11] She was at the time grappling with the last eight chapters of the finest work of the *cycle de la mère*, *The Marriage of William Ashe*. Willie's death coloured the sombre end of that novel.

For the plot of *William Ashe* Mary Ward had again borrowed from history. The main characters—Ashe and his wife Kitty—are based on William Lamb (Lord Melbourne) and his famously flighty wife, Caroline. The novel, however, is dominated by Ashe's mother, Lady Tranmore, whose ambitions for her politician son (and her terror lest his career be 'ruined' by any moral weakness) reflect Mary Ward's own anxieties about Arnold. Her outline of the first chapter in her notebook indicate clearly enough the maternal preoccupations at the core of the novel: 'The little house with its irregular bow windows in Park Lane—The mother waiting for him with her cousin Mary Lyster—*This* is her darling son. They talk of him, of his performance at Eton, of his idle time—His Election—his Prospects.'[12] Substitute Grosvenor Place for Park Lane and Dorothy Ward for Mary Lyster and it is the novelist's own fable. In the text of the novel, 'prospects' is further specified in Lady Tranmore's declaration that 'the first thing we've got to do is to marry him'.[13] Getting Arnold elected to Parliament and then well married was Mary Ward's order of priorities in 1904.

In the subsequent plot of *William Ashe* (initially 'Ashley') the hero in fact marries ill-advisedly, choosing Kitty Bristol, an 18-year-old girl of great beauty and spirit but with 'bad blood' in her veins. Kitty embarrasses her politician husband by her indiscretions as he ascends the rungs of the Liberal party to the post of Home Secretary. A physically handicapped child is born to the Ashes, whom Kitty cannot bring herself to love. Finally, as William stands on the brink of the premiership itself, Kitty betrays him doubly: by flirtation with a Byronic cad—Geoffrey Cliffe—and by writing a scandalous *roman à clef*, which divulges Cabinet secrets. The crisis of the marriage provoked by Kitty's unwifely acts is set against the background of a richly described Venice. (Humphry and Mary Ward had spent Easter 1904 there, at the Palazzo Barbaro on the Grand Canal.) Unable to live with her long-suffering husband's toleration, Kitty runs away with Cliffe to assist in the Balkan uprising against the Turks. She finally dies pathetically at Simplon in Switzerland, forgiven by William. The novel, which is often regarded as Mrs Ward's last significant work of fiction, is remarkable for the density of its socio-political descriptions. It embodies in its depiction of Ashe's career Ward's belief that modern England is ruled by 'government of country house'. And the strongest single scene is that at Ashe's own country house, Haggart, where Kitty mortally offends the Prime Minister with her thoughtless satire.

William Ashe marked another welcome increment in success for Mary Ward. In the period of negotiation with her publishers which preceded composition her American terms were again improved. She had £3,000 from Harper's for serialization of the story in their magazine and £1,500 advance on American royalties. Reginald Smith also relaxed his purse strings somewhat, with earlier releases of his 'advances' amounting to £6,000 in all.

Mary Ward began actual work on *William Ashe* with research into the life of Melbourne and a 'sketch' of her story as she foresaw it. The early chapters were done by February 1904. She was on chapter 11 and in Italy when news came in mid-May that Willie was dying. The manuscript was none the less complete by October 1904. *William Ashe* was then serialized in *Harper's Magazine* from June 1904 to May 1905. Reginald Smith shrewdly suggested that, for the book issue, Mrs Ward write a preface cueing the reader to the Lady Caroline Lamb identification, which she did. The novel was published in book form on 9 March 1905. Her mother was very happy about *some* of the reviews, Dorothy Ward recorded—although, for the first time, the *Manchester Guardian* was 'rather horrid'. There were some pleasant reactions to compensate, however. Sir Arthur Conan Doyle wrote to say that *William Ashe* 'kept me up till 2 a.m. It is really great. It has the rare merit of being so clear-cut as well as large. Personally I don't think George Eliot ever wrote anything to touch it.'[14]

Between 1900 and 1905, Mary Ward had produced three novels which between them earned some £30,000—possibly more. The novels all had features in common. Each was founded on a plot from history. Each was carefully calculated to appeal to American readers (or at least American book-buyers), who outnumbered their British counterparts by two or three to one. And—most important—each was written with the specific aim of dramatization. But Mary Ward's career in the theatre with adaptations of these works was short (effectively 1900–8) and tantalizing. Dramatic success was a jackpot apparently easily within her grasp but one that she could never quite lay hands on. It was the more tantalizing in that just one stage hit would have solved her financial problems at a stroke—elevating her to the ranks of those authors who earned hundreds not merely tens of thousands for their work.

Mary Ward had made some experimental stabs at writing plays in the 1890s, but her first serious 'wrestle' with drama was after writing *Eleanor*. The novel was hugely popular, and its American and English heroines seemed to make it an obvious candidate for stage adaptation. Mary Ward did not, however, trust her own abilities as a dramatizer. She was candid enough to realize that what came over as 'earnest' in a novel could be leaden in the theatre. She therefore approached Julian Russell Sturgis for help with adaptation. Sturgis was an American who had come to England as a child and gone through the

Eton and Balliol mill. Now naturalized and more English than the English, he nevertheless had a feel for what would work on Broadway. And his fiction was renowned for its light, frothy touch. Mary Ward gave Sturgis *Eleanor* to read in September 1900. He liked it and agreed to collaborate. In the spring of 1901 he duly came out to work with Mary at the Wards' villa in Rapallo. By the time he went home, two months later, 'the play stood up and lived'.[15]

At this stage, Mary Ward had grand ideas as to who might star in her new play. She made an appointment with the famous actress Eleonora Duse in April 1901 and was entranced: 'Heaven, what she would make of Eleanor!'[16] she told Humphry. Duse, it transpired, was not available for Eleanor. But Mary's sights remained high. 'Mrs Pat' [Campbell] and Ellen Terry were seriously talked about. But of course Mrs Ward was in no position to make casting or staging arrangements herself. In November 1901 she recruited the services of Addison Bright, the dramatic agent. He was—happy omen—Barrie's agent. Mary had high expectations, describing Bright as 'a very interesting and brilliant little man ... a sort of Napoleon of the stage'.[17] He certainly bustled about for his client and negotiated the lease of the St James's Theatre for the new play's London opening in matinée. But there followed a series of salutary shocks for the would-be dramatist. She discovered to her suprise that *she*, not the theatre proprietor, Mr Alexander, would have to foot the bills. For five matinée performances in May 1902, the cost was estimated at £500 and might rise as high as £800. If all the seats were sold (very unlikely for weekday afternoon performances) her receipts might theoretically go as high as £1,750. But the real return would be in an evening run, *if* the play pleased. It was a very big if.

It all came to nothing. On 14 March 1902 a distraught Mary wrote to Dorothy Ward telling her, 'Oh! my dear, your mummy is just now a distracted mummy! Last Wednesday Father came down to Stocks to break to me the news that after all, at the last moment, Alexander had thrown me over!—and the St James's was not to be had.'[18] Alexander's explanation was that he wanted to 'protect' the chances of another play, *Paolo*. Mary was not mollified. The sardonic Sturgis observed that he never trusted the word of any theatre person; 'pretending' was in their blood.

Over the early summer of 1902, Mary Ward was at Cadenabbia, and 'not very bright', as Arnold noted. Bright sent out copies of the play *Eleanor* (printed, for copyright protection) to New York and in London tried for the Lyceum and the Wyndham. Scenery for the play was a difficulty. So much depended on balconies and views of the Italian landscape. Bright finally secured a lease at the Court Theatre and Marion Terry (at least the surname was hopeful) promised to lead. Rehearsals began in October 1902, and were promptly set back by Miss Terry spraining a tendon in her leg (although, since she was playing the semi-invalid Eleanor, it was not disastrous). Mary was involved in constant rewriting of her text, something that rather stimulated her, and was forever being summoned to rehearsal by telephone.

Eleanor eventually ran for fifteen matinées, from 30 October to 15 November 1902. The reviews were goodish, but audiences (after the first afternoon) largely stayed away. *Eleanor* was never revived in the professional theatre. But if nothing else, Mary Ward now had some valuable experience. Rather than dramatize *Lady Rose's Daughter*, she wrote an independently conceived melodrama about glamorous adultery and illegitimacy, *Agatha*, which she finished between March and July 1903 with the collaborative assistance of Louis N. Parker (another established playwright). She had hurried the composition, being assured by Bright that the play would come off in London before Christmas. But this fell through. 'All these theatrical matters are like a big quicksand—forever shifting',[19] Mary told Dorothy Ward with growing bitterness.

Meanwhile in America a version of *Lady Rose's Daughter* (which Mary Ward had no hand in) was produced at the Garrick Theatre in New York on 15 November and *Agatha* in the same week in Chicago. Both flopped. In Britain, plans to stage *Agatha* hung fire. Various managers turned it down over the spring months. But in August 1904 Mary told Henry James (who had his own sad stories about the theatre) that Beerbohm Tree had accepted the play, which she was rewriting. Viola Tree would take the lead. But she noted with what was becoming her favourite image, 'so far the theatre has proved the usual quicksand that it seems to be for everybody but Mr Barrie!'[20]

The first performance of *Agatha* in London finally took place at a matinée on 7 March 1905 at Tree's His Majesty's Theatre. Dorothy recaptures the excitement of the event in her diary.

> *The first Performance of Agatha*, at last—at last!—and in spite of the fact that the audience, as we expected, was of the dullest and stickiest it went off with the happiest success. Viola Tree's praise was in everyone's mouth .... Mummy is *very* happy, on the whole .... She had the circle box, over [by the] Princess of Wales (who sent for her after Act I) and with her father, Constance Fletcher and Mrs Clifford. Arnie and I and Walter were together in the balcony stalls, near Maud and others of the family.[21]

Tree was 'not discontented', Mary told Louise Creighton. But the reviews were tepid or cutting (e.g. 'passably repulsive', the *Athenaeum*), and after another matinée on 13 March, *Agatha* went into oblivion.

When Mary Ward wrote her preliminary notes for the novel *The Marriage of William Ashe*, she had also sketched out how the narrative would look as a five-act play. She was determined, this time, to have a hit. And things looked hopeful. The dramatic rights were bought in advance in America by a Mr Brady in July 1904. For the subsequent adaptation of *William Ashe*, she hired the services of a young American woman, Miss Margaret Mayo. The young lady had, Mary Ward thought, 'not a tincture of literature, but an astonishing instinct for stage effect'.[22] They worked well together. But despite Ward's 'literature' and Mayo's 'instinct for stage effect', the result was again frustration. The resulting play of *William Ashe* did well enough for an American touring company, which had bought the rights in that country from Brady. But there was no money for the dramatist in this. A long run in a metropolitan theatre was needed. And try as Mary Ward and her agent might, there seemed no London or New York theatre prepared to back her as a dramatist.

21

Family Matters: 1900–1905

THE first years of the twentieth century were golden for families like the Wards. Their income was large—between £6,000 and £8,000 a year—and it went a long way. Domestic service was still cheap and abundant. Dorothy Ward, for instance, interviewed at least nine cooks before finding one she liked for Grosvenor Place in 1903. Kitchen maids, upstairs maids, laundry maids, serving maids could be had for £30 to £100 a year plus board. Stocks had its butler, head gardener, and their subordinate retinues. Mary Ward's own entourage included her personal maid Lizzie Smith, her secretary Bessie Churcher, and her companion Dorothy.

The Wards could comfortably afford to run a landed country house (for which the mortgage was only 3.5 per cent on a loan of £15,000 with apparently no fixed term) and a West End house (for which the rent was around £500 a year). Mary Ward bore the expense of Dorothy's 'cottage' (it in fact had seven rooms)—Robin Ghyll, in the Lake District. The Wards holidayed *en famille* (a group which included Lizzie and sometimes Bessie as honorary Wards) three or four times a year on the Continent. In season, a furnished villa with ten large bedrooms, four sitting-rooms, views of the Mediterranean, and three servants could be had for as little as £8 a week, all in. The pound was the strongest currency in the world.

For their swarms of domestic servants and their luxurious accommodation the Wards paid prices that had barely changed in a hundred years. And, after 1900, they enjoyed the amenities of modern technology. Some were merely knick-knacks. In 1898, for instance, the Wards had entertained the Aldbury villagers at Christmas and New Year with family theatricals; in 1905, the same entertainment took the form of a kinematographic show. Culturally insignificant in itself, the episode prefigured the advent of twentieth-century mass media. That primitive, flickering film projector in a dusty little hall in Hertfordshire was something that was for ever altering the old 'organic' relationships between the landowning and the labouring classes of England. Mrs Humphry Ward's huge sales figures, although she did not realize it, were part of the same phenomenon. As she reached over the Atlantic to her vast American readership, so Hollywood would soon reach back.

In numberless other ways, new technology made the Wards' lives not just glossier but less tiring and more efficient. In 1898, Dorothy Ward would spend the whole of her Saturday mornings writing business letters for the Passmore Edwards Settlement. Her right arm would be rigid with secretarial cramp by lunch. In 1905, Saturday mornings were spent on the phone. But access to telecommunication did not then mean—as it does now—that the postal service had deteriorated. For a few years, people like the Wards had the best of both worlds. They could afford Mr Bell's expensive new apparatus and still have full use of Rowland Hill's penny post. Mary routinely managed an exchange of letters the same day with Reginald Smith. So too with transport. By 1905, the Wards had finally got rid of their lumbering old brougham and bought a car. (The Stocks butler doubled as chauffeur; Humphry could not drive, although Arnold Ward learned soon enough.) But despite the spread of privately owned automobiles in England, the train service between Tring and Euston still remained as good as ever and much better than it is today. In addition to faster and more frequent trains and better upholstery (in the first two classes), the Edwardian passenger was assisted at every station by an army of porters who—unlike their misnamed descendants—were actually prepared to carry luggage. When she arrived in London, Mary Ward—who had been chauffeured from Stocks to Tring Station—would not infrequently take a motorized omnibus from Euston. (Dorothy seems to have been fonder of the underground, which was adding substantially to its lines in the Edwardian period.)

Living when they did, the Wards enjoyed the best of private and public transport; of traditional and new modes of communication; of personal servants and new labour-saving devices. Mary and Dorothy could have their hair 'waved' under electric blow heaters in a salon in Oxford Street—and yet they still had the assistance of maids to help them dress for dinner. Their style of life at Stocks was practically feudal. And yet—in town—the Wards enjoyed all the conveniences of metropolitan life. New department stores like Evans's, Maple's, and Harrods had raised the quality and reduced the price of good clothes, furniture, and provisions to startling levels. In 1905, totting up her household expenses, Dorothy reckoned that buying all the household foodstuffs (less milk and cream) for the week came to £13 6*s*. This represented a sizeable catering operation. There were four dining-in family members and three live-in servants. The Wards would routinely have dinner parties twice a week and Mary's regular 'Thursdays'. Harrods of course delivered all the provisions Dorothy ordered, in their distinctive green vans or by their uniformed delivery boys on bicycles. It was, as Dorothy noted, 'considerably better than in the days of tradesmen'.[1] And considerably better than in the days (still to come) of supermarkets.

The Wards enjoyed this lifestyle—traditional comforts and newfangled mod cons—until the First World War. Suddenly, in 1914, good servants became difficult to find. The men were all at the Front; the women in the factory. And the servant mentality was destroyed. Dorothy Ward during the First World War constantly complains about how undeferential ('insolent') the servants are becoming. The low point was reached in 1917 when a chauffeur actually interviewed the Wards, to see whether he liked them, their car, and the quarters they offered. No, he decided, he did not like the Wards and went elsewhere. The *coup de grâce* was income tax and the war supertax which stripped all Humphry's pictures off the walls, and obliged the sale of Stocks's finer china, plate and furniture. But this ruination was well in the future. The 1905 life-style—which the Wards complacently thought would last for ever—would at least give them a good decade.

At any point during the pre-war years the Wards seemed to have a stable, unchanging family life. But, with the years, profound changes occurred. By 1905, Dorothy had found her permanent role as spinster companion to her mother. With Bessie Churcher (another spinster) she supervised the smooth running of PES, the play centres, and the invalid schools. She wrote 'endless letters', spent whole mornings on the telephone, and superintended innumerable girls' clubs, expeditions, concerts. Her life was one long act of duty and service—principally to her mother, whom she idolized. There is an illuminating vignette in her diary for 2 March 1905: 'At 11.40 tonight. I dishevelled and tired from an evening at the [PES girls'] Club and just going to bed. Mother and Father came in from their dinner at the Reginald Smiths, and mother came towards me smiling and looking so beautiful in her new black velvet and lace gown—and held out to me the *first copy* of *William Ashe* and bade me look—for there was a surprise for me—and lo! I discovered that the book was dedicated to *me* [i.e. 'D.M.W. Daughter and Friend']!'[2] Dorothy Ward was the kind of woman who inspired such gestures. Janet Trevelyan similarly dedicated her life of Mrs Humphry Ward to her sister Dorothy, with the clear implication that this 'daughter and friend'—invisible to the public and to posterity—was the foundation on which her mother's achievements rested. Not least, Dorothy spared her mother 'dishevelling' nights at the Centre which now bears her name.

Janet Ward was also kept close to her mother until quite late in life. Mary was temperamentally possessive. The girls, even as young ladies, did not have their own clothes allowances. Their mother made dress choices for them. In other ways, she reined her girls firmly in. When Dorothy, 18 years old and in Paris being 'finished', timidly asked if she might go to a ball, Mary Ward replied that it was 'out of the question'[3] without her mother present to chaperone. Dorothy Ward finally got her own bank account at the age of 29. By this date she was—among much else—managing several invalid schools.

Janet Ward made her decision to marry in 1903, immediately on finishing off the translation of Jülicher her mother imposed. Janet's chosen—George Macaulay Trevelyan—was not entirely to Mary Ward's liking. The pedigree, of course, was impeccable (his father, George Otto Trevelyan had succeeded Uncle Forster as Irish Secretary in 1882). So too was the education (Harrow, Trinity College, Cambridge). The young man was only three years older than Janet, but had substantial private means. He was brilliant and would in fact go on to become the country's leading social historian. The problems arose not with George's mind but with his soul. He was a product of the new Cambridge and, as Mary Ward thought, utterly godless. He was not for that reason without social idealism. He had been a member of the Christian Social Union in the 1890s (a politicized body of which Mrs Ward disapproved) and was now, in 1904, working on a study (a series of books as it would emerge) on the Italian revolutionary Garibaldi.

George wrote in April 1903 for permission to propose. The Wards (less Janet) were at the time vacationing in their favourite Villa Bonaventura, Cadenabbia. Formally, the suitor's application was to Humphry, but in reality it was Mary Ward whose assent was at issue. Dorothy recorded in her diary the domestic tumult provoked by George's letter: 'This has been a memorable day, for talks between M and F, M and F and me, and tonight me and F.—now a month's suspense!'[4] There would be a month's suspense because the young couple had been summoned out to be interviewed in person before permission would be finally granted.

The nervous lovers arrived on 20 May, amid a hailstorm of of telegrams. Mary Ward had her 'serious talk' with George soon after. It went well but there was, as Mary told Dorothy, 'a touch of pain in it for me . . . he is so wholly non-Christian. He looks forward to a civil marriage'. Did Janet realize, Mary Ward wondered, 'how different his ways of thought are from those in which she has been brought up'.[5] (The notion that the young woman might *want* something different from Jowett and the Greens seems not to have occurred to her mother.) Young Trevelyan's godlessness Mary ascribed principally to bad parenting by the other side. It was the product of the 'absence [of] any English Church tradition *whatever*, combined with the Unitarian *provenance* of Lady Trevelyan and the scepticism of Sir George'. But young George was, she conceded, 'a dear fellow'.[6] Permission was granted.

There ensued an epic battle on the great question of the form of wedding service. Oxford, of course, would be the site. That victory Mary Ward won easily. But ritual was stickier. The conflict was impishly described by Lytton Strachey in a letter of February 1904: 'Bride and bridegroom wanted *Office*, Bride's mother wanted *Church*; compromise arrived at—An Oxford Unitarian Chapel with a service drawn up by Bride, Bgroom and B's mother—at present chiefly Emerson.' Strachey was invited to this hybrid ceremony a month later and described to Leonard Woolf, with malignant hilarity, the wedding party's journey to Oxford by special train and the nuptials that followed:

My mother said we were a 'cultured crowd' and we were. Mostly matrons, in grey silk and hair – Henry James, Sheppard, Hawtrey, Theodore L[lewellyn] D[avies], etc. filled in the gaps. The lunch was free, and at separate tables, but the whole train was interconnected, so that there was a good deal of moving about. A High char-à-banc, with a horn, drove us from the station; flags were waved of course, and there was some cheering .... Mr Edward [Estlin?] Carpenter officiated. He began with an address composed of quotations and platitudes, during which, as Miss Souvestre said, the bride and bridegroom looked at the windows as much as they could ... the bride and bridegroom were almost completely hideous. But I suppose one must let copulation thrive. The service practically all balls in both senses.[7]

After the wedding Mary Ward was soon reconciled to her new son-in-law. It was all the easier since the couple lived in London and frequently visited Stocks. Mary was actually taking tea with Janet in February 1905 when her labour pains began (or, as Dorothy artlessly put it, when she was 'taken ill'). Mary serenely telephoned for a nurse and doctor and bundled Janet off to bed. She then spent the whole night by Janet's side and assisted at the birth of her first grandchild (a girl, Mary, named after her). 'I was allowed to see her just once,'[8] Dorothy pathetically complained.

Janet Trevelyan remained devoted to her mother, while at the same time expressing an ineradicable scorn and hatred for her mother's set. (It is likely that she had felt something of the kind since adolescence, but stifled it.) In a letter to Mary Ward in March 1908, she pens a description of the duffers and dead wood at a party at Lady Vera Herbert's which is almost Stracheyan in its savagery. And she concludes, 'it makes me sick to think of you in their company'.[9] Janet (a suffragist) remained convinced that it was Conservative bigwigs of this kind who had played on her mother's goodness to recruit her as leader of the antis.

By 1905, the Arnold daughters were, in their different ways, happily settled in life. As she looked at Jan nursing her baby, or at Dots dashing out of the house to PES, Mary Ward could feel a sense of maternal duty well done. Her girls were a credit to her. But Arnold, for whose future Mary had the highest hopes, remained obstinately unsettled. It was very perplexing. He had graduated from Oxford in Autumn 1899, a prize-man and a double first. He was dark, handsome, six foot tall, an athletic fourteen stone in weight, and a first-class shot. The world should have been at his feet. But frustratingly, it wasn't. Arnold Ward had no clear idea of what he should do beyond his mother's dream that one day he should do something 'great' in public life. How he should get into public life was not immediately apparent. He was not a patient young man; and although capable of terrific bursts of do-or-die industry he could not apply himself. And he had a most unfortunate tendency to quarrel with his superiors. Neither his housemasters at Eton nor his tutors at Balliol had found him easy to keep in line.

Like his father before him, Arnold Ward's first instinct was to mark time by winning an Oxford fellowship. This would give him an opportunity to look around. There was, as it happened, a vacancy coming up at All Souls in a couple of years. That college, with its links into British political life, would be ideal. In the meanwhile, it was decided that Arnold should go to Egypt and the Sudan, the scene of Kitchener's great imperial victories in 1898. He himself had rather preferred the option of a year in Europe—but his mother was against that. Something more purifyingly strenuous than Paris was called for. Egypt would broaden Arnie's mind and would give him an area of expertise when he made the eventual great step into public life. It would also keep him out of London's flesh-pots (Mary Ward had surely noticed her son's fondness for cards and club life).

Humphry arranged it so that in Egypt Arnold could title himself 'special correspondent for *The Times*'. It sounded grand but all it meant was that anything he wrote would be considered for publication on a per-column payment. But it would give the enterprising young man an entry and with luck supply a couple of hundred pounds to defray his expenses. The *Field* also agreed to look at any freelance pieces he might write. Arnold left in November 1899 on the month-long sea trip to Port Said. In Egypt, things started going wrong immediately. He presumed too much on his letters of introduction in January and was, as he indignantly told his parents, 'kicked out of the Sudan by Colonel Maxwell'.[10] John Grenfell Maxwell (a hero of Omdurman) was the commander-in-chief and was damned if he was going to have some whippersnapper calling himself a 'special correspondent' poking around his military installations. The Colonel was 'unwarrantably insolent', Arnold declared. He also took against the 'old crock of a *Times* Man' (the Cairo correspondent) and insulted him. Not surprisingly, the old crock wrote some sharp things about the young puppy, Arnold Ward, to his superiors in London.

The Orient irritated Arnold Ward immensely. 'Everything in Egypt costs double what it ought to ... the expenses of the place are *ghastly*,' he complained to his parents, who were paying the expenses. Nevertheless, as he coolly told his mother, attendance at Cairo's race meetings was 'obligatory'. Mary Ward's horror at a country whose inhabitants were required to go to race tracks was as nothing compared to Arnold's next proposal. December 1899 had seen the 'Black Week' of British defeats in the Boer War. Arnold—who had plenty of spirit—felt dreadfully 'out of it'. South Africa, he correctly perceived, was where the real action was. He therefore wrote asking his father to arrange a reporting position for him at the Front.

The suggestion struck sheer terror into Mary Ward, who saw with dreadful clarity her dearest boy lying stretched out dead on the veld. And who knew what reckless steps the young man might have taken during the month it took his letter to arrive? Humphry was made to cable at once to Cairo (on 2 February) peremptorily forbidding Arnold even to contemplate South Africa, reminding him that he was the 'eldest and only son' of the Wards. It was his *duty* not to risk his skin. In retrospect, the parents' reaction was a mistake. This sense of having great things expected of him—yet not being able to make any decisions for himself—was having a deleterious effect on Arnold's character. It was probably a main factor in his gambling—there, at least, he was a free agent, if only free to ruin himself. Humphry Ward should not have allowed himself to be pulled along in the train of his wife's hysterical anxiety. Photographers were at risk in South Africa, but reporters were not being killed in any great numbers. The Boer War might have been the making of Arnold (as it was for his friend Winston Churchill).

Arnold wrote back on 3 February a petulant letter, but one which was not entirely unreasonable. What on earth did his parents want from him? 'It would be different if I were employed in a definite profession. I am supposed to be seeing Europe and working with a view to securing £200 a year for myself. You have yourself dissuaded me from repairing to France or Germany. The Fellowship object though important is not particularly sacred. I don't suppose it would be damaged by a short time in South Africa; it might even be aided.'[11] Nevertheless, he conceded that he would not after all go to South Africa. But, as he told his parents two weeks later, his concession would cost them: 'I foresee that I shall require two years abroad rather than one to acquire the valuable foreign knowledge which I believe myself fitted politically to use. This will of course be relieved by considerable visits home.'[12] Arnold knew his mother well enough to anticipate that any suggestion of his leaving her side for good would reduce her to jelly. She agreed he should have his two years. On his part, Humphry Ward was told that a 'formal connection with the *Times*' was 'indispensable'. Arnold must be a '*Times* Correspondent', not a lowly special correspondent. Hard as he tried with his particular friend Moberly Bell, this was something Humphry simply could not do. He was not powerful enough himself on the staff. Nor had Arnold Ward's columns made a particularly good impression with the foreign desk.

Arnold decided to stay in Egypt until July 'when the *saison morte* begins' as he complacently explained. He was, in other words, going to kick up his heels a bit, before returning to to England to cram for All Souls in November. Whether he won the fellowship or not (and he was confident he would) he would take another year seeing the world. Then 'I shall have no hesitation in coming home for a lifetime's work.'[13] It is hard to imagine a more arrogant ultimatum from a young man with no fortune, no private income, and no skills other than an ability to turn Latin couplets and keep his end up on the cricket field. Mary replied, feebly adjuring 'meekness' but—as always—giving her darling what he demanded. Arnold spent his four easy months in Egypt toying with writing a play ('The Sphinx'). He learned a smattering of Arabic, wrote a couple of things for the *Field*, and made a useful friendship with Lord Cromer, the consul-general in Cairo. Arnold certainly did not live the life of an anchorite during these months and one suspects he may have come across some interesting new vices unknown to Eton and Oxford.

Arnold Ward returned as he said he would in July. To everyone's despair, he then failed to get the All Souls fellowship in November. It went to his old friend—and junior—Raymond Asquith. It was very hard. Men like John Buchan, Hilaire Belloc, and 'Raym' Asquith seemed to edge Arnie out at every turn. For the rest of his life, Raymond (with his wealthier background) would have the invaluable All Souls connection to help him to success in the bar and politics. Arnie's way would be stonier. Every year he shot with the Asquiths, and it must often have occurred to him how deceptively close his and Raymond's situations were. In fact, Raymond Asquith had the silver and Arnold Ward—unfailingly it seemed—had the wooden spoon.

What then could young Arnold Ward do with his life? He did not want to be a career journalist. The long hours and hard work did not suit him. Nor was it a convenient stepping stone to 'public life'. The only option open to him, as far as he could see, was the bar. That had the advantage of being a gentlemanly sort of thing to do, and traditionally barristers (H. H. Asquith, for instance) did go into politics and win high office. The disadvantage—from Mary Ward's point of view—was that young barristers spent their early years waiting for briefs and had to be supported in gentlemanly style in the West End of London. Nor was Arnold willing to take his bar exams until winter 1901. First he wanted that other year abroad which his parents had promised him in return for not risking his life in South Africa.

Arnold's first rather melodramatic plan was to spend the winter of 1900 in Australia and the summer in Siberia. Pliant as she was, his mother was not having that. They compromised on India. Humphry arranged with Valentine Chirol—*The Times*'s new foreign editor—that Arnold could again designate himself the paper's special correspondent but that this time his pieces would get a more sympathetic reception. Arnold set sail in December and landed at Calcutta on 9 January 1901, where he took up residence at the Bengal Club and introduced himself to H. Heasman, *The Times* correspondent. His first impressions of the subcontinent were not favourable. 'Calcutta', he declared, 'is rather like Dublin only if possible dirtier and more untidy. There seems to be too many natives, and they are mostly very repulsive in appearance.'[14] The 'educated natives' on the other hand he found to be 'bumptious'.

As in Egypt, Arnold was singularly immune to the charm of the Orient. But he settled down to study the country and work up the subjects Chirol had suggested for his *Times* articles. Indian finance was one ('my utter incompetence to deal with the subject is a slight drawback', he noted with rare self-mockery). The Frontier, tea tax, and Assam's railways were others. India, as it turned out, rather agreed with Arnold Ward. Study of the country's trade sharpened his ideas on Tariff Reform, which was later to be his main 'line' as a Unionist politician. He dined at Government House and hit it off with Lord Curzon, Viceroy since 1898. Curzon was a man notorious for his pomposity and aristocratic hauteur and was constantly in hot water with London. It is tempting to think that he and Arnold were two of a kind. For whatever reason, he smiled on the young man, who was made welcome at Viceregal functions.

Arnold also enjoyed the constant dining out. 'India,' he told his mother, 'is the home of hard work and big dinners'.[15] The shooting was also good and—unlike Egypt—there was an English 'society', composed of the families which administrators and senior military personnel brought with them. Arnold was in India in late January 1901, when the Empress Victoria died, which made his connection with *The Times* seem important. He learned some Hindustani, and in April—as the heat and dust descended on Calcutta—he began his serious travels. His first trip was to the Malakand Fort, from where he journeyed the length of the North West Frontier to Quetta. August and September were spent in Poona, working on his Frontier articles and in October 1901 he was off to Hyderabad.

Arnold was making good use of his time. But how much time did he want? It was clear that he had every intention of extending his Indian sojourn beyond a year. All this cost money. Money was, as it happened, in peculiarly short supply with the Wards just now. They had bought Barley End, a small estate bordering Stocks. (It was to be rented to their good friends the Whitridges. Frederick was a rich American lawyer, Lucy the daughter of Uncle Matt.) By way of reassurance, Arnold wrote estimating that his Indian expenses would amount to around £600 (a highly optimistic forecast). Of this, £200 would be covered by payments from *The Times* (another optimistic forecast).

Arnold Ward had some plausible reasons for extending his stay in India. He was thinking of writing a book on the country, and needed more research. He also felt he had exciting prospects. As he told Humphry, if his *Times* articles hit the right note he would have a very good chance of getting the Private Secretaryship to the next Viceroy. It would mean £1,600 a year and a luxurious house. And it was clear that Curzon—who had made himself very unpopular with the government—must soon go. In the face of these dazzling likelihoods, Arnold was allowed to stay on an extra three months.

He returned in April 1902, joining Mary and Humphry Ward at Cadenabbia. He and his father then went on to London, leaving Mary (exhausted by dramatising *Eleanor*) to recuperate more fully in Switzerland. In London, things did not go well. Chirol informed Arnold that he was going to hold back the young man's Assam articles (of which he was particularly proud) until the dead months of August, when Parliament was in recess and no one would be in town to read them. It was infuriating. But the fact was, Arnold Ward was a clumsy writer whose words—even when he knew what he was talking about—came badly off the page. Even he now realized that he had no great future as a journalist.

In May he began to read for the bar. Exams had never presented much problem to him, and he was duly called in early 1903. Back in London, he picked up his old social contacts, most importantly with his college friend Raymond Asquith and his stepmother, Margot. Arnold Ward was a favourite with the Asquith family—despite their being staunchly Liberal and he a Unionist. He was a regular guest at their dinners and parties and every autumn went up to shoot and play golf at their Scottish estate. He and Violet Asquith would eventually form something of an attachment—although it was absurd that as a penniless barrister Arnold should think of marriage.

In 1903, Arnold joined the 14th Middlesex Inns of Court Volunteer Rifle Corps. He liked the outdoor exercises and mock warfare. 'It is rather like being a lower boy at school again,'[16] he told his mother. Although he spent his evenings at clubs, he lived with the family at Grosvenor Place and Stocks. He was particularly close to his sister Dorothy ('Dots', or 'Haughty', as he affectionately called her) who worshipped the ground he trod. The two of them would go to Lords to see the MCC play or to galleries, she hanging on everything her wonderful brother said and faithfully recording it in her diary. He on his part was happy to help the family out by organizing cricket matches such as 'Stocks House versus The World' in July 1903—the world being young lads from PES, who probably also went away thinking Mr Ward was wonderful. In April 1903 he had dutifully enrolled at LSE summer school—to keep himself in trim for 'public life', whenever it should come.

For a few weeks in the early summer of 1903 it looked as if public life might beckon Arnold Ward sooner than expected. The question of Curzon's private secretaryship came up, and Mary wrote a letter to Walter Lawrence (the departing private secretary to Curzon) urging Arnold's qualifications for the position when it fell vacant. But Curzon held on as Viceroy, and insisted that Lawrence's successor should be an 'Indian civilian' (i.e. an Englishman resident in India). Arnold took it very hard. And thinking about Calcutta made him realize how boring law was. He would, he decided, go back to India anyway, and take his chances. If he did not get anything at Government House, so be it.

The prospect of Arnold leaving England for good terrified his mother. She wrote a letter, pleading with him to 'give up the India plan altogether'. The plea was accompanied by the expected bribe: 'it is *quite* possible that next year I may be able to help you quite materially'.[17] What she meant was that she expected a windfall from her new play *Agatha*. All those visionary tens of thousands of pounds should be Arnold's, to fund his equally visionary political career. She granted that England and English prospects were 'dingy and dull' compared with the viceregal life he had tasted in India: 'But, dearest son, nothing can really be dull for you, when you have once given it your strong clear intelligence, and you have only to work as other men have worked at the bar to succeed as they have done.'[18] Only.

Arnold thought it over in September when he went to shoot grouse with the Asquiths in Scotland. In October he told his mother that his plan was now to stay at Grosvenor Place until Christmas 1903, after which—if the Indian private secretaryship was still out of reach—he would move to a Chancery man's chambers for six months. And his next goal—not too far in the future he expected—was to go into Parliament. He made a rather perfunctory show of being uneasy about committing his mother to the large amounts which a parliamentary career would involve, and even suggested that the future golden earnings from the stage 'should go to your capital account ... it is neither to be desired or expected that you should continue much longer to do the amount of most fatiguing work you have done for so many years, and it will be very much better for me to get an independent job of my own. Home private secretaryships are of course perfectly useless for livelihood, though very good for parliamentary careers. Meanwhile the Indian scheme is at any rate in abeyance until Christmas.'[19] It is a typically confusing letter, in which Arnold seems to be saying a number of irreconcilable things. Not least, that he will have a parliamentary career, but somehow—magically—his mother will not have to pay for it.

December 1903 came and nothing more was said about India. Nevertheless, Arnold could not but feel that he was being left behind. His Oxford contemporary John Buchan came to see him. 'He is very likely to get a good appointment in Egypt,' Arnold told his mother. 'I am probably going in for Criminal law after Christmas,'[20] he added, without much enthusiasm. She had encouraged him to write, but a long article which he wrote on Curzon in the early months of 1904 was not accepted. Meanwhile, he worked on compiling a 'speaker's handbook' on tariff reform, to be published under the auspices of the Compatriots' Committee.

It was dreary work. So too was the interminable 'note making' which as a pupil was his main professional business. He was at the Birmingham Sessions in May where he prosecuted some luckless felon who pleaded guilty to stealing a bag of tools. After expenses, the fee of £2 4*s*. 6*d*. yielded the junior barrister precisely 6*d*. profit. In September, his first engagement in a London court brought the princely reward of four guineas. Two weeks later, he was at Dalquharra Castle arguing the merits of protectionism with Asquith and shooting (as he told his mother) 'well above my "form"'.[21] He also played the fine golf course at Prestwick while he was in Scotland. When he returned to Stocks, he talked Mary into extending the house's links—so that he could have proper golf tourneys, and invite a more important class of golfing guest. The extension would not, of course, be paid for by his one-guinea briefs. By the end of the year, Arnie was heartily sick of law work and its 'drudgery', which—as he complained to his mother—only 'possibly' leads to higher things 'in middle age'.[22] He was still a pupil—the lowest of the low in the profession.

By 1905, Arnold Ward had made some small progress in his profession. In March of that year, he was at Nottingham on duty for the first time as the junior on the Circuit. But it was nothing to boast about a week later when he dined at Brooks's with Winston Churchill and John Buchan—two contemporaries who were very much on their way in life. He was over the summer preoccupied with the extension to the golf course at Stocks and the golf festival in November to inaugurate it. His annual military service enlivened his vacation and in May he was—finally—elected to the MCC. All in all, it was a good season for cricket. In June he scored 118 at Bletchley Park in a club game. But July found him 'distinctly demoralised' by the prospects of another term's law. He was now playing 'hated bridge' at clubs and balls sufficiently often for his mother to warn him. He seemed to lose all sense when he was at the table. There had been some sticky moments when he needed to pay his debts. He was again considering giving up the law altogether and going into business—possibly with some colonial company or bank.

In August 1905 came a real stroke of luck. Arnold was in chambers as a pupil with the partnership of Mathews and Stephenson. On Friday, 4 August, Humphry Ward rushed down to Stocks by the 6.44 train with some extraordinary news. Guy Stephenson, the junior partner, intended to get married and was leaving the profession. Charles Mathews might well offer Arnold the vacant rooms and the legal work that went with them. It would be a major leap forward, and would bring remunerative briefs within a year (as Mathews promised). It could only be seen as 'a most extraordinary fluke'.[23]

On the following Tuesday, Arnold got up at 5 o'clock to see the volunteers parade at Ashridge and then caught an early train to London. The interview with Guy Stephenson in chambers went well. He could have the rooms for no rent, and under no conditions, 'but some degree of responsibility and regular attendance is understood'.[24] Mary—as soon as she heard the news—penned an overflowingly maternal letter: 'I have so wanted you to be *happy*—and to have all power drawn out, on the one side—and on the other—shall I say it?—by love and marriage (I hope no one will see this letter!) that I have been too anxious about the bar work.'[25] But now the bar work seemed to be settled. Parliament, marriage—anything seemed possible. (Privately, Mary Ward told Louise Creighton that she hoped this stroke of fortune would end Arnold's 'hankering after India'.)[26] Two weeks later, on 20 August, Dorothy had 'a long midnight talk with A in his room about finances and mother'.[27]

He was 29 on 8 November; by the time he was 35, he would—he confidently expected—be a burden on his mother no longer. As he shot and golfed with the Asquiths in September 1905, everything looked more cheerful. Arnold could now regard Violet with the eyes of a suitor. His year was crowned in December, when he was adopted as Unionist candidate for the Cricklade Division of Wiltshire.

22

Mid-Edwardian: 1906

By 1905 Mrs Humphry Ward could plausibly claim to be the most famous living novelist in the world. Russian prisoners taken in the 1905 war with Japan asked for her novels above all others. Their compatriot Tolstoy concurred in ranking Mrs Ward as England's greatest artist in fiction—England's Tolstoy. Although it was not a source of much money for her, she was translated and read everywhere on the Continent, in Scandinavia, in the Balkans, and in the English colonies. Serving army officers had a particular fondness for her, and over camp fires in the Karroo or on the Deccan, soldiers read *Marcella* or *Eleanor* and dreamed of lush gardens and demure English maidens. Mrs Humphry Ward was Baron Tauchnitz's best-selling British novelist, regularly selling over 5,000 copies in his English editions for Europe. When he went to India in 1901, Arnold Ward was astonished to find how many natives had heard of *Robert Elsmere*. Presumably if the Eskimo in his igloo or the Hottentot under the burning African skies knew of any novelist, it was Mrs Humphry Ward.

Dominating all else was her extraordinary hold over the American market. No English novelist—not even Marie Corelli or Hall Caine—had so devoted a transatlantic readership. As the American critic William Lyon Phelps complained, 'Her prodigious vogue is one of the most extraordinary literary phenomena of our day. A roar of approval greets the publication of every new novel from her active pen, and it is almost pathetic to contemplate the reverent awe of her army of worshippers when they behold the solemn announcement that she is "collecting material" for another masterpiece.'[1] He and his smart New England kind could sneer as much as they liked. Mrs Ward's novels still sold by the ton.

During all the upheaval and sheer physical strain of her years of fame, Mary Ward drew strength from the fanatic loyalty of her family and from her unusually friendly relationship with her publisher, George Smith. The attachment to Smith went far beyond business. He had, in a sense, discovered her ('made her', he might have thought). He had indulged her through the long travail that finally produced *Robert Elsmere*. He had been her wisest counsellor, banker, and—with the peculiar stress that she put on the word—'old friend'. She could always rely on him for the reassurance that she was a genius. There had only been one serious rift between them, just after the writing of *David Grieve*, when her demands for 'advances' had become excessive.

Ideally, this strength-giving relationship should not have broken on George Smith's death in April 1901. His place at the head of Smith, Elder was taken by his son-in-law, Reginald Smith. 'Mr Reginald', as Mary invariably called him, was well known to her, and was if anything a greater admirer of Mrs Humphry Ward's genius than George Smith himself. But psychologically, he did not quite fit. Mary Ward needed patriarchal authority in her publisher, and Reginald—for all his adulation—lacked that quality. He was, for one thing, six years younger than her. And in 1905 there were other wedges being driven between Mary Ward and her publisher. For other novelists, they would have been routine professional friction. For Mary—with her complex emotional needs—they were more damaging.

The main such wedge was her insatiable need for more money. It was not that Mary was greedy—although this was commonly said of her. Her demands arose directly from even more urgent demands that were being made on her. She had to pay for Arnold's career and settle his debts; and she had to pay for Stocks, a house which positively ate money. A hundred other rats were constantly gnawing at her granary.

Had the dramatic experiment come off, Mary would (like J. M. Barrie) have had her hundred thousand in the bank. She could have pensioned all her dependants in the style they expected and still have had money left over for her 'Life of Christ'. But the dramatic experiment had failed utterly. The Wards had accumulated no capital over the fat years since 1888, and were constantly remortgaging Stocks—their one asset. Humphry Ward was 60. The only substantial source of income the couple had in 1905 was Mary's fiction. They were always short and she was consequently always thinking ahead to the next novel. She did the 'scenario' for her 'Romney book' (*Fenwick's Career*) in July 1904, when she was only three-quarters of the way through *William Ashe*. This was not, of course, for her benefit as an artist, but to drum up advance interest in her publishers so that there should be no hiatus in her income. And then in November of the same year—with the ink barely dry on *William Ashe*'s manuscript—she proposed yet another work to Reginald Smith. Provisionally entitled 'Since *Robert Elsmere*' this would be a sequel to her great hit of 1888. In some alarm, Smith pointed out to his author that she was apparently intending to bring out three novels in the year 1905–6. The market simply would not stand it—to say nothing of her health and his bank account.

Mary Ward yielded the point. But if she could not treble it, she would at least have to double her rate of production to one novel a year. A logical step was to shorten narratives, while maintaining her per-book payment. Consequently the new 'Romney novel' was submitted as a 'shorter work'—something that would conclude in around six, rather than twelve, numbers of an American magazine. What she had in mind was a half-weight product of some 100,000 words. But although half-weight, it was certainly not going to be half-price. In July 1904—with the 'scenario' approved—Mrs Ward coolly demanded a £10,000 advance for her 'shorter work'—£4,000 more than she had asked for *William Ashe* and twice what she had asked for *Eleanor*. The sum petrified Reginald Smith. But he took the plunge and passed the cost along to America. The *Century Magazine* was charged £3,000 for the eight-part (as it turned out) *Fenwick's Career*.

It was a great *coup*. But her exactions had not finished. Having secured the £10,000 and a 40 per cent rise in her serial fee, she further demanded that her royalty rate be increased from 20 to 25 per cent—in both Britain and America. Edwardian authors enjoyed enviably high royalties by the current 10 per cent standard. But 1*s*. 6*d*. on a 6*s*. novel was steep, even in 1904. Smith feebly remonstrated about how expensive an author she was, with her incorrigible need to compose on proofs. With consummate skill Mrs Ward granted the point—and sweetly countered that, to cover this extra cost, she would expect the new 25 per cent royalty only *after* the first 10,000 were sold in England. She was inflexible about America—despite publishers telling her that no English writer had ever got 25 per cent. Smith finally gave in.[2] But, in his heart, he must have wondered whether his father-in-law would have been so easily manipulated. Would she have *dared* squeeze him so?

There was another wedge being driven between them—the *Times* Book Club and the so-called 'Book War' that it provoked in 1905–8. The *casus belli* was the Net Book Agreement of 1900. By this trade treaty, the Publishers' and Booksellers' Associations had agreed to abolish all 'underselling', or retail discounts. 'Net books' (effectively all books sold in shops) would sell at the publisher's marked price. No reductions would be allowed for bulk purchase, no surcharge imposed for buying in inconvenient places (like the provinces). A book would cost the same in London, Land's End, and John O'Groats. The theory was that the NBA would foster the nation-wide network of British bookshops, by not giving metropolitan stores with their huge volumes of sale an advantage over small regional outlets. History suggests that the theory was sound but Mary Ward had for her part never liked the Net Book Agreement, any more than she liked the old three-decker and the circulating library system. With every book she published, she could see that America bought three copies for every one bought in England. Why? To her mind the answer was simple. America had no retail price maintenance and no big circulating libraries.

The Times Book Club (TBC) was the brain-child of C. F. Moberly Bell, business manager of the paper. The scheme originated as something of a stunt to boost the paper's declining circulation. But Bell was also a man of some vision. He perceived that the main weakness in the British book trade was in advertising and delivery. Bookshops were altogether too passive. 'Any fool of a bookseller', he pointed out 'can wait for customers.'[3] You must—Bell argued—go out and hunt for customers. What he had in mind was to use *The Times* as a service for book borrowers and buyers. Readers of the paper would form themselves into a club, the membership fee being a year's subscription to *The Times*. The club would have its own rooms in Bond Street, which would serve to loan out books and also to sell 'next to new' library copies at huge reductions—usually while the same books were still on sale at full price in neighbouring bookshops. Members could also do their business by mail if they preferred.

The British book trade had no quarrel with the library activities of the club; it was furious at its book sales. These, it was felt, could make a sizeable dent in the retail sector. When he began the TBC scheme in September 1905, Moberly Bell was bandying about huge figures: there would be 50,000 potential members he claimed.[4] He foresaw orders of 4,000 for every new book published and possibly triple that when the scheme really got going. He told Reginald Smith, for instance, that he confidently expected to order 2,000 copies of *William Ashe*. The fact that he actually ordered 158 was only one of many reasons that Smith—like most publishers—was very chary about the club.[5] He felt it was simply undercutting bookshops; a ruse for circumventing the NBA and the retail system it aimed to preserve.

Smith concluded, very early, that Bell was 'vox et praeterea nihil'—all talk.[6] He wanted nothing to do with the TBC and became leader of the book-trade forces in the ensuing Book War. This put Mary Ward in a painful dilemma. Reginald Smith was her publisher. Moberly Bell was a particular friend of Humphry's. Moreover, she was convinced that the present structure of the British book trade accounted for her meagre sales in her home country. As Bell—not a man to mince his words—told Reginald Smith in October 1905, 'If you want to sell in dozens and don't want to sell in hundreds you are quite right, but *authors* like to sell.'[7] Mrs Humphry Ward was certainly one of those authors. The Book War smouldered on for three more years, until a compromise was reached. During all this time, she was in the thick of the fray, painfully torn between loyalty to her publisher, to her husband, and not least to her bank balance. She was a main architect of the eventual amnesty. But the book club affair served to cool even further her relationship with Smith, Elder. The house remained her publisher and her friend. But the old intimacy and closeness—the sense that they were something more than just author and publisher—was gone, perhaps for ever.

The mid-Edwardian year, 1906, began auspiciously for the Wards. Janet was pregnant again (Theodore, born on 5 July). Her husband George Trevelyan was also big with book (on Garibaldi) and clearly destined for greatness as a historian. Dorothy Ward's play centres were booming. Eight new ones had been opened in February 1905, and attendances were now running at over half a million a year. The vacation and invalid schools were also growing explosively. While Dorothy and Bessie Churcher did the front-line work, 'bussing' and 'cabbing' to all points of London at all times of the day and night, Mary Ward—like some Commander in Chief at GHQ—planned operations. She lobbied and fêted donors and men of power. Augustine Birrell, for instance, who was introducing a new Education Bill in Parliament, was brought to PES and made much of. Mary Ward prevailed on him to insert a clause obliging Local Education Authorities to fund children's play centres, vacation schools, and 'other means of recreation during their holidays'.[8]

Even Humphry Ward was on the crest of a little wave. He had weathered 60 well and for a couple of years had been happily engaged on a book-length study of Romney. 'Darling I am so glad about the Romney [book],' Mary told him, 'I really look forward to that book's coming out. It will help your position as an art critic very much, I think.'[9] She was, of course, flattering. Humphry Ward was now the dinosaur of British art critics. The notion that his monograph (authoritative as it is) on the eighteenth-century portrait painter would stem the tide of hated Modernism was laughable. But Mary Ward had to keep her husband from feeling too keenly how much he was now in her shadow. As a symbolic wifely submission (as if she were still Humphry's 'pupil', as she had been in those far-off days of their honeymoon) her new novel—*Fenwick's Career*—was also a 'Romney book'. It took (in sanitized form) as its main plot the best-known facts about the artist's career: his leaving his wife and children in Westmorland to make his way in London; his becoming entrapped by Lady Hamilton; his final reconciliation with his wife before his death. In the novel, John Fenwick—a gifted painter of humble origins—allows it to be thought that he is a bachelor. In London, his career thrives and he becomes romantically involved with the beautiful, rich, and learned Eugénie de Pastourelles (a character into whom Mary Ward put much of herself). His simple wife, Phoebe, learning of John's treachery, runs away from England with their daughter Carrie. It all results in disgrace when Fenwick's married state is maliciously exposed to the world. His reputation as an artist slumps and he descends to theatre scene painting. Finally, after many years, Fenwick is reconciled with his wife and grown-up daughter in their native Westmorland, 'cast down but not destroyed'.

Mary Ward—who never trusted herself to write a historical novel after 'A Tale of the Moors'—transposed the action of *Fenwick's Career* forward to the late nineteenth century ('about 1895', she noted in her outline). The novel was begun early in May 1905 at Robin Ghyll (which becomes the Fenwicks' married home and the scene of their eventual reconciliation) and was finished on 6 December—thus making it one of the fastest novels Mary Ward had hitherto written. All her fiction was fiction with a purpose. She made this

Romney book, like Humphry's critical biography of the artist, a manifesto for their reactionary artistic beliefs. In the story, Fenwick is a painter who combines the traditional strengths of the Pre-Raphaelites with the technical skill of the French Naturalists. He embodies a 'healthy' art against the 'maniac' and depraved Impressionists.

Everyone liked the new novel. It was hailed as a triumph by Mr Reginald when he read early proofs in January. So pleased was he that in addition to the 6*s*. version he agreed to an *édition de luxe* of *Fenwick's Career*. Finely bound, printed in 'Walpole' typeface, and carrying Sterner's illustrations (which Mary admired as the embodiment of Humphry Ward's artistic theory) it would cost a guinea for the boxed, two-volume set. Such things fed her authorial vanity and reassured her that Mrs Humphry Ward was more than just a best-selling novelist. On their part, Harper's were delighted with the reception of the book-form version of *Fenwick's Career* and with the stamina of *William Ashe*, which by March 1906 had sold 75,000 for them. As a mark of their pleasure they advanced Mrs Ward in January 1906 £3,000 for a 'long novel' which she need not deliver until autumn 1907.

Now, at last, the way was clear for Mary Ward to write the longed-for non-fiction 'Life of Christ'. Smith had always been hot for this project and now Harper's were also willing—'big gamble'[10] as they saw the sacred biography. They undertook to publish Mary's magnum opus in 1909 (after receiving their 'long novel') and to give it limited serialization in *Harper's Magazine*.

It was all highly gratifying. Mary Ward and Mr Reginald began seriously to lay plans for what would be the biggest publishing venture of their joint careers. 1906 should be kept as clear as possible, they determined. This would be gestation time. Harper's £3,000 was of course not enough, even with a matching supplement from Smith, Elder. Mrs Ward should, they agreed, top up her income with a short (three-number) story for Harper's in late 1906. At her new upgraded rates of pay, this should yield a quick couple of thousand and could be done in a few weeks. 1907 would be earmarked for Harper's 'long novel'. But with that out of the way, 1908 could be devoted entirely to finishing the 'Life of Christ'. For the first time in her professional writing career Mary had a breathing space and a long-term, three-book plan. She was off the treadmill.

As usual, the main hopes of the Wards in early 1906 were focused on their young prince, Arnold. Here again, things looked surprisingly hopeful. Arnie had been adopted as a candidate for Wiltshire North in the forthcoming general election. The constituency was actually in Swindon, a town of some 45,000 inhabitants, dominated by the vast wagon and engine works of the Great Western Railway. It was normally a Liberal stronghold but the political situation was highly confused and no seat was reckoned safe in January 1906. The previous December Balfour had resigned and his Conservative/Unionist

Government had fallen. Campbell-Bannerman had contrived to put together a convincing Liberal alternative. The general election would determine the nation's will.

Swindon was on the face of it a hard seat for Arnold Ward to win with its tough radical element. Men who spent the day swinging sledge-hammers in steel forges would not instinctively take to the young toff from London. But the country had been volatile since the 'khaki election' of 1900 and political upsets were likely. Pundits gave Arnold a better chance than his contemporary Winston Churchill, who was daring to stand in Manchester—Balfour's stronghold.

Arnold was standing as a Liberal Unionist. Even in the fluid parliamentary groupings of Edwardian politics, this was a somewhat eccentric party. It had been formed in the mid-1880s, by breakaways from Gladstone on the Home Rule issue. Since 1900, Conservative (also confusingly called 'Unionist') administrations had held power by forming alliances and electoral pacts with the Unionists like Arnold, who were in the odd position of Liberals who opposed Liberals. On their side, the Liberals proper comprised what modern politics would see as a broad 'Lib–Lab' coalition. Out and out socialists like John Burns rubbed shoulders in their Cabinet with old-style patrician Whigs.

As a Liberal Unionist Arnold was—by definition—against Home Rule. But Ireland was not going to be a factor in this election. He was in all important things a loyal ally of Balfour and of Bonar Law. But where he deviated from mainline Conservatism was on the matter of Tariff Reform. On this anti-Free Trade policy, Arnold devotedly followed Joseph Chamberlain. Not all Liberal Unionists did, and the party was currently in a dangerously schismatic condition. Effectively, Tariff Reform planned to erect a protectionist wall around the British Isles, with strategic preferences for Empire trade. Mary Ward was also a firm convert to TR, and in January 1906 declared that Joseph Chamberlain had a great future ahead of him. It was one of her unluckier predictions. In fact, he suffered a crippling stroke six months later.

Female suffrage was a dead issue in 1906, and if anything Arnold Ward was rather in favour of women having the vote. More controversially, he was convinced of the necessity of Chinese labour ('slaves'—as Liberals called them) in the Transvaal. Remote as it now seems, this was a live issue in the 1906 election. Coolies had been introduced after the Boer War to get the South African mines working again. (It would be some time before Kaffirs were ready to be exploited.) Rand millionaires were not generally popular in England and objections to coolie labour were framed in humanitarian terms. Mary Ward, whose views Arnold dutifully echoed, called it 'the Chinese Labour Falsehood'. When local churchmen mildly denounced the mine-owners she retorted furiously, asking 'whether Aldbury realises how much English prosperity and English industry depends on a free supply of gold'.[11] It was God's wish that Britain should be great and that yellow-skinned men should sweat in the bowels of Africa for it.

Before the campaign at Swindon began, Mary Ward confessed to Louise Creighton that it was a 'forlorn hope ... but it will tend to development and growth'.[12] As things got under way, she became more hopeful. Arnold might, she thought, have a chance 'if the socialists don't vote'.[13] Mary Ward was, of course, paying all her son's expenses, £500, from 'my own special account'[14] and felt herself fully entitled to meddle. He must not, she ordered, descend to 'personalities' against his Liberal opponent, the 64-year-old John Massie. She was not sure that 'Arnie knows the Free Trade side as well as he ought'[15] and gave him long lectures on the subject. Then she told *him* not to lecture the voters—'you are a politician not a professor'.[16] Arnold Ward must—for tactical reasons—support the religious clauses in the 1902 Education Bill, so as to keep the churchgoing electors sweet. It did not matter what he and she personally thought on the matter. He must firmly disavow the Tory slogan 'Ward and Cheap Beer', useful as it might be with the drinking classes. She edited his speeches, and was 'hurt' when he did not submit the proofs of his election address for her final correction. From London, she kept up a ceaseless barrage of telegrams and political newsletters (some of which she told him to publish as leaflets). And—she instructed—Arnold must be particularly mindful to wear a warm coat when he was out electioneering. Not for nothing was the young man lampooned as 'the member for Mrs Humphry Ward'.

Meanwhile, as usual, Dorothy Ward was doing the donkey work. She took up her post at the Goddard Arms on 27 December—probably the first time she had been in a public house in all her thirty-one years. Practical as ever, she hired a bicycle to patrol the Swindon streets. Dots was at her beloved Arnie's side as he (very nervously) gave his public addresses in the Corn Exchange and municipal swimming baths, and she bridled fiercely at the hecklers who dared interrupt her wonderful brother. She ventured with him into the even more nerve-racking 'Radical strongholds'; 'low-class' villages around the city, where young labouring men with villainous faces 'that they had not troubled to wash and caps adorned with Radical colours'[17] got up to taunt Arnold Ward and met his imperial slogans and quavering patriotic songs with a lusty chorus of boos.

In the first week, Arnold and Dorothy had met a couple of 'delightful policemen', the eldest of whom had remarked 'in a most kind and fatherly concern—"I'm afraid, sir, you're not going to get in"'.[18] But Dorothy was convinced that no one who saw and heard him could resist her brother. She worked herself to the bone for him and in ways that suggest that she was the real political animal in the family. In the opening days of the month-long campaign, Dorothy organized a 'woman's committee'. Although unenfranchised, women were charmed by Arnold Ward's looks, air of breeding, and gentlemanly manner.

Dorothy posted her women strategically outside the GWR works, distributing election leaflets to the men as they came home for dinner. She and her women canvassed the homes of 'doubtfuls', often tramping the terraced streets until eleven at night, while Arnold conferred with his committee over whisky and cigars in the congenial upper rooms of the Goddard Arms.

The general election—as was normal at the time—was spread over two weeks and Swindon voted late. Half-way through, Arnie's campaign was rocked by the sensational news on 14 January that Balfour—his putative leader—had been defeated at Manchester by nearly 2,000 votes. A barrister whose name no one could recall had trounced the out-going Prime Minister. It was very ominous. Mary Ward at once sent a letter of sympathy to the Balfour ladies. Privately, she prophesied that ungrateful England would rue its little faith in Tariff Reform. As always, she was puzzled by the lower classes' recalcitrance. 'The labour vote is amazing,'[19] she told Arnold. Her amazement about such things was to be frequently evoked in future years.

Mary Ward consoled herself with the thought that 'our boy is fighting hard' and that he and Dorothy were '*learning*, both of you, day by day'.[20] Even if they failed at Swindon, they might still be a winning team elsewhere. Mary, Humphry, Bessie, and Lizzie descended on the constituency for the eve of the election, 26 January. Polling day was fine and warm and from 6.30 a.m. the family toured the streets in a convoy of two motor cars (one hired, the other the Wards' faithful old crock, placarded and decked out with flowers). At lunch-time, they all stood outside the GWR gates as the men filed out to vote. The workers were polite enough to the ladies but felt no such chivalry towards Humphry who got some nasty looks and surly remarks. He expected the worst, and his worst fears were confirmed when at the count Arnold Ward was defeated by a whopping 1,578 votes. He took it stoically; Dorothy was inconsolable.

Accompanied by Lizzie, Mary Ward was off to Liverpool immediately after the declaration, to lecture on her new vacation school programme. Arnold returned to London. He was—he told Dorothy—too busy with briefs to come down to Stocks with her. He was perhaps intending to drown his sorrows. His rooms at 1 Essex Court Temple must have seemed very dreary. For a month or two he was very low. At the back of his mind must have been the thought that as Arnold Ward MP he might have presumed to marry Violet Asquith. (Her father was, with the Liberal landslide, heir apparent to the premiership.) As a briefless barrister marriage was out of the question. He bounced listlessly around the country-house circuit that spring.[21]

Even in less lugubrious moments, his political future was cloudy. 1906 had seen a Liberal landslide of such proportions—9 per cent swing, country-wide—as to suggest that the Conservatives were finished as a party of government and the Liberal Unionists as a party of any kind. As one historian notes, 'the tale of unseated Unionist chiefs reads like a list of the French chivalry at Poitiers'.[22] Balfour, Bonar Law, Alfred Lyttelton: all had gone under. Arnold Ward it seemed had joined the fight too late and on the wrong side. The irony was that he had not the slightest sympathy for the Union and regarded Ulster loyalists as a bunch of odious thugs. Had his mother not got that bee in her bonnet about Home Rule in the 1880s he would have been in the Liberal Party where he belonged, and probably on his way to a junior Cabinet post under old 'Sicky' (as he and his young friends irreverently called Asquith behind his back).

Mary Ward returned from Liverpool in early February to some unpleasant news. Her sister Judy had contracted cancer. It was heart-breaking. She was only in her mid-forties and her school at Prior's Field had been going from strength to strength; so much so that she had taken on a partner—Mrs Burton Brown (whom the Wards actively disliked, as a busybody). Her favourite sister's illness brought back all sorts of depressing memories of that other Julia, their mother. An operation was performed in the second week of February, by the eminent surgeon Sir Victor Horsley. It was pronounced a success and Judy came to Stocks to convalesce. But Mary—remembering her mother's 'successful' surgery in 1877 and 1882—remained apprehensive. With the onset of depression came the familiar physical symptoms. First, insomnia. Dorothy Ward recorded on 10 February that 'M. had poor night and is not at all well—rather *grey* which I don't like and *very* tired tonight.'[23] A month later, the eczema erupted. With it came its inevitable partner 'dejection'. Mary Ward was again at her breaking point: exhausted, sleepless, inflamed, depressed. She saw a specialist on 29 March about her arms, which were a mass of white blisters, and allowed herself to be bundled off the next day to Paris. On 3 April at the 'Français' Theatre she was taken 'horribly faint and ill' and had to be carried out at the beginning of the second act. She contrived to drug herself sufficiently to endure the journey to Cadenabbia, where the family settled in at the Villa Bonaventura, five days later.

It was, however, just a flying visit. For once the Alfred Trenches, who owned the villa, were in residence.[24] Humphry returned to London, leaving his wife in the capable hands of Lizzie and Dorothy. After a few days stay and much lounging on the villa balcony, Mary Ward was capable again of walking short distances and doing some sketching. Her health was about up to the next leg of the holiday by train to Milan, then onward to Rome. (The very *dirtiest* journey the family had ever taken, Dorothy thought.) In the capital, they stayed in a four star hotel and mixed with embassy people, *marchesas*, writers, painters, and rich Americans. It was a tonic. The weather was perfect. They took 'heavenly drives' into the Campagna, reliving the pleasures of seven years earlier at the Villa Barberini.

On 30 April, they moved on to Assisi, where Mary Ward had a powerful religious experience, which she recorded at length in her diary. She and Dorothy had arrived in the town on a cold and dark day, about five o'clock. After settling into their rooms at the little Hotel Giotto, the two ladies made an expedition to the church. The outer precincts were absolutely dark, until suddenly, turning the corner of the vestibule, they came on the eastern end of the church, incandescent with light. An Exposition of the Holy Sacrament was taking place, and the service was about to begin. Mary and Dorothy waited until it was under way, when an old peasant deferentially ushered them into a side chapel, directly behind the altar. Hundreds of slender candles were arranged in a blazing pyramid, showing the low-groined roof, dimly glowing with Giotto's frescoes. It was for Mary Ward—in her weakened condition—an epiphany. Later they hurried back through the cold and deserted Assisi streets, too moved to speak.

Four days later, they moved on to Perugia. There it was on 3 May that they celebrated 'Fenwick's birthday'. On 4 May, the three women all moved on to Florence. Every step of this itinerary was, of course, organized and superintended by Dorothy, who invariably spent the best part of any holiday fussing about luggage, couriers, tickets, hotel rooms, and her mother's physical and mental comfort. Lizzie, meanwhile, did the packing, unpacking, laying out, ironing, and nursing. Mary Ward rested. At Florence, they read telegrams from Janet, Humphry, and Reginald Smith, all reporting 'reviews are excellent'. The newspapers themselves followed, and 5 May was a day-long 'happy orgy in reviews'.[25]

They were indeed designed to make any author happy. The *New York Times*, for instance, wrote about *Fenwick's Career* in terms that would have seemed rather overpitched for Shakespeare:

> [*Fenwick's Career*] attains a height hitherto unreached by Mrs Ward. She has poured into it the deepest thought, her ripest wisdom and *Fenwick's Career* stands to-day as the noblest expression of her genius. It has qualities finer than anything Mrs Ward has yet written. One marvels at the warm, vital humanity of its creation. We rise from the reading spiritually and intellectually illumined. Mrs Ward has written a book of rare power and beauty. She has enriched literature and interpreted life.[26]

Excellent reviews of *Fenwick's Career* were accompanied by excellent sales. On publication day, the English subscription was 18,000; the colonial was 8,500. In America 37,000 had been taken by booksellers, of which 5,000 went to Canada (a market in which the pro-imperialist Mary was increasingly interested). Getting on for 65,000 full-price hardback copies had been sold before publication. Stephen King would be respectful.

Mary and Dorothy Ward returned to England on Saturday 12 May, to be met at Victoria Station at 3.50 by a solicitous Humphry. He bustled them back to a redecorated Grosvenor Place. The indefatigable Bessie Churcher was there to report that the play centres and PES were all running smoothly. Mrs

Ward—ostensibly rested by her weeks abroad—plunged back into London business. Early summer was always socially hectic. 'If only there were forty-eight hours in the day!'[27] she told Humphry on 30 May who was currently in Spain, reporting on art for *The Times*. Her 'Thursdays' were now attracting thirty to forty people every week—many of them politicians who, it was hoped, could do something for Arnold.

The *Times* Book Club affair was still festering on, without any sign of a solution. Reginald Smith had been furious in early May when Moberly Bell prominently advertised *William Ashe* (a book which Smith, Elder now had in print at 2*s*. 6*d*.) for 11*d*. It was no help that *The Times* itself was in a bad way and Humphry Ward's days were possibly numbered. Mary had to tread very carefully lest by siding with her publisher she precipitated her husband's dismissal. There were other sources of pressure. Play centres were opening up all over London and all of them bursting at the seams. Raising cash for them could be called a Sisyphean task, except that Sisyphus at least had a rock that stayed the same size. Eight centres needed £900 running costs in 1905. By 1908, there would be twelve centres with running costs of £3,000. There was no end in sight and no one but Mrs Humphry Ward to drum up the ever larger amounts of charitable donation.

Overriding all else in the summer of 1906 was the thought that she *must* soon begin her 'short novel' for Harper's. (Obviously, the 'Life of Christ' would have to be put aside for a while). Arnold's election expenses had been substantial. The political situation was fluid and another election might come any week. 'If anything untoward happens', she told Arnold, 'the family resources are so slender alack! I realise that so much when I am ill.'[28] She *was* the family resources, and she was—as it gradually emerged—seriously ill; perhaps even dying.

The little remission Mary Ward had enjoyed between Assisi and her return to London quickly evaporated. All through May and June she agonized about money and Arnold, Arnold and money, money and Arnold. It wore her down and stole her sleep from her. On 11 June, Dorothy Ward returned from Robin Ghyll 'shocked to find how *ill* [Mother] looks with this horrible eczema. I am afraid it will be a long business again.'[29] The fugue of insomnia, eczema, drugs, and depression was set in motion. But this time, the outbreak of white blisters on her arms, shoulders, and face was worse than ever, and exacerbated by 'a plague of boils'. All the while, Dorothy and Bessie Churcher—infected by Mary Ward's anxiety—were themselves working to the point of breakdown. Bessie could not even be persuaded to take a day off to meet her sister, just back from China.

When he saw her in late June, Mrs Ward's London doctor, Henry Huxley, was 'shocked by her condition, especially the boils'.[30] He lanced the worst of them, freezing her skin first with a local anaesthetic. He then arranged a consultation with Sir Phineas Abraham (a specialist in dermatology) on 29 June. This great medical man prescribed a new remedy—boracic wrappings—but no drastic change of treatment. 'Harrogate may have to come,' he warned. Mary Ward, that is, might have to take a proper convalescent stay at some resort, where she would be confined to bed-rest and absolute quiet for months. The prospect terrified her. If she sank into permanent invalidism, who would pay the bills?

On Friday 13 July the London season was formally over and the family made their annual 'summer move' to Stocks. 'Poor Bessie', as Dorothy noted, was left alone in London to handle the opening of the vacation school at PES on 1 August. Dorothy was needed at her mother's side. Unable to face the train, a 'weak and languid' Mary Ward was chauffeured down to Hertfordshire. It added to her depression rather than otherwise that on the same night of the thirteenth, Arnold was adopted as Liberal Unionist candidate in Watford. It was for him a minor triumph. The local association was initially hostile; but after he addressed them for thirty-five minutes they accepted him 'with enthusiasm' (or at least so the faithful Dorothy thought). It was true that the constituency was local—just a few miles from Stocks—and that all sorts of strings could be pulled for Arnold. But the expense would be never-ending and with an electorate to nurse Arnold would be even less inclined to work at law. Already the committee had bluntly asked him, 'Are you prepared to pay your election expenses and to subscribe to the funds of the West Herts Unionist Association.'[31] Arnold was prepared—so long as his mother gave him the money wherewith to pay. Dorothy noted Mary Ward's despondency on Sunday 15 July: 'A[rnold] went off this evng. as he plays cricket at Woolverstone tomorrow. M. very low and depressed tonight, also about A. Mr Percy, the Assoc. Agent to luncheon and a long talk with A.'[32]

It triggered off the worst health crisis in Mary Ward's life. Natural sleep went altogether, and rest was only possible with dangerous overdoses of sleeping draughts. By day, she was often too weak and light-sensitive even to lie on the lawn at Stocks in her *chaise-longue* or be pulled in the lug chair. Styes formed in her eyes to add to the 'maddening irritation' of the eczema. Dorothy urged a night nurse; she and Lizzie were themselves ragged from lack of sleep. But Mary Ward would not have it. She associated nurses with deathbeds. On 22 August, Dorothy organized a nightmare journey up to London and back, to see a new specialist, Henry Crocker, a foremost expert on eczema. The next day (Thursday 23 August), Dorothy had again to go up to town to help a distraught Bessie, sinking under the strain of running the vacation school single-handed. She left her mother lying in the garden, looking 'very grey and tired', and rushed off to catch the 2.39 from Tring Station. Still rushing, she was back by the 8.13 that night. She found 'M. in great misery and *utterly* dejected'.[33] It was clearly dangerous for Dorothy to leave her mother, even for a few hours.

August 24 she recorded as 'the worst day poor M. has had since the arms were acute. This afternoon she could not *bear* the pain—which was made most acute of all by Dr Crocker's new ointment'.[34]

Mary Ward was losing weight rapidly (a pound a day at the end of August). She was getting two and three hours of sleep a night. More was only possible with doubled doses of veronal or 'by dint of a m.s.' (morphine syringe). Waking at three or four in the morning, her depression was suicidal. No less than six doctors were treating her. One of them had prescribed a high-milk diet, which disagreed with her. She was plastered with remedies for the eczema, none of which helped. Dr Crocker's lead-based lotions, Sir Phineas's boracic wrappings, and Dr Brown's 'hemlock ointment' all seemed if anything to make the irritation worse. She was thrown back on her old standbys—oleate of zinc and cocaine. These at least soothed if they did not remedy. Meanwhile, Stocks was sweltering in a late August heatwave.

On 30 August, unable to stand her mother's piteous moaning at night any longer and 'at the end of my tether',[35] Dorothy Ward took it on herself to hire a night nurse by telephone from London. Nurse Ferrall was at Stocks the same evening. There followed a series of days and nights which the frantic Dorothy recorded in her diary as 'the worst yet'. Mary Ward's misery was compounded by a 'wholly unexpected bout' of her old enemy 'side' on 31 August. A week later Dorothy recorded, 'a very bad day alas, after a *third* miserable night, barely two hours sleep. Irritation almost unbearable again. It fills me with rage and despair to see her suffer so, to hear her groan and *yet* be so sweet and gentle with us all.'[36] Nevertheless, they all went to Robin Ghyll for the month of September (Humphry's annual leave from *The Times*). In the small whitewashed cottage and surrounded by her beloved Lakes, Mary Ward slowly recovered. By early October, she was back at Stocks and two weeks later Nurse Ferrall could leave. And on 15 October after an unaccustomed two nights' good sleep Mary Ward again took up her pen.

She began tentatively to write 'A Story of Assisi', an account of her experiences in early May. Notes survive,[37] and suggest a rather interesting idea. Arthur and Anne Lewison are unhappily married: he is a bank clerk, she is a 'Newnham girl' with a second-class degree. They live in an 'ugly block of flats' in St John's Wood. Arthur has ritualist tendencies, and converts to Catholicism. There are violent scenes between the couple. He goes to Assisi. Eventually she follows him, and has the same epiphanic experience in the church that Mary and Dorothy Ward had on 30 April 1906. She continues to loathe Catholicism intellectually, 'but the artistic side of her is captured'. Out of pity, love, and greater understanding, Anne is thus reconciled with her Catholic husband. 'The wrinkles smooth out of his face,' and he dies. The story then peters out on the page. Mary simply did not have the energy to write it up, and perhaps she felt it opened up too many of her complex feelings about her parents' marriage. But the episode at Assisi was too good to waste, and she eventually wove it as a patch of Italian colour into *The Testing of Diana Mallory*.

However feebly, Mary Ward was back in harness and Dorothy rejoiced to see 'the old eager light in her dear old eyes'.[38] There were other flickers of life. Arnold had been conspicuously absent during the worst six months of his mother's illness and was just back from shooting with the Asquiths. Mary Ward wrote to him a long letter in October, complaining that 'I always feel with some trouble that Ireland is rather a gap in your mind—and you ought really to try and fill it.'[39] There followed the familiar lecture telling him exactly what he must do and think on the great question (on which he was, indeed, entirely lukewarm).

That Mary Ward could again worry herself about Arnold's career was another sign of recovery. She would, after all, not rest in peace. Not that she was out of the wood. There had been some alarming new heart symptoms, and she was prescribed a strychnine tonic, which she would take at times of stress for the rest of her life. She was also bothered with persistent 'internal troubles' (probably arising from the pappy diets her doctors recommended). Full recovery was out of the question. Her health was permanently impaired and she would never be an entirely well woman again. But semi-invalids—as many Victorian women had demonstrated—could still write novels. They could not, however, write lives of Christ, with all the travel and library research such a project demanded. Mary Ward's great enterprise was shelved and never taken down again.

Mary Ward had sufficient worries for the moment. The de luxe *Fenwick's Career*—unlike its humble 6*s*. and $1.20 brothers—lost money. Having pushed the thing on Reginald Smith, she felt obliged to recompense him from royalties on the ordinary edition. The sums were not large, but it was wounding to the *amour propre*. She hated any hint that her popularity was vulnerable. She must, she decided, find some new wrinkle for her next novels. It presented itself with the visit of Mackenzie King to Stocks on 21 October. He was a former resident of PES and a future Prime Minister of Canada. His evangelical zeal for things Canadian 'and about the enormous field for British immigration'[40] infected her. Mary Ward had for some time been edging towards the Liberal Imperialist party ('Limps', as they were called). She would—she decided—write a couple of 'Canadian novels' when her next long novel for Harper's was out of the way. And to research these new 'novels of empire' she would make her long deferred trip to the North American continent—perhaps as soon as autumn 1907.

Meanwhile, in late October 1906, the Book War smouldered on acrimoniously. Moberly Bell's huge 'jumble sale' in May infuriated the publishers. The sight of their in-print titles—good as new but mendaciously described as 'used' copies—confirmed that the TBC was just a dirty ruse to circumvent the NBA. An embargo on supplies to *The Times* was organized. Reginald Smith was a principal organizer. Moberly Bell wanted his copies of *Fenwick's Career* at the wholesale price so that he could, in a month or two, dump them as loss leaders at less than half what the bookshops were charging. Smith refused to supply. Of course, he expected Mrs Ward—whom his firm had published for eighteen years—to support him. Humphry as confidently expected his wife of thirty-four years to back him and his friend Moberly Bell. Mary Ward explained her dilemma to Mr Reginald. He sympathized but stood by his decision: the TBC should have no copies of *Fenwick's Career*—not one. But he compensated Mrs Ward with an *ex gratia* payment for the lost sales (2,000, as Moberly Bell estimated). It was a gallant act, but did not in any sense constitute a permanent solution to the problem.

Mary Ward assumed her usual role of conciliator. She arranged for a private meeting of authors to be held at her house on 6 November, followed by a public meeting the next day. The *Daily Mail* got hold of the story, and Grosvenor Place was besieged by reporters, interviewers, and photographers snapping England's famous authors on Mrs Humphry Ward's doorstep. After their discussions, the authors (with Mrs Ward as the acknowledged leader) dispatched an ultimatum to the *Times* Book Club and to the Publishers' Association via Viscount Goschen as emissary. It was not their wish to destroy the TBC, or any other means of actually *selling* their books. What they proposed was a fair-minded compromise. Publishers should withdraw their boycott in return for a guarantee by the TBC that no new net book would be sold by the club for at least six months from publication and that books advertised as second-hand should be genuinely used copies. The publishers' answer came back via Goschen on the morning of Thursday, 29 November: 'an emphatic non-possumus!—M is *very* angry',[41] Dorothy Ward noted.

All in all, the year which had begun so auspiciously limped out in anger, frustration, and gloom. Birrell's Education Bill passed the Commons but foundered in the Lords on the religious clauses. It was dead by December 1906. The Book War was further from solution than ever. Mary Ward and Reginald Smith were now very tense in their relationship. Arnold Ward was in the doldrums, picking up small fees at the Old Bailey and complaining incessantly at the boredom of it all. Mary Ward would never now write her 'Life of Christ' and indeed would have to set to and write a stream of romances if she wanted to keep the wolf from the door.

Among all this gloom, Dorothy Ward was profoundly grateful. Her mother had not died. She was sure in the summer of 1906 that she would lose her. To have Mary Ward alive, and to be with her was joy enough. 1906 was, she wrote, 'the happiest Christmas day I have spent for a long time—I have enjoyed my family so! and Mother has been so *well*—walking to and from Church, to which

A. and she and I all went.'[42] There was further cause for joy. On 28 December, after some days of preliminary note-taking and planning, Mary told the family at luncheon 'that she had actually written the first pages of the new novel "The Testing of Diana Mallory" '. She would, she told them, 'allow fifteen months to the book so as not to feel at all hurried. So America will again not come off in the Autumn but probably in March 1908.'[43]

23

The Testing of Diana Mallory: 1907

THE New Year of 1907 began at a gallop. By 2 January, Mary Ward had the first chapter of *Diana Mallory* off to Smith, Elder's manager (the loyal Aitchison—now celebrating his fiftieth year with Smith, Elder). She was her own woman again. On the advice of Curtis Brown—not yet her agent but her 'adviser'—she suggested to Reginald Smith that her old copyrights were still not fully exploited. A 7*d*. Library Edition was duly arranged with Nelson, to partner Newnes's less sturdy 6*d*. editions. The Scottish firm had higher standards than its Fleet Street counterpart and Nelson's 'sevenpennies' were (and are)—for their rock-bottom price—handsome little volumes. They sold not spectacularly but steadily. The publisher cleared around 26,000 of *David Grieve*, 50,000 of *Robert Elsmere*, and 70,000 of *William Ashe* in the first six months. Mrs Ward received on account of her 1*d*. per copy royalty a £100 advance for the two older novels and £150 for the still fresh *William Ashe*. It was a pleasant boost to an income still stunted by 1906's illness.

But the growing influence of Curtis Brown in his and Mrs Humphry Ward's affairs nettled Reginald Smith. Literary agents had only just appeared on the professional scene. Edwardian publishers saw them as the serpents in their Eden, destroying former intimacies and reducing everything to cash nexus—almost as bad as if Aitchison had joined a trade union. Like other best-selling writers, Mary Ward saw it differently. Smith, Elder were her 'old friends'—but none the less she could not but be seduced by Curtis Brown's silken promises of 25 per cent from America—that figure which had become something of an obsession with her. And—Brown pointed out—Harper's were skimming 50 per cent off her cheap edition receipts. Normal now, the half division seemed barefaced robbery then.

Hyperactive though she was in early 1907, Mary Ward was still not looking well; she was 'anaemic and not a good colour',[1] Dorothy thought. But she was at least moving forward on all fronts. On 4 January, she chaired a conference of LCC teachers on 'Organised Games for Girls'. This was one of her more recent causes. Like the feminists, she strongly resented the physical passivity imposed on little girls. Wreck that she now was, Mary Arnold had once herself been a boisterous little girl, all of whose energies were strait-jacketed by the decorous immobilities and cult of deportment of the Victorian school system for young ladies. Netball, hockey, ballet, or eurhythmics would have done her the world of good at Rock Terrace. All the young Mary Arnold had was evangelical Christianity.

At the end of January 1907 Mary, Dorothy, and Humphry Ward all went down to Cornwall, to see 'poor Uncle Austin',[2] Humphry's brother, on his deathbed. It was something that would disappear in the twentieth century, but Mary firmly adhered to the Victorian practice of the family congregating round a dying relative, to see them 'over the bar'. Uncle Austin Ward had been paralysed in the left leg since an attack of influenza. He died on 17 February, peacefully, comforted to the end by Mary, now sadly expert in such terminal ministration. As always, Cornwall with its remoteness and mild winters suited her. She loved to fantasize about her wild, Celtic, Penrose origins and—more to the point perhaps—there was no telephone. She decided to extend her stay out of town with a holiday at Ventnor on the Isle of Wight. There, ministered to by Lizzie and with the occasional supportive visit from Dots and Bessie, she could steam ahead with *Diana Mallory*.

Mary Ward realized the necessity of regularly gingering up her fiction with new ingredients. The new ingredient for *Diana Mallory* would be passionate crime. A number of things had put this in her mind—not least the violent romances of Hall Caine, her nearest English rival with the American public. Arnold was now a practising criminal lawyer and could help her explore this new territory. Indeed, he was eager to do so. He questioned his senior, Mathews, about 'distinguished female thieves'. Together, Arnold and Mary reviewed the cases of likely woman criminals. Arnold had never been allowed into such a collaborative relation with his mother, and he found that he liked it. In fact, he got rather carried away. 'As for the female crime,' he asked her in March, 'have you thought of corrosive acid? She might vitriolize her husband for infidelity. There is a most fascinating vitriol trial begins at the Old Bailey tomorrow morning, story of passion, name of Foucault.'[3]

Mary Ward was not attracted by the suggestion. Wives throwing acid in the faces of their husbands was too far out of her orbit. And crime itself was eventually relegated to the antecedent matter of *Diana Mallory*'s narrative. The story, set in 1905, mainly concerns the love affair between two wealthy and beautiful young people, Oliver Marsham and Diana Mallory. He is a radical Liberal, she a fanatic Liberal Imperialist. Diana, a wealthy orphan, has been brought up abroad by an explorer father with a strange antipathy to England. On the day of their engagement, by melodramatic contrivance, Oliver discovers that Diana's mother was convicted of murder. Diana herself was only four at the time and unconscious of the scandal. The family name was changed and she was never told of the family disgrace. Oliver is willing to keep his word to his fiancée but his mother, Lady Lucy, implores Diana to withdraw for the sake of her son's political career. With aching heart she does so and the match is broken off. Diana subsequently discovers from a friendly old lawyer, Sir James Chide, that her mother committed her murder in mitigating circumstances, in self defence against a ruthless blackmailer. Left alone at Brighton while her husband went on an archaeological dig to Asia Minor the unfortunate woman was lured into baccarat and thence into forgery by evil companions, Sir Francis and Lady Wing. Taunted by the fiend-like woman, she 'accidentally' stabbed her through the heart.

The second half of *Diana Mallory* is political in flavour, melodramatizing Mary Ward's recollections of Swindon and Arnie's (verbal) buffetings by the rude radicals there. Oliver—whose family has been enriched by coal mining—falls foul of the Labour voters at West Brookshire. He is struck while electioneering by a stone thrown by a disaffected colliery-worker. His spine is obscurely injured. He loses the election, sickens, falls into morphia addiction, and almost goes blind before being restored by reconciliation and marriage with Diana (who has meanwhile been at Assisi).

Diana Mallory has never been claimed, even by her partisans, as one of Mrs Ward's great novels. But it has undeniable points of interest. It discloses, under the veil of fiction, some of her deepest fears and anxieties: her apprehension about Arnold among the unruly proletariat of Swindon; her growing alarm about gambling; her ineradicable sense of personal shame at the parental scandals of her childhood (here projected on to Diana, the murderess's daughter); her still aching memories of Willie's last morphia-ridden years as he died from his spinal disease. More positively, *Diana Mallory* indicates how thoroughly nationalism and imperialism had replaced religion as her main cause. The liveliest sections of the novel are those describing Diana's enthusiasm for 'England'. Diana, the novelist asserted, is 'the instinctive Conservative among women, just as Marcella is the instinctive Liberal'.[4] Mrs Humphry Ward's own instincts were now well to the right. She 'despaired of democracies,'[5] she told Dorothy in March 1907.

Mary Ward stayed at Ventnor until 23 March. Her weeks on the Isle of Wight had done her good. On the Saturday morning of her departure, Dorothy exulted at finding 'M. looking so well and vigorous and having washed her hair and read *much* Homer before I went in to her at 7.30! Her dear hair was plaited in the Gretchen plaits and her eyes were dancing. The island bade her a smiling farewell, as if to say caressingly—"I have done you good and helped you with your book, haven't I?" '.[6] Installed at Stocks the next day, Mary Ward walked vigorously to church and back. The whole family (less Arnold) was off to Paris on 1 April, where Mary was capable of actually shopping in Printemps and promenading at Fontainebleau. It was an auspicious few days. She and Humphry celebrated their thirty-fifth wedding anniversary in better spirits than for years. On the day itself, Dorothy took the traditional posy of flowers to her mother in bed, Humphry walking in behind her. Then they all trooped happily off to the Louvre.

The family returned to Stocks on 7 April, where they saw George and Janet Trevelyan off on their bicycles for a strenuous tour of the Alps. The couple had adopted an austerely intellectual life-style. Mary and Humphry Ward cleaved to the old comforts and spent the Easter holiday period at Ascott with the Leopold Rothschilds. By the end of April, Mary Ward was well into her novel and the familiar spiral of exhaustion and over-stimulation had started up. She was, Dorothy noted, waking up too early—at four or five in the morning. The ominous lug and bath chair were again necessary at the weekends in Hertfordshire. In addition to the strain of 'wrestling' with the novel, her relations with Reginald Smith were deteriorating fast; something that gnawed at her emotional security. The wily Curtis Brown was one rift. Another was the still raging Book War. Mrs Ward continued to give aid and comfort to the TBC while between Reginald Smith and the TBC it was war to the knife.

Moberly Bell was forever swelling Mary Ward's head with wild estimates of what her earnings might be, if only the club got off the ground. She could, he argued, sell her novels at half a crown and still receive 1*s*. 6*d*. royalty. Where was all her income going? Bell asked with a meaningful nod at Smith, Elder and Co. The public paid *nine times* the production cost of one of her novels, he told her: 'that is to say that the writing of the book—the paper of it—the mechanical printing of it—the binding of it—in fact the complete production of the book costs 2/- and [for] the mere handling of the book someone gets 2/6d'.[7] That someone was Reginald Smith and his bookseller friends. At the end of April, Bell (who seems to have been a consummate twister) claimed to Mary Ward that the TBC was selling over 2,000 volumes a day—which would have made it the biggest book merchandiser in England. 'Sober-minded men', as Reginald Smith tartly responded, saw the TBC as a 'fiasco'—the last desperate stunt of a bankrupt newspaper.

Nevertheless, Mrs Ward listened to Bell; and, more importantly, she wanted to sell her novels through TBC and Smith, Elder would not supply them. Mr Reginald was adamant on the point. There was only one alternative, painful as it might be. On 29 April a dumbfounded Smith asked Mrs Ward if what he heard were true—that without consulting him she was 'entrusting to another publisher the publication of a shorter book in England as well as in America?'[8] She was, and the publisher she had in mind was Cassell, a house willing to deal with Moberly Bell. Smith was deeply hurt. He could not understand it 'after what you have said and—I will add—what I have done'[9] (i.e. his *ex gratia* payment for 2,000 copies of *Fenwick's Career*). She did not relent.

Mary Ward did not reserve this new toughness for her English publisher alone. In the United States she was similarly intending to desert Harper for Doubleday—a firm more amenable on the 25 per cent royalty issue and less rapacious about reprints. Harper were even more appalled than Smith, Elder by Mrs Humphry Ward's defection from their list. They had given her an advance of £3,000 in expectation of a short work followed by a long work. When she fell so ill in summer 1906, they graciously agreed to postpone the short work until *after* the long. Mrs Ward—in their view—was legally obligated to give them this short work, not shuffle it off to Doubleday. Harper's rage doubled when they discovered that she additionally intended to serialize the disputed story with their despised low-class rival, *Everybody's*.

There were a number of official festivities in London that 1907 summer, and at an address to a Colonial Wives' Luncheon in the Hyde Park Hotel on 2 May, Mary Ward suffered an alarming pain in the region of her heart. Had it anything to do with nervousness? Dorothy wondered—or was it the dreaded Arnold angina, come to claim another victim?[10] On other occasions that summer, Mary Ward felt curiously feeble and tired when making any special exertion such as trying on a veil or putting on her hat. Even when dressing she had sometimes to sit down and rest.

It was very ominous. She resorted more frequently to her strychnine drops, but could not slow down. There still remained half of *Diana Mallory* to write. And in May, the American Miss Margaret Mayo, who had earlier dramatized *William Ashe*, came across to discuss alterations for the English stage production. Mary Ward intended to have a last try at the elusive theatrical jackpot. Mayo made her suggestions, Bessie typed them in, and the script was sent to Cyril Maude of the Waldorf Theatre in the Strand.

Mary Ward paid the usual physical cost for these exertions. By the end of May, her eczema was back and Dorothy 'was suddenly filled with a horrible fear of this summer being as bad for her as the last'.[11] Nevertheless, Mary Ward managed (with 7,000 others) the Windsor Garden Party on 22 June. As always, the early summer was hectic. There was 'too much social racketting these days',[12] Dorothy noted, and it was with the usual relief that the family made their summer move to Stocks ('this blessed place') on 13 July. They were, however, less away from things than in most years. Curtis Brown was down in mid-July to discuss Mrs Ward's still stalled relations with America and to guide her negotiations with Smith for *Diana Mallory*. At Brown's direction, she succeeded on 30 July in arranging best ever terms with her still rather balky publisher. Smith, Elder would pay a 25 per cent royalty for the first 10,000 of the novel, 27.5 per cent for the second 10,000, and 30 per cent for all copies sold over 20,000. She would additionally receive 6*d*. per copy for the colonial edition. Illustrations were out of the question with these rates of pay. On the day of publication, Mrs Ward would receive a £2,000 advance on account of all payments except for Harper's. Smith, Elder would not, under any circumstances, supply the TBC; and they undertook not to bring out a 1*s*. or cheaper edition for two years.

Curtis Brown had advised well. Although the advance looked small, it was founded on substantially improved royalty rates. Coming into the twentieth year after *Robert Elsmere*, Mrs Ward was holding her own as an author. The Wards would, however, need every penny she could earn. When they came down from town on 13 July they found waiting for them the architect's plans and estimates for remodelling Stocks. This had been something Mary and Humphry Ward had dickered with for some years. They finally took the plunge. The estimate was a fairly modest £3,900. Arnold offered to sell his South African shares and contribute a thousand, which his mother thought gallant but unnecessary. In the back of her mind, she was again gambling on the play of *William Ashe* coming good. And, even if it didn't, the sum was not large. Unfortunately, once the builders started uncovering its 120-year-old fabric, Stocks was revealed to be structurally rotten. Drastically revised estimates came one after the other. Within a year, Mary Ward had another financial nightmare to keep her awake all night.

But in late summer 1907, all looked well; rosy even. Cyril Maude had not come through but on 25 July Arthur Bourchier accepted *William Ashe* for the Garrick Theatre in London with enthusiasm, 'promising to make it his first new London production in autumn [1907]'.[13] (In fact, auditioning did not start until November.) Mary Ward wired Dorothy the good news in London. A contract was signed on 22 August. Nothing could now go wrong, Mary consoled herself. She wrote to Arnold telling him that the money from the play would be entirely his, a war chest for his political career. After some demur that the money might perhaps go to the rebuilding of Stocks (which he would, after all, one day inherit) Arnie good-naturedly accepted the offer.

When the contractors made themselves obtrusive, Dorothy whisked her mother off to her cottage at Robin Ghyll for ten days. To have her mother to herself under her own roof was the purest bliss for Dorothy, who recorded the treasured moment in her diary for 11 August: 'Tonight in my little firelit sitting room, so cosy with its two lamps while a storm of high wind and rain blew outside—Mother read to me Chapters 8, 9, and 10 of "Diana" [the proposal scenes between Oliver and Diana]. She lay back on the sofa, the old green reading lamp beside her—wearing the pretty lace bodice we bought last year from Paris and her dear purple velveteen skirt, and her dear hair shone softly round her eager, beautiful face. I don't want to forget it.'[14]

Storms continued to rage until 19 August, when the two women returned to London so that Mrs Ward could receive the Duke of Bedford on one of his ceremonial visits to PES. She continued all this while to be afflicted by eczema and—to a torturing degree—by insomnia. Her intake of trional reached near toxic levels for someone who had not built up a massive tolerance to the drug.[15] Stocks was by now a gutted shell; the furniture had been put in store and Humphry's pictures lovingly stowed away in very safe places. On Monday 7 September Dorothy recorded their farewell to the old house: 'This is the last night which—DV—we shall spend under this roof for 8 or 9 months to come.' She added, 'dear Mother and Father mind going away from it a good deal, I think'.[16] While workmen started tearing off the roof the family settled in at Stocks Cottage. It was comfortable enough but rather cramped. Humphry spent a lot of time in town. Reginald Smith, however, thought the asocial life at the cottage was good for Mrs Ward's writing. By mid-September, thirteen chapters of the novel were finished.

With the glorious prospect of a parliamentary career ahead of him, Arnold Ward spent much of the spring and early summer 1907 working on his India Tariff Reform Plan; something with which he hoped to impress the Liberal Unionist ideologues. Unfortunately, Austen Chamberlain (Joseph's political heir) was unimpressed. Arnold was meanwhile increasingly disenchanted with his Old Bailey work, although his briefs now earned a little more money for him. White flannels, shooting tweed, and khaki suited him better than wig and gown. He got huge satisfaction from playing for the MCC during the summer of 1907 and in August he joined the Hertfordshire Yeomanry at their annual summer camp. The yeomen were now a more efficient fighting force, following the government Act of 1907 which had reorganized them as a regular Territorial Army. Training was nowadays rather more than fun and games. Learning that he was at his military exercise by the sea, his mother wrote in her usual alarm '*Please* don't run any risk bathing! ... you know you were never a great swimmer.'[17]

In September there was another pleasant interlude for Arnold—his annual shooting with the Asquiths in the north. He was now seeing a lot of 'wild Violet'.[18] Her brother, his old friend Raymond Asquith, had married earlier in the summer. Raym's career had run an almost exact parallel to Arnie Ward's: he was an undergraduate a year behind Arnold at Balliol; both men got firsts; both competed for the All Souls fellowship; both were called to the bar. And now, Raymond had married Katharine Horner. She was not rich and his parents spent all their large income in entertainment. But the young people were so much in love that somehow they would make do.

In October 1907, Arnold steeled himself to do the same. Here too, he would run parallel with his friend Raymond. There followed a period of high tension at Stocks when he informed Mary and Humphry of his plans. On Monday, 21 October an excited Dorothy recorded that: 'Arnold came down by the 7.10, as promised, to have a long talk with mother. He and M. and F. have been closeted for an hour and a half in Mother's sitting room. I feel out of it—but that is a role that the unmarried woman of my age must often fill—And really I don't care for anything but for his happiness, and that he should see clear to do the right thing about it all.'[19] As Dorothy Ward was writing, on the other side of the door Arnold was explaining his position. He was in love: his intended was highly suitable; he had little money and little prospect of making money at the bar for many years; he had immediate political commitments for which he would need heavy subsidy. But, he evidently told his parents, he could probably pick up some useful income in business speculation and playing bridge for money at his London clubs.

Mary Ward gave no direct answer. The matter was, after all, very important. She thought about Arnold's situation for a day or two, then wrote a letter:

My darling Arnold—This is just to wish you God-speed in whatever may await you at Blagdon—supposing indeed that any crisis may arise. Father and I are only anxious that you should be certain first of your own *heart*—and then that in any future calculation of income, you should not be too ready to leave the bar out of account—that you should include writing—and that you should *not* include other casual things, to reckon on which is, we believe, not good for a man's best and highest character. Is it not well to ask yourself always what the *best men* one has known would have thought? That I think was the advice of T. H. Green and I often think of it. Meanwhile my darling, there is nothing we would not do to help you and as you know we have given you leave to pledge our guarantee—we are ambitious for you, we always have been—but perhaps we care most about your happiness. If a man is once happily married—with brain and character—life does the rest! Well, it will be hard if we can't manage to help our only son through—supposing always, it has come to a question of helping through—of course I dread lest the demands—the very natural demands—put forward might be more than could be possibly met—then you would have to wait. But everything might still come right—Ever my dearest son,—Your most loving mother.[20]

In the first part of this letter, Mary Ward indicates her disapproval of Arnold's gambling and speculations—but is not too worried about them. It seems that in 1907 she was well aware of her son's love of play, but still hopeful that it would be curbed. And a main curb would be a good wife and home. On the question of whether she would actually pay for that wife and home, Mary remained rather enigmatic. A 'guarantee' that Mr and Mrs Arnold Ward would not want is as far as she was prepared to go.

In one sense we know exactly what happened in that last fateful weekend of October 1907. Arnold Ward did not win his bride. Violet Asquith did not marry him. Why she never became Mrs Ward remains mysterious. Perhaps Arnold could not pluck up his courage to ask. Perhaps her increasingly fanatic Liberal principles (she was to be her father's principal apologist) prevented her marrying a Unionist. Most likely, Violet's parents—having already sanctioned one love-in-a-cottage match in 1907—discouraged their 20-year-old daughter from making another. Arnold simply could not support an Asquith in the style expected. Arnold did not win his bride in 1907, and remained a bachelor all his life.

There were other excitements at this period. On Sunday 27 October, Mary Ward received a cable from Miss Mayo: 'Ashe performance in New York splendid success.' If he failed to win a bride, it at least looked likely that Arnie would have his parliamentary fighting fund. On 30 October, Mary Ward finished the manuscript *Diana Mallory* in the same gallop that she had begun, and duly received her £2,000 advance from Smith, Elder. It was deposited at Barclay's Bank in Oxford—the account which the Wards had kept ever since their early married days.

There was further good news the following week when, after a searching examination, Henry Huxley pronounced that Mrs Ward was not—as she had feared in the summer—dying of cardiac disease. There was, he found '*no* trouble in the heart itself, but [the] *muscles* there [are] flabby, [and] don't tighten up as they should. *Rest* every day [is] *the* thing to be desired.'[21] This was very reassuring, but doubts remained. Even at her best, Mary Ward was now a dangerously frail woman. Just how frail was noted by Dorothy on a walk at Stocks on 17 December: 'in her struggles to save Flip [their dog] from the onslaughts of a donkey [Mother] *ran* a little way and then felt out of breath and *faint*, and had to sit down and put her head between her knees. I don't like that much.'[22]

On the whole, 1907 had been not a bad year. There had been the usual crop of death and sickness. Most distressingly, Judy relapsed with her cancer, as Mary Ward feared she must. In September, she was having 'X-ray treatment' for neuritic symptoms. But it was some relief that Mary Ward's own health was much better than it had been in the depths of summer 1906. She had finished her novel in ten months—very fast by her normal standards. And she had a promising new subject matter in mind. By 20 December, she was sketching out two 'stories of Empire', as she called them. One would be set in the Canada 'of the old Régime' the other in the Canada 'of the Far West'.[23] To supply colour and detail for these stories, she would—at long last—make her pilgrimage across the Atlantic.

It was a happy Christmas, but ruined on New Year's Eve, when Humphry phoned from London with the appalling news that Bourchier had—like all his theatrical predecessors—thrown Mrs Ward over. He was, it emerged, nearly bankrupt. The contract for the stage production of *William Ashe* in which she had so trusted was not worth the paper it was written on. 'there is no believing these theatrical people',[24] she sadly concluded.

24

The New World: 1908

BOURCHIER'S treachery was a depressing start to the new year of 1908. An alternative arrangement for the play was cobbled together at desperately short notice. *William Ashe* would now be performed at Terry's Theatre in London from April to May. To reduce rehearsal time, actors already familiar with their roles were brought from America, where the play (whose rights had been bought) had been in repertory for some years. The accents of the 'third-class Yanks' (as Arnold called them) were a liability. But much hope was put in the sexual attractions of Fanny Ward, the 33-year-old American who would star in the part of Kitty Ashe. Married to an Englishman, Miss Ward lived in London, but regularly toured in the United States. Mary Ward had her strong suspicions that this namesake was 'vulgar'. Nor was she in the first flush of youth and would have to be made up younger to play the 18-year-old Kitty. But there was no denying the actress's vitality.

These dramatic arrangements put Mrs Ward more than ever in the hands of her agents, Addison Bright and Curtis Brown. Brown was by January 1908 her closest literary adviser, filling the vacancy left by Reginald Smith. Brown dazzled his client with promises of a still-higher 33 per cent royalty—if only she dropped her exclusive arrangement with Smith, Elder and allowed him as agent to auction her literary work. (Mary Ward did not appreciate at this stage that he would charge a 10 per cent commission for the service.) Brown—a true believer in the twentieth-century mass market—argued his client out of the élitist prejudices which would restrict her appeal to a relatively few middle-class English readers. There was, he told her:

> a vast and really admirable class of readers in the United States who have never had your work brought into their homes in serial form because they do not, as a rule, buy 6*s*. books or magazines for which the annual subscription is half-a-guinea or more. The publication of a serial by you in a periodical having a circulation of approaching a million copies a month would, I believe, introduce you to a new field of readers and have a pronounced effect on the interest in your books.[1]

The periodicals which Brown mainly had in mind for this grand popularization were the Philadelphia based *Ladies' Home Journal* and the New York based *McClure's Magazine*.

Brown's was good advice, if maximizing her income were all that mattered to Mrs Humphry Ward. But what she did not yet fully realize was the pernicious effect that writing for the million would have on her self-image and sense of artistic mission. This in turn would exacerbate the already pronounced romantic, melodramatic, and posturing elements in her writing. Mary Ward had always flattered herself that she was a superior kind of writer—someone like Henry James with the mysterious difference that her work seemed to sell more copies. When she eventually saw *Canadian Born* (1910) printed alongside the *Tit Bits*-style journalism of Edward Bok's *Ladies' Home Journal*, she was appalled by the company she was now keeping and the tawdry appearance of her fiction on luridly illustrated newsprint. She sent the editor, Edward Bok, a copy of *Cornhill* (which carried the British serialization of *Canadian Born*) by way of censure.

Bok, however, was wholly unrepentant. 'You send me a copy of the *Cornhill* magazine and call my attention to the fact that there your novel is "properly presented", but, my dear Mrs Ward, if we published a magazine like the *Cornhill* we would have what the *Cornhill* has—a few thousand circulation and a few pages of advertising, and we would pay you for your novel what the *Cornhill* pays you. Such a dull-looking magazine as the *Cornhill* would be impossible in America.'[2] George Smith might spin in his grave, but Bok's logic was unassailable.

As a sop to her artistic self-regard, Mary Ward (like Henry James before her) invested more and more of her hopes in the de luxe Autograph Edition (called the Westmoreland Edition in England). This, she determined, would be her monument. Houghton Mifflin had agreed to co-publish the work in America. She had to swallow a diminished 7.5 per cent royalty, but there were good trade reasons for this. The boxed sets would have to be sold largely by subscription, which was expensive. Moreover, preparation costs of the lavishly illustrated edition were extremely high. But, if the sixteen volumes retailed at their estimated £1 apiece, even a reduced royalty should produce some worthwhile income.

Mary Ward had *Diana Mallory* off her hands in February 1908. For the next twelve months she adopted a dual authorial strategy. First she would, at long last as she hoped, earn some of the rich theatrical pickings which had so far eluded her. On the literary front, she would continue to mix long and short novels. Her next long work was already committed to Harper's and she had received £3,000 advance for it in 1906. The story had already taken firm shape in her mind as 'Robert Elsmere II'. (It would, in the event, be called *The Case of Richard Meynell*.) A principal inspiration for the new novel was the series of lectures on the 'Modernist' movement in the French Protestant Church which Paul Sabatier gave at PES in February and March 1908. This movement Mary Ward saw (quite deludedly) as her long-awaited 'New Reformation'—the dawn of the new 'rational' Christianity which, with British imperialism, would usher in the twentieth-century millennium.

In the interim, before getting on to 'Robert Elsmere II', she resolved to write two 'Canadian stories'. Fiction of Empire was to be her new line of goods. But first she would have to see the Empire—or at least Canada—with her own eyes. Thus it was that Mary Ward's long-meditated and much-postponed trip to the New World came about. There were a number of other motives: by travelling via New York she would be able to examine the American Play Center movement—generally agreed to be years ahead of its British counterpart. Stocks was uninhabitable, and the cottage was—as spring came on—somewhat cramped. The ocean air would do her good. She could give eight lectures in America for her favourite charities. She could make apostolic contact with the North American masses, who were her most fanatic admirers. Above all, she could refresh her imagination and mind with new stimuli—things and places larger than the Europe which was already rather stale in her fiction.

The last proofs of *Diana Mallory* were delivered to Reginald Smith on 20 February. Three weeks later, on 11 March, Mary, Dorothy, Lizzie, and Humphry were off to the United States on the *Adriatic*, a White Star liner. Mary mentally compared the luxurious liner with the 'cockleshell' *William Brown* which her mother had endured on the family's 1856 crossing. Even twenty years after her death, Julia's memory was often with her. 'The discomforts my mother must have suffered,' she mused, 'my heart often aches when I think of it'.[3]

Her heart might ache, but nothing else did. Mrs Ward was sedulously protected against any possible discomfort or inconvenience on her great voyage. Royalty could not have been more pampered. And even the Atlantic remained 'marvellously smooth'.[4] Mary Ward spent hours watching the ocean's 'boundlessness and solitude', indulging a poetic reverie which she recorded in her shipboard diary: 'I only saw two ships all the way. We passed perhaps five or six. At one time we sailed midway between the poles—nothing but the primeval ocean on either side, between us and the Polar ice, the brightly lit ship at night, gliding alone over the fathomless depths.' It was not all solitude and poetry, however. There were some interesting passengers in first class, including Lord Dalhousie's young son, fair-haired and 'athirst for adventure'. He was going up to Alaska for the gold-mining, 'with glee!'. Mary Ward noted him down as a future hero. Captain Smith, master of the *Adriatic*, was another pleasant find, 'handsome, fair-haired [an] ardent Tory and protectionist'.[5] Just the kind of Englishman in whose hands Mary Ward felt safe.

They arrived in New York on 19 March on a bright day. To Mrs Ward's annoyance, on landing the decks were rushed by reporters taking snapshots and shouting questions. They possessed, she thought, a 'strange ugliness' and were rather like a pack of 'Pekinese spaniels'.[6] At the dock the Ward party were met by Frederick Wallingford Whitridge, who whisked them off by motor car to a luxurious suite of rooms in his New York house. Whitridge was the American lawyer who rented the Wards' house at Barley End. A 56-year-old New Englander, he had married Lucy Arnold (Matthew's daughter) in 1884. He was a partner in the New York firm of Cary and Whitridge and a director of innumerable companies. A cultivated magnate, Whitridge also contributed to the leading reviews on political and economic topics. He was as rich as Croesus, and loved his landlords the Wards. He knew every American who mattered—including Theodore Roosevelt, the President, to whom he gave Mary an introduction. It was his dearest wish to be Mrs Humphry Ward's American travel agent.

Another of her distinguished travel agents was the imperialist proconsul, Earl Grey, the Governor-General of Canada (1904–11). Grey invited Mrs Ward to stay as an honoured guest at Government House in Ottawa. Sir William Cornelius Van Horne—the former president of the Canadian Pacific Railway—was a third supremely useful contact. Van Horne was a collector of old masters (something of particular interest to Humphry). He made available to Mrs Ward and her party for as long as she wanted a first-class CPR car, to do whatever she wished with.

For a few days the Wards first remained at the Whitridges' house on East 11th Street. Frederick was on hand to conduct them about. Mary Ward was stimulated by what she saw and even by the air she breathed, which she discovered to be of 'an extraordinary clearness'.[7] From the first, New York was a maelstrom. Among the innumerable overcrowded dinners that Mary Ward was obliged to sit through in the city was one 'where everybody had a card containing a quotation from my wretched works'. This naked adulation upset her. Even more so when, at one of Mrs Whitridge's teas, she overheard an ecstatic New York lady exclaim: 'to think that I should have lived to shake hands with the authoress of *Little Lord Fauntleroy*!'[8] On 28 March, she was hosted at a zoo-like dinner by the Play Centers Association of America, at which 3,000 people were present. In Brooklyn, her lecture on the 'Peasant in Literature' was heard by an audience of 1,200.

On 4 April, Mary and Dorothy Ward took the train to Philadelphia, a city they found 'all very American and amusing'.[9] On the journey, they were boarded by a lady reporter. Out of her 'six words with Dorothy' (as Mary calculated) this hackette concocted a column of nonsensical stuff for the next morning's *Philadelphia Ledger*. That was very American and not amusing. In Philadelphia, the women stayed in the Belgravia Hotel, which was quiet and something of a relief after New York. The audiences for Mary Ward's two presentations of the 'Peasant' lecture were gratefully smaller and more 'cultivated'. Humphry Ward, meanwhile, was in Boston lecturing and hobnobbing with Sir William Van Horne about pictures. 'One feels as if the poor things were all in exile,'[10] Mary told her husband of the old masters she saw on

American walls. Four days later on 8 April—with four lectures of the eight now done—the Ward ladies were back in New York, preparatory to their Washington jaunt. There they would stay with another distinguished friend, James Bryce, since 1907 British Ambassador to America.

The three Wards arrived in Washington on 12 April. In the capital Mrs Ward basked in 'brilliantly warm and sunny' spring weather and she and Humphry strolled past the White House, admiring the magnolias and the statues, and drinking in the 'warm southern air'. At lunch in the restaurant of the Library of Congress the author of *Robert Elsmere* was charmed to be recognized by the waitress who served them. Having ascertained Mary was comfortably installed at the Bryces, Dorothy and Humphry Ward then returned to New York. The remainder of her stay in the capital was a high point. As the guest of the Ambassador, Mary Ward met 'most of the Cabinet and high officials'. She was guest of honour at a 'little round-table of eight'[11] given by Henry Adams, at which the President himself led her into dinner. She had also had dinner at the White House (an experience she subsequently incorporated into *Daphne*). Afterwards, as a mark of signal honour, the President indicated that she should sit next to him in the ballroom. 'When the music was over,' she told Arnold, 'he and I plunged into all sorts of things, ending up with religion and theology!'[12] Roosevelt—who liked big egos—took to Mrs Ward immensely and accepted her offer to visit at London or Stocks when next across the Atlantic.

Mary Ward went north again on 20 April and was met at Boston Station by Dorothy with a porter beside her and a cab ticking over in the street. Her suite at the Somerset Hotel was full of flowers from Charles Norton and Sarah Orne Jewett. There was a log fire in the drawing room, and—miraculously—only one reporter (male) in the lobby. Annie Fields—the publisher's widow and hostess of Boston's most influential literary salon—was at hand to receive her. Dinners with Parkman and other literary brahmins followed. This was the pleasant part of the visit. The ordeal was a series of receptions for 500 to 600 people. 'They are *too* ridiculous!' she complained, 'but the touching thing is the distance people come—one lame lady came 300 miles!—it made me feel badly.'[13]

She would give her lecture twice more, at Boston and at Smith College: 'and then I think the poor old "Peasant" may really go to a well earned repose!'[14] The poor old peasant drew 1,200 in Boston's Jordan Hall and was very hard work (although by now, Mary Ward had mastered thoroughly the acoustics of public speaking). Her lecturing experience at Smith College was both hard work and very 'odd'. After the lecture, she told Humphry (who by now was back in England), 'we were carried off to a kind of Town Hall, where Mr [George Washington] Cable, the writer, addressed me,—before an audience—for half an hour by the clock, turning his back to the audience and haranguing the luckless me, till the world became one vast nightmare, and

Sally [Norton] and Dorothy looking on were alternately bursting with laughter and wrath'.[15]

Physically, she stood up to the strain remarkably well, although it would have been too much to expect perfect health. On 30 April, she experienced a sharp touch of lumbago. It persisted and by 3 May it was a full-blown attack of generalized rheumatism. She and Dorothy were now at Fairholt, Burlington, Vermont as guests of the Henry Holts (yet another gilded age New England publisher, of the kind that seems to have found Mrs Humphry Ward irresistible). On 5 May Mary and Dorothy Ward (always with the faithful Lizzie in attendance) went north to Canada. This inaugurated the inspirational and—hopefully—recuperative phase of the trip. In Canada, Mary Ward would relax in 'the warm feeling of being amongst one's own people'.[16] Much as she liked individual Americans—especially the rich and famous ones who liked her—she retained an Arnoldian sense that the country's culture was at root depraved and doomed. Canada, by contrast, was 'a land with an almost boundless future'. It was a pity it did not have more readers of novels, but Mary Ward would overlook that.

They arrived at Montreal on the evening of 5 May to be met at the station by Lady and Miss Van Horne. The first impressions of the city were not favourable. Mary Ward found it 'not at all attractive ... ugly and smoky'. But there were compensations. The Governor General, Earl Grey, was up from Ottawa to attend a great horse show. Mary Ward was also there in the Van Horne's private box. During his progress around the arena, Grey stopped to talk with Mrs Ward and that night invited himself to the Van Hornes' party, giving as his reason, 'I like Van Horne, and I wanted to see Mrs Ward.'[17] On her part, Mrs Ward liked these great people who so liked her. But the party itself was something of a bore. There were 100 guests and she was placed at the head of the top table, between Sir William Van Horne and the Governor General. As she later jotted down, 'talk difficult and spasmodic—food very bad—of which poor Sir William was mildly and philosophically conscious'. The dreary affair dragged on until 2 a.m. Mary Ward remained depressed by the urban aspect of Montreal, which she found French-looking and 'unspeakably ugly'.[18] Spring had not yet come this far north. Canadian Gothic was not, she found, to her taste. This was not the New World she had foreseen.

By 11 May Dorothy and Mary Ward were residing in state at Government House, Ottawa, which was both physically and socially more to her taste. Earl Grey obligingly fed them gossip ('it *ought not to be repeated*') about 'these plotting Catholics'.[19] In Ottawa Mary Ward hobnobbed with the kind of politician she liked best, Tariff Reformers, Trade Preferentialists, and British Imperialists. Four days later, it was on to Toronto and, on 20 May, Buffalo. From here Mary Ward motored down to the Niagara Falls, which figure so strikingly in *Canadian Born*. As it happened, she visited them but did not actually *see* the Falls. On the day she was there, 'a Thames valley fog, thick, white, impenetrable [hung] over the whole gorge'.[20] All Mary Ward saw was the Clifton Hotel railing; but her imagination was quite capable of supplying the shrouded scenery.

Mary and Dorothy Ward returned next day to Toronto, where her private first class carriage on the CPR was waiting for her. 'The car is yours, the railway is yours,' Sir William loftily declared: 'do exactly as you like and give your orders.'[21] Van Horne had also put at Mary's disposal the Royal Suite at the Queen's Hotel—free of charge. (All this was eventually repaid by puffs in *Canadian Born*). On 23 May, the Wards (accompanied by Lizzie and Dorothy's bosom friend Sara Norton) set off on their great journey west. Two days later, their train was stalled at a small wayside station, six hours from Winnipeg. The track had collapsed into a sink hole. They waited sixteen hours while thousands of cubic yards of sand and gravel were shovelled in, to make the road firm. Meanwhile, 'Snell, our wonderful cook and factotum, being in want of milk, went out and milked a cow!'[22] It was frustrating, it brought on an attack of 'side', and Mary missed important engagements in Winnipeg. But she had the main event for her new novel, *Canadian Born*.

By the end of May they were in Vancouver. Here they were looked after by F. C. Wade—Agent-General for British Columbia—and by Mackenzie King. The Far West was what Mary Ward expected Canada to be and she found Vancouver 'amazingly interesting'. 'If we only had some money!' she told Humphry, they might buy some real estate. 'Arnold ought to come,'[23] she added, hopefully. On the way back to the East Coast, the Ward women stopped at Banff at sunset on 3 June ('more wonderful ... than Switzerland').[24] Mary Ward now had a special train all to herself. 'We shall part from the Rockies with a pang,' she told Humphry: 'one's physical eyes will never see them again, but it is something to have seen them once.'[25] By the middle of June, they were back in New York, ready to embark for Liverpool. Lest there be any misapprehension about the purity of her motives in making her trip Mary asked Humphry Ward to put a notice in *The Times* declaring that 'the proceeds of Mrs Ward's eight lectures in America [i.e. £400] have been divided between the Evening Play Centres' Fund (£270), the Children's Recreation School, Passmore Edwards Settlement and the Quebec Battlefield Fund'. This last was a project to buy back the battlefields as a memorial to Wolfe. Mary Ward admired its imperialist sentiment.

The American tour was one of the most unequivocally triumphant episodes in Mary Ward's life. She must be a great person, if all these other great persons (not to mention the thousands of commoners) paid such elaborate court to her. But meanwhile, back at home, her fortunes as a dramatist obstinately declined to mend. *William Ashe* had its first night on 24 April. Bright cabled that evening with the news 'Ashe well received. Ward made personal success.

Press divided.' Unfortunately, the 'Ward' who made the success was not Mrs Humphry, but her *bête noire*, Fanny. This lady did not endear herself by sending her own cable: 'Press unfriendly to play—*my* performance highly praised!'[26] Mary Ward raged at her lead actress: 'her ill manners and conceit make me pretty furious . . . . The picture of Miss Fanny Ward in the *Daily Mirror* makes me perfectly sick.'[27] Fanny Ward had—in its author's opinion—'ruined' Mary Ward's play and she was all for sacking the actress until Bright reminded her that Miss Ward had a contract. It was all academic, anyway. *William Ashe* was withdrawn from performance at the end of May. Notices had been uniformly lukewarm or hostile.

No play of Mrs Humphry Ward's was ever afterwards performed professionally in England. Her experience as a playwright had been painful, expensive, and ominous. She had around £400 in rights revenue; but had paid out thousands in bills for theatres, casts, and collaborators. The time she had spent writing and rewriting could—if devoted to fiction—have earned her income she badly needed. Worst of all, her experience in the theatre had given her the sharpest taste of failure since J. R. Green rejected her 'Primer' in 1874.

This final theatrical disappointment came at an awkward time. The Wards were under unusual financial pressure in 1908. At the beginning of the year, Mary Ward had asked Reginald Smith to deposit letters of guarantee at their Oxford bank, in order to secure an extension on the loans required for the never-ending renovations to their house. As 1908 dragged into August, they were still in the cottage. Smith had also come through with extra advances to cover 'unforeseen expenditure on the rebuilding of Stocks'.[28] £2,500 had been received in this way on 28 February. Mary Ward seriously discussed with Louise Creighton the possible necessity of their having to sell Stocks. In May she was writing to Humphry from America hoping that 'you have been able to devise something [about] Barley End and the loan'.[29] (The hope, presumably, was to mortgage this adjoining property.)

In fact, Humphry Ward had done something rather cleverer. In late April he contrived to sell a Rembrandt in which he owned a half share. According to the *Daily Mail*, the painting went for £8,000. And then, on 23 May, he wrote to Mary to tell her that he had bought another Rembrandt, 'Bathsheba at her Toilet', for £60 which should mean another £1,000 for the Stocks renovations. The buying and selling fever was on him strongly in 1908. Later in the year, Arnold reported that 'Father has done a good deal over a Bonington.'[30] Arnold thought this money should be used to restore the walls of the Stocks billiards-room to their original panelled condition.

There is no precise surviving record of how much the remodelling of Stocks cost Mary Ward. But on the back of a letter of 12 August 1908 from Reginald Smith, she doodled some revealing figures. Recognizable are the £18,000 original purchase price on Stocks and the £7,000 which the Wards had given for the adjoining farm. But two other higher figures are found in Mary's jotted calculations: £34,000 and £45,000. The second of these is apparently the estimated market value of the improved Stocks property; £34,000 is apparently what the Wards had paid, or committed themselves to pay, as total purchase plus renovation costs. When Mary died in 1920, the outstanding mortgage to Lord Grey was almost £14,000 and £8,000 was owed to the Oxford bank. There is every reason to believe that Mary Ward's indebtedness increased inexorably over her last ten years with Arnold's compulsive gambling. But it also seems the case that the Stocks improvements—initially costed at under £4,000—eventually rose to three or four times that, driving her deep into long-term debt.

Books would have, as usual, to clear the debt. The immediate literary task awaiting her on the return from America was correcting the proofs of *Diana Mallory*. She also intended to 'greatly improve the last two chapters'.[31] A main problem was Oliver's blindness. Mary Ward wrote to Humphry's distinguished surgeon cousin, Fleming M. Sandwith, on this tricky subject. In the novel as she had written it, Oliver Marsham has a stone thrown at him by a rowdy, is hit on the spine, and goes blind from the injury. On being married to Diana he begins to recover his sight. (This ending was, of course, a recollection of what happens to Rochester in *Jane Eyre*.)

Was this medically plausible? she asked Sandwith. 'Yes,' he replied, 'it is possible to have disseminated (or insular) sclerosis of the spinal cord following an injury, and this may produce blindness.'[32] But there was a catch. This condition was incurable. But Sandwith had some colleagues at St Thomas's who were testing ideas put into circulation by controversial continental psychologists. They would tell her that 'hysteria' could produce Oliver's blindness symptoms and their remission. 'You will perhaps not like the word hysteria,' Sandwith said, 'but all you have to do is call it neurasthenia or a nervous breakdown.' Hysteria would also, he pointed out, explain Oliver's 'strange subservience to his mother'.[33] Mary adopted all Sandwith's suggestions into her last chapters. She thus became, I suspect, the first novelist in England consciously to incorporate the theories of Sigmund Freud into fiction.

With *Diana Mallory* out of the way, Mary Ward could concentrate on her unwritten novels. Actual experience of North America had rather altered her 'stories of Empire' project. She would still write her novel of the Far West. Called *Canadian Born* it would closely trace her exciting CPR experiences. This work would show Mary Ward in her role as 'amateur emigration agent'. But *Canadian Born* would not be written for a year. The other Canadian novel of the 'old régime' would never be written (possibly because she did not feel she had time now to do the necessary historical research). Instead, Mary Ward resolved to write straight away a novel attacking American (not Canadian) divorce mores. This would show her in her character as reactionary matriarch.

Her change of subject was primarily dictated by a growing antipathy to suffragism: an antipathy which would distort almost everything she wrote for the next decade. Divorce and emancipation went together in Mary Ward's mind. As she told Louise Creighton, 'the whole idea of marriage is becoming radically transformed in that strange [American] nation, and part of the strong opposition to the suffrage comes there from the feeling that it is the suffragist women who are helping on the disintegration of the family'.[34] The family must be kept intact, the franchise must be kept male. Oddly, Mary Ward does not seem to have considered for a moment that this new anti-feminist militancy might threaten her loyal (and principally female) readership in America. Just as the women in her own family were pro-suffrage and still loved her, so she expected that American women would swallow her criticisms of them and continue to buy her books.

The anti-divorce novel was initially called *Marriage à la Mode*. It was sold as a serial in England to the *Pall Mall Magazine* for £700. In America, *McClure's Magazine* took it for $2,100. For a work of just over 50,000 words, this was handsome payment. Curtis Brown was particularly pleased with the arrangement—the first time that Mrs Ward had contrived to serialize a major work simultaneously on both sides of the Atlantic. Another first was Cassell's being the book's British publisher, with the specific aim that this house would undertake (as Reginald Smith would not) to supply the TBC. Cassell gave Mrs Ward her now usual 25 per cent royalty, but the firm's chief editor, James W. Smith, insisted that the title be changed to something less 'difficult' for British readers. He suggested *Daphne*. Mary Ward acceded to his wish. It was the only favour the publisher can be said to have done for her. Cassell's eventually sold lamentably few copies of a rather wretched-looking edition, and by the time the book came out in June 1909, the Book War was over—meaning no sales to TBC.

McClure—who wanted to begin his serialization in January 1909—needed copy of *Marriage à la Mode* to be going on with by October 1908, and Mary Ward had the whole of the novel completed by February 1909. Curtis Brown meanwhile contrived to sell *Canadian Born* (called *Lady Merton, Colonist* in the USA) to Edward Bok's *Ladies' Home Journal*, for £1,700. Doubleday would take the American book rights.

All this was very galling to Harper's, who still fondly imagined that they had a special relationship with Mrs Humphry Ward. But she was growing very impatient with the New York firm. Pleading trade depression, they were selling half to two-thirds of the normal quantity of her in-print novels. Having savoured her transatlantic popularity at first hand, Mary Ward could not believe that they were really trying. Wherever she had gone in America, she had been besieged by readers following *Diana Mallory* in *Harper's Magazine* who wanted to know how it would all end. Her relations with Harper had been particularly soured in February 1908, when they asked her to remove from *Diana Mallory* a reference to Catholicism as a dying Church. She wrote back in fury. If Harper changed her text 'whatever you suppress I shall probably publish simultaneously in a New York paper with my protest. That a periodical like *Harper's Magazine* should be to this extent under Catholic dictation seems to me perfectly incredible—and I will be no party to it.'[35] The publisher capitulated at once but Mrs Ward was unmollified. She offered them in February her two Canadian stories, but demanded an exorbitant sum she knew they could not afford—£7,000 for book and serial rights to both works, half in advance.

Paradoxically, Mary Ward's relationship with Reginald Smith mended over 1908. The main bone of contention between them disappeared when *The Times* was taken over by Northcliffe. The book club was quickly brought to heel; Bell was made to submit to the Net Book Agreement, and the three-year Book War was over. Mary Ward resumed her confidential relationship with Mr Reginald, entrusting him with her most urgent business and financial worries. But she did not dispense with Curtis Brown. Eventually, Smith found himself in the oddly reversed position of advising his author how to deal with her agent and his pernicious 10 per cent. She was astonished to discover how big a bite this commission took out of her earnings. Smith solemnly instructed her how to haggle Brown's rate down while Brown, at the same time, was instructing her how to drive her royalty rate from Smith up. It was part of Mary Ward's peculiarly conciliatory genius that she could get both sides to work for her in this way.

Diana Mallory was finally published on 17 September 1908. The English subscription was disappointing: 11,000, as against 16,700 for *Fenwick's Career*. In America, Harper's sent out an initial 44,000 on sale or return in America and Canada: more than *Fenwick's Career*, if slightly less than *William Ashe*. Despite the presidential election in November (whose outcome Mary had discussed in camera with Roosevelt at the White House), *Diana Mallory* had sold 54,000 in America in its first two months: 'satisfactory',[36] as Reginald Smith thought. By December, Harper's realized that, in losing Mrs Humphry Ward, they would, after all, be losing a major asset; they indicated, finally, that they were prepared to surrender on the 25 per cent royalty. It was a year too late. Mrs Ward's benchmark was already in the low thirties.

1908 brought the usual burden of bereavement. By early September, the symptoms of Judy's cancer had reappeared with new virulence. Although she was clearly terminally ill, it was resolved by the family that nothing must be publicly admitted, 'for the sake of the school'. She *was* Prior's Field. If it were known she was dying, many parents would remove their children immediately. Even Julia herself had the grim facts of her condition kept from her. When she was finally told, she bitterly cried 'Why do I have to die, and die so young?'[37]

All through October and November, she languished. On 26 October, Leonard Huxley wrote a letter to Mary Ward, denying the report that Judy was dabbling in Christian Science on her deathbed. She had, he confirmed, tried therapeutic 'suggestion' earlier, but it was unsuccessful. A Dr Finzi was now treating her. Judy '*doesn't* want frequent visits from Ethel', Leonard told Mary (he himself disliked his sister-in-law Ethel Arnold). Mary was welcome. His wife's last reserves of strength were ebbing away: 'you see the enemy has got so deep. It is no ordinary case—alas—alas!'.[38]

During October, Judy was looked after at a nursing-home in Leinster Gardens. Then, in November, she went back to Prior's Field to die. Mary Ward wrote to Dorothy: 'my feelings are so mixed. I rejoice to think of her in her own room and lying in her garden—and yet those 80 girls, and the responsibility to them and their parents. The whole thing weighs upon me terribly.'[39] 'Everything seems to be a nightmare!' she added. On 10 November Judy was having the 'Becker treatment' for her disease.[40] Finally, on 26 November, Mary Ward informed her daughter that 'the summons has come'. She rushed down to Godalming for the final gathering. (The boys were similarly summoned from Oxford and Eton.) In her notebook Julia left a poignant message for her children: 'It is very hard to leave you all—but after these weeks of quiet thought, I know that all life is but one—and that I am only going into another room "of the sounding labour-house vast of Being".'[41] As always, the spectacle of death fascinated Mary. On 29 November, she reported that her sister's last hours were 'peaceful I am thankful to say—more peaceful than yesterday—but most piteous. She looks very unlike herself—yet younger—and in many ways like my mother. Her pretty hands, so full of character, are crossed on her poor chest—the eyes of course are closed. Yet once this morning she lifted them—and I think she knew me.'[42] Thinking she would not last the night Mary sat up alone with her sister, who died the next day, 30 November. The body was cremated: the first of the Arnold children to be so disposed of.

The death of the last of the siblings to whom she was close focused Mary Ward's family ambitions and cares ever more firmly on Arnold. But, at 32, the no-longer young man seemed to be no nearer his destiny than ever. The Liberal Parliament dragged on interminably. He was not married. He was speculating rather too freely for Mary's taste in business. And—most worryingly—he had fallen out with his senior in his law firm, Sir Charles Mathews. In April 1908 Mathews made it very clear that he had no further use for Arnold Ward and told him to find new chambers. Arnold found a berth with another senior barrister, Marshall Hall, but remained gloomy about the whole business and his future as a lawyer. As he bluntly told his parents, 'from my experience of four-and-a-half years under the especial wing of the head of one great department of the bar, all absolutely useless, I [am] not anxious to embark on a second venture of a similar kind'. He now regarded the law as so much 'wasted slavery'.[43]

At Marshall Hall's, Arnold would work 'on an absolutely independent basis'.[44] His mother had no confidence in Arnold's self discipline in such a drudging matter as law, and had Humphry inquire into what had actually gone wrong with Mathews. The lawyer was perfectly civil. He thought Arnold 'has a power of grasping facts and putting them in a narrative form and coming to quick and sound decisions'.[45] But the young man would not work. And, although Mathews did not say so outright, Arnold was insubordinate. Arnold on his part was not reluctant to criticize Mathews, and allege breach of faith: 'what has it all come to? [Arnold] declares that it is unheard of that he should not have got lots of work by this time, after three years as [Mathews's] official assistant.'[46] Humphry, rather dispirited about the whole business himself, urged Arnold to keep his spirits up. 'Mr Hall will probably help him more',[47] he assured Mary. 'How strange it is he has not found an opening,' his mother mused. Perhaps, after all, he should go into commerce, since 'he has evident business and financial ability.'[48] Probably, however, the best thing would be to put their hopes in a political future.

As the summer drew on, Mary Ward became more and more worried about her boy, who seemed not to have recovered from his disappointed marriage hopes in 1907. 'One knows nothing about you,' she wrote in August. 'Do write a chatty letter to your mummy—that can't make your head ache.'[49] What she did know, that Arnold was devoting his main energies to nights in London clubs and golfing weekends, was not reassuring. The letters of Arnold which survive from 1908 make frequent reference to Violet Asquith, and one at least strongly implies that she too was under considerable emotional strain. From Slains Castle, Aberdeenshire, where he was staying as usual in September with the Asquiths, Arnold wrote his mother that:

> an extraordinary thing happened last night. I had been for a walk with Violet, we returned at 7.30, nearly dark. Five to ten minutes afterwards she left the house, saying she must fetch a book she had left on the rocks. At dinnertime she did not appear. Several of us went out to look for her at once, the rest—about half—went through dinner. We could not find her. The house is built on the indented cliffs. It was not known to which of the inlets she had gone. The night was pitch dark, the cliffs in many places sheer from the sea. The coastguards and fishermen were summoned, and came bringing rope ladders and great torches, while a life boat put out and cruised below, burning red lights, but could not come near owing to heavy seas running. The scene was marvellously romantic and weird. Archie [Gordon, Arnold's best friend] performed heroic feats of crag descent. All the gullies were searched and nothing found. Margot and Elizabeth stood on the brink, and the Premier paced up and down.[50]

Finally Violet was found, lying unconscious on a grassy slope, a quarter of a mile from the house. She had, allegedly, 'fainted'.

A more Machiavellian mother would have urged Arnold Ward to make a journey across to the Liberals, and so exploit his friendship with the Prime Minister, who clearly liked him. Such changes of party were common and honourable enough in Edwardian politics. Instead—in one of the most unequivocally misguided decisions of her career—Mary Ward resolved to throw in her lot with the most bone-headed lobby in British Conservatism. Asquith's declaration in May 1908 to a deputation of ultra-Liberals that he had no objection in principle to an extension of the suffrage to women caused something like panic in the bastion of male reaction. Lords Cromer and Curzon ('the Curzon–Cromer combine', as they were called) desperately looked around for some woman to oppose the pernicious demands of new womanhood. Obedient figureheads were easy enough to come by from the ranks of their own blue-blooded wives. But they needed someone who would command popular respect and who had a head on her shoulders.

They finally asked Mrs Humphry Ward to head their anti-suffrage protest group. As Janet Trevelyan puts it, 'Mrs Ward groaned but acquiesced.'[51] Why did she acquiesce? She was in a minority among the women in her family and among the women in her own class. (Of those few notable women who could be induced to declare a public opposition to the suffrage, the vast majority were peeresses.) It may be that Mary felt she owed favours to the House of Lords for the Education Acts and Childrens' Bills of 1907–8 which had been solicitous of the interests of her play centres and invalid schools. Possibly she thought the proconsuls of Empire who headed the 'antis' might pay back her services with favours to Arnold.

Whatever her motives, Mary Ward acquiesced. The opening move was an article in *The Times* for 12 June 1908 entitled 'A Counter-Movement' announcing that 'A National Women's Anti-Suffrage Association is being formed.' Mrs Ward's name proudly headed the list of commoner ladies, qualified by 'intellectual eminence'. The suffragists, it was promised 'will not be allowed to hold the field unchallenged'. The declaration finished ringingly that 'The association will represent those who believe that, though Mrs Fawcett is personally well fitted to have a vote, or many votes, that is a poor reason for admitting to the franchise two or three millions of women who know nothing whatever of politics, or parties, or the nation, or the Empire.'

Mary Ward wrote in July to her Aunt Fan (a 75-year-old staunch suffragist) at Fox How that 'You will see from the papers what it is that has been taking all my time—the foundation of an Anti-Suffrage League'.[52] The Women's National Anti-Suffrage League was duly launched at the Westminster Palace Hotel on 21 July. *The Times*, which Mary Ward had been using as her personal pulpit on this subject for some weeks, gave the event prominent coverage. Under the title 'Enthusiastic Meeting' it described the function as 'largely attended'. The Countess of Jersey was deputed to open the proceedings. But the manifesto was introduced to the meeting by 'Mrs Humphry Ward, who was received with cheers'. The document was clearly framed by her, and stated in rather more formal terms the arguments she had recently been pushing in the paper's correspondence columns. It finished with 'Women of England! We appeal to your patriotism and common sense.' She followed with a good punchy speech, which brought the audience to its feet.

Although Lady Jersey was appointed President, all the organizational work fell, inevitably, on Mary Ward and on Dorothy who worked through the league's executive committee. By December 1908 Mary Ward had got going the *Anti-Suffrage Review*, which—especially when she wrote for it—gave some intellectual respectability to the rejectionist cause. But one of the insoluble problems for the WNASL was the thinness of its ranks. By February 1910, ninety-seven branches were claimed, but most of these were simply inert lists of passively friendly ladies. The activity of the league resided in the *Review* and the immensely lively public meetings at which Mrs Ward was inevitably the main anti gun.

To give the anti lobby some muscle it was decided in December 1910 to amalgamate with the Men's League for Opposing Women's Suffrage (formed in 1909). Numbers were swollen but the masculine character of the movement was, after 1910, transparent. The Countess of Jersey was obliged to step down as President in favour of the Earl of Cromer. The main problem that the antis faced was that any honest explication of their position tended inevitably to make converts for the other side. They did not have a good case. Explicit opposition to the parliamentary vote and membership of Parliament was therefore avoided by the more intelligent antis. In public, they stressed the one 'positive' element of their programme, namely the wider representation of women 'on muncipal and other bodies concerned with the domestic and social affairs of the community'. The justification of this was that local politics constituted 'enlarged housekeeping'. But even this concession was hard to wring from the old men who were the real power in the anti bloc. They simply thought women too ignorant, too emotional, and too irrational to vote. And every time these patriarchs opened their mouths in public, support for the anti-suffrage cause haemorrhaged away.

25

Anti-Suffragist: 1909

CHRISTMAS 1908 and the New Year 1909 were gloomy; the Huxleys—still mourning Julia's death—spent the holidays at Stocks. From now until her death, Mary Ward was to be a particularly fond aunt to the three motherless boys Aldous (Dorothy's favourite), Julian (Mary's favourite), and Trev. January itself promised to be busy. Mrs Ward's Women's League to oppose suffrage had taken off and a full programme of lectures and debates had been arranged by the executive, of which she was the main strategist. Arnold Ward was now a dutifully convinced anti-suffragist (when the Men's League for Opposing Women's Suffrage was formed on 19 January he was prominent among its officers). Dorothy Ward was organizing women's meetings for the anti cause at Tring. Ominously, not all Arnie's local Liberal Unionist patrons were in favour of their candidate's illiberal line; among others, Lord and Lady Ebury were angry at him and already working to unseat him. And, by May 1909, there was a suffragists' group active in Watford. But in 1909, it still seemed wholly feasible that the antis might emerge ultimately victorious, or at least hold back the enemy another forty years. It was true that a large majority of Liberals in the Cabinet were on record as for—if tepidly for—the suffrage (the Premier, Asquith, and the Chancellor of the Exchequer, Lloyd George, among them). But so nervous were they of splits that a full-blown Reform Bill was not on the cards.

Frustration, meanwhile, was driving the suffragettes to reckless and arguably counter-productive stunts. There were violent plans to 'rush' Parliament, and physically mob the members. Mrs Pankhurst (who took some of her tactics from Russian anarchist *émigrés*) was advocating an axis between suffragettes and the unemployed, whom George Lansbury was mobilizing in Poplar. In prison, suffragettes were being treated ever more savagely and in September 1909 forcible feeding was authorized.

All this could be seen as playing into the hands of the antis and their 'Patriotism and Common Sense' doctrine. It was they who seemed to represent the sane and decent part of the sex. Not that they were above propaganda, and her desire to score points against the opposition was beginning to affect Mrs Humphry Ward's fiction. Suffrage was palpably shaping her novel-in-hand, *Daphne*, which she had begun on 27 July 1908. 'My grewsome tale of divorce iniquities in America', she ironically called it in a January 1909 letter to Louise

Creighton, adding, 'I am afraid [it] has grown into rather a tract.'[1] Louise, like many of Mary's friends, was already on the other side on the votes for women question.

On 29 January 1909, Mary Ward addressed a 'splendid' anti-suffrage meeting in the Victoria Rooms, Bristol. There were 1,100 people present and she spoke for forty-five minutes from typewritten notes. Dorothy recorded for posterity that her mother was dressed in a black satin dress (made by her maid Lizzie), with a black boa and hat. Sober black was the colour of her militancy. One of the effects of Mary's new political activism was in fact to bring her closer than ever to Dorothy. Dorothy was now with Lizzie the indispensable out-of-town companion; the arranger of cushions, tickets, shawls, and hotel rooms. Alone among the women of the family, Dorothy followed Mary Ward into the ranks of the antis, and could be confided in on matters of strategy. Dorothy was more and more responsible with Bessie Churcher for the running of the London playgrounds under PES management, whose number was now scheduled to rise to twelve. Dorothy was undertaking much of the arduous research work for the Autograph Edition of her mother's fiction. Her chores included finding Mary Ward's old contracts, and answering innumerable queries from Houghton Mifflin about the album of photographs which were needed for the project. 'What a brick you are! my best and kindest of daughters,'[2] Mary told Dorothy Ward in March, after she had succeeded in turning up her mother's lost records of the famous Gladstone conversation in 1888. Now that they were too small for Arnold, she informed Dorothy on 2 February that *she* should have the proceeds from all future American performances of *William Ashe*. (They amounted to £12 in 1909.)

At Stocks on 15 February 1909 ('as the snowdrops came out in the wood') Mary Ward finished *Daphne*. On the twenty-sixth of the month, there took place her long-arranged and postponed debate with Mrs Fawcett at PES. Dorothy Ward was sick with rheumatism and toothache, and could not go. (She stayed at home reading Mill's *Subjection of Women*, which her mother had prescribed for her.) The antis, expecting a home ground advantage, reserved only 150 tickets from the 500 issued, confident that Mrs Humphry Ward's eloquence would easily sway the don't-knows in the audience. Sir Edward Busk presided and Mary Ward had the privilege of opening the debate. She began by recalling her own activities as a feminist at Oxford in the 1870s. But the meat of her speech was the Imperial argument. The Empire was the creation of male brawn and male violence. The reason why women's suffrage had become a more dangerous leap in the dark than it was in the 1860s, she told her listeners,

> is simply because of the vast growth of the Empire, the immense increase of England's imperial responsibilities, and therewith the increased complexity and risk of the problems which lie before our statesmen—constitutional, legal, financial, military, international problems—problems of men, only to be solved by the labour and special

knowledge of men, and where the men who bear the burden ought to be left unhampered by the political inexperience of women.[3]

In reply, an unimpressed Mrs Fawcett calmly pointed out the disadvantages to women (in such areas as wages, conditions of work, and legitimacy laws) which arose from their having no direct presence in the legislature. 'Men were not intentionally neglectful, but they were so taken up by their own affairs that the claims of women had little chance of attention.' The vote was put: Ward received 74 votes, Fawcett 235. It was a slaughter. Mary Ward was furious for days after at what she saw as the dishonesty of Fawcett's arguments and the suffragists' unfair packing of the audience. But the experience was salutary. Hitherto, she had believed that her arguments—delivered with all her personal authority—would convince. Now she realized that, to the unbiased mind, the opposition had the better arguments. The antis would be wiser to proceed by judicious lobbying, by nobbling the men in power, and avoiding wherever possible the open forum.

For a couple of weeks after the PES debate, Mary Ward's voice was gone and she was racked with 'side' and eczema. But despite Fawcett's dirty tricks she felt her Anti-Suffrage League had made its mark and she prided herself that she had the suffragists on the run. Her friends in the Lords (notably Cromer) assured her that, even if the women's amendment got through the Commons, it would be voted out in the other place—like the Licensing Bill in October 1908. Mary Ward put great faith in the House of Lords—foolishly as it eventually turned out.

By 26 March, she had recovered sufficiently for another ordeal, the monster Anti-Suffrage Rally at Queen's Hall. Although the proceedings were 'much heckled by suffragettes of the most persistent type', they went off 'splendidly',[4] Dorothy thought. Mrs Ward spoke first and was wildly cheered when she got to her feet. She spoke for eleven minutes, was very well received (only one lonely voice could be heard exclaiming 'shame!'), and the servants (brought en masse from Grosvenor Place and Stocks) reported that they heard every word of their mistress in the top gallery. Mary Ward made much of the 250,000 signatures her league had collected, and bravely promised that before the campaign was over, they would have two million. Arnold Ward sat behind his mother on the platform, Dorothy and her aides were on the floor at the side, 'rather bad places', she felt. The hall was crammed and 'all the speeches went down admirably, *especially* Lord Cromer's. At the end there was chaos, raised by the Suffragettes who behaved abominably trying to get an amendment put .... I felt at the time that Mother's voice would not have held out for a *long* speech.'[5] That night, her mother needed an exceptionally heavy dose of trional.

At Easter, the Wards (less Arnold, who was suffering a suspicious number of 'colds', 'catarrh' and 'flu') retired to Cadenabbia. Mary was prostrated with a combination of ailments. But nowadays, she had to write, even when she was relaxing by Lake Como. After lunch on 20 May, Dorothy took out her presentation copy of *Daphne* (or *Marriage à la Mode*) and retired to the villa's cool loggia, to read it through in leisure. The book gave her a 'vivid impression'. Indomitably loyal, she predicted that 'for every one that it will anger and estrange I *can't* help thinking that it will capture and touch deeply and sympathetically too'.[6]

It didn't. *Marriage à la Mode* is the story of an impulsive American girl 'of Irish extraction',[7] Daphne Floyd, who falls in love with an Englishman, Roger Barnes. He is tall, fair-haired, and has 'the carriage of a young Apollo'. Roger has an impeccable background (Eton and Oxford) and good breeding but he also has 'an easy character without much strength of will'.[8] His family has fallen on hard times and he has come to America to seek his fortune. Daphne is an orphan, heiress to a lumber fortune, and a feminist who 'acts humility and simplicity [but is] in reality intoxicated with the sense of power'.[9] After a whirlwind courtship in Washington (an episode incorporating many of Mary Ward's 1908 experiences) the couple marry. On their return to England and Roger's country house, Heston, the marriage goes wrong. Daphne wrongly suspects that Roger is betraying her with his former fiancée, Chloe Morant, now Mrs Fairmile. Daphne corrupts her own judgement with feminist tracts and decides to gain a divorce by recovering her American citizenship. She takes flight to South Dakota, whose indulgent courts give her a decree against Roger (on the false grounds of his misconduct) and sole custody of their daughter Beatty. She is, of course, still married under British law. Roger meanwhile takes to brandy and consorts with a shop-girl. Beatty (whom he has tried vainly to abduct) dies of convulsions. Daphne finally returns to England, to discover that Roger has only two years to live. The novel's final image is of them parting for the last time. The unvarnished theme of the book is that divorce is an unmitigated disaster and a peculiarly American moral monstrosity.

Americans, particularly young female Americans, were not inclined to undo their liberal marriage laws for Mrs Humphry Ward. As the *Metropolitan Magazine* tartly observed in its review in July 1909, 'The fact is Mrs Ward does not know much about America or the American woman.' Sales of *Daphne* were anaemic on both sides of the Atlantic. Between May and August, Doubleday sold 23,056 in American and Canada—about half what *Diana Mallory* had sold. Between May and November, Cassell sold 15,268 in England and 6,072 in the colonies. These were the worst sales figures Mary Ward had had since *Miss Bretherton*.

The Wards returned from Cadenabbia on 26 May, via the Simplon route. The train was crowded and a 'Yankee' couple caused an ugly commotion in the restaurant car. Mary was still very 'brain and nerve-tired',[10] but physically recovered. At Stocks, they were met by Arnold, bronzed after his three weeks' annual yeomanry service at Aldershot. He and his father had some grand schemes to lay before Mary. They proposed to buy Aldbury Nowers, and an adjoining farm. Altogether, it would be 1,100 acres, lying to the side of Stocks. It was 'rather a big bite', Arnold laconically observed: 'price somewhere between £30,000 and £60,000'. Ten thousand down would secure the property, the rest could be raised on mortgage. Arnold—who had been doing well just recently in his business speculations—offered to chip in £3,000 'as I regard it as very important'.[11]

£60,000 was a sum to give Mary Ward nightmares—but, as always, she submitted to what her men wanted. She would buy them their thousand unnecessary acres. In fact, her earning prospects looked currently powerful. In January 1909, Smith had agreed to give £10,000 for 'Robert Elsmere II', whenever it should be written. Half would be paid on delivery of the manuscript, the other half on publication. After making their first £1,000 profit, 75 per cent of all subsequent profits would be Mrs Ward's. It was, quite simply, the most generous arrangement she had ever had. In gratitude, she gave Reginald Smith the serial rights of *Canadian Born* for nothing. (Smith did, however, give her £1,750 for the book rights.)

June 1909 was, as always, socially frenetic. Arnold wearied himself speaking at fêtes, playing cricket for the Gentleman of Hertfordshire and being seen at Ascot. Humphry and Mary Ward went to the Royal opening of the Victoria and Albert, he bursting out of his court suit. Mary motored all over London (always with cotton wool stuffed carefully in her ears) making her summer inspection of the play centres. As she did once a year, she laboured up the steps at Bow, where the playground was on the roof. This year, she took time off to show the dirtiest children how to wash their faces and hands, before herself returning to the Ladies' Athenaeum for tea and ices.

Over the summer of 1909 at Stocks, Mary Ward gathered her strength for the anti-suffrage campaign of the autumn. She was developing rheumatism in the neck—perhaps it was the draughtiness of the old car. She followed Arnold's and the chauffeur Larkin's urging and authorized the purchase of a new saloon. Humphry, meanwhile, got a touch of tonsillitis, and was taken off by Dorothy to Cornwall to get better. Arnold played much golf and tennis and waited for the shooting at the end of September. Bessie Churcher took her annual two weeks' holiday. Dorothy (now returned from Cornwall) took her favourite cousin—the immensely gangling Aldous Huxley—up to Robin Ghyll with her.

On 21 September (without Dorothy by her side, for once) Mary Ward argued the anti-suffrage case at Newnham and Girton colleges in Cambridge. It was a dispiriting experience. At Newnham, 'the hostile though quite polite audience was really too chilly', she told her daughter. At more conservative

Girton, a third of the women in the audience 'made a tremendous noise and clapped every point on our side'. But there was cold antagonism and unpleasant heckling from the majority of her listeners. 'The fire and the rage were immense,' Mary recorded in some wonderment, 'and considering that the "staff" is hotly suffrage at both colleges I should think discipline will soon suffer.'[12] Somerville, meanwhile, had broken off relations with her altogether. Mary had to face the sad fact that she—who was once a role model for the intellectual young womanhood of England—was now a figure of scorn and absurdity in their eyes.

Through all this distraction, Mary Ward soldiered on with her next 'short' novel, *Canadian Born*, and had the fifth number (of eight) written by the end of August. The story was entirely finished by October. Her speed was commendable; but nothing else about the novel is. *Canadian Born* is a work of blatant inferiority and marks a clear downward skid in Mary's process of artistic selling out. The novel was, as she told Reginald Smith in June 1909, 'framed to embody the vivid impression which I received last year in our swallow flight from Toronto to Vancouver and back to Quebec—a sojourn of eighteen days in a [railroad] private car, with occasional nights in a hotel'.[13] It was her holiday snaps and diary, clumsily transformed into fiction. The novel's title echoes a favourite line from her pro-imperialist speeches of the period, 'I am colonial born.'

The flimsy narrative of *Canadian Born* centres on a fine-boned English widow, Lady Elizabeth Merton, who is bringing her delicate (i.e. potentially dipsomaniac) younger brother, Philip Gaddesden, to Canada 'on a journey of health'. The couple are given a private car by the CPR, in recompense for services earlier rendered the railroad by Elizabeth's father. On their luxurious journey west, the train is stopped by a sink-hole. Elizabeth meets a young Scots Canadian engineer, George Anderson, who asks for some fresh milk for the ailing child of some Galician immigrants in third class. George is intended to represent 'the energy, the independence, and the success of Canada'. He has, however, a guilty secret in his past in the shape of a drunken, vicious father now thought dead. George's father returns and sets up a robbery of the train in which he is killed. At the inquest, George is obliged to admit his disgraceful parentage, thus ruining any future career he might have in Canadian public life. Meanwhile, Elizabeth is torn between two suitors: George, the energetic Canadian, and a middle-aged, over-civilized Englishman of her own class and background, Arthur. This dilemma forms the core of *Canadian Born* as Elizabeth speeds across the great continent in her luxurious car:

> Could she—*could* she marry a Canadian? There was the central question, out at last!—irrevocable!—writ large on the mountains and the forests, as she sped through them. Could she, possessed by inheritance of all that is most desirable and delightful in English society, linked with its great interests and its dominant class, and through them

with the rich cosmopolitan life of cultivated Europe—could she tear herself from that old soil, and that dear familiar environment?[14]

Of course she can. The 'epilogue' of the novel finds plain 'Elizabeth' and George setting up as farmers at the foot of the Rockies, 'first invaders of an inviolable nature, pioneers of a long future line of travellers and worshippers'.[15] Their 'frugal meals', however, are prepared and served by 'Indian half breeds'. The pioneer life, as conceived by Mrs Humphry Ward, is not without its amenities.

The notion that Lady Merton would adapt so easily to the rigours of life in a wilderness cabin was implausible. Even Mary Ward's most loyal readers felt the idea of the work was somewhat forced. Arnold, who as a literary critic had a blunt turn of phrase if nothing else, pointed to 'the enormous gulf between the man and the woman, so that you almost feel she is marrying a nigger—magnificently but against nature'.[16] He did not like the novel. Nor did other English readers; in its first five months it sold just 15,129 copies. Her previous major novels had easily cleared the 20,000 threshold in three months.

Canadian Born is very unnatural, as Arnold Ward rather gracelessly pointed out. But this is not the novel's main fault. What most offends is the author's assumption that she *knows* Canada—knows it, that is, as someone might who had passed the best years of her life there. The arrogant assumption of inwardness with a country which she had, after all, only glimpsed on a fortnight's trip from a first-class carriage window hardly conferred the right to celebrate the hardy pioneer's life. And anyway, as Edward Bok pointed out, her constant reference to the CPR and the free hotels they had put her up in made *Canadian Born* look like a crude pay-off to her hosts. Doubleday recommended mentioning the railroad firm's name a *little* less often in the text. To make things worse, they warned, 'Canadian matters are not especially popular in the United States.'[17] Mary was, in short, killing her market with novels like *Marriage à la Mode* which were offensive to Americans and works like *Canadian Born* (renamed *Lady Merton, Colonist* to avoid the dreaded word 'Canadian') which bored them to tears. Nor did Mary Ward's American publishers much like the sound of 'Robert Elsmere II'; Anglicanism rated with Canada as a topic of less than burning interest to the average American reader. Unfortunately, Mrs Ward paid no heed to her American publisher's shrewd advice, and went her suicidal way.

Autumn 1909 saw the resumption of the Wards' increasingly interdependent social activities. Arnold's business interests were not prospering.[18] On 21 October, at a concert for the Herts. Yeomanry, he evidently drank too much and made injudicious political remarks. He managed to restrain some reporters, but others filed the story. That weekend at Stocks, he was 'distrait'.[19] When she wrote her birthday letter to him on 7 November, Mary Ward sounded unusually alarmed. 'There could be no greater happiness for me', she told him, 'than to see you in Parliament, disciplined and stimulated by your intellectual equals, and working for your country.' But was he *disciplined*? 'Dearest,' she went on, 'we hope you are going back to Grosvenor Place to sleep when you are in town. Father and I are unhappy at the amount you must be spending at the Portland [Club].'[20] Spending, that was, not just at the dining- but at the gaming-table. The Portland Club, in St James's Square, was (and is) the home of British bridge. The game had actually been invented there. It was not, Mary and Humphry thought, a taste to be encouraged in their son. 'Do for all our sakes cut down expenses wherever you can. It is not right—my dear old boy—to behave as if we were rich people when we have nothing but what we earn.' She would, she promised, give him as before £500 for his forthcoming election expenses. But, she told him, 'keeping the house gay, paying the remainder of the alteration expense and buying Aldbury Nowers will tax all our resources'. Frederick Whitridge had been offered Barley End but had declined to buy. Everything, as usual, would have to come from Mary Ward's pen. But—if Arnold would only be good—she did not despair: 'take me into your confidence—live frugally—and let us see what the family pot can produce for next year',[21] she told her son.

On 26 October 1909, Mary Ward was in Manchester for a great debate on suffrage in the Free Trade Hall. Her actual departure was delayed by an agitated series of telegrams, when she learned that the organizers of the debate had changed the order of events to allow the suffragists to speak first. She refused to budge until her right to address the audience first were restored. She won the point. The debate itself played out before 5,000 people. Mrs Ward was well pleased with her performance. The next day, 27 October, there was another meeting in Sheffield. 'Dear Mummy felt her speech a great effort,' Dorothy reported: 'the audience was very flat at first but in time she roused them well.'[22] After Sheffield it was on to Edinburgh, for which Mary Ward wrote an entirely new speech. 'We have to show', she told her audience of 2,000, 'that neither sentimentalism nor reaction has anything to do with our opposition to the suffrage; that it is we who in these days of differentiation and division of labour are the scientific and modern party.'[23]

By early November, Mary Ward was back at Stocks. The shooting was extraordinarily good this year. On Saturday 13 November, the Ward men and their guests bagged a record 281 pheasants and 81 hares. When not shooting, Arnold Ward was now spending all his time motoring around his constituency. It was clear that a general election could not be far off. It was provoked on 30 November when the House of Lords rejected the Government's Budget by a majority of 275, thus making a January 1910 election inevitable. Arnold's agent, Mr Blount, got the machinery working and Dorothy, as usual, organized her Women's Committee. Mary Ward looked over his shoulder as Arnold penned his election address. The campaign proper began on 6 December, with a meeting of Unionists in Tring. Lord Rothschild made a speech, endorsing Arnie. The candidate himself 'did not speak his best,' Dorothy thought, 'being rather nervous'.[24] His nerves were further affected on 16 December, when he received the news that his best friend, Archie Gordon, had been killed in a car accident. In later life, he was to mark this as a turning-point and would sometimes melodramatically claim that he never made another real friend again.

26

Arnold Ward, MP: 1910–1911

THE general election of January 1910 offered Arnold Ward's best chance of securing the parliamentary seat that had so far eluded him. The Liberals were out, on the Lords' rejection of their budget. The country seemed ready for a change. On 4 January, Arnold was formally adopted as the Unionist candidate for West Herts. at a '*stirringly* unanimous and enthusiastic meeting'[1] at the Bucks Inn, Watford.

Mary Ward threw herself into her son's ensuing campaign with a series of 'Letters to my Neighbours', of which the obliging Reginald Smith printed 100,000 gratis. The first of these letters was a paean of praise for the House of Lords, and more specifically for such local aristocrats as Baron Rothschild, Lord Brownlow, and the Duke of Bedford, all of whom were eulogized by Mrs Ward as enlightened landlords. 'There is not a tenant', she declared, 'that would not rather a hundred times be a tenant of the Duke than a tenant of Mr Lloyd George.'[2] Other letters attacked the land tax and argued for the benefits of Tariff Reform. By now, Mary Ward was an adept propagandist and her letters have a reactionary punch which has lasted well. But it was rather hard on Arnold that he could not write his own position papers. More and more, he was looking like the member for Mrs Humphry Ward.

His mother was helpful in other ways, paying for a male secretary to assist Arnold, giving him complete use of the Ward automobile and loaning him Dorothy. As usual, his sister performed prodigies with her corps of women helpers despite having fallen painfully on her knee in early January. She would have had the leg sawn off, if it might help her brother to Westminster. During the election campaign—which was spread out over a leisurely three weeks—the family all ostentatiously attended church together. Arnold's speeches attacked the Liberals' 'black record' and he made much of the impious atrocities of the suffragettes.

Despite racking rheumatism, Mary Ward was playing a central part in the election, using all the resources of her considerable local popularity. On the night of 22 January, she and Arnold were tugged in a carriage by thirty local supporters, a band marching ahead, torchlight-carrying Unionists in a militant host around them. Resounding speeches from the steps of the Unionist Club rounded off the evening. On the eve of the poll, Arnold spoke at five meetings, his mother and sister by his side, except where there was fear of rowdies. From eight o'clock on Friday 28 January, the day of the poll, the Wards' placarded car patrolled the Watford streets. Dorothy and Humphry nervously paced the counting room. The result, when it came at noon the next day, was a triumph—Arnold Ward was home with a majority of 1,551.

It was an unalloyed moment of bliss for Mary Ward. She was physically wrecked but her son was at last in Parliament. On 21 February, she and Dorothy (who had thoughtfully brought a camp stool for her mother) were at the momentous opening of the new session. It was, perhaps, a fly in the ointment that—although they had sustained grievous losses—the Liberals and Asquith were still in power. But Mary Ward's excitement at witnessing the opening debate from the Serjeant at Arms's gallery was unbounded. (The episode was used as the vivid opening scene of *The Coryston Family*, 1913.) Over the next few months, she and Dorothy were often at the House: lunching on the terrace, taking tea, using Arnold's tickets to hear debates. In other ways, it was a hectic spring season. There were, in March, no less than five secretaries at Grosvenor Place; all furiously typing either for Arnold, for the Anti-Suffrage executive, or for Mrs Ward's play centre petition to candidates in the spring LCC election. By mid-March, Mary Ward looked 'white and tormented',[3] the irritation of her 'side' and eczema having come on unbearably. Her eyes were bothering her. Dorothy was administering up to a sixth of a grain of morphia nightly to get her mother off to sleep.

With a breakdown imminent the family left for a holiday on the Continent at the end of March. They had rented a new villa at Valescure which turned out something of a disappointment, being hemmed in by trees (Mary Ward always craved views) and by a nearby unused English church. It all 'rather depressed'[4] Mary Ward, who was in full physical and emotional reaction from the excitement of January. On 8 April, the Wards adjourned to Cannes and their favourite Hotel Californie. From there, they moved on by *train de luxe* to Florence, before returning to England in early May.

May and June in London were even more full of business than usual. The King's funeral was on 20 May. And ten days later, President and Mrs Roosevelt made their promised visit to Grosvenor Place. It was, as Dorothy recorded, 'a *succès fou*. The Roosevelts arrived at twenty-to-six, very soon after they were due, and did not leave till quarter-to-seven. For the last half-hour, the President was holding forth and everybody else listening! M. was wonderfully ingenious about bringing everybody up and introducing them and it really *did* go well.'[5]

The summer was dominated politically by the suffrage question. On resuming power, Asquith's Government set up an all-party 'Conciliation' Committee, which drafted a bill satisfactory to the Women's Social and Political Union, which called a truce in its militant campaign. June and July were dominated by monster rallies in London. *The Times*, with Mrs Humphry Ward well to the fore, was as ever virulently hostile to the women's cause. Her tactics had, however, changed somewhat over the year. She now played the American card much less than she had previously. It was clear that emancipation was not, after all, in full retreat over there, as she had boldly claimed in 1908 and 1909. Her 1910 line was that the violence of the suffragettes had left an 'indelible' impression; that women in the main knew the issues perfectly well and had no wish for the vote; and that the dramatic growth of her anti-suffrage ranks was confirmation of these trends. In a *Times* article of 4 June, she claimed 15,000 paying members and 110 branches. A petition had been delivered to Parliament, containing upwards of 320,000 signatures 'of women of all classes, but especially of the working class, protesting the concession of the Parliamentary franchise to women'.[6] Against this, the suffrage societies claimed a paltry membership of 290,000 out of a female population of seven-and-a-half millions. Mary Ward concluded her article with a plea for 'Information, vigilance, an earnest local government policy—let me then once more press these three needs upon the vast multitude of women throughout England who resent the suffrage campaign. If we keep them well in mind we have no more to fear from the latest suffrage Bill than from those which have preceded it.'[7] Mary also pressed her demands on the Prime Minister himself, heading an anti-suffrage deputation on 22 June. Asquith, who had swung round to an overt anti position, took notice of what was made to seem like a damaging breaking of ranks among the women of England.

Mary Ward, who had a virtual freehold on the columns of *The Times*, penned another polemic for 11 July, 'The Eve of the Bill'. She stayed up until midnight on the ninth writing it, and 'consequently only slept four hours *with* sulphonal, poor darling',[8] Dorothy noted. The article itself was more defensive than usual. Pressed by the ultra-logical Mrs Fawcett as to her 'slanders' on women Mrs Ward stated the essence of her resistance to the suffrage—namely women's 'necessary' and wholly irreversible ignorance. 'It is, of course, replied', she concluded, 'that many men in a democracy are politically ignorant. True; it is the great risk of democracy. But men are not *necessarily* ignorant.'[9] It was always the weakest part of her anti-suffrage case that she herself was such a poor example of what she argued were the incapacities of her sex.

Despite a successful protest meeting by antis in Queen's Hall on 11 July, the first reading of the Conciliation Bill passed by 179 votes. A couple of days later, the second reading passed by 109. Mrs Ward and her supporters chose to be encouraged by the drop; and at a rally in Trafalgar Square on 13 July, she cheered her troops on. Meanwhile, in the House, Arnold Ward was speaking to order. In the debate on 12 July, he made a strangely frank admission of his own puppet status. As he told his audience of MPs, 'There was something in their situation that night which powerfully illustrated the indirect influence of women in politics. He did not know whether it was indiscreet to mention it, but in the course of making inquiries among members as to their attitude on this question he had found nothing more striking than the number of instances in which, he would not say the opinions but the course of action of a given member, had been influenced by a particular individual woman. (Hear, hear.)'[10] There is something piquant in the idea of Arnold skulking round the Westminster bars, asking whether *all* members were as henpecked as he was. He was quickly snuffed out by the suffragettes, who all during his parliamentary career would send him postcards after his speeches in the House, saying such things as 'Mother *will* be pleased!'[11]

Mary Ward had in fact been in the gallery of the House, watching her son play his part in the destiny of his country, and bursting with maternal pride. She thought he should have made his speech longer, but on the whole Mother *was* pleased. That evening (12 July) Arnold returned to Grosvenor Place at just after midnight, as Dorothy recorded, 'and came up to Mother's bedroom and talked it all over with her and me—looking so bright and handsome and animated and having so evidently and keenly enjoyed the work and excitement of the last days—his first real "job" in the House.'[12] As it happened, the antis had nothing immediate to fear from the majorities which the Conciliation Bill received. Asquith expertly fudged the committee stages of its forward progress, and moved the House on to the more urgent business (as he saw it) of the Lords' Veto. The most uncertain phase of the women's struggle began, in which they had a theoretical assurance of a victory, which—in practical terms—never got any nearer.

As usual, the Wards made their summer move to Stocks in the second week of August, and Mary Ward began to write her long-delayed *The Case of Richard Meynell* (i.e. 'Robert Elsmere II'). By 21 August she had the first two chapters written, and read them to Dorothy in the summer house. They made the younger woman 'weep irresistibly'. By 1 September she had chapters 3 and 4 of *Richard Meynell* written and read them to a rather more distracted Dorothy, who as it happened was late for an interview with a new cleaner for Aldbury school, on whose management committee she sat. By 9 September, Mary Ward was again on the brink of physical breakdown. She had lain out in the north east wind, writing *Richard Meynell*, and brought on a bad eczema and insomnia attack. She was put on a yoghurt diet, which was given up after an extremely uncomfortable fortnight. She experimented with a new sulphur-based lotion for the eczema, which also turned out disastrously.

The Curzon–Cromer combine was all the while making insatiable demands on her time, and putting her book back months. On 27 October, Dorothy noted that 'M. finished her 7th. Chapter. How she *can*? with Anti-Suffrage bothering by every post.' Her face was 'grey and drawn and yet a spot of colour on each cheek'.[13] Heart symptoms were now added to her afflictions. After a couple of turns on the terrace at Stocks, she was 'puffed and breathless'. New tonics were prescribed. Nevertheless, she motored to Croydon on 31 October, where she spoke vigorously in the Public Hall on 'The Woman's Sphere'. She clinched her remarks (now a familiar litany) with the result of a postcard canvass of women householders which purported to show that antis outnumbered pro-suffragists in every region of the realm by five to one. Four days later, she motored to Cambridge to speak on play centres.

Weighing on Mary Ward all through the autumn was Arnold. He was beginning to fall out (as he always did) with powerful Unionists in his constituency. His mother was obliged to smooth things over on a couple of occasions. In August, he alienated his old (but easily offended) friend Margot Asquith by public criticism of the Prime Minister. His behaviour was becoming dangerously unstable as it invariably did when there was no new stimulus in his life. On 25 September, Mary Ward wrote to her son in some nervousness, complaining that she had had no line from him 'for weeks'. She had, she told him, 'thought often and anxiously about you, for Father gave me some unpleasing news of your money matters. Could you not arrange with Geoffrey Robinson for some regular *Times* work?'[14] Something, that is, which would keep him occupied of an evening. His late nights at the Portland Club were proving expensive, and Arnold now had no income beyond what his parents allowed him. And while his party was in opposition (for ever, as it wearily seemed) there was no prospect of paying office.

In fact, Arnold Ward was about to make still more demands. All through the year, the Lords' Veto issue had been festering on. A Constitutional Conference had been called to find some solution. On the night of Friday, 11 November, Arnold missed the last train from Boxmoor to Tring, and had to bicycle all the way through driving rain to Stocks, not arriving until 2 a.m. He learned, to complete his sodden misery, that the Conference on Lords' Reform had broken down; there would probably be a general election. It was a wretched weekend. The following Friday, Arnold telegrammed from town to announce the imminent dissolution of Parliament: 'so we knew that the worst was true,' Dorothy wrote. 'It is really a momentous and anxious state of things. The financial burden is great—and Mother and Father are much oppressed.'[15] Arnold too was 'nerve-y'. The election was set for early December. Mary was too anxious to attend the count or the declaration, and the result had to be telephoned to her. Arnold was in again, but by a much reduced majority—883. 'We breathe again,'[16] Dorothy noted.

1910 was a bleak Christmas. The Wards had been involved in a car accident. There was a court case, which—although the jury came out in their favour—caused Mary Ward nightmares all through December. And Arnold's election expenses provoked a bad row. He actually threatened to boycott Christmas at Stocks. His mother wrote imploringly on 22 December: 'Christmas won't be

Christmas at all till you come—so please come down early tomorrow if you can. Nobody will say a word about money affairs till next week.'[17] She gave Dorothy a small cheque for underwear by way of Christmas present.

All through January 1911, Mary Ward was in a 'tremendous grind' and 'depressed' over *Richard Meynell*. The book was simply not coming. By 22 January, she was still tinkering with chapter 5. Six weeks later, on 4 March, she had not progressed beyond chapter 8. Her health was meanwhile fragile. On Friday 3 February, Dorothy recorded that 'Mother went, already tired, to [the] Islington Play Centre display and returned with bad pain in chest, which she said she has had two or three times lately; she felt collapsed and had brandy and rested in dressing room before going to drawing room where she rested again. Took some strychnine drops before dinner and felt much better and got through all right—but I don't like it.'[18]

Arnold Ward was meanwhile overwhelming the house at Stocks with local politics. Even Dorothy was driven to wish 'that our constituency were in Northumberland'.[19] She and her mother spent some days in the Lake District, staying with Mary Cropper. But even in this northern remoteness, Mary Ward was obliged to speak at the local anti-suffrage association. In early April, the Wards got away for their real Easter break to Italy, where they rented a villa at Pisa. Here on 18 April they had the bad news that Janet Trevelyan's 5-year-old son Theo had appendicitis. On the following day at one o'clock came a terrible telegram announcing his death—together with a more cheerful letter written four days earlier. They prepared to go back at once. Then came another telegram, at ten at night, to say that the little corpse was being taken to Langdale in Westmorland 'and we should be too late.'[20] Dorothy hurried away to intercept Humphry, who was rushing back on the Rome express. The family decided to stay on in Italy. Arnold represented them at the funeral, helping the tiny coffin on to the carriage for its final journey.

This death, of all she had suffered, affected Mary Ward more profoundly than any since her mother's. She spent hours designing Theo's headstone. For the rest of the vacation she was 'unutterably sad' and 'suffering in an unhelpable way'.[21] She lay for hours on the loggia, or in her darkened room taking large doses of quinine. All travel in carriages brought on 'side'. Her eczema flared up. Humphry felt he could do nothing, and left for Venice. On 7 May, Dorothy recorded her mother's diet: at 7 o'clock, tea and biscuits; at 11.15, broth; at 3.30, tea, biscuits, and butter; at 7.30 broth, a little sieved chicken, and custard pudding. The weather was, throughout May, '*Awful* . . . almost ceaseless rain'.[22] They returned, disconsolately, on 20 May.

On 22 June, the new King was crowned. They had lived through the Victorian and Edwardian and were now in the Georgian era. Mary Ward had a seat in the Abbey. The rest of the family watched the procession from the balcony of the Athenaeum. By the end of the month, she was within a couple of pages of finishing *Richard Meynell*, and the work was out of her hands by 9 August. The novel had taken more out of her than any piece of writing since *Robert Elsmere*. Worse, and unlike that earlier novel, it had never come to life as she wrote it and she feared—not without reason—that it had turned out 'a woodenish *plaidoyer*'.[23] *Richard Meynell* came out on 26 October. Dorothy went in early that morning to read the reviews with her mother. They were mixed and sales were sluggish. The English pre-publication subscription was 9,488. By the end of January 1912, sales had stuck at 14,418. (*Lady Rose's Daughter*—not one of Mary's greatest hits—sold twice as many over the same length of time in 1903.) Smith quickly realized that he would make no profits on the new novel. In America, the story was even grimmer. The book was not reviewed and it did not sell.

The problem with *Richard Meynell* lay not in the writing—which is brisk enough—but in its conception as *Robert Elsmere*, twenty years on.[24] Richard Meynell is the rector of Upcote, a rural coal-mining parish around Lichfield. An old Jowett man (the novel is set around 1908) Meynell has made himself the leader of the Modernist movement which is sweeping through the Anglican Church. Catharine [*sic*] Elsmere and her grown daughter Mary are meanwhile in the area visiting Rose Flaxman (Catharine's married sister). Meynell and Mary fall in love. The spirit of Robert Elsmere appears to Catharine with the ghostly injunction 'Forbid him not.'[25] Meynell is tried by his church authorites for heresy. On the eve of the verdict he gives a powerful sermon, presenting his 'New Reformation' as the embodiment of Thomas Arnold's vision of an ecumenical English Church. The novel ends inconclusively with Richard united with Mary but with the outcome of the trial and its consequences for the Church unclear. On to this main plot Mary Ward attached a complex network of melodramatic sub-plots. One concerns Richard's 'ward', Hester Fox-Wilton. A malevolent local squire spreads the rumour that she is in fact the hero's bastard daughter. Hester (who is indeed illegitimate but not Richard's daughter) is seduced by a bigamist and returns from Paris to die at Long Whindale. Another linked sub-plot concerns the squire's son, who is discovered to be a thief.

Mrs Ward's publishers had paid through the nose for *Richard Meynell*'s tedious apologetics. Smith, Elder gave a £3,200 advance, Doubleday £3,000, and McClure another £4,000 for the serial rights. With bits and pieces for European translations and Tauchnitz's £250, Mary Ward cleared nearly £11,000 for *Richard Meynell*. Doubleday reckoned they needed sales of 66,000 to cover their costs. They sent out 22,500 to American and Canadian retailers on publication day, 26 October. A month later, sales had stuck at 24,600. 'I confess to being somewhat distressed about *Richard Meynell*,' Doubleday wrote to the author on 1 December 1911: 'I feel as strongly as you do the great merit of this book. We have gone at the job with enthusiasm and spent a lot of money.

As a matter of fact, at the moment, including interest on our advance, the book shows an unearned investment of about $15,000, and I may tell you frankly that it looks as if we were going to make a considerable loss.' It was the more harrowing, the publisher added, 'as the book business is really good in this country this fall.'[26] They spent another $2,000 on a sales campaign in January 1912, but *Richard Meynell* obdurately refused to move. Final sales were well under 40,000. Samuel McClure's opinion of *Richard Meynell* is not on record. But he certainly did not keep it to himself. Mrs Humphry Ward's novels were for the rest of her writing life virtually unsaleable to the best and best-paying American magazines. In short, after autumn 1911, she was a second-rate author as far as America was concerned.

Arnold Ward's thirty-fifth birthday on 8 November 1911 coincided with momentous political events. The morning papers contained Asquith's announcement that in 1912 his Government would bring in a Manhood Suffrage Bill on to which a Womanhood Suffrage clause might be grafted if the House so wished. In the evening papers of the same day, Balfour announced his resignation of the leadership of the Unionists. His place was taken by Bonar Law—a man who had little time for Arnold Ward.

Dorothy noted over the next few days that her brother did not look well, 'poor troubled dear'. There was nothing from his party leader to cheer him up. On 24 November, Dorothy recorded, 'Alas! it is in the papers today that Bonar Law has chosen Mr [John] Baird and [the] Hon. Arthur Stanley to be his Private Secretaries. I am keenly disappointed.'[27] So was Arnold's mother. It seemed that as in the university and law, her son simply could not get beyond the first threshold of success, despite his clear talents.

His Watford constituency was also proving difficult. And his party was becoming increasingly pertinacious on the question of money. On 5 December Arnold Ward was obliged to explain 'his financial position deliberatively and with great care and tact ... saying he "was not a rich man"'.[28] (Nor a rich woman's son, he might have added.) The suffragists—such a joke two years before—were now a formidable force in the area. At a 'Grand Anti-Suffrage Meeting' at Berkhamsted on 13 December, Arnold and Mary Ward contrived to carry their vote only by a small majority. Both of them were a 'little ragged'. Arnold discovered that his women helpers were melting away from him. 'Suffrage is worse than morphia,' he told Dorothy, 'and ruins women for all party purposes'.[29] Nevertheless, the antis were still fighting hard. On 14 December, Mary Ward led an influential deputation to Asquith, and elicited from him the admission (which infuriated suffragists and suffragettes both) that the inclusion of Women's Suffrage in his forthcoming Reform Bill would be 'a political mistake of a disastrous kind'. It was the signal for a new outbreak of militancy—including the Pankhurst-led raid on the window-panes of 10 Downing Street.

All in all, 1911 had proved a bad year for all the Wards. There was tragic death; the anti-suffrage cause lost ground; Arnold got nowhere in his career; *Richard Meynell* was a mighty flop; Humphry's picture dealing incurred losses. Money—always money—was at the root of their problems. They were constantly living a few notches above their means. And what means would they have if Mary Ward's popularity waned? On 24 December, she wrote a rather despairing letter to Arnold: 'I have been anxious my darling and depressed. So much of our family income depends on my work, that any real diminution of vogue and popularity for my books must be an anxiety to us all.'[30] That same night she began to pen her next novel.

27

Calamities: 1912–1914

THE early months of 1912 saw the anticlimax of the Conciliation Bill. Mrs Ward, who was the antis' only star woman speaker, motored all over England to rallies. That at Colston Hall in Bristol on 16 February was memorable. Lord Cromer—now a doddering old man—was in the chair; Charles Hobhouse, MP for Bristol, was the main male speaker. Hobhouse harangued the audience—through the usual interruptions—for an interminable fifty minutes on women's mental and biological inadequacies. When Mrs Ward finally got up to speak, a suffragette emerged staggering from the organ pipes behind the platform. Still deafened from the recital which preceded the speeches, this militant attempted vainly to address the audience before being forcibly ejected. A number of 'our men' (Dorothy's phrase) took the occasion to attack the recognizable suffragettes in their midst. It was very distressing and degrading. Mary's speech was curtailed by half its length, and, as Dorothy observed, 'it sounded like a great effort for her dear voice'.[1] Increasingly, public meetings like this were nightmarish occasions for Mary Ward, who had a visceral fear of physical violence. At the giant Albert Hall meeting on 28 February, she declined to mount the platform, terrified as she was of well-publicized suffragette threats of blaring dictaphones concealed in the hall, of stink bombs, of the lights being cut off.

Whether they won or lost on the suffrage issue, the Wards' bills had to be paid, and Mary was embarked on a new novel in January 1912. From the start, it was a vexatious thing. Even the title gave inordinate trouble. It was first called 'Fortune's Wheel' (already used, it emerged); then 'Lydia' (also already used); then 'Lydia's Lovers' (too like Mrs Gaskell's *Sylvia's Lovers*, Reginald Smith thought); then 'Lydia Victrix' (likely to suggest a pro-suffrage theme, as Smith pointed out). Finally, publisher and author settled on the infelicitous title *The Mating of Lydia*.

As usual, Mary Ward had the main scenario (her term) clearly mapped out before she started, to which she adhered reasonably closely.[2] The narrative begins with a prelude set thirty years before. Edmund Melrose arrives at Threlfall Tower in the Lake District (near Ambleside) with an Italian wife, Netta, and a little daughter, Felicia. There is a domestic argument and Netta absconds, stealing a valuable Florentine bronze wherewith to pay her way. Over the years Melrose becomes a recluse, obsessed with the collection of *objets d'art*.

Claude Faversham, an idle young barrister on vacation, has an accident on his bicycle outside Threlfall's gates. Despite his habitual misanthropy, Melrose takes to the injured young man, because—it emerges—Claude is heir to some gems that he covets. Once recovered, Claude becomes Melrose's secretary and eventually his heir. He also becomes involved with Lydia Penfold, a young girl who lives in the area with her widowed mother. Lydia has a 'professional' background (her father was a naval officer) and is 'a "modern" of a charming type [with] an idealist contempt for wealth'.[3] Lydia is Claude's good angel, Melrose his bad. The novel comes to a climax with the murder by shooting of Melrose, of which Claude is first suspected then found innocent. Uplifted by Lydia's idealism, the young man passes the wealth he inherits on to Felicia, whom Melrose had cut out of his will. Threlfall Tower becomes a museum, with Claude (now Lydia's husband) as its curator.

The Mating of Lydia has some interest as an analysis (not without an edge of comedy) of the collecting mania which Mary had observed over the years in her husband, Humphry Ward. But like much of the author's later fiction it is damaged by an over-reliance on plot contrivance, something that had increasingly disfigured her fiction since *William Ashe*. There are too many surprises, too many strokes of theatre, too many forced coincidences.

Nevertheless, Reginald Smith and Doubleday were relieved at the intellectually unambitious quality of *Lydia* when they saw the scenario and made better offers than the *Richard Meynell* figures warranted. Doubleday offered £2,000 in guaranteed advances; Smith £1,200 against 'all your receipts for the new book other than from America'.[4] Curtis Brown bustled about to find a taker for the American serial rights. Mrs Humphry Ward's price had effectively been halved. But, given the fact that it had cost her publishers heavily to put out her previous book, she could hardly complain. She set to work and the first chapter of *Lydia* was read to a dutiful Julian and Aldous Huxley on Sunday 25 February 1912.

It was in other ways a busy spring. On a Sunday at Stocks in March, Dorothy observed with some wonder the combined industry of her family: 'between us all we posted *85* letters and postcards to-night—I think a record!—and Arnold and I wrote about 9 or 10 which are to go tomorrow. Mother has been doing [her] book hard and Father wrote about a column for the *Times*.'[5] Much of this Stocks correspondence was generated by the festering suffrage business. The debate on the Conciliation Bill extended over Wednesday 27 and Thursday 28 March. Mary and Dorothy Ward were entertained on the evening of the final debate in Room D by Arnold and various anti grandees, who regaled themselves over their wine with 'amusing' stories about the outrageous doings of Christabel Pankhurst—currently directing the suffragette fight from exile in Paris while her mother, Emmeline Pankhurst, bore the brunt of government persecution in London.

The division took place at 11.15, and to the Wards' immense relief the measure was defeated by fourteen votes. It was, as an astonished Dorothy recorded, contrary to all their expectations. 'The Irishmen all went with our lobby,' she added, and generously gave credit for the astonishing turnabout to her brother's oratory: 'I heard Arnold make a really excellent speech—25 minutes—but alas! alas! Mother missed hearing him and by only two minutes .... It was terribly disappointing for her. Our boy's speech caught, roused, amused and thoroughly stirred the House, and his was the *only* one to do so.'[6] The next day, Dorothy, Mary, and Humphry Ward left victorious for Paris and Italy. The parents celebrated their fortieth wedding anniversary (6 April), drinking each others' health in Asti Spumante at a little hotel at Bogliaco, near their beloved Lake Como. 'We don't intend to be old a day sooner than we need,'[7] they gaily wrote to Arnold.

Their good spirits were forced. March and April 1912 were in fact gloomy months. Mary Ward had been plunged into the deepest depression by the fortunes of her Autograph Edition in America. For three years, everything had seemed to go swimmingly. Deferential photographers had come from America to take pictures for the edition; Houghton Mifflin had printed and bound the sixteen volumes to the highest standard. Since 1909, volumes had been sold as they were printed to prepaying subscribers. There had been much publicity in the papers. Mary Ward understood, as a matter of course, that such an edition would have immense production and distribution costs (£4,000, Reginald Smith estimated) which would need to be covered before any of the profits came her way. But she had always expected that the Autograph Edition would be a long-run money spinner, and three years was surely a long run. Heaven knew, she needed the money. She herself had spared no effort to make the venture a success. *Richard Meynell* had been written specifically to revive interest in *Robert Elsmere*, which would be the edition's flagship volume.

She received her first statement from the Boston firm on 23 March 1912. It was devastating: her total profit from three years' sale of the Autograph Edition in America was £8. Mary Ward exploded with wrath against the treacherous Americans:

> I have not yet received the account for my collected edition from your firm, and only—as news—the cable from you of ten days ago giving the incredible figure of £8 as due to me up to January 1 1912. I desire to protest in the strongest way against the manner in which your firm have handled the edition. It is all very well to say—as no doubt you will say—that Houghton, Mifflin and Co. have spent a large sum of money upon it and are threatened with a heavy loss. I regret this as much as you do. But I also have a large interest pecuniary and other in this edition, which involves the whole of my previous work, the new prefaces I have contributed and my general reputation. I cannot think that Houghton, Mifflin and Co. have acted with due consideration for these interests, or with the proper courtesy towards myself.[8]

The letter thundered on in this recriminatory vein for several pages more. But, however she protested, the facts were immutable. The American public had gone off Mrs Humphry Ward. They had hated *Marriage à la Mode*, yawned over *Canadian Born*, and were wholly baffled by the Anglican apologetics of *Richard Meynell*. All this returned as no sales for the Autograph Edition.

All through late April and May in France and Italy, Mary Ward struggled to complete *The Mating of Lydia*. Again the novel would not come easily. Nor—despite frantic efforts—could Curtis Brown secure a taker for the serial rights. She was paralysed with depression and exhaustion of spirit. In Paris they stayed in a hotel near the Tuileries, from where she wrote Arnold an anecdote which reflected her own demoralized state of mind: 'Someone said to the Empress Eugénie—"Madame, is not the outlook on the Tuileries too *triste* for you?" To which she replied quietly, "I have nothing in common now with the woman who once lived in the Tuileries. I look at the place where the palace stood and I think, 'poor woman! how she suffered!' But I do not seem to be thinking of myself".'[9] Mary Ward might well have thought, looking at Houghton Mifflin's sumptuous *Robert Elsmere*, that she too had nothing in common with the woman who had written that novel.

Still, the duties of tourism—like all duties—had to be undergone. On 30 April, Mary Ward saw the inauguration of the newly repaired Campanile at St Mark's in Venice. The brick tower glowed under its night illumination 'as if red hot'. A choir of children sang, and the lights over the lagoon composed 'the poetry of illumination'.[10] Yet, she found herself longing to be away in the mountains.

As the holiday progressed, Mary was increasingly stricken with spearing attacks of 'side', worse in their intensity than even Dorothy could remember. The 'unending mysteriousness of it' harrowed her, almost as much as the pain which clenched her whole back in its spasms. On 17 May, she collapsed in the street at Lugano. Humphry was also chronically depressed and looking his age. They trailed back to England on 19 May, glummer than when they had left six weeks before. In London, even worse news met Mary Ward as to the fate of the Westmoreland Edition—the English version of the Autograph. Smith, Elder had contrived to sell only fifty sets out of the 250 published. The unsold bulk would be remaindered. It was not—as Reginald Smith bluntly put it—a way of making money but 'only a loss lessened'.[11]

The subsequent summer in London was preoccupied in a never-ending round of meetings of the sundry committees, societies, leagues, guilds, clubs councils, and unions that Mrs Ward headed or helped run. With the Conciliation Bill buried, the anti-suffrage tempo had slackened; but she was working harder than ever on her Local Government Advancement Committee, for which a major campaign was planned in the autumn. Play centres were catering to 52,000 London children a week. Mary Ward would have to raise £1,200 in donations for them by Christmas. In order to keep her charitable causes in the public eye, she had to make herself available for speeches to such bodies as the Churchmen's Union annual meeting or the Booksellers' Association annual meeting, and scarcely a school within fifty miles radius of Stocks but did not recruit her for their prize day.

As always, she did her duty and showed the flag at innumerable public appearances over these summer months. She was now never entirely fit. But it was clear to the family that this year she was more than usually unwell. The trouble, it finally emerged, was gynaecological. On 17 July, Dr Mary Scharlieb called by appointment to see Mrs Ward at Grosvenor Place. Her decision to be examined by this distinguished woman obstetrician and gynaecologist (five years older than her patient) was, as Dorothy recorded, 'a great relief to us and will be a great help in guiding Mother and us in the future. But it is by no means altogether cheerful—for it means never being really *cured* of the evil, and just always keeping pain and weariness at bay. Still, evidently, she [Scharlieb] believes in her [Mother] leading an ordinary life as far as she *can*.'[12]

September and October saw a further decline into semi-invalidism as Mary Ward completed *The Mating of Lydia*. The novel had become a positive curse. Curtis Brown could not find an American magazine willing to take the serial. The word had got abroad that *Richard Meynell* had been a disaster for McClure. 'The Americans', Smith warned her in June, 'have to be won over again.'[13] But come-backs for novelists are as difficult as for prize-fighters—and Mary Ward was now sixty-one years old. In the short term, she was being throttled financially. Scheduled publication of the novel had to be pushed back. Clearly it could not come out as a book until after the serial run. This meant her on-publication payment for *The Mating of Lydia* was tantalizingly deferred. The novel was completed in manuscript by the third week of October 1912 and then put into cold storage. Finally, Curtis Brown made a deal with Hearst's *Good Housekeeping*. This monthly magazine had a constituency of middle-aged conservative women who remained in sympathy with Mrs Humphry Ward. It was the humblest periodical any of her serials had run through, but money was the only consideration now. The magazine gave $7,500 (£1,400) for *all* the English language rights in America. Serialization would run from November 1912 to November 1913, which would hold up publication of the novel in book form (and the main instalment of the author's money) six months until March 1913.

It was vexation on vexation. The pain in Mary Ward's side continued inexorably as she fretted. In September, she used quantities of laudanum to ease it and paid for numbness at night with lingering stupefaction by day. On 19 October (with twenty chapters of *Lydia* written) Dorothy reported her mother as 'feeling rather heart-y. Strychnine drops have done her good. This last grapple with the book is very exhausting for her.'[14] 'Nobody knows how depressed and anxious I have been of late,'[15] she told her daughter, despairingly, on 22 October. Henry Huxley in the medical check he gave her after every book reported Mrs Ward's heart action to be 'rather weak',[16] and packed her off to Paris for a recuperative fortnight. If you had gone to Huxley with gangrene he would have prescribed Paris or the Riviera. On 17 December, Mary Ward accepted Harper's proposal that she write a short novel, to run eight numbers in their magazine from April 1913. It would be called 'The Conisborough Family' (later changed to *The Coryston Family*) and would have as its subject 'the tyrannical sway of a political woman over her husband to whom she denies all freedom of action'.[17]

Humphry Ward had had an up and down year. His losses in art dealing had been a heavy drain on the funds and spirits of the family over 1911. His judgement—always prone to be over optimistic in attributions—had become dangerously erratic. But on 4 February 1912, Mary informed Arnold Ward that 'Father has had great luck with his Rembrandt which is sold. It *is* a mercy!—for what with his losses last year, and my losses on *Meynell* we were in a bad way. But this straightens things out.'[18] (From the going price for a Rembrandt, one gathers that Humphry's losses in 1911 must have run into thousands.) On 12 July he sold a Holbein at Christie's for £1,071. And on 27 August, his spirits had another fillip when his de Hooch was authenticated as genuine. In early September Mary told Dorothy Ward that 'Father has sold his de Hooch ... so a good many things can be paid,' which was 'a relief'.[19]

But other worries about Humphry had cropped up. Over the summer of 1912, Dorothy and Mary Ward had noticed that he was prone to 'mysterious little attacks of forgetfulness'. They had gone to Henry Huxley, who made light of these spells. He suggested 'thin blood' might be the cause and prescribed raw steak sandwiches and Paris. The attacks, however, persisted; recurring every six weeks or so. That on Monday 2 November was particularly frightening. Humphry had come back during the afternoon to Grosvenor Place, to dictate his *Times* article to their part-time secretary, Miss Salome. In the evening, he would—as usual—adjourn to the Athenaeum before dinner. Arnie came in a little late for the meal, 'a bit seedy with this *chill*' as Dorothy noted ('chills' were one of her many euphemisms for her brother's hangovers). Arnold went to bed at about 9.30, 'after lying against Mother's knee for her to stroke his head'. The two of them had been at a 'very crucial' executive meeting of the Anti Suffrage League that afternoon and were feeling close. As Dorothy then recorded,

Father fell asleep while reading at five-to-eleven, and I came in from [the] back drawing room to ask them if they could think of another man for the shoot [at Stocks] on the 13th. He woke up with a start and dazed look and asked us what we meant, what party it was, had he been consulted about it at all—and didn't remember anything about it. He exclaimed that it was an extraordinary thing, he couldn't take it in, would we write down in the morning a list of the people coming etc.—looking worried and distressed as he spoke. We tried to soothe him and assured him he was just tired and we had woken him up too suddenly . . . . We persuaded him to come up to bed with Mother, and she made him lie on her sofa with his feet higher than his head, and then he went to undress after a few minutes and came back soon. Mother came to tell me 'quite all right'.[20]

He was not, in fact, quite all right. The attacks continued over the following months, the intervals between them becoming insidiously closer.

Early 1913 was dominated by bad-tempered bickering about suffrage questions. In November 1912, Mrs Ward had been forced off the National Union of Women Workers—a body composed of women in social work, philanthropy, and education. For some years, Mary Ward had contrived to keep the NUWW neutral and 'above' politics. Now the executive could not be restrained from taking a position. It was a particular grief that her oldest friend, Louise Creighton, had precipitated the break. 'You are as always my dear old friend,' Mary told her, 'but I confess that you should be moving—as I am told—the suffrage resolution after the compromise of last year does rather surprise me.'[21] Nevertheless, the vote went forward, the NUWW identified itself with the suffrage cause, and Mary Ward was ejected into the wilderness. She would 'soon be sick of the name of "woman" ',[22] she told Louise. Nevertheless, she formed her own separate 'Joint Advisory Committee' (JAC) in April 1914, to liaise directly with Parliament on social matters relevant to her work with children.

In Parliament, the Liberals' franchise proposals were similarly mired in bad temper and confusion. Despite Mary Ward's attempts outside and Arnold Ward's attempts inside the House to redirect attention to the 'positive' issue of municipal franchise, general suffrage simply would not go away. However they stalled and obstructed it, the issue kept coming back. Even the most gross outrages of the suffragettes did not seem to turn the tide against them. They broke windows, burned down stately homes, immolated themselves under the hooves of their monarch's racehorses, chained themselves to London railings and defaced Sargent's portrait of Henry James with meat cleavers. And still sensible people listened to the wild women's arguments. It was most perplexing.

The debate in Parliament on 27 January 1913 was particularly frustrating. Mary and Dorothy Ward were in the gallery for five hours, but Arnold never got his call, though he sprang to his feet twenty times, speech in hand. 'He was absolutely disgusted,'[23] Dorothy noted. Their local government clause was treated, as Mary Ward thought, most shabbily and she wrote some very strong letters to erstwhile friends. Worst of all, although the suffragettes got no immediate satisfaction, the Government promised facilities for a Suffrage Bill early in the next session 'and the whole weary fight will have to come on again—in fact cannot be dropped for a moment.' It was becoming increasingly clear that anti-suffrage was going to be the losing party; it was merely a question of how long they could fight their rearguard delaying action. And when women got the vote, what would they do to Arnold Ward, MP?

Arnold was, in fact, becoming a major problem again. On 12 February 1913 he introduced his (in fact his mother's) Municipal Franchise Bill and then was not seen for a few days. Such disappearances invariably meant trouble. Mary Ward sensed something was unusually wrong and asked Dorothy to track down her brother and persuade him to come on holiday with the family later in the year. She must bind her son close to her—it was his only hope of salvation. Arnold consented to the holiday proposal. But there remained a strange tension in the family, at whose origins Dorothy (whom no one ever confided in) could only guess. All through March, Humphry Ward was ill and irritable. Arnold had an everlasting cold, and cried off innumerable engagements. Mary Ward was depressed and subdued.

Things were not helped by the reception of *The Mating of Lydia*, which came out in England on 15 March 1913. 'The story *must* be popular',[24] Mary Ward had told Mr Reginald. It wasn't. Subscription was poor and after three months Smith, Elder had sold only 11,430. Ten years earlier, novels like *Eleanor* and *Lady Rose's Daughter* had sold 25,000 or more in the weeks after publication. In those days, her copyrights had been gold in Smith, Elder's pocket. Like its predecessor, *The Mating of Lydia* would never make its luckless publisher money. In fact by March 1915, Smith reckoned that—despite the much lower payment his firm had given—he was still out of pocket £242 on *Lydia*. Advance orders in America for what Doubleday's publicists pointedly called 'this brilliant love story' were similarly poor—24,000 copies were sent out, which shrank with first month returns to 22,800. By midsummer, it was evident that the novel would probably not clear 30,000 copies in the United States.

The cloud that was lowering over the Ward family finally burst on 5 April 1913, as Dorothy returned from a week's holiday in Robin Ghyll. She arrived back at Grosvenor Place on the Saturday afternoon, at 6.25 p.m. She was met on the stairs by a solemn Mary and Humphry Ward 'and then followed a conversation which I can never forget'.[25] There was another solemn conversation with Arnold after dinner. The news was awful beyond belief. Her brother had run up gambling debts of many thousands of pounds, and had no way of paying them. He had brought them to his parents to settle for him.

The next day, Sunday 6 April, was Mary and Humphry's forty-first wedding anniversary. The house was plunged in sadness and desperate calculations.[26] Mary and Arnold Ward went to St Margaret's Church in the morning. Arnold and Dorothy made some desultory social visits in the afternoon, and they had supper without Humphry, who went off to bed complaining of a cough. On the

Monday, Arnold and his mother had another long talk after breakfast. 'It is, and *must* be, all this week, a time of constant consultation between us all,'[27] Dorothy declared. That evening they had arranged a party for forty French, German, Austrian, Dutch, Finnish and Russian historians. The function—with its surreal jabber—must have been one of the most grisly experiences of Mary's life.

On Wednesday, Dorothy wrote Arnold a letter before breakfast, and after ten they had another long talk: 'He very simple and direct and trusting. But we are embarked on a wrestle—a wrestle in which there is so much love—which cannot be ended or decided just yet, and we must go carefully every step of the way.'[28] A little later, Arnold saw his father, whose rages probably frightened him more than did his sister's tearful sympathy. Humphry had consulted the family solicitors, and there were melancholy business matters to decide. Her father seemed calm in his mind, but tired, Dorothy thought. That evening, she spoke to her sister Janet for hours on the telephone. The following day, Arnold was off to the House early and Mary Ward went down to Stocks 'alone—by preference—looking very tired and strained'.[29]

The following Sunday, Dorothy learned how the crisis was to be met. In the bastardized German–French–Italian patois that she used to keep her maid from reading her diary, she recorded the startling news that Barley End had been bought that day by the Whitridges: 'Es ist alles so schnell angekommen, ich kann's jetzt noch kaum glauben. Es ist in *ihrem* Namen gekauft, and she [Lucy] was very sweet to me about it, walking home together after they had lunched with us. Es ist eine grosse Erleichterung—doch macht es mich auch fast ersticken'.[30] That day, she and Humphry took a walk on the hill at Barley End. Arnold was meanwhile off at Ascot playing golf.

In her private notes, Mary Ward made some detailed estimates of their family worth. All their Hertfordshire property came to £46,486, on which they owed £18,900 in mortgages and some £3,500 in other debts. Barley End was bought by the Whitridges for £7,500 and another parcel of land called Tim's Spring was sold at the same time for £1,500.[31] On Wednesday 16 April, Mary Ward had a long interview with Mr Reginald, who was, as Dorothy recorded, '*most* kind and considerate and generous, but it is all more difficult and time-taking than we hoped'.[32] Smith evidently agreed to buy all Mrs Ward's old fiction copyrights for £5,000 and to advance her some £7,000 on work in hand and future novels.[33] Humphry Ward was at the same time making arrangements to sell the most valuable of his pictures.

On Thursday 17 April, Dorothy recorded that 'oggi è stata pagata la prima parte dei debiti, in quale è un certo triste rilievo. Non ho visto Arnold io stessa.'[34] All the while, Humphry Ward was very tired and depressed and beyond his family's power to rouse him. The following Wednesday, Mary Ward had yet another long talk with her son, after which she passed a wholly sleepless night. The next day, Dorothy recorded that 'Mother continues to sleep *so* badly alas—and only at all by dint of trional and aspirin. Father she tells me était epuisé de fatigue hier soir et d'autres choses aussi, mais mon frère, qui retournait diner avec eux, était de son mieux avec lui, bon et aimable: puis, après, lorsqu'il a reçu de lui le second grand ch[eque], lui qui ne témoigne jamais de l'émotion, était troublé de l'état de mon père et s'épanchait à ma mère, et ça lui a donné du bonheur quoique ce fût un triste bonheur.'[35]

Although all the precise financial records of this sorry business were destroyed, one can reconstruct plausibly what happened. Arnold Ward had been playing and losing at bridge, baccarat, and billiards for some years. His IOUs had amassed, and finally his creditors had threatened to expose him, if he did not reach some accommodation with them. There seems to have been one catastrophic loss—which required the two large cheques Dorothy refers to in her diary. To cover this Humphry and Mary sold their property at Barley End and Tim's Spring for £9,000. According to Julian Huxley, writing in his memoirs, they also sold the wooded plot of land called Aldbury Nowers, which Mary valued at £8,000; but it is not clear whether it was at this date, or to cover some of Arnold's subsequent gambling losses. From Reginald Smith in April and May 1913 Mary Ward raised something over £12,000. And over the following months, Humphry sold many of his best pictures. It seems likely, then, that Arnold had one way or another run up debts as much as £20,000. Clearly his parents could not negotiate directly with their son's creditors. Arnold's IOUs were bought up by his Portland Club friend Moy Thomas, who became his 'banker'. Mary and Humphry engaged to channel payments through Thomas, who was presumably in a position to buy Arnold's paper at cut price. Arnold evidently swore to his parents never to game again.

It was extraordinarily noble of the Wards to stand by their feckless son, who was clearly an incurably compulsive gambler. (His vow was not kept.) And the disaster was in fact kept fairly secret. Nevertheless, the family knew all about it and some of them were less than discreet. On 17 October 1917, Aldous Huxley met Virginia Woolf at Heal's. As she recalled, 'we walked up and down a gallery discussing his Aunt, Mrs Humphry Ward. The mystery of her character deepens; her charm and wit and character all marked as a woman, full of knowledge and humour—and then her novels. These are partly explained by Arnold, who brought them near bankruptcy four years ago [i.e. 1913], and she rescued the whole lot by driving her pen day and night.'[36] Aldous and Julian Huxley, like Janet Trevelyan, conceived after 1913 a violent dislike for Arnold,[37] and evidently blamed him for blighting the last eight years of Mary Ward's life.

During all this calamity, Mary Ward was writing *The Coryston Family*.[38] She had begun the novel in December 1912 and finished it by May 1913. The tragic climax of the narrative, in which Lady Coryston dies, shattered by her politician son's intransigence, was clearly coloured by the calamities of April. Originally, it had been intended to be a novel about the tyrannies of a wife over a husband; in the event it became the story of a mother destroyed by the fecklessness of her sons—a subject on which the author was now bitterly expert. In the circumstances, the speed with which she wrote *The Coryston Family* was extraordinary. But never had she needed money more. Harper's had agreed to advance £1,600 for the book and serial rights of *The Coryston Family*. Smith had offered a £1,000 advance for the English edition. The *Revue des deux mondes* gave £250 for the French serial translation. Mrs Humphry Ward's price was slipping fast, but she was in no position to haggle.

The Coryston Family is, in the context of what the author was going through while writing it, a remarkably fine novel, and one which most movingly records the woes of motherhood. The novel opens strikingly, with the matriarchal Lady Coryston watching her two MP sons (one obedient, the other disobedient to her) from the House of Commons's secluded Ladies' Gallery. She is a widow, her husband having been a 'weak amiable man of letters—absorbed in Spanish literature'.[39] Lady Coryston despised him and took her male ideal instead from her father, a politician and a 'comrade of Dizzy'. Lady Coryston has four children, three of them sons. The eldest defies her by becoming a socialist at Cambridge and goes into Parliament as a radical. Her husband has left everything to his domineering wife and Lady Coryston cuts the heir out of the will. The second son will have nothing to do with the inheritance. It goes to the third son, Arthur, also an MP. But this young man also defies his mother by falling in love with Enid Glenwilliam, the daughter of his mother's political *bête noire*. The combined strain of her sons' intransigence finally worries the elderly heroine into a stroke and death. It is hard not to see *The Coryston Family* as the author's own story. And it broke Dorothy's heart afresh to read the end of the novel, written as it was in the extremity of her mother's grief at Arnold's disgrace. Dorothy took the book to her bedroom and wept alone.[40]

Dorothy Ward was a partisan and wholly sympathetic reader. One of the features that makes *The Coryston Family* an interestingly complex work is its lack of self pity. At times, in fact, it reads like self-criticism. The novel does not disguise that it is Lady Coryston's 'political tyranny' which has turned her sons into rebels. She is as much victimizer as victim. Is this, one wonders, how Mary Ward saw herself? In chapter 16, in the course of a blazing row between Lady Coryston and Arthur, the young man is given a number of self-justifying speeches, which it is tempting to think echo Arnold Ward's in his painful exchanges with his mother in April 1913:

Have you ever let me, in anything—for one day, one hour—call my soul my own—since I went into Parliament? .... I intend to leave Parliament after this session. I do! I'm sick of it .... You've ruined me!—you've ruined me .... I suppose you've been fond of me mother, in your way—and I suppose I've been fond of you. But the fact is, as I told you before, I've stood in fear of you!—all my life—and lots of things you thought I did because I was fond of you I did because I was a coward—a disgusting coward![41]

The anti-suffrage battle raged on throughout 1913. On 5 May, Arnold moved the rejection of the Suffrage Bill in the House. Humphry and Dorothy Ward were there to hear him. He gave a speech of eighteen minutes, rather marred by nervousness, to a sparse House. The bill was rejected by forty-seven votes. 'Alas, Mummy was not here,' Dorothy noted, 'she decided yesterday that she could not spare the time to come up, she having to rewrite the greater part of the chapter written lately under such up and down conditions.'[42] It should have been her proudest moment. It was, in the event, dust in her mouth. Moreover, it was clear as could be that however many times the suffragettes were defeated constitutionally they would not give up their struggle. And who knew to what lengths these 'wild women' might go against those they conceived their enemies. On 15 May, Aldbury village was alarmed by the report that 'two mysterious women with rubber shoes were prowling about the churchyard last night'.[43] The scare—while adding another obstruction to Mary Ward's sleep—gave her the *donnée* for her 1915 novel of suffragette atrocity, *Delia Blanchflower*.

On 8 May, the Wards' beloved Titian was taken away from Stocks to the sale-room. Mary and Dorothy had one last long look at the Madonna and babe as the picture was on the floor, before being packed up. It gave rise, as Dorothy put it, 'to an ergreifendes little gespräch'.[44] Arnold meanwhile was skulking around the house, complaining of ill health and going to bed early, or staying up moodily drinking by himself. His mother thought he was 'making his way—slowly. If only we can guard and help him.'[45] But Arnold resented being guarded. And in his resentment, he went out of his way to vex his over-solicitous mother. On 13 June, Dorothy recorded that 'Arnold told us in *high* spirits just before lunch of his sitting up half last night and spending part of this morning with a group of MP friends maturing plans for helping Ulster. This rather terrifies us, and—alas, depresses Father—but what can one say?'[46] It seems clear that Arnold—by moving round to the Home Rule compromise—was rebelling against the prime article of his mother's political faith.

Humphry Ward suffered his worst ever attack of forgetfulness the next day (a Saturday), losing consciousness of himself for quarter of an hour before lunch. Dorothy learned that on the night before he had suffered another attack at the Athenaeum, during his game of billiards. But he had carried on with the game quite serenely, not having the faintest idea where or who he was, only knowing that he was supposed to pot the red. It was terrifying. Huxley was summoned, and admitted that three such attacks in a week was 'not normal'.

He prescribed arsenic, bromide, long rest, a 'very good nerve man, Risien Russell', and foreign travel.[47]

It was therefore arranged that Mary and Humphry Ward should go to the Tyrol to repair his nerves. Hopefully stress was the culprit and he would recover in the sweet Alpine air without Arnold around to madden him. Departure was fixed for 18 July. But then, the poor parents were driven distracted by Arnold's again going underground just before they left. 'Pas de nouvelles de mon frère,' Dorothy wrote despairingly in her diary for 15 July, 'il est ni venu à Stocks ... ni revenu ici [Grosvenor Place] pour dormir. Sehr angstvoll, alle heute nachmittag und abend. Mutter so traurig und weisse.'[48] She feared the desperate young man had killed himself. Finally, on 16 July, a telegram came explaining that he was so 'fatigued' on Monday (14 July) night, that he fell asleep on the last train and simply booked into a hotel at Bletchley. On his return, there was the inevitable long conversation, 'bien triste, bien anxieux'. On the Friday afternoon (18 July) Dorothy saw her parents off from Charing Cross. Arnold claimed pressing business in the House, and did not come. Humphry heavily dosed with bromide—was placid enough, but Mary Ward looked to her daughter 'horribly tired—and complained of malaise and indigestion pain in the chest which made her feel rather faint in the train'.[49]

Their Tyrolean trip would be no holiday for Mary Ward. In addition to her fretfulness about Arnold being left on his own in London, she had to look after Humphry, who was paralysed with depression and required constant attention. And she had somehow to get started on a new novel. It would, she had decided, be 'a feminist novel' with its early scenes set in the Tyrol. That way she might get some return on her expenses. She had suggested the idea to Smith on 1 July and—fanatic anti that he was—he liked it. On the strength of a renewed life insurance policy, he agreed to loan £2,000 on the spot—the money they needed for Humphry's Alpine convalescence.

The Wards' sleeper arrived at Basle on the morning of Monday 19 July. The next day they went to Mendel bei Bozen. The hotel was luxurious and 3,500 feet in the mountains. On 24 July, Mary Ward described in her diary 'our little sitting room ... Humphry and I sitting wrapped up with windows open. The Brenta and Presanella ranges visible to the west through the clouds.'[50] It was invigorating and Humphry sent back a prematurely optimistic telegram on 29 July, 'Both much better', and informed Dorothy (holding the fort at home) that he had left off his bromide medicine and was 'walking briskly'. Mary also wrote reassuringly two days later, 'I am earning about £250 a week, and that keeps my mind at rest! also I am beginning to enjoy the new *donnée* and take an interest in the new puppet play [i.e. *Delia Blanchflower*].'[51] (The novel opens at a hotel in the Tyrol and the line 'Not a Britisher to be seen!') They had by now moved to the Grand Hotel des Alpes and had the luxury of a suite with a balcony and full room service. 'I shudder to think of this week's bill!'[52]

Mary Ward told her daughter. But she lived in hopes that the *Century* might offer £3,000 for the serial rights of her new novel and several of her works had recently been optioned for the cinema at £20 apiece; this 'cinematographic thing' could turn out profitable, Curtis Brown had told her. Grosvenor Place was rented for £64 a month; Reginald Smith had promised another £1,500 for the new novel on 1 October. True the Wards only had £400 in their Barclay's bank account, but that was imminently due to be supplemented by the £700 Humphry had earned at Christie's from the sales of Stocks furniture, clocks, and china.

Unfortunately, Humphry's early recovery proved short-lived. By the first week in August, he had again his 'frail hollow-eyed look ... and his clouds of depression'.[53] He was again dependent on his bromide, and spent hours in the hotel room, moping. The weather was 'horrid'. 'Sometimes I wonder', Mary wrote to Dorothy Ward, 'whether home might not be as good as all this wandering!'[54] They nevertheless wandered on to a cheaper (£12 a week) hotel on 9 August. Humphry showed no signs of improvement and his wife fretted over the news that Arnold had flatly refused to go and stay with his sister Jan while they were away. She knew what evenings by himself in London meant. Mary Ward herself was now totally dependent on trional at night.

Dorothy Ward came out to join her parents at Gossensass in mid-August. She was shocked at Humphry's appearance: 'he seems disinclined to or afraid of anything but the mildest walks and turns back if they get at all too steep or seem to be getting longer than he thought'.[55] Complaining of 'lumbagoish pains in his neck' he would spend whole days listlessly doing nothing in his hotel room. Mary was also exhausted by the slightest physical effort, and would complain of being 'heart-y' after walking only a few hundred yards. She was, however, on the third chapter of *Delia Blanchflower* and by now had the story clear in her mind.

In fact, the scenario for this suffrage novel had been unusually difficult. She had begun with a marriage-in-public-life idea: 'Minister with suffragist wife—converted after taking office—missing one night—next day in prison—bails her out—and then he tells her that he shall resign—and they must part—He insists on the child.'[56] This novel, she thought, might introduce a gallery of 'suffragist types'. But having toyed with this idea, she diverged into something quite different. As reconceived and written, *Delia Blanchflower* is the story of a 22-year-old orphan left by her father, Sir Robert Blanchflower, in the care of a male guardian twice her age, Mark Winnington. The custody will last until she is 25. Earlier in life Winnington, a bachelor lawyer, saved the Blanchflowers' marriage by his wisdom and good sense. Delia has been misled into suffragette militancy by an evil angel, Gertrude Marvell (based on Christabel Pankhurst). This portrait of feminist fanaticism was to be a main attraction in the story. Delia's good angel is, of course, Mark—a sturdy, sensible, Conservative, cricket-playing English gentleman of impeccable stuffiness who sees it as his guardian's duty to curb his ward's juvenile infatuation with Gertrude's 'Daughters of Revolt'. Delia is also misled by a male feminist (and co-respondent in a recent sordid divorce case), Paul Lathrop. Out of sheer mischief, Lathrop helps Delia dispose of some jewels to raise money for the suffragette cause. Over time Delia comes to see the wisdom of Mark's views and repudiates militancy. The novel ends with the expected marriage. Gertrude perishes apocalyptically in the flames of a Cabinet Minister's country seat, Monk Lawrence, which she has set ablaze. With her perishes the guiltless and crippled child of the house's caretaker, Daunt. This climax was not melodrama. There were hundreds of such outrages in the early months of 1914 and Mary Ward evidently recalled the blaze at Lloyd George's country house in February 1913, which was generally thought to have been incited by the Pankhursts. Shrill and polemical as it is, *Delia Blanchflower* has an energy lacking in much of Mrs Ward's later fiction.

During the their first weeks in the Tyrol, it seemed blessedly that Humphry's 'attacks' were cured. But they returned with a vengeance in August, after a two-month remission. There was one particularly painful episode, in which he lost consciousness while teaching Mary her morning Greek in bed. He was taken to a specialist in Munich. The diagnosis was again '*liver* mainly'[57] and more of the odious raw beef sandwiches were prescribed. The three Wards eventually returned to London on 28 September out of pocket and out of spirits.

The reviews of *The Coryston Family* on 10 October were not good and 'curiously point-missing',[58] as Dorothy Ward thought. But why on earth would critics associate the tortured Lady Coryston with the ostensibly happy and wealthy Mrs Humphry Ward? The novel was unsuccessful in other more material ways. The English subscription had been a meagre 8,051 (some 500 less than even *The Mating of Lydia*). Sales were under 12,000 by the end of the year. Harper's had publication day orders for only 16,000. Transatlantic reviewers wholly ignored the book—there were more interesting, young, and *American* writers, Harper's brutally informed Mrs Ward. By December *The Coryston Family* had sold only 20,000 and was clearly going to be another flop.

Humphry Ward's attacks continued. And now they came at ever shorter intervals. Dorothy would sometimes catch him looking white-faced during meals, trying desperately to make sense of conversations that suddenly meant nothing to him. Typically, he would forget who he was. Dictating letters to Miss Churcher he would forget who he was writing to. On two more occasions, he lost consciousness playing billiards at the Athenaeum—once going on to win his game as he later discovered. On another distressing occasion, he was taken ill on the train from London, and was found by Dorothy disconsolately wandering Tring platform, uncertain of who he was and presumably taken by other passengers to be hopelessly drunk. By November and December, the attacks were coming every other day. His doctors now prescribed ether drops, brandy, strychnine, and valerian.

By 1913, Humphry's regime was that of the confirmed invalid. His day would begin with a dose of Sanatogen. Before breakfast (raw beef sandwiches) he would take further doses of anti-sclerosin and sulphur. After breakfast, he would take a supra-renal cachet. These medications would recur regularly through the day. The one indulgence was a 'small cup of coffee—half milk and *good* coffee'[59]—at 4.15. Supper would be brown bread and butter. If he seemed tired before going to bed, there would be what Mary Ward called the 'pink *dragée*'—a sweetened sedative. Humphry Ward, in short, was a shattered man. The strain of nursing him told on Mary, who confessed to Dorothy that 'the constant sleeping drugs are having a bad effect on her too'.[60] She was having occasional giddy spells during the day.

Her peace of mind was not helped by Arnold Ward's stout championing of Home Rule and Irish compromise in the House on 26 October 1913. Mary herself had never wavered an inch from Uncle William Forster's conviction that coercion was the only way to deal with the Irish. But Arnold obstinately defied her. On 28 October he came late to dinner at Grosvenor Place, and had clearly been drinking. It was, as Dorothy recorded, 'une soirée assez triste et difficile'.[61] There was the inevitable row about Ireland. Arnold was also kicking against the parental pricks in other ways. On Wednesday 19 November, there was a gambling scare. Humphry had looked tired and edgy all day, Dorothy noted: 'Il a été troublé parce que mon f[rère] est resté au P[ortland] C[lub] pour la nuit, quoiqu'il a fait téléphone de bonne heure pour nous le dire.'[62]

Even if he was not backsliding on his solemn promises (which he may well have been) Arnold was clearly determined to assert himself by breaking bounds. Most of the records have been destroyed. But there is a note from Arnold, dated '1913', from the Portland Club telling his mother, 'I feel very tired tonight because I stupidly played several games of billiards after my day's work. No rules broken.'[63] The 'rules' refer to Arnold's promise never again to play cards or billiards for money stakes. But his even *being* in the Portland Club, if only for a nightcap whisky and soda, was the stuff of Mary Ward's nightmares.

That Christmas the Wards, the Huxleys, and the Trevelyans gathered at Stocks. Humphry and Mary had just contrived to remortgage the property. It had been, from the point of view of health, wealth, literary reputation, and political causes, an unmitigatedly wretched year: But still Dorothy Ward thought that Christmas 1913 was 'on the whole happy, because of the sense of family unity underlying all and of A[rnold]'s being with us and caring to be. Outwardly it has been a *very* happy day.'[64] She and her brother went to church on Christmas day, enjoying the brief sense of comparative peace. Mary was exhausted, but nevertheless wrote her ten pages on Boxing Day. If she did not get her book out, they might well not have a roof over their heads by Christmas 1914.

The old battles were renewed in January. Humphry Ward continued to sell off his pictures whenever dealers could be persuaded that they were genuine and would offer a good price. As an art critic, however, he could not but think the world had gone mad. Modernists ('freaky people' as Dorothy called them) were everywhere spouting their nonsense. And suffragettes were assaulting Velazquez paintings in the National Gallery with hammers. On his part, Arnold Ward continued to defy his mother by supporting Home Rule and by defying that other home rule which specified that he should sleep at Grosvenor Place. Meanwhile, his constituency support was eroding fast, and he only narrowly survived a vote of confidence from his party supporters on 22 April. The Die-hards, the suffragists and those (an increasing faction) who simply resented Arnold Ward MP's high-handed ways were all combining against him. It was clear that he was going to be unseated at the next election, even if the newly enfranchised women voters did not have the pleasure of kicking him out.

Mary Ward was labouring to finish *Delia Blanchflower*. And then, when it was finished on 2 March 1914, everything went wrong for the feminist novel. Her arrangements were, as she thought, 'radically mismanaged by Curtis Brown'. But the truth was, no one wanted *Delia Blanchflower*. The book and serial rights could not be sold anywhere in America despite Curtis Brown's 'desperate campaign'[65] on its behalf. Magazine editors looked at the proofs, but they simply would not buy at any price. All this was extremely embarrassing for Reginald Smith who—in sympathy for Mary's domestic financial crisis—had made a series of unwarrantably large advances on the novel. 'The thought of my father in law [i.e. George Smith] is always present in my mind as I work for and try to aid you,'[66] he told Mrs Ward. By early 1914, his account with her stood at £4,557 overdrawn. Until an American publisher could be secured, there was no question of publishing *Delia Blanchflower* in England and Smith was becoming increasingly anxious. Mary Ward was already (in April 1914) making notes for a new novel provisionally called 'the Holland House novel' (later titled *Eltham House*). But Smith firmly declined to make any arrangement until something was worked out for *Delia Blanchflower*.

These delays were intolerable for Mary Ward, who needed money urgently and at once. Her annual household and personal expenses, as she estimated, were running at £6,031. After 'Food' (£750) the largest single item—£600—was cryptically labelled 'Arnold'. It was followed by £468 on the Stocks mortgage and £495 for the Grosvenor Place expenses. Other hefty items were £250 for keeping their Daimler on the road and the same amount for gardeners' wages; £400 life insurance; £150 for Dorothy's allowance; £100 for Bessie Churcher's wage (presumably supplemented by PES for her work at the settlement) and £40 for Lizzie Smith's annual wage. All these had to be paid and any delay in income from the novel in hand was excruciating. Mary Ward could of course write *Delia Blanchflower* off, and simply take whatever she could get for it. But that went against all her instincts and pride as an author. Finally, in July, she resolved to ask her 'kind and rich cousin F. Whitridge to lend me £1,000 against the American receipts to prevent my sacrificing the book'.[67] She was thus in the odd position of having taken advances in 1913 from Smith and an advance in 1914 from Whitridge for a book which would not, in the event, be published until 1915 and which would never come near meeting the cost of these advances.

Whatever else, it was clear that the feared thing had at last happened. Mrs Humphry Ward was no longer a best-selling author. And less-than-best-selling authors could not live in large country houses. On 3 May, Dorothy Ward recorded that 'Mother and Father made today the great decision—dass wir dieses Haus verkaufen müssen—and told Arnie of it first, before luncheon (he and F. only arrived 11.55) and me after luncheon.'[68] The triggering event in their decision to sell Stocks had been a 'disheartening' letter from Curtis Brown, enclosing one from Doubleday, who were no longer willing to be Mrs Ward's American publisher. All day, Mary Ward was grave and *distraite*. Humphry, however, seemed placid and even oddly 'cheerful'. Dorothy and Arnold Ward gloomily mulled over the latest crisis the next afternoon, 'walking up and down the path that crosses the cricket ground'.[69]

Mrs Ward confided the news in Reginald Smith a week later. He did his best to raise her spirits: 'To say that I am distressed in realising your distress is inadequate. And yet the courage with which you already face the pang of parting from Stocks—I can understand something of how dear the house is to you after all these years—is quite wonderful.' But he told her that selling Stocks was now her only option. She must not expect a return of popularity for her novels: 'I feel sure your work is as good and as greatly appreciated by *our* generation as ever: but we are passing away, and the tastes change.'[70]

The final arrangements made for *Delia Blanchflower* were—in orthodox publishing terms—bizarre. Curtis Brown finally managed to make an arrangement with Hearst Books in America to take rights for £1,400. In Britain, Smith and Curtis Brown between them cobbled together a sale to Ward Lock of £2,000 for the book rights and a consolation £150 for the serial rights in *Lady's Realm*. All this money would of course go directly to Reginald Smith, who (having advanced £4,500 plus interest) was effectively paying a thousand pounds for other publishers to make money from Mary's novel. Nominally he owned all copyright in *Delia Blanchflower* for five years. But his prospects of getting back his money were non-existent. (Apart from anything else, within five years suffragettism would be ancient history.) Even his father-in-law would have told him it was publishing madness.

In their straitened conditions Europe was out of the question for the Wards that fateful summer. They would holiday frugally in Fife, on the east coast of Scotland. One attraction was that they could call in at Fox How on the way north. But even this went wrong. On the Rydal Road, their Daimler (driven by the chauffeur-cum-gardener, Larkin) was involved in a serious collison with another vehicle belonging to the Red Lion Hotel at Grasmere. 'The attacking car' (as Mary Ward rather melodramatically called it) skidded, went up the roadside bank, and rolled 'gently over on its side'.[71] Miraculously, no one was injured except Larkin, whose nose was scraped. But Mary Ward was horribly shaken up. And their Daimler was extensively damaged. It had to be left in Grasmere for repair and the family continued their way north by train.

Fife itself proved wretched. The house they had rented was shabby. The east wind blew incessantly; the coast was shrouded in the oppressive grey sea mist which the locals call 'harr'. They had no car, and were confined to their draughty quarters. In early August, Humphry Ward had three attacks in two days. 'It was hard indeed for mother to keep up,' Dorothy recorded. 'Her face and her look as she sat by him as he lay on the sofa, chafing his hand, I don't think I can forget—but she was splendidly brave. She whispered to me that Lizzie must go with her to Edinburgh to see Mr Wallace (the surgeon recommended by Dr Brough).'[72]

One good thing came out of the wretched business. The 'young and clever' Edinburgh doctors quickly diagnosed what was wrong with Humphry Ward and pooh-poohed Huxley's theory about 'anaemia'. He had arteriosclerosis—hardening of the arteries and a consequent constriction of blood to the brain. It was a shock but at least Humphry's family now knew the truth. The next day came another shock: England had declared war on Germany.

28

The Wards and War: 1914–1916

LIKE other unhappy families, the Wards found world war to be something of a relief. Cataclysmic worry about the end of civilization took their minds off their own problems. Their concerns about money seemed positively petty as the Boche hordes swept over Belgium. And a number of painful decisions were mercifully postponed. There was now no question of selling Stocks—the bottom had dropped out of the property market and they would not get a fair price for it. Instead they made arrangements to let the house.

There would be no general election—so Arnold Ward would not after all be unseated. He was an officer in the Herts. Yeomanry (commissioned 1908, promoted lieutenant in 1910), and would certainly be removed from the flesh-pots of London. The War Office would relieve his mother of the responsibility of providing his bed and board. Whitehall would not of course pay his gambling debts, but with luck he would have no time for cards. When he returned bemedalled (or lightly wounded) from the wars, all might be forgotten by a grateful electorate.

All considerations of the suffrage were of course relegated indefinitely. The European war also took Humphry's mind off his ailments, and he gallantly offered his services to *The Times* in this hour of need. They allowed him to write some virulently anti-German leaders, which was wonderful therapy. Austerity, which would have been forced on the Wards anyway by Arnold's debts, now became a patriotic duty. Their male servants would be (eventually) conscripted. Foreign travel was out of the question. All expensive entertaining was dutifully postponed till after hostilities, whenever that might be. They need not even buy new clothes. One could economize *and* keep up appearances.

Lieutenant Arnold Ward joined his yeomanry unit at Watford on 5 August, the day war was declared. The regiment had been reorganized into three squadrons with the prospect of foreign service. In Dunfermline (suddenly swarming with soldiers) the Wards made their 'anxious plans for an uncertain future' and brooded over 'a nameless sick fear about Arnie—of whom we hear nothing'.[1] A telegram came on 12 August. Together with all his fellow officers, he had volunteered for the service abroad (something not required of the Territorial Force, who could opt for 'Second Line' duty at home). It was awful, but what they expected. 'Mummy is splendidly courageous,' Dorothy Ward wrote, 'yet the look in her face goes to the heart.'[2]

The family's emotional girding of themselves for war was complicated by a wholly unexpected tragedy. On 24 August—still in 'ugly' Fife—they received a telegram from Leonard Huxley: 'Read II Samuel 33. Dont reply'. Mary Ward rushed down to console her brother-in-law and his boys. It emerged that Trevenen Huxley—Mary's nephew—had gone missing from the nursing home in Surrey where he was being treated for depression. A week later his body was found; he had hanged himself from a tree. There had been a wretched business with a servant-girl, whom he had supposedly got pregnant. On the day of the funeral Janet divulged some of the details to her sister Dorothy. It was 'worse than I ever guessed'[3] she wrote in her diary.

The exact details of Trevenen Huxley's death were not publicized until his brother Julian wrote about them with touching frankness in his autobiographical *Memories* (1970). But the event is rendered mysterious by the use his other brother Aldous made of it in *Eyeless in Gaza* (1936). In that novel, Trevenen's hapless love and suicide forms the basis of Brian Foxe's story—the central strand of the narrative. Huxley made Brian an exact physical likeness to his dead brother, even down to Trev's stutter. Huxley also involves Mrs Humphry Ward in the drama, as Mrs Foxe. This lady is clearly intended as his Aunt Mary; allusions to Stocks, PES, cripple schools, and Mrs Foxe's physical mannerisms make the identification—which is satirically edged—unmistakable. Mrs Foxe 'adopts' the narrator, Anthony, after his mother dies early in life (as Julia Huxley had). Mrs Foxe is even made to talk like Mary Ward, her favourite epithet being 'splendid'. Mrs Foxe sums up Mary's theism when she tells Anthony, 'The wonderful thing for us [is that] Jesus was a man.'[4] In the novel as it climaxes, Mrs Foxe's smothering love is represented as a main cause of Brian's suicide by throwing himself over a cliff. Brian—a sensitive young man, who loves bird-watching—cannot withstand the force of Mrs Foxe's goodness. Anthony concludes: 'She had been like a vampire, fastened on poor Brian's spirit. Sucking his blood .... If anyone was responsible for Brian's death, it was she.'

Given Aldous's well-testified affection for his aunt—at least during her lifetime—it is not easy to know what one should make of *Eyeless in Gaza*. It is true that Mary Ward took an intense interest in her Huxley nephews' careers and upbringing. But they seem to have loved their aunt for her attention to them. No biographical account of Aldous Huxley has indicated what indirect part, if any, Mary Ward played in Trevenen's suicide or what the inner family meaning of the character of Mrs Foxe is. Does Huxley imply that guilt fostered by Mary Ward's goodness drove his luckless brother to hang himself when he fell so far below his aunt's moral standards? If so, it seems very hard.

Arnold Ward was not, after all, whisked up the line to death in August 1914. It emerged that his volunteering for the Front was something of an impulsively romantic gesture. Letters written to his family indicate that his more considered opinions were eminently sensible and unbelligerent. He believed that 'the right course for us is neutrality .... No British interests are concerned.'[5] Arnold's war strategy would have been to strengthen the Royal Navy for domestic defence and let the Europeans slaughter each other at their pleasure. The current surge of British jingoism he thought 'idiotic ... I think the world has gone mad. I do not want to shoot Germans and positively object to shooting Austrians.'[6] But his position as an MP was delicate. 'My political life is hanging by a thread,' he told his mother in August 1914: 'if I vote against the War I shall of course be ruined. And it is a particularly cheerful prospect after that to go down and join my regiment on mobilisation.'[7] Wisely, he seems not to have spoken out against the war. His fellow officers would probably have lynched him if he had.

On 31 August 1914, Arnold's regiment was ordered to Egypt to replace regular units needed for France. Dorothy Ward stayed up all night marking her brother's underclothing and darning his socks. She took them down in person to Suffolk, where the Herts. Yeomanry were encamped before embarkation. Humphry and Mary Ward had half an hour with their soldier son at lunchtime in London on 4 September. He then went off to say goodbye to Violet Asquith. The Prime Minister was too busy to see him, which was something of a blow. There was just time for a hurried last tea at Grosvenor Place, and then he was off to Liverpool Street at five. Mary, Humphry and Dorothy Ward sat among the teacups afterwards feeling very 'flat and choky'.[8] Then the ladies went for a short tearful stroll in Hyde Park while Humphry adjourned to his 'dear Athenaeum'.

Egypt was, in fact, an active theatre of war in 1914–15, although Arnold invariably had the bad luck to miss whatever fighting was going on. By October 1914, he was comfortably installed in barracks three miles from Cairo, and his old haunts the Casino de Paris and the Bohemian night club. Here he could gamble, drink and meet various young ladies living immoral lives.[9] Some of their lurid stories he actually transcribed and sent back to his astonished mother—apparently thinking they might supply her with valuable raw material for her fiction.

It was the regiment's hope that they would be transferred to some fighting front in three months. But nothing happened, and by January 1915 Arnold was complaining about being 'pretty bored ... helplessly watching the battle from afar'.[10] It was oddly similar to what he had felt in Egypt fifteen years earlier, during the Boer War. In his boredom and remembering his 'Special Correspondent' days he conceived a scheme 'for a rather grandiose job.'[11] He would write a report on Egypt and the war for the Government. They could release him from duties and he could rove all over the Middle East. He sent his enthusiastic proposal, together with a request for £75 motoring expenses, to the Prime Minister. Asquith did not reply.

Cairo in the morally relaxed atmosphere of wartime was just the wrong place for Arnold Ward. Almost at once the old vices reasserted themselves. In September 1914 his mother received an insolent letter from a Turk called Tueni Bey, who claimed Arnold owed him gambling debts—in fact incurred in London just before war broke out. Arnold wrote back instructing his mother to take no notice of the 'disgusting beast'. He went on to explain 'he deserves no consideration and won't get any from me. There was a general crisis in the card world when war broke out. I have over £500 to collect—perfectly good money in normal times—and of course have not dreamed of asking to be paid during the war.'[12] No sooner was this sordid little affair settled, than Moy Thomas was down at Stocks—evidently assuring himself that there would be no hiatus in *his* payments during the war.

On 8 January 1915 Barclay's Bank wrote to Mrs Ward that they would not honour a cheque which Arnold had drawn on them, for café expenses in Cairo. Mary had a 'bad night' and four days of 'grande peine'[13] before an explanation arrived from Arnold. It is one of the very few letters on his gambling which survived burning, and as such is worth quoting at length. Arnold explained to Mary that he had just joined the Mohamad Ali Club. As if it would mollify his mother, he added that it was 'the fashionable Cosmopolitan Club in Cairo'. He went on to confess that he had

> played a good deal of bridge there in the evenings at half-piastre (penny-farthing) and piastre (twopence-halfpenny) points. They are a very nice lot of members, native, English, Greek, etc. They haven't the slightest idea how to play bridge, but they defeated me for some time as I held very bad cards. On the evening of the 21st. or 22nd. [December] I met there the celebrated Omar Pasha Sultan, the original of Baroudi in [Robert] Hichens' *Bella Donna* .... By this time every one was anxious to back me at any card game; Omar is not a bad player but I can beat his head off—Eventually a contest at picquet was arranged at 7 piastre points (1*s* 5*d*). I took a quarter (fourpence-farthing). This was playing rather too high for me as it turned out. The contest lasted three hours and consisted of 18 games of which he won 15 and I won 3. The total damage came to 3,045 points, £220. My partner had gone to bed, and rather out of swagger I gave him an English cheque for the lot the same night. One of the partners was slow in paying up and the result was I did not send the remittance of £170 to Barclay's till the morning of January 6, by telegraph transfer... the last result I wanted it to produce was that you should be worried for funds, after your letters of the same period.[14]

It is a curiously stupid letter. He can beat Omar Sultan's head off, but mysteriously ends up losing fifteen games to three. No one in the club can play bridge, but they all contrive to beat Arnold Ward. There also survives a scrap of a letter from Mary Ward to her son from this period, which reads:

> I enclose this little statement of what has been paid you or for you since the outbreak of war:
>
> Allowance—£585-0-0

For charge at Alexandria—£20-0-0
Conservative Association—£50-0-0
Debt to Tueni Bey—£350[15]

The cosmopolitan café society of Cairo was a world away from the life the Wards were living in wartime Britain. They had let both Stocks and Grosvenor Place to the American author Edith Wharton, and had moved into a small flat in Queen Anne's Mansions. Later in the war (as in 1908) they would return to Stocks Cottage. But, although they were living modestly, huge wartime tax and supertax assessments (£800 in 1915) began to arrive every half year. Mary Ward's large income and their property holdings made for crippling liabilities. She took sole responsibility for dealing with these worrying matters, although the Inland Revenue correspondence was addressed exclusively to Humphry. 'She takes all *his* burdens on her,'[16] Dorothy Ward noted. Her father was now having attacks almost every day. He passed his sixty-ninth birthday (9 November 1914) 'confused and on the verge of an attack' all day. But he managed a glass of champagne at dinner and later that evening Dorothy read him a Kipling story, at which 'his brow cleared and he got interested and placid'.[17]

Mary Ward's health had meanwhile actually improved. In early October, she resolved after all to have an operation for her troubling gynaecological condition. It was performed on 16 October by Miss (later Dame) Louisa Aldrich Blake—the country's most eminent woman surgeon—and was pronounced a great success. There was, as Dr Blake told Dorothy, no malignancy. Almost immediately, Mary Ward became more sprightly. By December, she was finishing off *Eltham House*. The novel—conceived before the war in spring 1914—was actually completed on the last day of the year (the 'Annus Terribilis', as Mary Ward called it in her diary).[18] At 7.35 that evening, she emerged from her study to the hall, where Dorothy (in a hushed voice) was playing with her beloved 'chickies'—Janet's two children. She had 'the completed chapter in her hand and a look of happy and blessed relief in her dear and beautiful face.'[19] After the chickies were put to bed, she read the chapter to her daughter, by the hall fire.

Eltham House is another investigation of divorce, although less strident in tone than the ill-fated *Daphne*. The idea for the new novel was again taken from history. In 1796 Lord Holland eloped with Elizabeth, the wife of Sir Godfrey Webster. Webster was induced to divorce his spouse by the surrender of her fortune and children to him. Subsequently the runaway couple married and returned to England, where she set up a salon at Holland House; respectable women did not visit but men in public life put themselves out to receive her invitations. Holland, meanwhile, pursued a successful political career. He suffered no set-back and was included in several Whig ministries.

Mary Ward brought this plot forward to the present day posing the question 'how are things modified in a hundred years?'[20] How would the Hollands' drama re-enact itself in 1912–13, in the world of gramophones, telephones, cars, and new journalism?

Eltham House began, like many of Mary Ward's novels, with the vision of a grand house. Like Holland House, it is situated in the middle of fashionable London. As the story begins Alec (later Lord Alec) and Caroline Wing have returned to London 'to defy the world'.[21] Unlike their historical predecessors the Wings fail dismally in their attempt to reconquer English society. Caroline's salon is disreputable and she is snubbed by the Establishment (he, for instance, is invited to the Royal marriage, she is not). Alec is shunned by a Liberal Party currently undergoing a 'moral' revival; the Prime Minister will not appoint to his Cabinet a man who has been a co-respondent in a notorious divorce case. There is a quarrel between the Wings about his consorting with a former lover, Madge Whitton. He leaves England on his yacht for South America. Left in London, Caroline discovers that she has a fatal heart condition and is united briefly with her daughter Carina. Knowing nothing of his wife's illness, Wing returns unexpectedly to find Eltham House ablaze with light and company. In her plan for the novel, it was Mary Ward's intention to have the couple reunited in 'a passionate, though tragic love scene'[22] and then return for her to die in the Italian villa where they had first been lovers. But in the event her writing energies were not up to this climax and Caroline is made to die quietly in London. In keeping with its ending the novel is unusually morbid but interesting for the author's growing scepticism about conventional morality. Her views on divorce are markedly more sympathetic than they had been in the tract-like *Daphne*.

Mary Ward's relief at finishing *Eltham House* was short-lived. *Delia Blanchflower*, which finally appeared in the bookshops in January 1915, received reviews of chilling contempt. In the *Athenaeum* (6 February 1915) the reviewer dismissed the novel as beneath notice: 'We see no necessity to concern ourselves with this story as such, except to say it is quite unworthy of the author.' It was, the reviewer continued with a sneer, '*Victorian*'. As a term of contempt in 1915 it ranked with 'Prussian'. But Mary Ward could not afford to be put off. She writhed and wrote. By the middle of January she was racing through a new novel, *A Great Success*, at the rate of ten typed pages a day. The first instalment was due in the *Cornhill* in May 1915. She was meanwhile distracted by a new crisis at PES. The settlement could expect few men Residents during the war and it had been resolved by the management committee turn it into a women's settlement, under Hilda Oakeley (a first-class graduate of Somerville, 1898) as warden. Mary Ward was opposed to the reform but was eventually voted down. By March 1915, the first women residents were installed, and the centre was organized for all-out war effort.

After her initial disappointment, Mrs Ward took the transformation in good part, although she well knew that the new occupants of her settlement were suffragists to a woman.

The oppression of war, as often happens, made Mary Ward nostalgic for the far off days of her youth in Oxford. On 24 May she and Humphry carried out an 'old plan' (as Dorothy called it) and motored to Oxford for the day. It was delightful. They called on old friends like the Johnsons, and visited the college gardens 'in their blossoming glory'.[23] Out of this experience were to come two of her best later works: *Lady Connie*—a story of Oxford in the 1880s—and the autobiographical *A Writer's Recollections*.

The family was now (May 1915) mainly based at Stocks Cottage, tucked away among the woods on a hill near the great house. Stocks itself was meanwhile rented for £1,100 a year by a wealthy family called Blackwell. Whenever Dorothy peeked in she saw their house 'swarming with light-hearted, rich-looking young ladies in tennis dress'. The Blackwells rather disgusted her. But they were interested in buying Stocks, and that might one day soon be necessary. Mary Ward did not seem to mind the interlopers in her house. Nor did she mind not having Grosvenor Place at her disposal (it was also rented out, for £300 a year). But she remained curiously obsessed with events of half a century earlier, and had strange yearnings for Fox How. In June, after a 'last *orgie* of committees',[24] the Wards went off for their summer holidays in a rented house at Kelbarrow, near Penrith. By their standards they travelled light, taking with them two maids, a cook, twenty-three pieces of luggage, a chauffeur-butler, a motor car, and two bicycles.

As with Oxford, the immersion in places associated with her childhood was having a profound emotional impact on Mary Ward. She spent her sixty-fourth birthday (11 June) at Fox How with Aunt Fan. Dorothy described the afternoon in her diary:

> After luncheon (which we took out in the 'green parlour' amongst the rhododendrons on soft moss and grass) mother and I climbed up through the top garden up to the fell and she clambered a little way along the fell to the left delighting in the grey forms and beautiful falls and undulations of the fell and the exquisite peeps of lake and mountain beyond. We sat for long on a rock and talked of grave and serious things—of Stocks and the possibilities of Mr B[lackwell]'s offer—of her book-earning payments, of Father's health and of Arnold—M[other] always with marvellous sweet courage.[25]

The Wards did not, in the event, take Mr Blackwell's offer and repossessed their 'beloved Stocks' in mid-August 1915. True, it was not Stocks as in the days of its glory. The mains electricity was now cut off. They depended on a generator and when that broke down—as it often did—on lamps and candles. The golf links, cricket ground and every arable spot of land were ploughed up for vegetables which would be requisitioned by the government. Dorothy

Ward, practical as ever, was effectively running Stocks as a farm, aided by a whole corps of land girls.

On the whole, the non-combatant Wards were having a good war. The same could not be said for Arnold. He was 38-years-old and had served in the Yeomanry for almost a decade. Yet he was still in mid-1915 a lieutenant and there was never any suggestion that he might rise above this lowly rank. He had clearly been passed over. A man of his age, background, and military experience ought to have been a lieutenant-colonel or better. In November 1914, he was appointed assistant adjutant of his regiment, a junior post with the responsibility of keeping up the men's musketry. These amateur Hertfordshire soldiers were not yet regarded by the Staff as a reliable fighting unit, and had been put to routine garrison duties, guarding the Suez Canal. No incursion by the enemy was anticipated this far into Egypt. Nor did Arnold feel any great enmity towards Johnny Turk: 'It seems odd for me to be off shooting Turks . . . and I am even less enthusiastic about it than about shooting Germans. I wish we had a really congenial, palatable foe—such as Yankees, for instance.'[26] One wonders how such letters ever got past the censor.

More ominously, Arnold had fallen out with his Commanding Officer, Major Hugh Wyld. It was the old story. He simply could not help quarrelling with those put in authority over him. Wyld and his second-in-command, Captain Archie Clayton (who had been promoted over Arnold), were both stockbrokers in peace time and regarded themselves a cut above Lieutenant Ward. A fellow officer in the Squadron, Wilfrid Holland-Hibbert (a son of Viscount Knutsford), recorded in an unpublished memoir that Arnold Ward was 'a Member of Parliament, fat, uncountry and a bad rider'. He did not fit. The Ward–Wyld feud had simmered on for some months, and climaxed in June 1915. Dorothy Ward later destroyed all the letters written that summer, but it is clear that Arnold wrote a memorandum (which survives) complaining about conditions in his squadron. This memorandum which ran to twelve pages of extraordinary insubordination, was sent by Arnold to his regimental commander and eventually forwarded to the War Office in July.[27]

At this time, and indeed all through the war, the authorities were excessively nervous of anything that looked like mutiny. Lieutenant Ward was promptly posted well away from his regiment, to Cyprus, where he was made Officer Commanding an empty convalescent camp in the Troodos mountains. There were all of sixty white men and women in the local township. (Arnold did not consider the Cypriots to be white.) And the total garrison strength of the island was under 1,000. Most of these were recovering wounded soldiers—'crocks' (Arnold knocked up 101 not out against their bowling, which was one small bright spot). He had—as a mark of his CO status—been promoted to local or temporary captain. He would revert to the lower rank of lieutenant on rejoining his regiment if he ever did so. It was humiliating beyond belief.

Meanwhile, to cap it all, the Herts. Yeomanry were at long last in August 1915 thrown into action against the Turks in the Dardanelles. Arnold volunteered to join them, but was peremptorily refused leave to do so. They did not want him. He had been told that his memorandum to the War Office would be dealt with by mid-October and that if cleared he might then rejoin his unit. 'By that time', as he bitterly told Dorothy, 'most of my regiment will have been killed and wounded. You can foresee what will happen to me.'[28] He would be branded a coward, as well as a trouble-maker. It was horrible luck. Moreover, he had contrived to run up yet more gambling debts, which he had forwarded to his mother to settle. (She telegrammed back to say that she would need time to do so.)

In desperation Arnold bombarded the Commander in Chief with requests to let him back to fight with his comrades. Finally in early September he was ordered to rejoin his regiment in Gallipoli. He arrived at Alexandria on 4 September, to learn that the Herts. had already been in battle. There were forty casualties, including the Colonel, who died of shrapnel in the neck with the stirring words 'Cheers for the Herts. Yeomanry'. There were no cheers for the returned Lieutenant Ward, however, who was informed on the island of Lemnos that his regiment simply would not receive him back. He prevailed on his mother to write an appeal to the Commander in Chief in Egypt, but that did no good.

After a weary four weeks cooling his heels on Lemnos, Arnold was instructed to return to England. He arrived on 26 October looking, as Dorothy thought, 'his most vigorous old self'.[29] He was made much of and even his sister Janet was mollified, kissing him outside the porch at Stocks when they met. On Sunday 31 October, he and his mother went to church together in Aldbury. That afternoon, she wrote a letter to General Sir Ian Hamilton—Commander in Chief of Mediterranean Forces—asking for an extension of leave for her son. She also daringly suggested that he should be appointed to a cavalry regiment in the Balkans as a staff captain: 'It seems to his mother he has earned it.'[30] It did not seem so to GHQ. Nor to Janet, who later (in 1921) wrote across the head of the letter the caustic comment, 'scrounging a soft job for Arnold'. But he was given extended leave—probably because the War Office simply did not know what to do with him.

The following few weeks were all downhill for Arnold Ward. He pottered around the House of Commons, but there was really nothing for him to do there. On 30 November, there was Violet Asquith's wedding to Sir Maurice Bonham-Carter. Mary and Humphry Ward went, he didn't. On 8 November, he had celebrated his thirty-ninth birthday. And what was there to show for all those years? The old danger signs began insidiously to reappear. On Saturday, 11 December, he was engaged for political meetings with his party supporters at Tring and Berkhamsted. He simply did not turn up, nor did he telephone or telegram apologies. It was 'a nightmare of an evening'[31] for Dorothy, desperately phoning every possible London address she could think of. She eventually dismissed the disgruntled audience with white lies. Humphry Ward had a bad attack. After a sleepless night and a nerve-racked Sunday, Mary Ward phoned Moy Thomas the next evening. He undertook to get a message to Arnold. A telegram arrived on Monday. At least their son was alive.

Arnold Ward remained in London until 15 January 1916. Apparently, he was cleared in the matter of the July memorandum and was posted back to the Eastern theatre. 'He has gone!' Dorothy wrote in her diary, 'he said goodbye to Father at the top of the stairs in Queen Anne's [Mansions] and mother and I went with him to Charing Cross'.[32] He looked so tall and handsome in his British Warm she thought, loyal to the end. And it was she who remembered to buy him a pair of field-glasses.

The outbreak of war had posed some tricky authorial problems for Mary Ward. Traditionally, wartime is initially damaging to fiction sales; then—if the conflict is protracted—the civilian population tends to turn to novels for comfort and escape. But it was not clear how Mrs Humphry Ward's romantic, feminine melodrama might adapt to national emergency. On 24 October 1914, Curtis Brown negotiated the sale of the five-year lease of *Eltham House*'s copyright to Cassell's for £2,000. It was the lowest price she had ever had. Reginald Smith, meanwhile, was making ominous noises about how war had changed publishing conditions in England. But perhaps these changed conditions might be turned to her advantage. In October 1914, the American publisher Appleton contacted Mary to suggest that she write a romance of an aristocratic English lady who goes as a nurse to Belgium. With the assurance of some 'heroic sacrifice and inevitable brutalities',[33] the American publisher could see themselves advancing £1,200 on the spot.

Mary Ward tucked Appleton's proposal away. (It would re-emerge, somewhat altered, in *Missing*, 1917.) But she could not just now write a war novel. The interminable delays which made *Delia Blanchflower* twelve months late had a knock-on effect on all her writing. *Eltham House*, a pre-war novel—at least in its conception—did not come out until October 1915 and had a particularly poor sale in the US, where its publisher Hearst did not clear 15,000. And she had another novel conceived in peacetime to unload, *A Great Success*. Reginald Smith, who intended publishing *A Great Success* himself (unlike its two predecessors) made an agreement on 10 March 1915. He would pay £2,050 for *Cornhill* serialization and five years' possession of the British and colonial copyright. Mrs Ward would have an extra £200 if peace were declared before the novel was published.

A Great Success is, to put it charitably, a slight effort. The story, conceived in six short scenes, centres on a resourceful, university-educated heroine, Doris Meadows. Doris is married to a lecturer on great political figures, Arthur. His lectures have made a hit in (pre-war) London and he unexpectedly discovers himself a literary lion. Arthur is 'full of fine idealisms'[34] but needs his wife's good sense to keep him steady. The Meadowses live in a semi-detached Kensington house and she supplements their income by illustrating children's books. After the great success of his lectures, Arthur is taken up by a patroness, Lady Dunstable, who invites him to her magnificent country house, Crosby Ledgers. The haughty Dunstable (in her forties, but still a strikingly handsome woman) brutally snubs Doris who none the less encourages Arthur to cultivate his new acquaintance for his career's sake. In the plot complications that follow, Doris discovers that Lady Dunstable's disaffected son and heir Herbert has foolishly engaged himself to a vulgar poetess, Madame Vavasour (née Elena Flink). Doris has no reason to help Lady Dunstable, who is actively destroying her marriage. But native goodness impels the young woman to go uninvited to Crosby Ledgers and warn her rival of Herbert's danger. Her virtue triumphs and she and Arthur leave London to begin a new life in Scotland. *A Great Success* is probably the poorest novel Mrs Ward ever wrote and was mercifully ignored by reviewers in both Britain and America.

A Great Success was serialized in *Cornhill* from May to July 1915. But there was another interminable delay and it did not come out in book form until March 1916. Such delays were costing the author dear. Mary Ward was desperate to get more novels out and more money in. It might also be advisable, she realized, to devise subjects more appropriate to wartime than *A Great Success*. In March 1915, Mrs Ward tried Reginald Smith with two very tentative war scenarios. One was a story about German spies which he rather liked. The other was a story called 'The Pacifist', which he definitely didn't like.

Having made these proposals, Mary Ward two months later decided *against* a novel of war. On 5 May 1915 she sent Smith yet another scenario of a novel provisionally called 'A Cuckoo'; subsequently it was called 'The Intruder' and then 'The Newcomer'[35] before settling down as *Lady Connie*, a novel of Oxford life and undergraduates, set in the 1880s. It would be a nostalgic voyage back to the might-have-been world of Mary Ward's young married life, had she and Humphry decided not to come to London in 1881. Mrs Ward asked Reginald Smith on 1 September to give her £3,500 for English *and* American rights. He declined flatly to do so. In October 1915, he countered with an offer of £1,500 for five years' lease of the British and colonial copyright and serialization in *Cornhill*. The sum would be payable in full on completion of the manuscript. Hearst took it in America (presumably for around £1,000).

Mary Ward wrote *Lady Connie* over autumn and early winter 1915 among all the upheaval of reorganizing PES for its women residents. Nevertheless, she had the work finished on 11 November, 'two days after my dearest Humphry's 70th birthday',[36] and it ran as a serial in *Cornhill* from December 1915 to October 1916, coming out in book form in October 1916. After the disappointing ordinariness of *Eltham House* and *A Great Success*, *Lady Connie* is surprisingly good and has been rated as 'the best of Mrs Ward's late novels'.[37] The heroine of the title is Lady Constance ('Connie') Bledlow (the 'cuckoo', 'intruder', and 'newcomer' of the early titles). Brought up on the Continent, Connie is orphaned when both her parents are killed in an influenza epidemic. As a wealthy, worldly, but still inexperienced young woman she comes to Oxford to live in the uncongenially academic household of her uncle, Dr Ewen Hooper, a Reader in classics. There are two Hooper girls, Nora (closely based on Julia Arnold, Mary's clever sister) and Alice. Their love lives form sub-plots to that of Connie, which dominates the narrative.

Connie has three Oxford suitors: Douglas Falloden, a clever but violent undergraduate and leader of the college 'bloods'; Otto Radowitz, a sensitive Polish pianist of genius; and Alexander Sorell, an older man, a classics don, and by temperament a 'Hellene'. Mark Pattison makes a brief appearance as Mr Wenlock, Master of Beaufort. Out of malicious rivalry against Otto (whom he thinks Connie prefers) Douglas leads a 'ducking' expedition in which the young Pole's right hand is permanently injured. In remorse Douglas leaves Oxford and Connie. Meanwhile his father, Sir Arthur Falloden, has ruined himself by business speculation and is forced to sell his priceless collection of paintings. Mary Ward invests with great bitterness the scene in which a loathsome dealer makes calculatedly low offers on behalf of a vulgar German millionaire. Presumably it recalls the painful business of Humphry's selling his favourite pictures in 1913. Eventually, all tangles are untied and all necessary knots tied. Connie marries a regenerate and chastened Douglas (now a lord, his father having died of a heart attack). Otto—although he can never play again—becomes a great composer and Douglas's comrade. Alexander Sorell is united with Nora, now transformed from a pudgy-faced undergraduate to a striking intellectual woman.

Despite a somewhat cluttered story line *Lady Connie* is a fine novel. Its power largely resides in its reconstruction of the Oxford of Mary Arnold's young womanhood. This reconstruction was to be continued in the most enduringly valuable of her later books, the autobiography *A Writer's Recollections*, a work whose composition flowed naturally on from the peaceful world evoked by the novel.

29

Soldier in Skirts: 1916–1917

NOT surprisingly, the Wards who did most to help Britain win the war were Mary and her eldest daughter. The ever-willing Dorothy buckled to, and organized a force of land girls for the Hertfordshire Agricultural Committee. They pulled on trousers and tilled the land at Stocks. The Wards' work in London was no less vital to their country. With fathers at war and mothers in the factories, PES vacation schools and play centres (weekly attendances now running over 60,000) were highly valued by the wartime authorities. There had been a panic about juvenile delinquency and after consultation between Mary and the Board of Education over 1916–17, the Treasury agreed to fund 50 per cent of the 'approved expenditure' of play centres. By 1919, the proportion of the cost assumed by public authorities was raised to 75 per cent.[1]

More dramatically, Mrs Humphry Ward became the first woman journalist to visit the Allied Front. Her subsequent dispatches to the American people are plausibly credited with doing much to bring that country into the European fight. For a sick woman in her sixties, it was an extraordinary feat: something to rank with her self-education, the establishment of Somerville College, the building of PES, or *Robert Elsmere*.

Mary Ward's career as a war correspondent began with a florid invitation from one of her 'old friends'. Former President Roosevelt wrote to her on 27 December 1915, saying that in his view the English 'side' of things had not been at all well presented in America. He wished 'that some writer like yourself could, in a series of articles, put vividly before our people what the English people are doing, what the actual life of the men in the trenches is, what is actually being done by the straight and decent capitalist'.[2] Roosevelt made no secret of the fact that he wanted to recruit Mary's help in swinging American public opinion away from Wilson's 'even-handedness', which saw the war as nothing more than the clash of British navalism and German militarism. President Wilson saw no reason to plunge into the European blood-bath. Ex-President Roosevelt believed that the English were fighting a just war. The sinking of the *Lusitania* in May 1915 had usefully blackened the German image. What was needed now was something to glorify the image of the English. Mary Ward's pen could do the job, he thought.

It was a bolt from the blue. But it was also a powerful tonic. Mary was horribly depressed by Arnold Ward's continuing failure to make an honourable military career for himself. Her fiction was apparently never going to be really popular again and she could only hold on to those readers she had by writing below her best. This invitation to use the accumulated fame of thirty years as a Rooseveltian 'bully pulpit' held out the heartening prospect of actually doing something worthy of her best self. She knew the former President had once had an immense regard for her, and this regard was amazingly undiminished. In 1908, Roosevelt (then in office) had coolly informed the German Kaiser that he could spare him only twenty minutes, because he had an appointment with Mrs Humphry Ward.[3] Now, it seemed, he believed that Mrs Ward could again put Wilhelm in his place.

As Janet Trevelyan records, Mary at once 'consulted her friends C. F. G. Masterman and Sir Gilbert Parker at "Wellington House" (at that time the Government Propaganda Department) and found that they took Mr Roosevelt's letter quite as seriously as she did herself'.[4] They were immediately responsive: 'we must of course do it',[5] Masterman observed to Parker. But—patriot thought she was—Mrs Ward was not going to *give* her services to her country. The Wards had, as it happened, just had an income tax assessment for £1,240 and Humphry had his annual repayment of £750 to make to the picture dealer Agnew. 'I frankly said that I could not do it for nothing,'[6] she recorded herself as telling Masterman and Parker. They agreed and Parker undertook to make enquiries about American newspaper openings. A couple of days later, Mrs Ward wrote to Roosevelt saying that she would accept the commission 'if an opening could be found equivalent to what I was giving up, but that I must think of those dependent on me, and that the losses from the war and the pressure of war taxation were too heavy to be ignored'.[7] On this basis, the enterprise went ahead.

Of course, GHQ did not want Mrs Ward to *report* what was going on in France or in the munitions factories. That would be far too dangerous. But with Haig's great offensive coming on in summer 1916, American moral and material assistance might be very useful. The War Office's attitude to publicity had changed significantly since 1914. Kitchener loathed reporters, and thought that the Press ought to be throttled in time of war with more ferocity than even the enemy. But with the New Army of 1915, the need to keep the Home Front actively involved overrode the British Army's ineradicable distrust of the Press. Five British and two American war correspondents had been accredited in 1915. These men were accompanied everywhere by censors, who read what they wrote before the ink was dry, and who were directly accountable to GHQ. The seven-man Press corps were resourceful, decent and generally appalled professionals. Mrs Ward would not be one of their number. She would be a VIP visitor. War correspondents proper like Philip Gibbs despised amateurs like Mrs Humphry Ward or Harry Lauder who spent a day in the trenches as one might visit Madame Tussaud's. Probably the serving soldiers were not enamoured by these famous tourists to their fields of death. But after ten years of arguing the anti-suffrage cause, Mrs Ward was, without question, the most accomplished woman propagandist in England and, more importantly, she was the best-known Englishwoman in America. In total war Britain would have been foolish not to use her as a weapon once she had been primed by Roosevelt.

On 20 January, Mary Ward went to see the Foreign Minister, Sir Edward Grey (her old landlord at Stocks), in his office at Eccleston Square. She found him 'a little stouter',[8] almost blind, but as affable as ever. Over half-an-hour's conversation in his upstairs library sitting room she learned that Kitchener was 'keen' on the articles. It was agreed that she should have permission to visit the factories and the Fleet. Gilbert Parker would 'outline' the articles for her; and they would of course be reviewed by the censor before publication. Their theme should be England's initial 'unpreparedness' and her current mighty industrial and military effort. It was the authorities' expectation that Mrs Ward would restrict her investigations to the Home Front. She, however, had different and more ambitious plans: 'I said that I should like to go to France, just for the sake of giving some life and colour to the articles—and that a novelist could not work from films, however good.'[9] Grey conceded, somewhat reluctantly. She might go to France.

From 31 January to 15 February, Mrs Ward toured munitions works in the north of England. Lizzie was with her to look after her physical needs, Bessie Churcher was there to take down her notes and observations, all of which were duly passed to the censor to blue-pencil. On 15 February, she spent a couple of days at Invergordon, where she reviewed a battle squadron of the Fleet as the guest of Admiral Sir John Jellicoe. On 16 February, on a 'bitterly cold' morning and huddled in her 'American fur coat' she was taken by open car and barge to the flagship in the Firth of Forth. Her notes capture the excitement of her whirlwind tour: 'Thrilling to see a ship in war-time, that might be in action any moment. The loading of the guns—the wireless rooms—the look down to the engine deck—the anchor held by the three great chains—the middies' quarters—the officers' ward-room. The brains of the ship—men trained to transmit signals from the fire-control above to all parts of the ship, directing the guns.'[10]

On 23 February, Mary Ward and her two lady companions motored back to London in their War Department car. Five days later, she and Dorothy were in France, touring the army's rear echelons. They saw the 'miles of sheds [and] forests of cranes and derricks' in the dock area. 'What an example of sea-power and the command of the seas!' Mary Ward exclaimed in her diary. Inside the sheds they saw field bakeries, boot shops, and the pathetic heaps of dead soldiers' belongings, awaiting identification and return to bereaved families. All the while the weather was bitter and snowy. On 25 February the

Wards visited the *Oxfordshire*, a hospital ship. The wounded were being loaded aboard and Mary was struck by a 'poor, poor boy, who had lost his eyes, God be with him; he is quite cheerful, the nurses say,—and he doesn't realise'.[11]

On 29 February came news from GHQ that the ladies were to prepare themselves for a two-day visit 'to the place which no one here [Abbeville] mentions but with bated breath, and which I will not write'.[12] Gas masks were duly issued. The unmentionable place was St Omer with an en route stop at the combat training camp, Etaples. There followed a series of 'interminable' car journeys along roads crowded with lorries before they arrived at the 'Visitors' Chateau'. Its comfort was, in the circumstances, something of a surprise. Mary Ward's bedroom was 'extraordinarily pretty with its Louis Quinze carvings and its old Armoire and its windows open to the old neglected garden'.[13]

Over the next few days Mary and Dorothy Ward visited the rear lines, often within range of the enemy artillery. On 2 March, after some 'excellent coffee', they started out by car at 9.30 for the Front, 'after hearing that there had been a successful little fight that morning [in which] we had retaken the International Trench on the Ypres–Comines Canal.'[14] They were, at last, 'within actual sight of the *Great Aggression*'.[15] Their escorting officer, Captain Roberts, noting their keenness, laughingly congratulated them on their 'bloodthirstiness'.[16] From the Scherpenberg Hill, at 1.30, two English ladies watched through field glasses the repulse of the great aggressors by British artillery.

As they had started out that morning, their escorting officer handed Mary some telegrams from England. One announced the death of Henry James on 28 February. He had only a few months before become a naturalized Englishman. The news of her old friend's death heightened Mary's perception of what followed to a sacramental pitch:

> All through that wonderful day, when we watched a German counter-attack in the Ypres salient from one of the hills south-east of Poperinghe, the ruined tower of Ypres rising from the mists of the horizon, the news was intermittently with me as a dull pain, breaking in upon the excitement and novelty of the great spectacle around us.—'*A mortal, a mortal is dead!*'—I was looking over ground where every inch was consecrated to the dead sons of England, dead for her; but even through their ghostly voices came the voice of Henry James, who, spiritually, had fought in their fight and suffered in their pain.[17]

After this exaltation it was all banal bustle. She and Dorothy were back at Boulogne by eleven o'clock the next night, and fighting their way through Victoria Station at an ungodly quarter to four the following morning.

Bessie, Dorothy, and Mary all set to, putting the notes of the great adventure in order. Mary Ward then assembled and polished them into six 'letters to an American friend'. By 19 March she had finished the first letter, by 26 March the third. By 8 April, she adjourned to Stocks 'in despair' about the fifth letter. On 20 April, she had Joseph Choate, the former American Ambassador, to tea ('most delightful') and on 11 May she had the whole book finished and was asking Balfour to write the preface for the English edition. He refused. There were other impediments. On 18 May, they read in the evening paper that Mary Ward's nephew, Arthur Selwyn, had been killed flying. 'M. is minding terribly,'[18] Dorothy noted. Humphry Ward was, however, in such a state of nervous anxiety about his attacks that Mary could not leave him to go the funeral. Instead, she went to church by herself at Aldbury on 20 May, at the same time that the shattered body of her nephew was being buried. On 26 May, the Wards lost their chauffeur to the Army (despite Mary Ward's attempts to get him exemption). The useless Daimler was put on the market and later sold (in July) for £210. The Wards reverted to horse-drawn and public transport.

Despite everything, the *England's Effort* 'letters' were finished at breakneck speed, syndicated to American newspapers, and came out from April onwards. They were published in book form with Scribner's at the end of May. There was a preface by Joseph Choate. It contained, among much compliment to Mrs Humphry Ward, the alarming misstatement that 'Her only son is a member of Parliament and is fighting in the trenches in France, just as all the able-bodied men she knows are doing.'[19] Arnold's exploits—dangerous enough to be sure—were currently confined to the card-tables of Cairo's clubs. The English edition, with a preface by second-choice Lord Rosebery, came out on 8 June 1916.

England's Effort is a shrewd mixture of eye-witness reportage, statistical bombardment, telling anecdote, and 'beastly Hun' propaganda. Its theme is that England was indeed hopelessly unready in 1914. But there was no shame in that unreadiness. The country had nothing like the Prussian war machine because it meant no harm to its European neighbours. But now, like a giant awoken, 'this old, illogical, unready country' was formidably girded to smash the aggressor. Mary Ward played down the 'drunkenness, trade union difficulties', and the 'small—very small revolutionary element among our people'. Ideologically, England was sound as a bell. The adjective which echoes through her account of women working in the factories, of serving sailors and soldiers, is 'splendid'. England was making a splendid effort. England but not, unfortunately, its neighbouring island. The epilogue contains some bitter words on 'the squalid Irish rising . . . the Irish outrage paid for by German money'.[20] Ever since 1882 she had hated the Irish; now, in April 1916, they had showed their true, indelibly treacherous colours.

Odious as her war-mongering seems to later readers, *England's Effort* was extremely well done. Propaganda is judged by results and Mary Ward's book certainly got results. As she boasted to Roosevelt, its godfather, 'the little book has [been] translated into every European language, and there has been a long and very civil review of it even in the *Preussische Jahrbücher*!' As an Englishwoman she rejoiced in doing her bit; but she was also personally grateful to Roosevelt for suggesting the enterprise. 'It brought me a wonderful experience,'[21] she told him. Part of that experience was feeling useful again. The whole of her life had been an attempt to move from the edge to the centre of things: to the centre of her family, the centre of Oxford, the centre of the London literary world. Now she felt at the centre of the war. It was exhilarating.

Aside from the exhilaration of *England's Effort*, 1916 was a horrible year for Mary Ward as for the whole British population. The war ground on, with the murderous Somme offensive in July. While his fellow officers died by the tens of thousands in Flanders fields, Arnold went from bad to worse. He had returned to Egypt in late January. But his regiment did not welcome him. He was considered a bad influence on his junior officers (junior in years, that is, not in rank; most lieutenants in 1916 were in their teens—Arnold Ward was 40, the two pips on his shoulder a brand of incompetence). Again he had been gambling, losing, and not paying. Again, he began to make official complaints of unfairness against him by his superior officers. In April, Mary Ward besought her friends in the High Command to get him a Foreign Office appointment in Athens. Nothing came of it. Yet another posting back to England was arranged for this soldier that no one wanted. There would be another semi-official hearing on his grudges.

Bad reports had meanwhile begun to circulate the constituency about the conduct of their MP. In mid-May, Dorothy Ward recorded posting a letter to Arnold 'in which I for the first time suggested that si les affaires ne marchaient pas mieux au moment où il retournerait en Angleterre—il ferait peut-être mieux de donner sa démission Parlementaire en avance'.[22] In other words, his forced resignation might be extremely ugly, if angry electors brought up his gambling, drinking, and alleged scrimshanking while their brothers, fathers, and sons were fighting and dying.

Arnold Ward arrived back in England on 21 June, and booked into the Hotel Curzon. Mary and Dorothy went to visit him there the next day, and in a long and sticky four hours' conversation had 'the whole story of the regimental difficulties'.[23] On Wednesday 28 June, Dorothy recorded 'a day of almost unbearable waiting, but at 7.15 [Arnie] rang up and—thank God—with good news! He had been interviewed by General Bethune himself, the Director-General of the Territorial Force—at 6! Il n'y a rien "against you" in the Report—he had said—and the way is straight now, with honour. Mother talked to him on the telephone, sitting at my table.'[24] The next day, Mary Ward went to see her son by herself, lunching at his hotel.

One can put the story together. Gambling was at the bottom of the sorry business. On 29 June, after lunch with her son, Mrs Ward had a long talk with Arnold's 'banker' Moy Thomas, who agreed again to help. Clearly she could not deal directly with her son's gaming creditors, but she could supply the wherewithal. Over the following years, Moy Thomas was paid regular (apparently six-monthly) instalments of £500. At Mary Ward's death in 1920, some £1,700 was still owing him, and a cryptic note of Dorothy's (who destroyed the relevant correspondence) confirms that in summer 1916 Moy Thomas advanced £6,000 on Arnold's behalf to a creditor called only 'M.L.'[25] Army records show that Lieutenant Ward was transferred on 18 August 1916 to the Reserve Territorial Force, which suggests that his regiment would have no more to do with him. If he was to serve his country, it would have be elsewhere than in the Herts. Yeomanry.

That same evening (29 June) Dorothy Ward was sitting at home at Stocks reading a story-book to her little niece when her mother returned at seven, from her harrowing day in London: 'as she came in, round the corner of the sofa, by the screen where we were sitting, I saw that what I hoped yet dreaded had taken place.—Elle a été si douce et si bonne to J[an] and me—but oh—what it is.'[26] What it was, of course, was the undertaking yet again to clear Arnold's honour. She had given her assurance to pay the £6,000. It had also been agreed that Arnold must resign his seat in Parliament. Too many people locally—people with relatives serving with honour in the Herts. Yeomanry—knew the whole ugly business. The next day, looking 'very strained and white', Mary Ward wrote letters requesting face-saving favours for her son. On 5 July, they were told that the Unionist Executive (out of respect for Mrs Ward, presumably) had accepted Arnold's offer to resign without fuss at the end of the present Parliament—the disgrace of public ousting was thus spared the family. The next week was, as Dorothy put it, 'critical . . . quant à les affaires'.[27] And on 14 July, she wrote, 'On this night, about 10, we got a pencil note from A., sent back by the car, to say the affair was settled in principle and pagamenti would be made early next week. He gave details—but the horrible size of the amount seems in a way lost in the relief of his being cleared. We were all rather emoted. Poor Father!'[28] 'Gli affari', as Dorothy called the agreement with Moy Thomas, was signed and sealed and left in the keeping of the family solicitors, Burton.

Under no circumstances would his county regiment take Lieutenant Ward back. There was some hope of what Dorothy called a 'French job'—i.e. a post on the staff of Field-Marshal Sir John French. Mary told Dorothy Ward that 'It seems to me *possible* that French might *privately*—through [Lord] Haldane—ask some pledge from [Arnold] with regard to cards—as a condition of taking him on his general staff.'[29] But the War Office was not in the habit of making private agreements with superannuated lieutenants as to the conditions on which they would consent to serve their country. Their decision was to bury the obstreperous Lieutenant Ward where he could do least harm, in a training camp without promotion. 'The stupidity of this from the WO point of view strikes me almost more than the indignity for Arnold,' Dorothy Ward wrote, with a loyalty to her brother bordering on obtuseness. Mary Ward—as loyal as her daughter—continued to write a stream of letters to Bethune, appealing for something better. His replies 'depressed her'.

Meanwhile, Arnold Ward himself was reacting as he usually did to stress, by going to ground. On 2 August, Dorothy wrote that 'It has been a bad day for at 10 we heard the bewildering news by wire that A[rnold] was at St Anne's on Sea till the end of the week.' Dorothy telephoned endlessly in vain attempts to track him down. Finally, on 3 August, she went to St Anne's herself. She found her brother there 'with a whole party of his Portland [Club] friends. Moy Thomas, Sir H[erbert] Leon, etc.' 'It was', as she cryptically noted, 'a difficult afternoon.'[30] Mary Ward meanwhile was writing her 'Somme Letter', an appendix to the latest edition of *England's Effort*, saluting the ranks of British soldiers dying by the hundred thousand in the mud for their country, while her son Arnold drank cocktails at St Anne's on Sea.

By the end of September, it was clear that Arnold Ward was not going to get his General Staff appointment with French. He lingered on dissipating himself in London. On 23 September, he told Dorothy that 'he may not run up any other "debiti"[presumably mess bills and gambling debts] without the Major's permission'.[31] The Major was evidently the luckless officer appointed as his nominal commander. Mary Ward was of course still pulling whatever strings she could for him. There were certainly strings for her to pull. On 6 October, she had received a three-page letter from Douglas Haig himself, thanking her for her book.[32] Two days later, she went to London, and had an interview with Lord Haldane, imploring him to do something for the son of the author of *England's Effort*.

Favours were in order, and something was eventually done for Arnold. On 25 October came the order that he was to join the Third Battalion of the King's Own Scottish Borderers, in Edinburgh. The unit was due to join its other battalions in France, and Arnold might at last see action. He was 'pleased and relieved'.[33] It was his hope that (with his mother's influence) he might eventually—after a bit of fighting in the trenches—get something in intelligence at Salonika. But in the meantime, this posting to an active service unit was very welcome. As she told Dorothy, Mary Ward felt 'a sort of serious joy because it is just *everything* that he has got back into the great machine, into a definite niche again'.[34]

Before Arnold left for the KOSB depot, there was another unpleasant crisis about the 'Parliamentary Salary business'. Local Unionists ('the Watford people', as Arnold called them) were appalled that he was still drawing his Westminster salary. Their understanding was that he had withdrawn from politics. There was a heated discussion in the billiard room at Stocks. After his mother left in despair, Dorothy Ward 'persevered, but in vain. It is an utter conviction with him—he thinks it is *the* thing in which he has recht gethan.'[35]

That night, Arnold Ward, in a fury, destroyed his parliamentary papers and constituency materials. Dorothy, meanwhile, went to a memorial service in the village for three local lads, killed in the recent offensive. Two days later, the 'Watford people' delivered a 'memorandum' for Arnold to sign, guaranteeing that he would resign at the end of the current Parliament.

There was one melancholy comfort in all this. There was now no point in leaving Stocks. Any money they got from the sale of the house would simply be eaten up by Arnold Ward's incorrigible gambling. Mary felt that she and Humphry could not live long anyway. There would be enough left for Dorothy who would be left the property and who—unlike her parents—would have no responsibility for Arnold's debts. He could declare bankruptcy. From now on, Humphry and Mary Ward would sell off the contents of the house, piece by piece, and live there until the end. Dorothy wrote on 3 December that 'It is past all doubting now that [Mother] would *rather* stay here at any rate for several years more, whatever the struggle.'[36]

For the family as for the country, 1916 was a grim year. The Wards lost their last male servants—including their gardeners and chauffeur. The replacements they tried for—invalid men excused military service—contemptuously disdained them after one look at Stocks, which now resembled a rather run-down farmyard. There were plenty of soft-faced profiteers to work for. Not having a car (after 21 July 1916) was fiendishly inconvenient. The transfer of PES to a woman's settlement under Hilda Oakeley went smoothly enough, but Mary could not but feel excluded from her own building which was swarming with suffragists. PES ran itself very happily without its founder. Even the omnicompetent Bessie Churcher was out of it, having suffered a heart attack in winter 1916. (She was never to be a vigorous woman again.) Humphry's arteriosclerotic attacks were as frequent as ever, and often more severe. On 6 December, he fell on Tring Station platform, and seriously banged his head. On 12 December, the Wards' faithful old Flip was worried to death by farm dogs. It seemed a symbol of something. 'His poor little body was buried at the bottom of the garden,'[37] a distraught Dorothy Ward recorded; she could not get out of bed all that day.

On 28 December, Humphry Ward brought from town news of a more calamitous death. Reginald Smith had fallen from the second-floor window of his house. Mary Ward, as Dorothy observed, was 'quite stricken. Apart from all else it will be a very *serious* matter to us. Never has author met with kinder consideration from publisher than she from him.'[38] Dorothy and Mary went up to Mr Reginald's funeral on 30 December, at St Mark's Church, North Audley Street. After the service, they watched from their taxi in Oxford Street as the little funeral procession started on its long way to Hampstead Cemetery, with the motor hearse in front. Mary Ward—involuntarily—broke out with the cry 'Goodbye old friend!'[39] That same day, another old and generous friend,

Frederick Whitridge, died. He had always been their last and unfailing resort in the financial crises that had piled on them since 1912.

In one respect only—the authorial—1916 was an exceptionally good year. It saw the publication of *England's Effort*, which was an unalloyed triumph of *useful* writing. Mary Ward also published a novel as good as anything she had done for twenty years—*Lady Connie*. The strength of that work is in its overpowering, almost intoxicated, nostalgia for the Oxford of the author's girlhood. In other ways, she was drawing on those funds of long-past strength and happiness. From mid-July to mid-September 1916, the family had holidayed in the Lakes. The return to Fox How, conversations with Aunt Fan (still indestructibly hale) and the turning over of old letters, inspired Mary Ward at last to write her autobiography. Through the summer months, she lived in the 1850s, 1860s, and 1870s and meditated what became the opening reminiscential chapters of her *Writer's Recollections*. There was considerable publishers' interest in this venture, and Curtis Brown actually got Collins and Methuen to wage a little bidding war for the work. Collins won, with their offer to take the three novels which Mrs Ward would write after her autobiography. Harper's eagerly bought the serial rights of the autobiography for the United States. Miraculously, Mrs Humphry Ward was a desirable literary property again.

The past was much with her, as it often is with people who expect to die soon. But in another compartment of her mind, Mary Ward continued to engage herself in England's present crisis. Following the success of her first foray into war reporting, she began to agitate in November to be allowed to write another book on the same propagandistic lines as *England's Effort*. She got an assurance from Roosevelt that the old warrior himself would write a preface for any such work. On Monday 6 November 1916, she duly went to see Lord Onslow at the War Office, to apply formally for permission to revisit the French lines. He wrote three days later acknowledging 'the valuable effect which your first book produced on the public' and indicated General Charteris's approval of her application, adding ominously: 'but it would have to be clearly understood that in the event of your being allowed to go, it would not constitute a precedent as regards any other ladies'.[40]

The arrangements for what became *Towards the Goal* were made at the War Office in February. On the fifteenth of the month she had a conversation with Balfour, who confided that 'the French are often very difficult' and made it clear that he felt that Mrs Ward should 'tactfully' indicate to the American reader that the English army was 'confronted with the bulk of the German forces'.[41] On 23 February, she had a 'pleasant half-hour's talk' with Sir William Robertson, 'a powerfully built square-jawed man with a broad though low forehead and bushy eye-brows',[42] after which, all was clear for the expedition. Before she and Dorothy left for France the Wards had their first dinner party 'for some years',[43] as Dorothy recorded. Once the most convivial family in London, they had almost forgotten how to serve guests.

Dorothy and Mary Ward then spent 'fourteen most wonderful days'[44] in France. They left Charing Cross Station at 1.40 on 28 February on a train crowded with soldiers. They arrived in Boulogne at six o'clock that evening to be met by their officer escort in a motor car. On 1 March, to their delight, the Wards slept at Agincourt. The routine of their visit was much the same as before but the machinery of war had changed significantly. At St Omer they saw hundreds of aeroplanes, 'like dragonflies'.[45] Nissen huts had sprung up everywhere. And motoring to the Front Mary Ward saw her first tanks. 'What luck!' she recorded in her diary. 'We stop and watch one ascending a field bank eighteen or twenty feet, and quickly descending the next—the plunge of the monster—quiet, deliberate, but irresistible—the ascent the same—And there, while we watch, on the horizon, a ploughman with his plough—primitive like those of the Odyssey—and in front the Tanks!'[46] The vivid conjunction was later worked up as a set piece for the second letter of *Towards the Goal*.

On this occasion Dorothy and Mary Ward spent two days at the Front, including a visit to the Somme 'where [Mrs Ward] stood in the midst of the world's uttermost scene of desolation'.[47] It was clear that she was now even more of a VIP than in 1916. At Abbeville on the day of her departure (14 March), she had exclusive use of General Asser's Rolls Royce and several tête-à-tête conversations with top brass, including General Charteris. Maps were shown her and hush-hush plans divulged. It was, Mary Ward thought, much more 'worthwhile ... than last time',[48] almost as if she were a brevet-General herself. In France, GHQ was careful to show her scenes of German atrocity against the civilian French population. (England was becoming more expert in the arts of propaganda.) Mrs Ward was duly appalled, and stored away numerous evidences of the invaders' savagery.

Returned to England, Mary Ward was taken on 'a secret visit to the Navy' at Harwich on 16 March. She examined a submarine and, as Dorothy put it, 'might apparently have been seen climbing up and down the sides of ships on ladders with the agility of a squirrel'.[49] It fell on Dorothy to write up the notes single-handed, Bessie Churcher having collapsed with cardiac pains on 23 March. Mary Ward was again—as with *England's Effort*—buoyed by an extraordinary surge of energy. The first letter of *Towards the Goal* was sent for censoring to Wellington House on 26 March. The next day, she sent twelve chapters of her war novel, *Missing*, to Curtis Brown. By 11 May, the last censorship corrections were incorporated, and *Towards the Goal* was cleared for publication. The ten letters, addressed to a mysterious 'Mr R.', were syndicated in American newspapers through the summer and the book version came out in late July with a preface by Mr R. (i.e. Roosevelt) declaring that 'England has in this war reached a height of achievement loftier than that which she attained in the struggle with Napoleon.'[50]

In tone, *Towards the Goal* is less confident than its predecessor and dwells more on the statistics and technology of total war, as if to blind the reader with numbers and military science. In 1916, Mary Ward had shared the Staff's confidence that the Big Push would end the war by Christmas. In 1917, the war looked interminable. Her moral emphasis is consequently less on the heroism of Britain than the bestiality of the enemy—it is, in short, a book designed to stir up hate (particularly in letters 7 to 10). And it ends on a peculiarly desperate note. The military set-backs of spring 1917, the Russian Revolution, the Irish uprising, the submarine blockade, the delay in American reinforcements, the air raids on London all suddenly made it horribly plausible that Britain might actually lose the war or be fought to a standstill. Mary Ward's letters tail off in uncertainty as to what the goal might after all be, echoing only the harsh slogan 'Reparation, Restitution, Guarantees!'

For its author *Towards the Goal* was, like its predecessor, a success. It sold well, earned well, and continued the refurbishing of Mrs Humphry Ward's reputation. Six thousand of Murray's first edition of 10,000 were subscribed on publication day, 15 July. Scribner's had paid £360 apiece for the ten letters. On the coat-tails of *Towards the Goal* her war novel *Missing*, which came out in November 1917, sold 20,000 copies in its first two months in America. No novel of Mrs Ward's had done that for five years. On the strength of her new fame, Brown contrived to sell film rights to *Missing*, opening the possibility of a whole new kind of international fame.

Part of *Missing*'s appeal was its strong war-centred plot for which, after 1917, America was eager. Mary Ward's scenario for the novel began with a vivid vignette: 'Two young lovers—A husband and wife—just married—The outbreak of war—The lodging beside Rydal Water—The poignant happiness, with the shadow of death behind it.'[51] The honeymooning couple are Lieutenant George and Mrs Nelly Sarratt. It is 1915, and he must soon go to the Front. In the Lake District, where she remains, Nelly is befriended by Sir William Farrell. Prevented from serving by age and lameness, he has turned his vast country house into a hospital for the war wounded. Sir William falls desperately in love with Nelly. His attachment is noted by Nelly's older and ruthlessly ambitious sister, Bridget. George is meanwhile reported missing. On a mission in France Sir William comes across a 'nameless, speechless' wounded man who he thinks may be Nelly's husband. He summons Bridget, who recognizes George but denies the fact, thinking that he will soon die and release her sister to marry Sir William. But George does not die immediately and recovers sufficiently to speak. Nelly rushes out to be reconciled with her dying husband. Widowed, she devotes herself thereafter to being a military nurse.

Missing was the first of Mrs Humphry Ward's novels published in wartime to be directly concerned with the war. And like everything she wrote in those circumstances it was a great success. For a writer well into her sixties it was a remarkable accomplishment. To hear critical applause again was very pleasant. But not the least rewarding aspect of Mary Ward's war writing was that it took her mind off problems that worried her at night even more than the end of European civilization: problems such as Arnold's disgrace, the take-over of Smith, Elder by John Murray, income tax, supertax, servant problems, government requisitions of Stocks timber and potatoes, Humphry's health, the poor bids for his pictures in the sale-rooms, the similarly poor prices they were getting for the sale of their Stocks cottages.

On 23 March, Dorothy recorded that 'Mother is very troubled and worried today over her affairs and said to me tonight sadly that she felt so "rested" in France because there she did not feel beset by these worries.'[52] Compared to what she went through nightly in her bedroom at Stocks, the trenches would have been a rest cure. Much of her agony came from her reluctance to share her miseries. As always, confession was hateful to her. She must—whatever the cost—keep the pain within. On 12 September she wrote pathetically to her oldest friend, Louise Creighton: 'I thought of telling you something of what your old friends have gone through in the last four years and how changed our life has been. But on the whole it seemed better not—Silence is best. Work has been my best help .... One must bear it patiently. And beside the war, what matters?'[53]

The Wards were involved in two unpleasant court cases in the spring and early summer of 1917. On 4 April, their houseboy, Fred Mogg, was caught forging a cheque for £10. There were 'painful scenes' and the family felt obliged to charge the lad. He got two years probation and three years in youth custody. The thought of her Arnold's recent £6,000 must surely have crossed Mary Ward's mind as she put in motion the unforgiving machinery that would ruin Mogg's young life and shame his mother in the eyes of Aldbury village. And on 17 May, Humphry Ward was served a subpoena to give evidence in a disputed Romney case. Despite a medical certificate from Frank Arnold, the sick old man was made to give evidence in Justice Darling's rooms for fifty minutes, his wife Mary standing by him all the while. Over the next two days, he had a devastating series of attacks in which he lost consciousness entirely.

There was a bitter knock for the family on 11 October 1917 when Frank Arnold's eldest son, Tom, was killed. Mary Ward had already lost two Selwyn nephews but Tom was particularly dear to her. He had been her father's namesake and stood in the family's main male line of descent. But, inevitably, the main difficulties in 1917 were with Arnold. He had not after all been sent to France. His subsequent application to join the tank corps had been turned down. He was usually to be found on vague and indefinite 'leave', hanging round London. Fallen as he was, he was still lovable and could still occasionally elicit the old all-forgiving tenderness from Mary and Dorothy Ward. On Sunday 11 February they sat *à trois* in the billiard room after an unwell Humphry had gone off to bed early. As Dorothy pictured the scene, 'M. leant back in the red chair and I sat on the window side of the fire, knitting and A. half lay, half sat on one of the pink sofas which he had moved down from the raised ledge, and put opposite the fire; and he was very "easy" and friendly; ready to discuss his prospects and possibilities and we tried together to thrash out the next steps. He is very keen on his "New Europe".'[54]

If only it had been forever Sunday evening in the Stocks billiard room, Arnold Ward's life would have been quite manageable. It was when he ventured into the outside world that his problems began. The King's Own Scottish Borderers did not want him, and he was sent in mid-February to the Machine Gun Corps, Cavalry Section, on a training course at Uckfield, in Sussex. Again, he seems to have misbehaved in the mess. The fiction was subsequently concocted that he was 'requested' by the anti-suffragists to return to the House of Commons to assist their struggle. It is true that the Representation of the People Bill was preoccupying Parliament all that spring and early summer. It was also true that with Lloyd George at the head of the Liberal Party, things looked fair for the women's cause. But it can hardly have been the War Office's policy to second serving officers to oppose the government of the land. The truth was that Arnold was yet again in dispute with his commanding officers.

On 18 April, he came back on indefinite leave and 'on the understanding with the [KOSB] officers at [Edinburgh] that he is *not* to go back! pending final sanction from the W.O.'[55] He reverted to unemployment on the Herts. Yeomanry list in May 1917. In April, there was another crisis with his 'banker' Moy Thomas 'à cause des remboursements' (as Dorothy delicately put it, in her diary code). It was all sorted out by June, but it was a clear indication that Arnold was still gambling. On 12 August (at the height of Haig's 'Great Push') Arnold confessed to Dorothy that he had not been able to keep his promise:

> Mi diceva stamattina che non avesse potuto sein versprechen halten—piuttosto sa résolution, che non sarebbe possibile di tenerla. A nightmare of a whole morning's talk first alone, then with Mother and him and then I left them soli. Encore des conversations ce soir mais nous ne pouvons le détourner—et il pense qu'il a le droit on his side. Non ho mai visto mia madre tanto commossa, e veramente non è sopportabile, questo stato degli affari.[56]

Insupportable or not, the state of affairs dragged on. Arnold could clearly no more stop gambling than he could stop breathing.

Despite her earlier resolution, Mary Ward had again to consider the 'vente' of Stocks. If Arnold gambled, she must lose money. All through October it was impossible to talk to her son. On Friday 9 November, Dorothy recorded in her diary patois a poignant exchange with her mother as she set out on her little daily constitutional stroll. What oppressed the old woman most was 'la pensée insupportable que lui, A[rnold], ne pouvait rejoindre l'esercito fino a che siano pagate le fatture(?). C'était la chose qui lui pesait le plus lourdement sur le cœur. Et l'expression de douleur avec laquelle elle me regardait!—I entirely feel with her—but à quel prix doit elle libérer mi fratello! Le prix de sa propre santé, de ses meilleurs forces.'[57]

Until his debts were paid Arnold was forbidden to rejoin the army. And he was running up debts faster than his mother could settle them. But, whatever else, he had time enough in 1917 to play a leading part in the debates in Parliament on the suffrage. Few other MPs were prepared to put themselves in the firing line. But Arnold Ward had nothing to lose—he was after all bound to resign his seat at the next general election. His executive in Watford sent him a resolution passed on 2 May 'dissociating themselves from him as to his action re suffrage'. But he ignored them.

By summer 1917 the antis were a sadly depleted rump. Their main delaying tactic was a campaign for a referendum. This had no validity other than that it would postpone the awful day that women got the vote. In the House, the antis were now treated with open hostility. The Speaker routinely cut Arnold off in mid-speech, or ruled him out of order when he leapt to his feet, waving his order paper. He was jeered by suffragist women as he entered or left the House. On 20 June, Arnold's amendment to the Woman's Suffrage Clause was defeated by 300 votes. Now everything depended on the Lords. But, ominously, Curzon was turning tail. He refused Arnold's request to bring in a referendum bill in the Upper House. Arnold and Mary felt, with good cause, that their leader—the man who had got them into this mess in the first place—was about to rat.

The *coup de grâce* for the antis came on Thursday, 10 January 1918. The suffrage clause was carried in the Lords, 134 for, 71 against. Curzon abstained. Mrs Humphry Ward, who had passed a few chivalrous words with her great opponent Mrs Fawcett in the corridor outside, heard the debate from the visitors' gallery. Dorothy Ward recorded their subsequent dismay in her diary: 'We have been betrayed by our leader Lord Curzon. Coward! After making a long A[nti] S[uffrage] speech, with every appearance of believing in the arguments against the Vote which he was advancing—he suddenly announced that in view of the gravity of the conflict with the Commons at this moment *he was not going to vote* and advised noble Lords not to vote.'[58] Arnold Ward, standing at the bar, went white in the face with rage as he heard Curzon's pompous treachery. Afterwards the three Wards walked down Whitehall in the dark together, sharing their defeat. It was all Dorothy could do to prevail on her mother to come back with her on the only available vehicle, a number 11 bus. Two days later, the West Herts. Unionist association adopted a new candidate—by 22 to 5 votes. Arnold Ward's parliamentary career was over.

On 16 January, the referendum amendment was easily beaten in the Lords, and the suffrage was assured.

There was some small consolation for Mary Ward in the spring of 1918. Her Joint Advisory Parliamentary Committee (which had been formed in 1914 when the antis broke away from the NUWW) managed to get a clause on behalf of physically defective children inserted into the current Education Bill. It was the outcome of a massive lobbying by mail to all the country's Education Authorities and MPs. The whole operation was masterminded from Stocks. As Janet Trevelyan records, 'all this took place during those agonising weeks when the British Armies were being hurled back in France and Flanders, and when Mrs Ward, of all people, realised only too clearly the peril we were in. But the task was accomplished and the clause added to the Bill, so that a new charter was thus provided for the 30,000 or so of crippled and invalid children who still remained throughout the country uneducated and uncared for.'[59] It was her last great intervention in Parliament. It is a poignant thought how much other valuable social change she might have accomplished had she not been preoccupied for ten years with the futilities of the anti-suffrage struggle.

Curzon's tergiversation marked the end for Arnold, who was now to all intents a civilian and an ex-MP. On 1 April 1918, he evidently came to Stocks the worse for drink. Dorothy felt obliged to write him a stern letter, 'the first of the kind I have écrit for some time', chiding him for 'manque de considération pour ma mère'.[60] Janet, meanwhile, passed on to her innocent elder sister some of the 'more horrible'[61] details of Arnold's behaviour. Dorothy was appalled, but her love for Arnold was inextinguishable. She could not, like Jan, hate the poor fellow. On Mary Ward's birthday in June, Arnold only wired congratulations, claiming he was too busy in town to come down to Stocks. On other occasions, he pleaded illness or his everlasting cold as reasons for keeping out of his parents' way. In October he submitted himself to the Preston Unionist Association as a candidate, but they would have nothing to do with him. On 7 October, the Wards paid over £500 and took excited possession of a cow, which happily grazed on what had been Arnold's beloved cricket pitch and golf links. Who would have thought, twenty years ago, as he delighted his mother in flannels and plus-fours, that it would come to this? What was to blame—that Ward propensity for gambling, sheer bad luck, or Mary's smothering possessiveness? Arnold was 42 on 8 November. Three days later, the war was over.

On 19 December 1918 Mary, Humphry, and Dorothy Ward watched Haig's triumphant entry into London from their usual places on the balcony of the Athenaeum. That 'Peace Christmas' they all got together at Stocks—including Janet and Arnold who had their own private armistice. There was a tree, many presents and the Trevelyan children. Christmas Day itself was brilliant, sunny, and warm. 'We have all enjoyed it and been happy and more serene than on the last two or three Christmases,' Dorothy wrote. 'I can't explain why, because goodness knows there are plenty of anxieties and sadness and dreads in our hearts all the time.'[62] Three days later, the new member for Watford (Dennis Herbert) was elected, and Arnold was an MP no longer.

1918 was authorially one of the more productive years of Mrs Humphry Ward's writing life. She finished writing her autobiographical *A Writer's Recollections* on Christmas Eve 1917. The work was serialized in *Harper's Magazine* from January to October 1918. In April 1918 she finished writing *The War and Elizabeth* (called *Elizabeth's Campaign* in America). The novel opens in war-time with Sir Henry Chicksands, Chairman of the Inspection Committee of the Corn Production Act, calling on the eccentric Squire Mannering (a widower) to tell him that part of his park must be ploughed up, by the authority of the new Defence of the Realm Act. Mannering, a connoisseur and scholar, conceives himself as standing for civilized values during the barbarism of war and intends to barricade his gates against the government's tractors. Mannering's oldest son, Aubrey, is meanwhile a decorated and much-wounded serving officer who has mortally offended his father by falling in love with Sir Henry's daughter, Beryl Chicksands. Mannering's younger son, Desmond, is also about to go to the Front as an artillery officer. Desmond's twin, Pamela, has as her governess the clever, university trained Elizabeth Bremerton—a never-failing source of good sense and intelligence.

Squire Mannering is eventually reconciled to the war, to the edicts of his government and to his alienated son by Elizabeth's patient efforts. The climax of the story is precipitated by Desmond's protracted death at Mannering Park from wounds sustained in battle. The squire proposes to Elizabeth. She refuses but they resolve to work on together, shoulder to shoulder, for England's victory, if victory it is to be: 'A few days—or weeks—or months will decide.' The mood of *The War and Elizabeth* is by no means confident as to the outcome and, as Mrs Ward wrote in her preface, 'This book ... represents the mood of a supremely critical moment in the war' (i.e. November 1917–March 1918). Published as it was in the victory winter of 1918, its dubiousness fell rather flat. But it remains highly readable for its reconstruction of grim wartime determination.

Mrs Humphry Ward's last war novel, *Harvest*, was begun in August and finished on New Year's Eve, 1918. The narrative centres on the last summer of the war, 1918, and a farm run by two young women, former comrades at agricultural college. Rachel Henderson and Janet Leighton are 'scientific farmers'. Captain Ellesborough, an American forestry battalion commander working in the area, falls in love with Rachel. He is about to join the Air Force. Rachel, although she does not tell Ellesborough, has been earlier married and divorced in Canada. Her former husband, Roger Delane, reappears and blackmails her. He is dying of tuberculosis and is morally worthless. The action of the novel climaxes during Armistice week and the national celebrations. Rachel and Ellesborough are shot—she is killed, he is wounded. Delane, the murderer, creeps off to an abandoned hut in woods, where he eventually kills himself: 'of such acts', the narration concludes, 'there is no real explanation. They are the product of that black seed in human nature which is born with a man, and flowers in due time, through devious stages, into such a deed as that which destroyed Rachel Henderson.'[63]

Harvest is an extraordinarily interesting late work. It represents Mary Ward's first extended analysis of violent crime and the relationship of the independent young women on the farm recalls—as critics have noted—D. H. Lawrence. Socially, *Harvest* offers an astute analysis of how scientific farming was altering the old ways of rural England—something that Mary Ward had been in touch with at Stocks since 1892. In one aspect, the story is a homage to what Dorothy Ward and her corps of land girls had achieved on the Wards' land over the war years. But as with other of her wartime novels the timing of *Harvest* was unfortunate and damaging to its sales prospects. In late 1918 and 1919 the reading public did not want to remember the austerities of the last years of the war. Consequently the novel was shelved. Finished on 31 December 1918 *Harvest* did not appear until April 1920, just after Mary Ward's death. As a result it has often been analysed as her last work of fiction, which it is not. That place is claimed by the far inferior *Cousin Philip*.

30

The End: 1918–1920

THE WARDS discovered like the rest of the British middle classes that world wars do not come cheap. They were in fact peculiarly vulnerable to sky-high tax assessments. Mary Ward had an extraordinarily large income from her books and the family possessed what looked on paper like vast property holdings. They leased a fine house in the West End of London; they owned a country house and estate in Hertfordshire, and a clutch of adjoining farms and cottages. On 7 August 1918, Humphry Ward received from his accountant their supertax assessment—£4,634. On 7 October came the income tax demand—£4,403. The bulk of this was evidently paid by consolidated remortgage of their Hertfordshire property. But all through 1917, 1918, and 1919 Humphry sold off the contents of Stocks piecemeal. Even after five years of pensioning Arnold's vices, the house was amazingly well appointed. Dealers fell on the Wards' living rooms like a flock of vultures. The silver went to Carrington; ceramics and jewellery were sold at Christie's; Parson's took the rare books; Edmond's took the art books; Maple's took the carpets; Block took the antique clocks (which Humphry had specially collected), the satinwood cabinets, and even the mantelpieces. By 1920, Stocks was picked clean.

It is clear that over the years Mary Ward had allowed her husband to buy without any check the *objets d'art* which he loved. In his way, he may well have been as much a drain as her son with the difference that his collection—unlike Arnold Ward's IOUs—had some resale value. Judging by the prices he got in the depressed post-war market, Humphry Ward had over the years a good eye for books, silver, china, and furniture. But clearly he was at best an erratic judge of pictures. From mid-1917 until the last sale on 5 December 1919, he sold off his art collection *en masse*. Although he loyally listed the pictures in his diary as by a dazzling array of old masters, it seems that he had often been hoodwinked or hoodwinked himself. The prices rarely reached more than £200, and the bulk of his pictures offered in the sale-room had to be bought in when they failed to reach their reserve. For someone who prided himself on his artistic expertise, it was a constant humiliation. On 12 July 1919, Dorothy noted that 'F[ather]'s "Velazquez"—rather ill-fated picture, the one he bought from the man who bought it from Lord Clarendon—[was] sold at Christie's for 160 guineas.'[1] A real Velazquez would have fetched a hundred times as much. Humphry Ward had been had. Most of his treasures could only be cleared in job lots; thus, in early 1919, he sold at Christie's four pictures for 149 guineas on 3 February, three pictures for £192 on 21 February, and nine pictures for 290 guineas on 14 March.

Clearly the Wards could not justify their pre-war ménage of a large house in town and a place in the country. They began tentatively to investigate disposing of their property. Mary Ward began, almost on Armistice Day itself, to make her calculations. On 19 November 1918, she recorded that Dolly Thursfield had written 'saying that Mr Cohen, head of the Shell Motor Spirit Co., wanted to buy Stocks, and was prepared to give whatever we asked'. It might well solve all their problems at a stroke: 'supposing we sold Stocks for a sum that would give us £30,000 clear above the mortages [currently around £18,000] there would be in addition sales of pictures and furniture £3,000'.[2] Against this £33,000, however, was £2,300 which Humphry owed the art dealer Agnew's and £3,000 to Moy Thomas. There were also overdrafts at the bank (against Stocks as collateral) and personal loans outstanding to friends like Lucy Whitridge. And if Arnold lost more money—as he surely would—the nest-egg would begin to look very small.

In the event she decided to hold on to Stocks. But in summer 1919, the Wards decided that they must let Grosvenor Place go; the £600 a year which the lease cost them could not be justified. There had been yet more demands from Moy Thomas, and Arnold signed a new agreement (presumably guaranteed by Mary) with his 'banker' on 21 August 1919. She spent her last night in the London house on 29 August and described her sombre feelings to her daughter Janet: 'The poor old house begins to look dismantled. I had many thoughts about it that last night there—of the people who had dined and talked in it—Henry James, and Burne-Jones, Stopford Brooke, Martineau, Watts, Tadema, Lowell, Roosevelt, Page, Northbrook, Goschen and so many more—of one's own good times, and follies and mistakes—everything passing at last into the words, "He knoweth whereof we are made, He remembereth that we are but dust".'[3]

Humphry Ward stayed on until 4 September, packing things away. 'Father minds it a great deal,' Dorothy noted sadly. The Wards had had a London address for forty years. The next day, vans took the remaining possessions down to Stocks. Mary Ward recorded in her notebook on 7 September '[Humphry] has borne the fatigues and worries of the move wonderfully, though no doubt it is a great wrench to him .... Furniture and pictures from G.P. will have brought in by the end of the year substantially more than £4,000. And we must have sold about £1,000 worth before—out of it we shall have paid £1,700 of debt ... and we are putting by £2,000—we hope. Not so bad in these hard times.'[4] From November to December 1919, the remaining Grosvenor Place pictures were sold at Christie's. On 14 November, ten went for 440 guineas and what Humphry obstinately called a 'Bonington' for £189. That same week in the same sale-room, someone else's genuine Carpaccio went for 32,000 guineas. On 28 November, Humphry's 'Rubens' went for £225, his 'Wilkie' for 40 guineas, and his 'Watts' for 32 guineas. On 5 December, there was the final sale of Grosvenor Place pictures. Eight were sold for £430; six were bought in by Humphry. And what he persisted in calling a 'Tintoretto' went for £162. It seems highly unlikely that he got anywhere near the £4,000 Mary Ward projected in September.

While they were selling themselves up, Mary Ward continued to write. It had been decided in the flush of November 1918 that she would make up a trilogy of non-fiction war books with a final eyewitness account of the battlegrounds. The new book would be called *Fields of Victory*. Permission to visit was granted by the military authorities and on 3 January 1919, she and Dorothy set off for France. There was a comic episode when they discovered that since it was peacetime they would need civilian passports—a nicety they had forgotten all about. Once this detail was sorted out, they embarked on a strenuous 900-mile motor journey along what had been the British, American and French front lines. Escorted by a young officer (the Allies were still in France in force) they examined at leisure the eerily silent battlegrounds and 'ugly wreck' of Ypres, Lens, Bapaume, Verdun, and Arras. They were all 'cities of the dead'.[5] Mary Ward climbed the Vimy Ridge, 'scrambling up over trenches and barbed wire and other *débris* to the top'.[6] 'I have never yet came home with such an overwhelming impression of wreck and wrong,'[7] she later told Barrett Wendell.

Not that she had lacked deference and compliments to herself. In Paris she had 'a personal contact' with President Wilson and the British Ambassador, Lord Derby, whom she found 'plebeian' and 'rather liked'.[8] The man of the hour, Sir Douglas Haig, gave Mrs Humphry Ward an audience at his small and modernized Montreuil Château. 'I sat at his right hand, Dots opposite', she recorded in her notes. They talked at lunch about demobilisation and the workers' disturbances that were rocking peacetime England. Haig was bitter against the press, particularly the *Daily Mail*, for 'setting soldiers against their officers'. Mrs Ward 'pleaded for *The Times*—but he was not very cordial'.[9] During what seems to have been a thoroughly bad-tempered occasion Haig was sarcastic about 'the women', the League of Nations, the general election and (vehemently) about General Pershing, his American 'ally'. Things were more cheerful in Strasburg where Mrs Ward was ceremonially received by General Henri Gouraud, who was all praise for *The War and Elizabeth*, which had just been serialized in the *Revue des deux Mondes*.

On 30 January 1919 Dorothy brought her mother home to London. Mary had been sick for a week and 'on the verge of bronchitis'. Throughout the long car trip, she had been constantly chilled and rheumatic and had got through it all only with the help of heavy doses of her pain-killers. It was 3 February before she could leave her sick-bed for an easy chair in the living room, and she remained weak and 'very depressed'.[10] But by 17 February, she had the first letter of *Fields of Victory* in draft form for the Foreign Office to censor. On 11 March, she was barely well enough to go to the Palace with Humphry to receive her CBE from the King. It was a poor reward for someone who had done what she had for the children of the nation and for the war effort. But the powers in the land had evidently not forgiven her for her 1918 outburst against Curzon.

Septic throat, decayed teeth, and Arnold's creditors afflicted her through the rest of March. But she had *Fields of Victory* complete by mid-April. It was syndicated in various newspapers in late summer and appeared in book form in July in Britain and a couple of months later in the US. In volume form it was accompanied by photographs, a coloured map, and a folding statistical chart of fearsome complexity. The book was dedicated to the 'Allied Armies' and struck a defiantly pro-imperial tone. It foresaw with European victory the rising up again of a reinvigorated British Empire. And in an epilogue which is unintentionally comic Mrs Ward invited America to join Britain's commonwealth of nations as an apprentice world power.

Mary Ward's private world was falling apart as she wrote *Fields of Victory*. Arrangements were being made to sell up Grosvenor Place. At Stocks every day the walls were barer with pathetic light patches where the loved pictures had once been. Her health was degenerating fast. Well into May, she was confined to a bath chair. She was all the while chronically afflicted with an array of complaints that she knew would never mend: neuritis, rheumatism, bronchitis, giddiness. Her heart and lungs were giving way. Good servants were impossible to keep and bad servants caused her infinite vexation. They had a new chauffeur and a new car (a Renault, purchased for £260), which was cheaply made and whose back axle promptly collapsed two weeks after they bought it. Arnold (now living at Sunningdale) expected an allowance and his 'banker' Moy Thomas was forever pestering them with demands for payment.

There was little consolation to be had in the state of the nation. After all England's effort, victory turned out to be strangely hollow. Throughout late 1918 and 1919, the country was paralysed by a never-ending series of strikes. Was this the 'goal'? In February 1919, Mary Ward told her American professor friend Barrett Wendell that

> Just at present we in England are descending again into the trough of the wave and no-one can tell when and how we shall emerge. Some two million men—miners, railway men and transport workers—are plotting to hold up the whole nation in their interests, deaf to argument or council, or any other considerations whatever. Yet these are the men who were fighting heroically in France a year ago. There is indeed no fathoming the ironies of this strange world.[11]

One thing was crystal clear: she must continue writing until she dropped.

Her new novel, *Cousin Philip* (called *Helena* in America), was finished on 10 September and rushed off by registered post to New York, where it was published by Dodd, Mead. They had advanced £1,200 for the work in February. The novel is—unsurprisingly—a tired effort. In her pre-composition sketch she foresaw the work (then provisionally called 'Lord Buntingford') as a light-hearted anatomy of bachelorhood. The hero would be Philip Buntingford, 'a man of nearly 50, handsome, melancholy, charming, undecided, a great favourite with women, and generally inclined to make love to them—up to a point'. Philip was to find himself guardian to a young ward, Helena, who inevitably falls desperately in love with him. At the same time he is beloved by Lady Cynthia, '35, cultivated, clever and eccentric'. Buntingford was almost to succumb to Lady Cynthia 'but finally escapes, and she at last, seeing that it is hopeless, resigns herself to the second best. As to pretty Helena, Lord B. when he recognises to his dismay that he has become her hero, has an even harder task. But he finds a lover for her, and invents a story about himself which chills and disillusions the girl.' At the end of this projected narrative 'Buntingford is left triumphantly unwed—the friend of everybody and the slave of none.'[12]

In *Cousin Philip* as it was written Mary Ward preserved the above triangle more or less intact. But what prevents Philip's giving himself to either of the women is not bachelor's reluctance. He has—buried deep in his past—a wife whom he married on the Continent, and who returns to disgrace him, before dying. This turn of plot was cannibalized from a 'spy story' which Mary Ward had contemplated writing in lieu of the romantic 'Lord Buntingford'. The amalgamation of plot elements does not work. All that *Cousin Philip* has to recommend it are some lively class-war scenes in which middle-class protagonists do battle with strikers. The novel appeared in November 1919 and was largely ignored by reviewers.

With the writing of *Cousin Philip* out of the way and Grosvenor Place disposed of, the three Wards motored up to Fox How in mid-September 1919. As she came to the end of her life, the Lake District called more and more irresistibly to Mary Ward. Of all the homes she had occupied, this was the one that meant most to her. They returned to Stocks on 10 October. She was if anything slightly better in health, although her left hand had become mysteriously paralysed. X-rays showed gouty deposits to be the problem. It would get better gradually.

That autumn, for the last time, Mary Ward played her part in public life. On 13 October 1919, she motored to Oxford to give a speech at Manchester College. On 23 October, she gave a speech for the Save the Children Fund. The children in Germany should not be excluded from British charity, she asserted: 'we have no war with children'.[13] On 29 October, she gave out the prizes at Berkhamsted Girls' School speech day. Her 15-year-old grandchild, Mary Trevelyan, won a prize for literature and history and Mrs Ward's address was on 'The Right Use of Leisure'. On 9 November, Humphry Ward was 74, and there was a little shooting at Stocks. That December, Mrs Ward embarked on her very last campaign against the Enabling Bill. This Act would, she felt, exclude free-thinkers like herself from the Anglican congregation by setting up participation in Holy Communion (and assent to the biblical miracles) as a necessary test. In early December, she organized a collective letter to *The Times* and in January 1920 she issued a fighting leaflet from Stocks on the subject. It was curiously appropriate that her last public intervention should have been a plea for the liberalized Anglicanism that she had first promulgated in 1881 with her pamphlet 'Unbelief and Sin'.

Over Christmas 1919, Mary Ward's health began finally to crack up. She was afflicted with crippling neuritis in the arms and shoulders. This was worse than anything she had previously experienced. On 6 January, Dorothy recruited a nurse to look after her mother while she (Dorothy) looked for a new house in London. It was essential that Mary should now be near good doctors and specialists. Dorothy eventually found affordable quarters to rent in Connaught Square and Mary Ward was installed there for various treatments, none of which worked. She was, by February, wholly invalid. Arnold, Dorothy, and Janet would take turns reading to her and old friends dropped by to chat and keep her spirits up. 'Those who came to the house to sit with her left it, usually, with aching hearts,'[14] Janet Trevelyan remembers. There were some good things to distract Mary, however. In February she was asked by the Lord Chancellor to be one of the first seven women magistrates in England. She accepted although she can hardly have expected ever to sit on the bench in Hertfordshire. She would, one feels, have made a good JP.

On 11 March, Humphry Ward suddenly fell ill. He was examined at noon by Sir Humphry Rolleston, whose measures failed to relieve the condition. Rolleston was a specialist in the liver, and this presumably was what ailed Humphry. He was taken to a surgery in Henrietta Street, in the early evening. There he was examined at six in the evening and rushed by car to a nursing home three-quarters of an hour later. Mary—herself scarcely able to walk—was by his side in the vehicle, Arnold Ward riding in the dicky seat behind. Humphry was operated on that same night at 9 p.m. by Sir Alfred Tripp. While she waited outside the theatre, Mary Ward's right hand was seized by crippling spasms. But the forty-minute operation was a complete success. On the next two days, Mary Ward managed to visit her husband in his nursing-home. But she was too weak to climb the stairs to his bedroom unaided. And, on returning from the second visit, she was so breathless that she had to be carried up to her bedroom. She remained in her room the rest of the day. She would, in fact, never come downstairs alive. This was Sunday, 14 March 1920.

Mary Ward was suffering from bronchitis and heart disease—both in an advanced state. On 17 March at 11 a.m., Dorothy became frightened by a sudden bout of breathlessness in her mother and telephoned their physician, Dr Williams. Mary Ward was herself more bothered by agonizing internal pain (the 'old enemy' was vigilant to the end). On the doctor's instructions Janet—who had been called in—administered chloroform, which 'quieted her wonderfully'.[15] On the next day, Dorothy hired a twenty-four-hour nurse.

By Saturday 20 March, Mary Ward was much worse. Williams was 'for the first time alarmed'[16] and gave an injection of digitalis. Sir Humphry Rolleston came the same day to examine her. On Sunday 21 March, her pulse was 125, and she managed to spend a little time on her bedroom sofa, where she was visited by her old childhood friend and cousin, Mary Cropper. On the Monday, Williams reported that 'she seemed very nervous about herself, to have lost confidence in us and in herself for getting well'.[17] On going up to see her mother that evening, Dorothy found her 'quieter and more serene', and she remained stroking her until midnight. Occasionally, Mary Ward would seem shaken with strong internal emotion, and at one point she 'suddenly quoted two E. Brontë lines to me in a kind of passion of feeling'. The lines were from the stanza

> O god within my breast,
> Almighty, ever-present Deity!
> *Life—that in me has rest,*
> *As I—undying Life—have power in thee!*[18]

After a brief recovery on the Tuesday, Mary Ward died on Wednesday 24 March at 9.35 in the morning. Dorothy had brought her breakfast at just after eight, and was the last person to see her conscious. As she recalled, her mother 'opened her dear eyes wide, her dear brown eyes, like the eyes of a young woman, and looked out into the unknown with a most wonderful look on her face'.[19] There followed an hour of gradual heart failure. Arnold arrived just in time to see her alive. The three children then went to Humphry's nursing home at noon, to tell him the news. He had a heart attack three hours later, but responded well to restorative measures.

On Friday 26 March, Dorothy Ward, Bessie Churcher and Lizzie Smith accompanied the body on its last journey to Stocks by hearse and train. It was fitting that she should have this company. They had always been her companions in life's many travels. Arnold had gone on ahead. He, Janet, and George Trevelyan and the servants received the body at the door of Stocks. They laid the coffin in the sitting room, decorated with spring flowers from the garden outside.

The funeral was at 3.15 p.m. the next day, a Saturday. There was a private ceremony first at 12.30 in the Stocks sitting room, led by Mary's old PES comrade Dr Estlin Carpenter. Relations arrived from London by train at 1.20 and lunch was taken in the house. The family walked then to Aldbury churchyard, behind the hand-bier, which was guided on its way by the Stocks gardeners. Janet carried red roses for Humphry, who was too ill and too overcome to attend. A detachment of Hertfordshire constabulary escorted the coffin, as was proper for a Justice of the Peace. The coffin was lowered into the ground under the shadow of the church tower. It was evidently what Mary Ward herself had desired as her last resting place. In addition to the villagers who knew her, there were many London friends already waiting in the church. 'I was conscious of being surrounded by love and sympathy,'[20] Dorothy wrote. Mrs Ward's friend, Canon Wood, performed the ceremony and another friend, Dean Inge, gave the address in which he called Mary 'perhaps the greatest Englishwoman of our time'. There was a message of condolence from Buckingham Palace, declaring that 'Their Majesties believe that Mrs Humphry Ward's distinguished literary achievements, her philanthropic activities and her successful organisations to promote the health and recreation of children will endear her memory to the hearts of the English-speaking people.'

Virginia Woolf, who had always had her doubts about Mary Ward, remained obdurately unendeared, writing in her diary a week later, 'Mrs Ward is dead; poor Mrs Humphry Ward; and it appears that she was merely a woman of straw after all—shovelled into the grave and already forgotten.'[21] Unfeeling as Woolf's comment is, it has a grain of truth. Even Inge, who had publicly called her the greatest living Englishwoman, privately doubted in his diary that evening whether her 'great reputation as a novelist will be lasting'.[22] Nor on the whole, has it lasted. There was no bust in the Abbey, no great memorial. Edinburgh in the last days of her life gave her an honorary degree; but Oxford, which meant much more to her, has never done anything for Mrs Humphry Ward's memory. And as soon as she died, the world stopped reading her fiction. *The Times* gave her a two column obituary, and then she slipped into an oblivion where—for all literary purposes—she still resides.

Surprisingly perhaps, Humphry Ward survived the loss of his wife and did not die in March. He came down to Stocks by borrowed car on 8 April with Dorothy. He was still effectively bedridden. But two weeks later, on an exquisite spring day, he was well enough for his daughter to bring him down to the kitchen garden in a bath chair. He was subdued and morose-seeming as he sat in the cool sunshine. But after Dorothy brought him back to the drawing-room sofa and the nurse had left, 'he broke down with a bitter cry—Vanity of vanities—all is vanity—that garden, it was all for her!'[23] Over the next few weeks, he gradually conquered his despair and gathered his strength for the unbearable business of selling Stocks, proving the will, sorting out jewellery and personal mementoes. The house was advertised in *The Times* on 20 May 1920. An auction was held on 11 July, with a reserve of £35,000. But Stocks did not meet the price.

It was not until December 1921 that the great house was finally sold.

Humphry Ward left some surviving notes of the financial details.[24] Stocks and its grounds went for £18,000 (exactly what they had paid for it in 1895). It was shockingly little. The adjoining farm, which the Wards owned, went for £7,000. The attached cottages fetched £400 and Stocks Cottage £2,700. Furniture was knocked down for £1,500. After seller's commission, the grand total was £29,000. Against this, there was a daunting set of outstanding debts and obligations. Moy Thomas, Arnold's 'banker', was still owed £1,733. Lord Grey—the original owner of Stocks and their principal mortgage holder—was owed £13,800. Lucy Whitridge was owed £2,300. The total of these debts came to £17,833. To them, Humphry added £3,892 and £350, without any indication of who the creditors were. There was another £1,500 mysteriously listed as owing to 'W.C.W.' All the debts came to a total of £23,675.

With these debts deducted from the sale total of £29,000, Humphry and Dorothy were left with a surplus of £5,325. On the same scrap of paper Humphry made some preliminary calculations as to how he and his daughter would support themselves. '£5,325 invested means, say, £180 a year,'[25] he calculated. To this could be added £120 for royalties (from his *English Poets*) and £400 from his pension from *The Times*. They would, in short, have £700 a year. Exactly—that is—what he had earned as a tutor when he married Mary Arnold fifty years before. Humphry did not include in his reckoning the copyrights of Mary's books, since these had all been sold off to Reginald Smith in the terrible spring of 1913. Dorothy Ward's calculations—on another slip of paper—were £80 'D. dress and locomotion', £200 for 'rent', and £120 'wages 3 servants'.[26] It would be very tight; but they would manage.

On 13 February 1922, Humphry and Dorothy took their last farewell from Stocks. Dorothy left her father to go up to the grave by himself, while she paid some last visits to local friends. Later she went to the churchyard herself and decked her mother's grave with Stocks snowdrops. It now had its simple, crossless headstone. Then, 'on that very radiant and exquisite morning, Father and I left our beloved Stocks for good'.[27] Lizzie had gone senile in January 1921 and been put into care. Dorothy and Humphry Ward went to live in a small flat in Eccleston Square. He died on 6 May 1926, 'very, very peacefully, my hand holding his',[28] as Dorothy recorded. He was buried at Aldbury beside his wife. He left £8,850, all to Dorothy except for £100 to his wife's 'faithful maid' Harriet Elizabeth ('Lizzie') Smith. Humphry had destroyed many of his wife's letters in 1922. After his death, on 2 July 1926, Dorothy destroyed many of his letters, including those referring to his speculations in pictures. She also destroyed many of her mother's papers, including the 'to me infinitely sad and significant financial calculations on odd bits of paper'.[29] After his mother's death, Arnold went to live at the coast, where he apparently earned a living as a journalist. On 1 January 1950 he died. Dorothy was with him. He too was buried at Aldbury, alongside his mother, on the very edge of the churchyard's consecrated ground. In February, while listening to the election results on the radio, Dorothy methodically went through the brown leather suitcase her brother left her. She burned in her garden incinerator all the surviving records of his financial and career difficulties. Fourteen years later, Dorothy herself died, aged 89, and her ashes by her own wish were buried in her mother's grave at Aldbury. Half a mile away, Stocks became by turns private residence, girls' school and—for a wild interlude—the country mansion of Victor Lownes, English manager of the Playboy empire. It is now a country club owned by the former test cricketer Phil Edmonds who has restored the pitch on which Arnold Ward once played.

Notes

Abbreviations used in Notes

ASW	Arnold Sandwith Ward.
BC	Ms holdings of Brasenose College, Oxford.
BL	Ms holdings of the British Library.
DW	Dorothy Ward.
EHJ	Enid Huws Jones, *Mrs Humphry Ward* (London, 1973).
GS	George Smith.
Honn	Ms holdings of the Honnold Library, Claremont, California.
Hou	Ms holdings of Houghton Library, Harvard.
Hunt	Ms holdings of the Henry E. Huntington Library, San Marino, California.
JA	Julia Arnold.
JPT	Janet P. Trevelyan, *The Life of Mrs Humphry Ward* (London, 1923).
LC	Louise Creighton.
LTA	*The Letters of Thomas Arnold the Younger, 1850–1900*, ed. James Bertram (Auckland, 1980).
MA	Mary Augusta Arnold.
MAW	Mary Augusta (Arnold) Ward.
MPA	Mary Penrose Arnold.
NZL	*The New Zealand Letters of Thomas Arnold the Younger, 1847–51*, ed. James Bertram (Auckland, 1966).
PH	Ms holdings of Pusey House, Oxford.
PWL	Thomas Arnold, *Passages in a Wandering Life* (London, 1899).
RS	Reginald Smith.
SRI	*Studies of Roman Imperialism, by W. T. Arnold*, ed. Edward Fiddes, with a memoir of the author by Mrs Humphry Ward and C. E. Montague (Manchester, 1906).
TA	Thomas Arnold.
THW	Thomas Humphry Ward.
UCL	Ms holdings of University College London.
VH	William Peterson, *Victorian Heretic* (Leicester, 1976).
WTA	William Thomas Arnold.
WR	*A Writer's Recollections*, 2 vols, (London, 1918).

Where editions are not otherwise specified, reference is to the Westmoreland Edition of Mary Ward's novels, 16 vols (London, 1912).

1. The Girlhood of Mary Arnold: 1851–1860

1. The dream, dated 5 Apr. 1902, is in Honn notebook 13.
2. *WR* i. 4. Mary Ward paraphrases a well-known passage in A. P. Stanley's *Life and Correspondence of Dr Arnold* (London, 1846), 434.
3. MAW, *Marcella* i. 7
4. WTA, 'Thomas Arnold the Younger', *Century Magazine*, May 1903, p. 128.
5. Stanley, *Life and Correspondence of Dr Arnold*, p. 432.
6. *LTA*, p. xvii.
7. WTA, 'Thomas Arnold the Younger', p. 118.
8. *NZL*, p. xxvii.
9. TA to J. C. Shairp, 11 Jan. 1848, quoted in *NZL* 218.
10. WTA, 'Thomas Arnold the Younger', p. 122.
11. *SRI*, p. iv.
12. TA to MPA, 20 Aug. 1849, quoted in *NZL* 131.
13. *PWL* 129.
14. *The Australian Dictionary of Biography* (London, 1966, repr. 1986), ii. 462.
15. JPT 1.
16. *The Australian Dictionary of Biography* i. 30.
17. Mary Ward may never have known the full story. In *WR* i. 7, she asserts that William Sorell, her great-grandfather, retired from his governorship 'in 1830 or so'. In fact he was dismissed for moral turpitude seven years before.
18. TA to MPA, 16 Dec. 1852, quoted in *LTA* 31.
19. TA to MPA, 15 Feb. 1853, quoted in *LTA* 33.
20. This passage, dated 16 Feb. 1853, is from a sheaf of extracts taken from Thomas Arnold's diaries and letters to his mother, made up for Janet Trevelyan's biography and now held at Pusey House.
21. Ibid. No clear date is attached to this observation.
22. Ibid.
23. Extracted notes from Thomas Arnold's diaries and letters, 5 Jan. 1855, PH; quoted in JPT 6.
24. *WR* i. 27.
25. TA to MPA, 21 Nov. 1854, quoted in *LTA* 56.
26. TA to MPA, 23 Feb. 1855, quoted in *LTA* 59.
27. JA to TA, 23 June 1855(?), quoted in *LTA* 62.
28. TA to MPA, 21 Feb. 1856, quoted in *LTA* 68.
29. Ibid.
30. TA to MPA, 17 Sept. 1855, quoted in *LTA* 66.
31. TA to Mary Arnold, 5 Mar. 1856, notes PH.
32. TA to MPA, 21 Feb. 1856, quoted in *LTA* 68.
33. *WR* i. 5.
34. TA to MPA, 19 Oct. 1856, quoted in *LTA* p. 80.
35. MAW, *Milly and Olly* (New York, 1907), 99.
36. *PWL* 38, quoted in *VH* 27–8.
37. *WR* i. 31.
38. JPT 6.
39. J. H. Newman to Miss M. R. Giberne, 11 Feb. 1868, quoted in *The Letters and Diaries of John Henry Newman*, eds. C. S. Dessain and T. Gornall (Oxford, 1973), vol. xxiv, p. 34. James Bertram notes that Newman's is the only such comment about Julia on record.
40. *SRI*, p. v.
41. Ibid.
42. *Marcella* i. 10.

43. *WR* i. 118.
44. *WR* i. 96.
45. *PWL* 38, quoted in *VH* 27.
46. *SRI*, p. vii.
47. MA to TA and JA, 13 Feb. 1857, PH.
48. MA to WTA, 16 Feb. 1857, PH.
49. MA to TA, 29 May 1859, PH.
50. MA to JA, 12 Aug. 1860, PH.
51. MAW, *Milly and Olly*, p. 64.
52. Margaret L. Woods, 'Mrs Humphry Ward: a Sketch from Memory', *Quarterly Review*, July 1920, p. 148. Woods suggests that Mary's rebellion was staged at Eller How. Miss Clough's assistants were grandly called 'governesses'.
53. TA to JA, 'September 1859', quoted in *LTA* 101.
54. Quoted *VH* 45.
55. MA to TA and JA, 3 Apr. 1859, PH.
56. T. C. Down, 'Schooldays with Miss Clough', *Cornhill Magazine*, June 1920, p. 62.
57. Ibid., 682.
58. Ibid., 678–81.

2. Schooldays: 1860–1867

1. MA to TA and JA, 1 Mar. 1861, PH.
2. *The Letters and Diaries of John Henry Newman*, eds. C. S. Dessain and T. Gornall (Oxford, 1973), vol. xxiv, p. 34.
3. See *Marcella* i. 10, where Mary Ward recalls this period of her early life.
4. Ibid.
5. MAW, *Marcella* i. 7.
6. Ibid.
7. MA to TA, 19 Nov. 1863, PH.
8. MAW, *Marcella* i. 8.
9. *SRI*, p. ix. In *Robert Elsmere*, chap. 4, Mary Ward puts this statement in the mouth of the hero.
10. JPT 11.
11. MAW, *Marcella* i. 12–13.
12. MA to JA, 4 June 1862, PH.
13. MA to TA, 'October 1862(?)', PH.
14. MA to TA, 19 Nov. 1863, PH.
15. TA to MA, undated, presumably 1862, Honn.
16. MA to TA, '1862(?)', PH.
17. MA to TA, 24 May 1862. PH. Robert A. Slaney, Liberal member for Shrewsbury, died 19 May 1862, aged 71.
18. Honn notebook 1.
19. MAW, *Marcella* i. 13.
20. Honn notebook 1.
21. 'A Tale of the Moors' survives in Honn notebook 1. William Peterson analyses the tale in *VH* 48.
22. Anthony Trollope, *An Autobiography*, (first pub. London, 1883, repr. 1950, ed. Frederick Page), 42–3.
23. MAW, *Marcella* i. 9.
24. Honn notebook 1.
25. *WR* i. 26.
26. JPT 13.

27. *WR* i. 181.
28. MA to JA, 'Summer 1865(?)', PH.
29. MA to JA, 'Summer 1865(?)', PH.
30. MA to JA, 1865(?), PH.
31. MA to JA, 20 June 1865, PH.
32. MAW, *Marcella* i. 21.
33. JPT 15.
34. MAW, *Marcella* i. 21.
35. MAW to Mrs Cunliffe, winter 1865, quoted in JPT 15 (my italics).
36. *WR* i. 135–6.
37. The partial text of the story survives in Honn notebooks 2, 3, and 4. Peterson gives a close and informative account of 'Lansdale Manor' (among other juvenilia) in *VH* 50–1 to which I am indebted.
38. This comment appears in the *Dictionary of National Biography* entry on Charlotte Yonge.
39. MA, diary, Honn, 14 Jan. 1866. See *VH* 51–2.
40. JPT 16.
41. *WR* i. 134.
42. JPT 11.
43. MA to JA, 20 June 1865, PH.
44. *WR* ii. 180–1.
45. MAW to TA, 20 Apr. 1896, PH.
46. *WR* ii. 259.
47. T. C. Down, 'Schooldays with Miss Clough', *Cornhill Magazine*, June 1920, p.675.
48. Honn notebook 1.
49. *WR* i. 129–30.

3. Oxford: 1867–1871

1. Before rejoining her family, Mary Arnold went for a short holiday to Glasgow, commemorated in 'A Journal of my first Visit to Scotland, June 7–July 8, 1867', Honn notebook 3.
2. *WR* i. 136–7.
3. MAW, *Lady Connie* (New York, 1916), 133.
4. Honn notebook 5.
5. MAW, *Helbeck of Bannisdale* i. 38.
6. *WR* i. 137. Ward actually dates this encounter '1868 or 1869'. Peterson in *VH* 67 notes that her name is first mentioned in Pattison's engagement diary on 31 December 1867.
7. *WR* i. 138.
8. Mark Pattison, *Memoirs* (London, 1885), 292.
9. See Vivian Green, *Love in a Cool Climate* (Oxford, 1985), 19–20.
10. Quoted in *VH* 66.
11. Honn notebook 5, dated 21 June 1869.
12. *Robert Elsmere* i. xxi, quoted in *VH* 68.
13. Pattison, *Memoirs*, p. 1.
14. Pattison, *Memoirs*, p. 290.
15. Quoted by Anthony Grafton, 'Mark Pattison', *American Scholar*, Spring 1983, pp. 229–30. Grafton's article contains a convincing critique of the essential sterility of Pattison's scholarship.
16. *WR* i. 141.
17. MAW, *Lady Connie*, p. 117.

18. Honn notebook 5.
19. TA to T. B. Collinson, 15 Sept. 1851, in *LTA*, p. 12.
20. *WR* i. 150.
21. *WR* i. 144–6.
22. J. R. Green to Louise von Glehn, 6 Mar. 1871, quoted in *VH* 81.

4. Stabs at Fiction: 1867–1871

1. Honn notebook 5.
2. Quoted in *VH* 53. Peterson dates the letter 21 Sept. 1870(?).
3. Honn notebook 3.
4. MAW, *Sir George Tressady*, ii. 222. The corrected manuscript is at the Honnold Library.
5. MA to Smith, Elder, 1 Oct. 1869, quoted in JPT 25.
6. Edith C. Rickards *Felicia Skene of Oxford*, (London, 1902), 132–3.
7. Rickards, *Felicia Skene of Oxford* p. 133. In this same recollection of Skene, Mary Ward says that she wrote *A Westmoreland Story* when 'I was seventeen or eighteen'. Her eighteenth birthday was on 11 June 1869.
8. MAW, *Sir George Tressady*, ii. 222.
9. TA to MPA, 16 June 1870, in *LTA* 167.
10. MAW, *Sir George Tressady*, ii. 222.
11. The fragment is found in Honn notebook 5.

5. Marriage: 1870–1872

1. THW to E. H. Cradock, 15 June 1870, BC.
2. THW, diary UCL, 7 May 1871.
3. These were figures reported by the Duke of Cleveland's Commission, in 1871. Other sources suggest an even wider discrepancy.
4. This crucial episode in Oxford's history is dealt with authoritatively in A. J. Engel's *From Clergyman to Don* (Oxford, 1983). See particularly p. 100.
5. Quoted in Richard Ellmann, *Oscar Wilde* (New York, 1984, repr. 1987), 84.
6. A. C. Benson, *Walter Pater* (London, 1906), 26.
7. *The Letters of J. R. Green*, ed. Leslie Stephen (London, 1901), 53.
8. *The Letters of J. R. Green*, p. 193.
9. This guidance is given in A. M. S. Methuen, *Oxford, Its Social and Intellectual Life* (London, 1878).
10. THW, diary UCL, 'Memorandum of 1870'.
11. Ibid.
12. Ibid.
13. Ibid.
14. Ibid.
15. Ibid. See also MAW, *Robert Elsmere*, ii. 237.
16. THW, diary UCL, 'Memorandum of 1870'.
17. THW was alarmed to discover that J. R. Green had put his shipboard flirtation into a *Saturday Review* piece, 'Winter Flittings', which appeared 4 Feb. 1871.
18. THW, diary UCL, 25 Jan. 1871.
19. THW, diary UCL, 30 Jan. 1871.
20. THW, diary UCL, 1 Feb. 1871.
21. THW, diary UCL, 2 Feb. 1871.
22. THW, diary UCL, 9 Feb. 1871.
23. THW, diary UCL, 14 Feb. 1871.
24. LC, *The Life and Letters of Mandell Creighton*, 2 vols. (London, 1905) i. 77.

25. THW, diary UCL, 15 Feb. 1871.
26. THW, diary UCL, 16 Feb. 1871.
27. THW, diary UCL, 17 Feb. 1871.
28. THW, diary UCL, 20 Feb. 1871.
29. THW, diary UCL, 1 Mar. 1871.
30. THW, diary UCL, 10 Mar. 1871.
31. THW, diary UCL, 20 Mar. 1871.
32. THW, diary UCL, 3 May 1871.
33. Quoted JPT 82.
34. THW, diary UCL, 13 June 1871.
35. Rickards, *The Life of Felicia Skene*, 134.
36. THW, diary UCL, 15 June 1871.
37. THW, diary UCL, 16 June 1871.
38. THW, diary UCL, 18 June 1871.
39. THW, diary UCL, 23 June 1871.
40. THW, diary UCL, 12 July 1871.
41. 'A Morning in the Bodleian' was published pseudonymously ('By Two Fellows') and privately, dated 21 Aug. 1871. The Wards republished it anonymously in the magazine *Dark Blue* in Feb. 1872.

6. Marriage and Oxford: 1872–1878

1. Mandell Creighton to Louise von Glehn, 19 Apr. 1871, quoted in LC, *The Life and Letters of Mandell Creighton*, 2 vols. (London, 1905), i. 78.
2. MAW to TA, 11 Aug. 1873, PH.
3. John Wordsworth to Walter Pater, 17 Mar. 1873, *The Letters of Walter Pater*, ed. Lawrence Evans (Oxford, 1970), 14.
4. See A. J. Engel, *From Clergyman to Don* (Oxford, 1983), pp. 156–256.
5. Mandell Creighton to Louise von Glehn, 12 Sept. 1871, quoted in LC, *The Life and Letters of Mandell Creighton*, pp. 106–7.
6. *WR* i. 159–60.
7. MAW, *Lady Connie* (New York, 1916), 132.
8. *WR* ii. 120.
9. *WR* i. 191.
10. MAW, *Sir George Tressady*, ii. 222.
11. JPT 21.
12. Contract between MAW and Macmillan, 19 June 1874, PH.
13. MAW to Macmillan, 19 June 1874, BL Add. MS 54928.
14. *WR* i. 193-4.
15. A sympathetic exposition of T. H. Green's theism is given in W. H. Fairbrother's entry on the philosopher in the *Encyclopaedia Britannica* (11th edn.).
16. MAW to TA, 13 Jan. 1874, PH.
17. T. H. Green to MAW, quoted in JPT 9. Peterson (*VH* 77) points out that in *Robert Elsmere*, Ward gives these words to Mr Grey, a character who was clearly modelled on T. H. Green.
18. *WR* i. 177.
19. LC to MAW, 17 Sept. 1878, PH.
20. *WR* i. 203–4.
21. *WR* i. 205–15.
22. The earlier project is mentioned as 'the French history' in a letter, THW to Macmillan, 26 Apr. 1880, BL Add. MS 54928.
23. MAW to TA, 16 Feb. 1876, PH, quoted in *LTA* 178.

24. TA to THW, 14 Oct. 1876, quoted in *LTA* 182. The Rawlinsonian professorship of Anglo-Saxon was formerly tenable only for five-year periods and in 1876 had been vacant for some time when it was decided to make the post permanent. In the absence of Tom Arnold's candidacy the election went to John Earle.
25. MAW to TA, 23 Oct. 1876, PH, quoted in *LTA* 185.
26. MAW to TA, 21 May 1880, PH.
27. Ibid.
28. *SRI*, p. xx.
29. MAW to TA, 5 July 1873, PH.
30. *SRI*, p. xxiv.

7. Fighting Back: 1878–1880

1. See *WR* i. 206. JPT gives Oct. 1877 as the date Wace recruited Mary Ward; Peterson gives 1879; Mary Ward gives 1878. The volume in which her first contributions appeared came out in 1880.
2. *WR* i. 217–18.
3. *WR* i. 202.
4. *WR* i. 218.
5. Honn notebook 8.
6. *WR* i. 202.
7. *WR* i. 222.
8. MAW to TA 13 Jan. 1874, PH, quoted in *LTA* 175.
9. JPT 28.
10. MAW to JA, 22 Feb. 1883, PH. In *WR* ii. 14, Mary Ward dates her examining appointment as 1882.
11. *WR* i. 191.
12. MAW to Macmillan, 29 Oct. 1880, BL Add. MS 54928.
13. MAW to Macmillan, 20 Oct. 1880, BL Add. MS 54928.
14. MAW to Macmillan, 6 Feb. 1881, BL Add. MS 54928.
15. THW to Macmillan, 6 Dec. 1880, BL Add. MS 54928.
16. MAW to Macmillan, 18 Oct. 1882, BL Add. MS 54928.
17. See *VH* 108: 'Of the 3,000 copies printed by Macmillan, nearly 900 were still unsold in 1888.'
18. *WR* i. 233–4.
19. *The History of The Times: 1841–84* (London, 1939), 518.
20. MAW to TA, 7 June 1880 and 21 May 1880, PH.
21. MAW to THW, 20 July 1880, PH.
22. MAW to THW 25 Aug. 1881, PH.
23. *WR* i. 234–5.
24. *WR* i. 235–8.
25. MAW to TA, 7 May 1882, PH, quoted *LTA* 202.
26. See JPT 40–2.
27. MAW to THW, 29 Dec. 1880, PH.
28. MAW to Macmillan, 6 Feb. 1881, BL Add. MS 54928.
29. *WR* i. 224.
30. *WR* i. 286.
31. Mark Pattison to MAW, 'March 1881', quoted in JPT 34.
32. MAW to TA, 26 Aug. 1881, PH.
33. MAW to THW, 12 June 1881, PH.
34. MAW to THW, 19 July 1881, PH.
35. MAW to JA, 31 July 1881, PH.

36. MAW to JA, 6 Aug. 1881, PH.
37. MAW to WTA, 10 Sept. 1881, PH.
38. Ibid.
39. MAW to TA, 31 July 1881, PH.

8. London: 1881–1886

1. Memorandum, Aug. 1881, PH.
2. MAW to WTA, 10 Sept. 1881, PH.
3. *WR* ii. 12.
4. MAW to WTA, 24 Dec. 1881, PH.
5. MAW to THW, 11 Oct. 1882, PH.
6. MAW to TA, 29 Nov. 1882, PH.
7. MAW to WTA, 14 July 1887, PH.
8. MAW to TA, 25 Nov. 1882, PH. £5 may not seem much. But in a letter to THW, 11 Oct. 1882, PH, MAW noted that they had £188 in the bank and debts of £143—'not much to go through November with my Darling', she added.
9. MAW to TA, 4 Aug. 1883, PH.
10. JPT 39.
11. MAW to JA, 23 Jan. 1883, PH.
12. MAW to JA, 14 Feb. 1883, PH.
13. MAW to JA, 16 Apr. 1883, PH.
14. MAW to TA, 2 Dec. 1884, PH.
15. MAW to JA, 11 June 1886, PH.
16. MAW to JA, 'October 1887', PH.
17. JPT 249.
18. MAW to JA, 28 July 1886, PH.
19. MAW to TA, 3 Oct. 1886, PH.
20. *WR* ii. 10.
21. MAW to JA, 23 Apr. 1886, PH.
22. THW, diary UCL, 6 Aug. 1887.
23. MAW to THW, 7 Aug. 1888, PH.
24. Mary Ward uses the phrase in a letter to Arnold in 1905, PH.
25. MAW to WTA, 10 Sept. 1881, PH.
26. *VH* 87.
27. See *VH* 113–14.
28. *WR* ii. 2.
29. MAW to TA, 29 Nov. 1882, PH.
30. J. Morley to MAW, 22 Mar. 1883, quoted in JPT 42.
31. *WR* ii. 83.
32. See *Letters to Macmillan*, ed. Simon Nowell-Smith (London, 1967), 195. Shorthouse's novel was first published privately in a limited edition in 1880 before Mary Ward introduced it to her publisher.
33. JPT 38.
34. MAW to Henry James, 9 Dec. 1882, Hou MS Am. 1094 (477).
35. Henry James to Edmund Gosse, 22 Aug. 1895, quoted in *Selected Letters of Henry James to Edmund Gosse, 1882–1915*, ed. Rayburn S. Moore (Baton Rouge, La., 1988), 131.
36. *WR* II.17.
37. MAW to JA, 21 May 1885, PH.
38. THW, diary UCL, 6 Mar. 1886.
39. THW, diaries UCL, 1886–88.

40. Oscar Wilde, 'The Decay of Lying', in *Intentions* (London, 1891, repr. 1913) 10, in reference to *Robert Elsmere*. In the first version of the essay, published in *Nineteenth Century*, Jan. 1889, Wilde omitted the 'meat tea' comment.
41. The Marquis of Queensberry's spelling.
42. MAW to JA, 29 Apr. 1885, PH.
43. *WR* ii. 23. The odd grammar of this sentence is Mrs Ward's.
44. Pat Jalland, *Women, Marriage, and Politics, 1860–1914* (Oxford, 1986, repr. 1988), 103.
45. *WR* ii. 24.
46. *WR* ii. 26.
47. Ibid.
48. *WR* ii. 27–8.
49. *WR* ii. 31.
50. MAW to JA, 26 Apr. 1886, PH.

9. The Right Book: 1883–1884

1. MAW to A. Macmillan, 19 July 1883, BL Add. MS 54928.
2. Ibid.
3. MAW to A. Macmillan, 22 July 1883, BL Add. MS 54928.
4. MAW to Frederick Macmillan, 20 Nov. 1888, BL Add. MS 54928.
5. MAW to Macmillan, 6 Sept. 1883, BL Add. MS 54928.
6. *Amiel's Journal*, trans. and with an introduction by Mrs Humphry Ward, 2 vols. (London 1885, repr. New York, 1928), 388–9.
7. *WR* ii. 38.
8. MAW to Macmillan, 6 Sept. 1883, BL Add. MS 54928.
9. *Amiel's Journal*, 'Introduction', p. xxx.
10. *Amiel's Journal*, 'Introduction', p. xiv.
11. *Amiel's Journal*, 'Introduction', p. xii.
12. MAW to A. Macmillan, 3 Feb. 1884, BL Add. MS 54928.
13. MAW to G. L. Craik, 27 Apr. 1885, BL Add. MS 54928.
14. MAW to A. Macmillan, 1 Sept. 1885, BL Add. MS 54928.
15. MAW to A. Macmillan, 14 Dec. 1885, BL Add. MS 54928.
16. See *Matthew Arnold: The Last Word*, ed. R. H. Super (Michigan, 1977), 460.
17. MAW, *Sir George Tressady*, ii. 222. *Miss Bretherton* is appended to this volume in the Westmoreland Edition.
18. Honn notebook 9.
19. MAW to JA, 19 May 1883, PH.
20. Quoted in EHJ 63.
21. Quoted in JPT 43.
22. J. Morley to MAW, 8 Oct. 1884, Quoted in *VH* 108.
23. G. L. Craik to MAW, 12 Oct. 1884, BL Add. MS 55418.
24. F. Macmillan to MAW, 21 Nov. 1884, BL Add. MS 55418.
25. MAW to Macmillan, 4 Nov. 1884, BL Add. MS 54928.
26. A. Macmillan to MAW, 6 Nov. 1884, BL Add. MS 55418..
27. G. L. Craik to MAW, 28 Nov. 1884, BL Add. MS 55418.
28. MAW, *Sir George Tressady*, p. 327.
29. MAW, *Sir George Tressady*, pp. 392–3.
30. The manuscript version of *Miss Bretherton* is in Honn notebooks 9 and 10.
31. MAW, *Sir George Tressady*, ii. 224.
32. The comments are quoted in Macmillan's advertisements for *Miss Bretherton*.
33. *WR* ii. 15.

34. JA to MAW, 14 Dec. 1884, PH.

10. The Elsmere Ordeal: 1884–1888

1. MAW to G. L. Craik, 5 Feb. 1885, BL Add. MS 54928.
2. MAW to G. L. Craik, 26 Feb. 1885, BL Add. MS 54928.
3. G. L. Craik to MAW, 27 Feb. 1885, BL Add. MS 55419.
4. MAW, *Sir George Tressady* vol. i, pp. xiv–xv.
5. GS to MAW, 2 Mar. 1885, Honn. Smith, whose memory was not good in his later years, put down in his memoirs a fallacious account of the episode: 'One afternoon [in the autumn of 1886] Mrs Humphry Ward called on me in Waterloo Place, and said she had "come on business". I put on my business expression of countenance, and prepared to listen to her. She had written—or partly written—a novel, had asked Macmillans to publish it for her, and to pay her £200 for it. As her previous book—which they had published—had not been a success, Macmillans had declined her proposal. She was very friendly with Macmillans, she said, but had need for the £200, and she had come to ask if I would purchase her book on her terms. I said "yes". But she said in a tone of surprise, "You have not even seen it yet." "No," I said, "but I read your previous tale. It is a poor tale, but it shows that you can write, and I am willing to take the risk." She then said the book would be in two volumes, and she had written one already. Would I let her have £100 on account? I remarked that it would save both her and me trouble to make it one transaction, and I wrote her the cheque for £200.' See L. Huxley, *The House of Smith, Elder and Co.* (London, 1923), 191–2. When Mary Ward made her overture in March 1885 there was nothing of *Robert Elsmere* written. She asked for £250. George Smith made over his £200 advance to the author a year later, in March 1886.
6. MAW to G. L. Craik, 2 Mar. 1885, BL Add. MS 55419.
7. G. L. Craik to MAW, 27 Apr. 1885, BL Add. MS 55419.
8. See C. Morgan, *The House of Macmillan* (London, 1943), 133–4.
9. MAW to Frances Power Cobbe, 15 Aug. 1890, Hunt. I take the observation from *VH* 121. Peterson gives an authoritative account of the composition of *Robert Elsmere*.
10. *WR* ii. 12.
11. Gertrude Ward's Diary, 'May 1885', quoted *VH* p. 114.
12. THW, diary UCL, 1 May 1885.
13. The notebook (which is in the Honnold Library) is transcribed with annotations by Peterson, *VH* 213–20.
14. THW, diary UCL, 21 June 1885.
15. THW, diary UCL, 30 Sept. 1885.
16. THW, diary UCL, 25 Nov. 1885.
17. *The Autobiography of Margot Tennant*, ed. Mark Bonham Carter (London, 1962), 76–7.
18. MAW to WTA, 16 Dec. 1885, quoted in *VH* 117.
19. MAW to JA, 30 Dec. 1885, PH.
20. THW, diary UCL, 9 Jan. 1886.
21. THW, diary UCL, 12 Mar. 1886.
22. MAW to TA, 5 Feb. 1886, PH.
23. THW, diary UCL, 5 Feb. 1886.
24. THW, diary UCL, 5 Mar. 1886.
25. MAW to TA, 'Spring 1886', PH.
26. MAW to JA, 18. Apr. 1886, PH.

27. MAW to JA, 23 Apr. 1886, PH.
28. MAW to JA, 26 Apr. 1886, PH.
29. MAW to JA, 29 June 1886, PH.
30. MAW to JA, 11 June 1886, PH.
31. Ibid.
32. MAW to JA, 4 Aug. 1886, PH.
33. MAW to GS, 20 Oct. 1886, Honn.
34. MAW to TA, 3 Oct. 1886, PH.
35. Ibid.
36. MAW to GS, 20 Oct. 1886, Honn.
37. GS to MAW, 21 Oct. 1886, Honn.
38. MAW to GS, 20 Oct. 1886, Honn.
39. GS to MAW, 21 Oct. 1886, Honn.
40. MAW to JA, 16 Dec. 1886, quoted in *VH* 120.
41. MAW to TA, 27 Dec. 1886, PH.
42. THW, diary UCL, 31 Dec. 1886.
43. THW, diary UCL, 1887.
44. MAW to JA, 23 Jan. 1887, PH.
45. MAW to WTA, 'February 1887', quoted in *VH* 121.
46. MAW to JA, 7 Mar. 1887, PH.
47. MAW to TA, 'February 1887', PH.
48. MAW to JA, 29 Nov. 1886, PH.
49. MAW to JA, 7 Mar. 1887, PH.
50. MAW to TA, 9 Mar. 1887, PH.
51. *WR* ii. 64.
52. MAW to GS, 24 Oct. 1891, Honn.
53. MAW to JA, 7 Mar. 1887, PH.
54. MAW to JA, 8 Apr. 1887, PH.
55. For instance a letter, MAW to JA, 6 Apr. 1887, PH, recounting a dinner party conversation with Henry James on how beastly the Irish are.
56. MAW to WTA, 'March 1887', PH.
57. Ibid.
58. MAW to Ethel Arnold, 12 Apr. 1887, PH.
59. MAW to JA, 11 Apr. 1887, PH.
60. MAW to Ethel Arnold, 12 Apr. 1887, PH.
61. MAW to GS, 29 Apr. 1887, Honn.
62. MAW to GS, 7 May 1887, Honn.
63. GS to MAW, 13 May 1887, Honn.
64. GS to MAW, 17 May 1887, Honn.
65. MAW to GS, 17 May 1887, Honn.
66. THW, diary UCL, 27 May 1887.
67. THW, diary UCL, 10 June 1887.
68. MAW to WTA, 14 July 1887, PH.
69. MAW to GS, 29 July 1887, 11 Aug. 1887, Honn.
70. THW, diary UCL, 6 Aug. 1887.
71. MAW to GS, 5 Sept. 1887, Honn.
72. MAW to GS, 7 Oct. 1887, Honn.
73. MAW to GS, 5 Nov. 1887, Honn.
74. *VH* 137.
75. See *VH* 123. I am here indebted to Peterson's chap. 7, '*Robert Elsmere*, an Analysis'.
76. JA to MAW, 5 Feb. 1888, PH.

77. MAW to TA, 6 Feb. 1888, PH.
78. MAW to TA, 2 Mar. 1888, PH.
79. MAW to JA, 19 Feb. 1888, PH.
80. GS to MAW, 15 Mar. 1888, Honn.
81. MAW to TA, 25 Mar. 1888, PH.
82. MAW to DW, 3 Apr. 1888, PH.
83. MAW to THW, 6 Apr. 1888, PH.
84. MAW to THW, 6 Apr. 1888, PH.
85. MAW to WTA, 7 Apr. 1888, PH. For another account see *LTA* 217–18.
86. MAW to THW, 8 Apr. 1888, PH.
87. TA to Frances Arnold, 7 Apr. 1888, in *LTA* 217. Thomas Arnold gives a moving account of Julia's death in this letter.
88. MAW to TA, 16 Apr. 1888, PH.

11. *Elsmere* Mania: 1888

1. *The Times*, 7 Apr. 1888. Peterson describes the British press reaction in detail, *VH* 169–75.
2. GS to MAW, 18 Apr. 1888, Honn.
3. MAW to GS, 11 May 1888, 16 May 1888, Honn.
4. MAW to TA, 29 May 1888, PH.
5. *WR* ii. 87.
6. *VH* 172.
7. MAW to GS, 14 Mar. 1888, Honn.
8. Priscilla Metcalf, *James Knowles* (Oxford, 1980), 322.
9. Quoted in *VH* 165.
10. Mary Ward made notes of her meeting with Gladstone, and transcribed them as a narrative in the appendix to the Westmoreland Edition of *Robert Elsmere*. Her actual notes are transcribed, with annotation, in W. Peterson, 'Gladstone's Review of *Robert Elsmere*,' *Review of English Studies*, Nov. 1970. Peterson also discusses the encounter in *VH* 166–7.
11. JPT 60.
12. *WR* ii. 78.
13. *WR* ii. 91.
14. JPT 77.
15. JPT 78.
16. MAW to WTA, 3 Jan. 1889, PH.
17. MAW to TA, 'October 1888', PH.
18. GS to MAW, 21 Dec. 1888, Honn.

12. The Fiction Machine: 1890–1900

1. MAW to WTA, 'September 1894', PH.
2. MAW to WTA, 'April 1892(?)', PH.
3. MAW to TA, 30 Nov. 1888, PH.
4. MAW to GS, 25 Feb. 1889, Honn.
5. MAW, *David Grieve*, vol. i, p. xxiv.
6. MAW to unknown correspondent, presumably a bookdealer, 27 Sept. 1888, Hou bMS Am. 1622 (199).
7. MAW to TA, 5 Oct. 1888, PH.
8. MAW to TA, 'October 1888', PH.
9. MAW to ASW, 19 Nov. 1889, PH.
10. MAW to GS, 11 Dec. 1889, Honn.

11. GS to MAW, 8 Mar. 1890, Honn.
12. MAW to GS, 29 Sept. 1890, Honn.
13. MAW to GS, 30 Apr. 1891, Honn.
14. MAW to WTA, 11 Oct. 1891, PH.
15. GS to MAW, 12 Oct. 1891, Honn.
16. Ibid.
17. MAW to GS, 24 Oct. 1891, Honn.
18. MAW, *David Grieve*, vol. i, p. xxxvi.
19. GS to MAW, 8 Apr. 1890, Honn.
20. MAW to GS, 14 Feb. 1889, Honn.
21. GS to MAW, 9 Dec. 1890, Honn.
22. MAW to GS, 30 Apr. 1891, Honn.
23. GS to MAW, 13 June 1891, Honn. In his memoirs Smith claims that he originally asked for £8,000.
24. GS to MAW, 16 June 1891, Honn.
25. MAW to GS, 30 Dec. 1891, Honn.
26. MAW to GS, 1 Feb. 1892, Honn.
27. MAW, *David Grieve*, vol. i, p. ix.
28. GS to MAW, 20 June 1892, Honn.
29. MAW to GS, 20 June 1892, Honn.
30. MAW, *Marcella*, vol. i, p. ix.
31. Ibid.
32. MAW, *Marcella*, vol. i, p. x.
33. Ibid. The 'circumstance' included a ten-year feud between the keeper and the principal poacher, which Mary Ward wove into her story. A full account of the episode is given by Ruth Crauford, 'The Aldbury Double Murder' (privately printed: Aldbury, 1963).
34. *WR* ii. 146.
35. The working notes and much of the manuscript and proofs of *Marcella* are held at Princeton Library, New Jersey.
36. MAW to Macmillan, 3 May 1892, BL Add. MS 54928..
37. MAW, *Marcella*, vol. i. pp. 253–4.
38. MAW, *Marcella*, vol. i, p. xviii.
39. MAW to GS, 1 June 1892, Honn.
40. MAW to GS, 7 Aug. 1892, Honn.
41. GS to MAW, 16 Aug. 1892, Honn.
42. Ibid.
43. MAW to GS, 19 Aug. 1892, Honn.
44. MAW to GS, 19 Oct. 1892, Honn.
45. MAW, *Marcella*, vol. i, pp. xiii–xiv.
46. MAW to GS, 26 Jan. 1893, Honn.
47. GS to MAW, 21 July 1893, Honn.
48. MAW to GS, 21 July 1893, Honn.
49. MAW to GS, 24 July 1893, Honn.
50. MAW to GS, 8 Sept. 1893, Honn.
51. Ibid.
52. MAW, *Marcella*, vol. ii, p. 343. Toynbee was similarly jeered at, just before his death.
53. MAW to GS, 8 Jan. 1894, Honn.
54. MAW to GS, 18 Jan. 1894, Honn.
55. GS to MAW, 21 Jan. 1894, Honn.
56. MAW to GS, 22 Jan. 1894, Honn.

57. MAW to GS, 23 Jan. 1894, Honn.
58. GS to MAW, 25 Jan. 1894, Honn.
59. MAW to GS, 1 Feb. 1894, Honn.
60. MAW to GS, 2 Feb. 1894, Honn.
61. GS to MAW, 5 Feb. 1894, Honn.
62. G. Brett to MAW, 13 Mar. 1894, Honn.
63. *WR* ii. 162.
64. MAW to GS, 24 May 1894, Honn.
65. MAW to an unknown American, 29 May 1894, Hou bMS Eng. 132 (600).
66. MAW to GS, 24 May 1894, Honn. Mudie initially took only 300.
67. R. A. Gettmann, *A Victorian Publisher* (Cambridge, 1960), 258.
68. GS to MAW, 5 June 1894, Honn.
69. MAW to GS, 12 July 1894, Honn.
70. GS to MAW, 13 July 1894, Honn.
71. GS to MAW, 16 Aug. 1894, Honn.
72. MAW to TA, 5 Sept. 1888, PH.
73. MAW to GS, 6 Jan. 1895, Honn.
74. MAW to GS, 8 Jan. 1895, Honn.
75. MAW, *Sir George Tressady*, vol. i, p. x.
76. MAW to GS, 14 Jan. 1895, Honn.
77. MAW to GS, 1 Feb. 1895, Honn.
78. MAW, *Sir George Tressady*, vol. ii, p. 215.
79. GS to MAW, 4 Jan. 1897, Honn.
80. MAW to TA, 19 May 1896, PH.
81. MAW to WTA, 30 Oct. 1896, PH.
82. *WR* ii. 179.
83. *WR* ii. 180.
84. MAW to TA, 15 Nov. 1896, PH.
85. MAW, *Helbeck of Bannisdale*, p. 534.
86. *WR* ii. 180.
87. MAW to DW, 7 Mar. 1897, PH.
88. MAW to THW, 16 Mar. 1897, PH.
89. MAW to THW, 12 Mar. 1897, PH.
90. MAW to WTA, 24 Mar. 1897, PH.
91. MAW to THW, 26 Mar. 1897, PH.
92. MAW, *Helbeck of Bannisdale*, p. 4.
93. MAW to TA, 9 June 1897, PH.
94. MAW to WTA, 19 June 1897, PH.
95. MAW to TA, 20 Apr. 1898.
96. GS to MAW, 11 Jan. 1898, Honn.
97. W. Addis to MAW, 8 Apr. 1898, Honn.
98. MAW to TA, 20 Apr. 1898, PH.
99. MAW to THW, 3 May 1898, PH.
100. MAW to TA, 27 May 1898, PH.
101. RS to MAW, 4 May 1898, Honn.
102. DW, diary UCL, 10 June 1898.
103. John Wordsworth to MAW, 10 June 1898, PH.
104. GS to MAW, 8 Aug. 1898, Honn.

13. Families—the Arnolds: 1890–1900

1. TA to JA, 27 Mar. 1877, quoted in *LTA* 189.

2. TA to J. H. Newman, 13 Jan. 1879, quoted in *LTA* 197–8.
3. MAW to WTA, 28 Jan. 1889, PH.
4. MAW to TA, 18 Dec. 1893, PH.
5. MAW to TA, 10 Jan. 1894, PH.
6. TA to General Collinson, 15 Jan. 1899, quoted in *LTA* 242.
7. MAW to WTA, 17 May 1896, PH.
8. MAW to TA, 25 May 1899, PH.
9. MAW to TA, 'June 1890', PH.
10. MAW to WTA, 22 Feb. 1892, PH.
11. MAW to WTA, 11 Oct. 1891, PH.
12. *SRI*, p. xlii.
13. MAW to WTA, 12 Feb. 1897, PH.
14. MAW to THW, 12 Mar. 1897, PH.
15. MAW to WTA, 4 Apr. 1897, PH.
16. Ibid.
17. MAW to WTA, 9 Oct. 1903, PH.
18. MAW to WTA, 19 June 1897, PH.
19. For Carroll's interest in the Arnold girls, see *The Letters of Lewis Carroll*, ed. Morton Cohen, 2 vols. (New York, 1979).
20. See *The Letters of Lewis Carroll*, i. 209.
21. The fact of the adoption is listed in Carus Selwyn's *Who's Who* entry.
22. MAW to TA, 19 Dec. 1890, PH.
23. MAW to GS, 5 Oct. 1894, Honn.
24. MAW to TA 2 Oct. 1894, PH.
25. MAW to THW, 3 Oct. 1894, PH.
26. Ibid.
27. Ibid.
28. Ronald W. Clark, *The Huxleys* (New York, 1968), 101.
29. Clark, *The Huxleys*, p. 133.
30. JA to TA, 22 Mar. 1884, in *LTA* 211.
31. MAW to JA, 19 May 1883, PH. Ethel had the role of Portia in the the Philothespian production of *The Merchant of Venice* in November 1883. Dodgson thought her performance 'a great deal better than I had expected'; see Cohen, *The Letters of Lewis Carroll*, i. 516.
32. MAW to JA, 18 Apr. 1886, PH.
33. MAW to THW, 9 Aug. 1888, PH.
34. MAW to TA, 23 Oct. 1889, PH.
35. MAW to TA, 22 June 1890, PH.
36. MAW to THW, 30 Sept. 1896, PH.
37. MAW to THW, 22 Apr. 1899, PH.
38. MAW to THW, 20 June 1899, PH.
39. Ibid.
40. Ibid.
41. MAW to THW, 16 July 1889, PH.
42. MAW to TA, 23 Oct. 1889, PH.
43. MAW to TA, 29 June 1890, PH.
44. MAW to TA, 5(?) Nov. 1891, PH.
45. *WR* ii. 219.
46. MAW to THW, 13 Aug. 1896, PH.
47. MAW to TA, 18 Dec. 1898, PH.
48. MAW to TA, 12 Feb. 1890, PH.
49. MAW to TA, 26 Dec. 1888, PH.

50. MAW to TA, 7 Apr. 1898, PH.

14. Families—the Wards: 1890–1900

1. THW to MAW, 5 Mar. 1895, PH.
2. MAW to THW, 9 Feb. 1895, PH.
3. Richard Ellmann, *Oscar Wilde*, (New York, 1984, repr. 1987), 133.
4. MAW to WTA, 27 Oct. 1895, PH.
5. Dorothy Ward's diaries are held at the library of University College London.
6. DW, diary UCL, 25 Nov. 1890.
7. DW, diary UCL, 9 June 1890.
8. MAW to DW, 11 Aug. 1888, PH.
9. Gertrude Ward's diary, 20 Mar. 1890, quoted in EHJ 98.
10. MAW to LC, 22 Sept. 1895, PH.
11. MAW to ASW, 28 Nov. 1892, PH.
12. Bessie Churcher to MAW, 24 Aug. 1895, PH.
13. DW, diary UCL, 25 Mar. 1898.
14. MAW to DW, 21 Sept. 1901, PH.
15. MAW to LC, '1905', PH.
16. MAW to JA, 27 Mar. 1886, PH.
17. MAW to JA, 3 Apr. 1886, PH.
18. MAW to JA, 14 June 1886, PH.
19. MAW to JA, 4 Aug. 1886, PH.
20. ASW, diary UCL, 1887.
21. Mary Ward informed her husband of her intention to do this in a letter from Borough Farm, 28 Apr. 1888, PH.
22. MAW to LC, 24 Jan. 1889, PH.
23. MAW to TA, 2 June 1889, PH.
24. Ibid.
25. MAW to THW, 16 July 1889, PH.
26. MAW to TA, 12 Apr. 1890, PH.
27. MAW to ASW, 15 Feb. 1892, PH.
28. MAW to TA, 14 July 1895, PH.
29. MAW to ASW, 'August 1891(?)', PH.
30. MAW to ASW, 25 Nov. 1891, PH.
31. MAW to ASW, 24 June 1894, PH.
32. Ibid.
33. MAW to ASW, 10 May 1893, PH.
34. MAW to ASW, 25 July 1894, PH.
35. ASW to MAW, 5 Oct. 1895, PH.
36. ASW to MAW, 7 May 1899, UCL.
37. ASW to MAW, 6 Mar. 1896, UCL.
38. ASW to MAW, 'November 1897', UCL.
39. ASW to MAW, 6 Mar. 1898, UCL.
40. John Jolliffe, *Raymond Asquith* (London, 1980), 36–7.
41. THW in a postscript, MAW to TA, 18 Dec. 1898, PH.
42. MAW to TA, 1 Aug. 1899, PH.

15. Homes: 1888–1900

1. THW, diary UCL, 18 Aug. 1888.
2. JPT 78.
3. MAW to THW, 30 Aug. 1888, PH.

4. MAW to TA, 24 Apr. 1890, PH.
5. MAW to TA, 29 June 1889, PH.
6. MAW to TA, 30 Nov. 1888, PH.
7. DW, diary UCL, 18 July 1890.
8. MAW to TA, 30 Mar. 1890, PH.
9. JPT, 93–4.
10. ASW to DW, 6 Jan. 1892 (misdated 1891), UCL. The former tenant of Stocks, Mrs Bright, died in the influenza epidemic of winter 1891.
11. JPT, pp.78–9
12. MAW to THW, 23 July 1889, PH.
13. MAW to Frances Arnold, 28 June 1889, PH.
14. JPT 72.
15. Ibid.
16. MAW to TA, 29 June 1890, PH.
17. MAW to TA, 14 July 1892, PH.
18. MAW to TA, 9 Nov. 1890, PH.
19. THW to MAW, 22 Mar. 1895, PH.
20. MAW to THW, 13 Mar. 1895, PH.
21. MAW to THW, 19 July 1897, PH.
22. MAW to THW, 20 Apr. 1899, PH.

16. Respectable Genius: 1890–1900

1. MAW to TA, 19 Feb. 1894, PH.
2. MAW to TA, 15 June 1891, PH.
3. See Katherine L. Mix, *A Study in Yellow* (New York, 1960), 146.
4. MAW to ASW, 21 June 1889, PH.
5. MAW to WTA, 11 Sept. 1890, PH.
6. MAW to TA, 26 Feb. 1898, PH.
7. MAW to LC, 22 Sept. 1895, PH.
8. MAW to ASW, 18 May 1891, PH.
9. EHJ 92. Jones quotes from *Fenwick's Career* and notes THW's influence on MAW's ideas about art.
10. DW, diary UCL, 11 Jan. 1911.
11. MAW to ASW, '1891(?)', PH.
12. JPT 153.
13. MAW to Stopford Brooke, 25 Aug. 1895, PH.
14. MAW to DW, 15 Mar. 1897, PH.
15. Mary Cropper to MAW, 24 Jan. 1920, PH, quoted in *VH*, pp. 189–90.
16. MAW to Gladstone, 16 Sept. 1895, PH.
17. MAW to LC, 13 Apr. 1897, PH.
18. MAW to W. Addis, 4 Feb. 1898, PH.
19. MAW to Mandell Creighton, 9 Aug. 1898, PH. The letter is reproduced and discussed in *VH* 189.
20. *The Times*, 5 Sept. 1899.
21. MAW to DW, 20 May 1903, PH.
22. MAW to Mandell Creighton, 9 Aug. 1898, PH.
23. See *VH* 189–90 for a discussion of this theme in Mrs Ward's fiction.
24. DW, diary UCL, 3 Mar. 1907.
25. Mary Cropper to MAW, 24 Jan. 1920, PH.
26. MAW to TA, 7 July 1894, PH.
27. MAW to WTA, 29 Apr. 1899, PH.

28. MAW to TA, 14 July 1892, PH.
29. MAW to TA, 20 Sept. 1890, PH.
30. MAW to TA, 5 May 1895, PH.
31. MAW to A. J. Balfour, 10 Feb. 1897, PH.
32. ASW to MAW, 19 Feb. 1899, UCL.
33. MAW to THW, 5 Sept. 1898, PH. 'Recessional' was the poem Kipling wrote in celebration of Queen Victoria's Diamond Jubilee, published in *The Times*, 13 July 1897.
34. MAW to TA, 14 Sept. 1899, PH.
35. JPT 175.
36. MAW to WTA, 8 Mar. 1900, PH.
37. JPT 176.
38. MAW to TA, 21 Jan. 1889, PH.
39. MAW to TA, 7 Feb. 1897, PH.
40. JPT 229.
41. See Chap. 8, n. 40.
42. Michael Holroyd, *Lytton Strachey*, 2 vols. (London, 1968), ii. 330, i. 443.
43. Elaine Showalter, *A Literature of their Own* (Princeton, NJ, 1977), 110.
44. Victoria Glendinning, *Rebecca West* (New York, 1987), 48.
45. Arnold Bennett, *Books and Persons* (London, 1917), 47.
46. Rebecca West, reviewing *A Writer's Recollections*, *The Bookman*, Dec. 1918, pp. 106–7.
47. EHJ 163.

17. Health: 1890–1900

1. TA to MPA, 19 Dec. 1852, in *LTA* 31.
2. MAW to THW, 2 Feb. 1895, PH.
3. DW, diary UCL, 24 Sept. 1898.
4. MAW to TA, 13 Jan. 1874, PH.
5. DW, diary UCL, 1890 *passim*.
6. MAW to JA, 16 Apr. 1883, PH.
7. DW, diary UCL, 2 Feb. 1903.
8. MAW to GS, 29 Dec. 1895, Honn.
9. MAW to C. S. Olcott, 27 July 1909, Hou bMS Am. 1925 (1860).
10. DW, diary UCL, 14 Mar. 1919.
11. Mary Cropper to MAW, 24 Jan. 1920, PH.
12. DW, diary UCL, 16 Mar. 1917.
13. MAW to GS, 29 Dec. 1895, Honn.
14. MAW to TA, 29 Nov. 1895, PH.
15. JPT 118.
16. MAW to WTA, 12 Aug. 1902, PH.
17. MAW to TA, 23 Dec. 1892, PH.
18. MAW to TA, 19 May 1896, PH.
19. MAW to Sarah Orne Jewett, 23 Mar. 1895, Hou bMS Am. 1743.1 (106).
20. DW, diary UCL, 29 Aug. 1918.
21. DW, diary UCL, 21 Aug. 1919.
22. DW, diary UCL, 24 Mar. 1903.
23. JPT 119.
24. MAW to GS, 19 Oct. 1892, Honn.
25. JPT 105.
26. MAW to WTA, 9 Oct. 1903, PH.

27. MAW to WTA, 'February(?) 1897', PH.
28. DW, diary UCL, 7 Mar. 1903.
29. DW, diary UCL, 27 Apr. 1905.
30. DW, diary UCL, 29 July 1898.

18. The Passmore Edwards Settlement: 1892–1900

1. My account of the growth of the settlement movement draws on J. A. R. Pimlott, *Toynbee Hall* (London, 1935).
2. Quoted in Pimlott *Toynbee Hall*, p. 42.
3. JPT 85.
4. MAW to TA, 29 May 1888, PH.
5. JPT 79.
6. JPT 80.
7. MAW to TA, 12 Feb. 1890, PH.
8. JPT 82.
9. JPT 85.
10. DW, diary UCL, 29 Nov. 1890.
11. MAW to TA, 30 Nov. 1890, PH.
12. JPT 88.
13. Quoted from an 'In Memoriam' pamphlet, published by the Passmore Edwards Settlement in June 1921.
14. MAW to ASW, 5 Nov. 1891, PH.
15. MAW to ASW, 31 May 1892, PH.
16. MAW to TA, 9 Nov. 1890, PH.
17. MAW to THW, 21 Aug. 1894, PH.
18. MAW to THW, 9 Mar. 1895, PH.
19. *Manchester Guardian*, 11 Oct. 1897.
20. MAW to WTA, 11 Oct. 1897, PH.
21. DW, diary UCL, 11 Feb. 1898.
22. JPT 123.
23. MAW to WTA, 8 Oct. 1898, PH.
24. Janet Trevelyan gives a book-length account of the growth of the play centre movement in her *Evening Play Centres for Children* (London, 1920).
25. JPT 135.
26. Estlin Carpenter to MAW, 29 Mar. 1895, PH.

19. *Eleanor*: 1900

1. MAW to WTA, 30 Oct. 1896, PH.
2. MAW to TA, 28 Feb. 1899, PH.
3. The poem was published in *Cornhill Magazine*, March 1900.
4. MAW to WTA, 29 Apr. 1898, PH.
5. MAW to Mandell Creighton, 27 July 1900, PH.
6. *WR* ii. 230.
7. JPT 157.
8. MAW to ASW, 27 Mar. 1899, PH.
9. MAW to THW, 20 June 1899, PH.
10. *WR* ii. 219.
11. *WR* ii. 218.
12. *WR* ii. 221.
13. GS to MAW, 26 July 1900, Honn.
14. MAW to WTA, 9(?) Nov. 1900, PH.

15. MAW to WTA, 10 Nov. 1900, PH.
16. Ibid.
17. MAW to Mandell Creighton, 15 Nov. 1900, PH.
18. Ibid.
19. MAW to LC, 18 Jan. 1901, PH.

20. Bestselling Novelist, Failed Dramatist: 1901–1905

1. RS to MAW, 26 Jan. 1903, Honn.
2. DW, diary UCL, 5 March 1903.
3. Smith, Elder and Co. to MAW, 26 Oct. 1904, Honn.
4. MAW, *Lady Rose's Daughter*, p. 3.
5. MAW, *Lady Rose's Daughter*, p. x.
6. F. Rives, *Mrs Humphry Ward, Romancière*, 2 vols. (Lille, 1981), ii. 868.
7. MAW to ASW, 9 Aug. 1903, PH.
8. DW, diary UCL, 9 Nov. 1905.
9. DW, diary UCL, 8 Sept. 1903.
10. MAW to Henry James, 15 Aug. 1904, Honn. MAW recalled Willie's last words to be: 'God only knows what I have suffered .... I love God', *SRI*, pp. cxxi–cxii.
11. Ibid.
12. Honn notebook 14.
13. *The Marriage of William Ashe*, p. 6.
14. RS to MAW, 3 Apr. 1905, Honn.
15. JPT 177.
16. MAW to THW, 25 Apr. 1901, PH.
17. MAW to Corinne (Roosevelt) Robinson, 2 Oct. 1903, Hou bMS Am. 1785 (1446).
18. MAW to DW, 14 Mar. 1902, PH.
19. MAW to DW, 25 Aug. 1903, PH.
20. MAW to Henry James, 15 Aug. 1904, Honn.
21. DW, diary UCL, 7 Mar. 1905.
22. MAW to LC, 29 Sept. 1907, PH.

21. Family Matters: 1900–1905

1. DW, diary UCL, 1 Feb. 1905.
2. DW, diary UCL, 2 Mar. 1905.
3. MAW to DW, 7 Jan. 1892, PH.
4. DW, diary UCL, 20 Apr. 1903.
5. MAW to DW, 20 May 1903, PH
6. Ibid.
7. Michael Holroyd, *Lytton Strachey*, 2 vols. (London, 1968), i. 189–90.
8. DW, diary UCL, 19 Feb. 1905.
9. JPT to MAW, 28 Mar. 1908, PH.
10. ASW to THW, 30 Dec. 1899, UCL.
11. ASW to MAW and THW, 3 Feb. 1900, UCL.
12. ASW to MAW and THW, 18 Feb. 1900, UCL.
13. Ibid.
14. ASW to MAW and THW, 9 Jan. 1901, UCL.
15. ASW to MAW, 'February 1902', UCL.
16. ASW to MAW, 'Summer 1903(?)', UCL.
17. MAW to ASW, 9 Aug. 1903, PH.
18. Ibid.
19. ASW to MAW, 18 Oct. 1903, UCL.

20. Ibid.
21. ASW to MAW, 'September 1904', UCL.
22. ASW to MAW, 'Easter 1904', UCL.
23. DW, diary UCL, 6 Aug. 1905.
24. Ibid.
25. MAW to ASW, 5 Aug. 1905, PH.
26. MAW to LC, 'August(?) 1905', PH.
27. DW, diary UCL, 20 Aug. 1905.

22. Mid-Edwardian: 1906

1. W. L. Phelps, 'Mrs Humphry Ward', *Essays on Modern Novelists* (New York, 1910, repr. 1921), 91; quoted by Esther M. G. Smith in *Mrs Humphry Ward* (Boston, 1980), 143. Smith gives a useful account of Mrs Ward's reputation in the twentieth century.
2. RS to MAW, 25 Nov. 1905, Honn.
3. Moberly Bell to RS, 16 Oct. 1905, Honn.
4. Moberly Bell to RS, 20 Oct. 1905, Honn.
5. RS to MAW, 23 Oct. 1905, Honn.
6. RS to MAW, 10 Oct. 1905, Honn.
7. Moberly Bell to RS, 16 Oct. 1905, Honn.
8. JPT 195–6.
9. MAW to THW, 'April 1904', PH.
10. Harper's to RS, 19 Jan. 1906, Honn.
11. MAW to Canon Wood, 22 Jan. 1906, PH.
12. MAW to LC, 14 Dec. 1905, PH.
13. MAW to DW, 15 Jan. 1906, PH.
14. MAW to DW, 11 Jan. 1906, PH.
15. MAW to DW, 29 Dec. 1905, PH.
16. MAW to ASW, 14 Jan. 1906, PH.
17. Gertrude Ward to MAW, 20 Jan. 1906, UCL.
18. DW, diary UCL, 8 Jan. 1906.
19. MAW to ASW, 19 Jan. 1906, PH.
20. MAW to DW, 17 Jan. 1906, PH.
21. ASW to MAW, 27 Mar. 1906, UCL.
22. R. Shannon, *The Crisis of Imperialism, 1865–1915* (1976, repr. 1986), 378.
23. DW, diary UCL, 10 Feb. 1906.
24. DW, diary UCL, 10 Apr. 1906.
25. DW, diary UCL, 5 May 1906.
26. *New York Times*, 5 May 1906.
27. MAW to THW, 30 May 1906, PH.
28. MAW to ASW, 'May 1906', PH.
29. DW, diary UCL, 11 June 1906.
30. DW, diary UCL, 27 June 1906.
31. ASW to MAW, 12 Apr. 1906, UCL.
32. DW, diary UCL, 15 July 1906.
33. DW, diary UCL, 23 Aug. 1906.
34. DW, diary UCL, 24 Aug. 1906.
35. DW, diary UCL, 30 Aug. 1906.
36. DW, diary UCL, 7 Sept. 1906.
37. Honn notebook 16.
38. DW, diary UCL, 19 Oct. 1906.

39. MAW to ASW, 16 Oct. 1906, PH.
40. DW, diary UCL, 21 Oct. 1906.
41. DW, diary UCL, 29 Nov. 1906.
42. DW, diary UCL, 25 Dec. 1906.
43. DW, diary UCL, 28 Dec. 1906.

23. *The Testing of Diana Mallory*: 1907

1. DW, diary UCL, 2 Jan. 1907.
2. DW, diary UCL, 25 Jan. 1907.
3. ASW to MAW, 20 Mar. 1907, UCL.
4. MAW, *Diana Mallory*, p. ix.
5. MAW to DW, 4 Mar. 1907, PH.
6. DW, diary UCL, 23 Mar. 1907.
7. Moberly Bell to MAW, 23 May 1907, Honn.
8. RS to MAW, 29 Apr. 1907, Honn.
9. RS to MAW, 1 May 1907.
10. DW, diary UCL, 2 May 1907.
11. DW, diary UCL, 27 May 1907.
12. DW, diary UCL, 28 June 1907.
13. DW, diary UCL, 25 July 1907.
14. DW, diary UCL, 11 Aug. 1907.
15. DW, diary UCL, 14 Aug. 1907. DW records her mother taking large doses of trional at Robin Ghyll during this period.
16. DW, diary UCL, 7 Sept. 1907.
17. MAW to ASW, 'May 1907', PH.
18. ASW to MAW, 'September 1907', PH.
19. DW, diary UCL, 21 Oct. 1907.
20. MAW to ASW, 23 Oct. 1907, PH. Blagdon was the seat of Viscount Ridley in Northumberland. The surviving correspondence has been weeded out so as to conceal the name of Arnold's intended wife. I identify her as Violet Asquith on the prominence she has in his letters, and from references in DW's diaries.
21. DW, diary UCL, 5 Nov. 1907.
22. DW, diary UCL, 17 Dec. 1907.
23. MAW to RS, 20 Dec. 1907, Honn.
24. DW, diary UCL, 31 Dec. 1907.

24. The New World: 1908

1. Curtis Brown to MAW, 26 June 1908, Honn.
2. Edward Bok to MAW, 20 Oct. 1909, Honn.
3. JPT 208.
4. Honn notebook 18.
5. Ibid.
6. Ibid.
7. Ibid.
8. JPT 209.
9. MAW to THW, 4 Apr. 1908, PH.
10. Ibid.
11. MAW to ASW, 13 Apr. 1908, PH.
12. Ibid.
13. MAW to THW, 24 Apr. 1908, PH.
14. MAW to THW, 8 Apr. 1908, PH.

15. MAW to THW, 24 Apr. 1908, PH.
16. MAW to RS, 12 May 1908, Honn.
17. JPT 214–15.
18. MAW to THW, 5 May 1908, PH.
19. MAW to THW, 11 May 1908, PH.
20. MAW to THW, 20 May 1908, PH.
21. JPT 217.
22. MAW to THW, 25 May 1908, PH.
23. MAW to THW, 29 May 1908, PH.
24. MAW to THW, 4 June 1908, PH.
25. Ibid.
26. JPT 179.
27. MAW to THW, 16 May. 1908, PH.
28. RS to MAW, 28 Feb. 1908, Honn.
29. MAW to THW, 11 May 1908, PH.
30. ASW to MAW, 10 Dec. 1908, UCL.
31. MAW to RS, 12 May 1908, Honn.
32. Fleming Sandwith to MAW, 6 July 1908, Honn.
33. Ibid.
34. MAW to LC, 23 Aug. 1908, PH.
35. MAW to 'Mr Alden', 19 Feb. 1908, Honn.
36. RS to MAW, 11 Nov. 1908, Honn.
37. Julian Huxley, *Memories*, (New York, 1970), 70.
38. Leonard Huxley to MAW, 26 Oct. 1908, Honn.
39. MAW to DW, 'November 1908', PH.
40. MAW to DW, 10 Nov. 1908, PH.
41. Julian Huxley, *Memories*, p. 20.
42. MAW to DW, 29 Nov. 1908, PH.
43. ASW to MAW, 23 Apr. 1908, UCL.
44. Ibid.
45. THW to MAW, 23 May 1908, UCL.
46. Ibid.
47. Ibid.
48. MAW to THW, 3 May 1908, PH.
49. MAW to ASW, 14 Aug. 1908, PH.
50. ASW to MAW, 'September 1908', UCI..
51. JPT 230.
52. MAW to Frances Arnold, quoted in JPT 223.

25. Anti-Suffragist: 1909

1. MAW to LC, 31 Jan. 1909, PH.
2. MAW to DW, 31 Mar. 1909, PH.
3. *The Times*, 27 Feb. 1909.
4. DW, diary UCL, 26 Mar. 1909.
5. Ibid.
6. DW, diary UCL, 20 May 1909.
7. Honn notebook 18.
8. Ibid.
9. Ibid.
10. DW, diary UCL, 26 May 1909.
11. ASW to THW, 'June 1909', UCL.

12. MAW to DW, 21 Sept. 1909, PH.
13. MAW to RS, 25 June 1909, PH.
14. MAW, *Canadian Born* (New York, 1910), 223.
15. MAW, *Canadian Born*, p. 329.
16. ASW to MAW, 19 Apr. 1910, UCL.
17. Doubleday to MAW, 9 Nov. 1909, Honn.
18. DW, diary UCL, 14 Aug. 1909.
19. DW, diary UCL, 21 Oct. 1909.
20. MAW to ASW, 7 Nov. 1909, PH.
21. Ibid.
22. DW, diary UCL, 27 Oct. 1909.
23. *The Times*, 15 Nov. 1909.
24. DW, diary UCL, 6 Dec. 1909.

26. Arnold Ward, MP: 1910–1911

1. DW, diary UCL, 4 Jan. 1910.
2. Quoted in *The Times*, 13 Jan. 1910.
3. DW, diary UCL, 12 Mar. 1910.
4. DW, diary UCL, 25 Mar. 1910.
5. DW, diary UCL, 30 May 1910.
6. *The Times*, 4 June 1910.
7. Ibid.
8. DW, diary UCL, 10 July 1910.
9. *The Times*, 11 July 1910.
10. *The Times*, 13 July 1910.
11. ASW to DW, 8 Apr. 1912, PH.
12. DW, diary UCL, 12 July 1910.
13. DW, diary UCL, 27 Oct. 1910.
14. MAW to ASW, 25 Sept. 1910, PH.
15. DW, diary UCL, 18 Nov. 1910.
16. DW, diary UCL, 8 Dec. 1910.
17. MAW to ASW, 22 Dec. 1910, PH.
18. DW, diary UCL, 3 Feb. 1911.
19. DW, diary UCL, 11 Mar. 1911.
20. DW, diary UCL, 19 Apr. 1911.
21. DW, diary UCL, 26 Apr. 1911.
22. DW, diary UCL, 4–10 May 1911.
23. Reginald Smith's description. The French means 'speech for the defence'.
24. For the earlier and very different 'scenario' of the novel, see *VH* 192–3.
25. MAW, *The Case of Richard Meynell* (London, 1911), 352.
26. F. N. Doubleday to MAW, 1 Dec. 1911, Honn. Harper's had originally contracted to publish *The Case of Richard Meynell* and advanced £3,000 in Jan. 1906. Mrs Ward evidently did not return the advance, leaving it to be paid off by receipts from her novels which Harper's still had in print.
27. DW, diary UCL, 24 Nov. 1911.
28. DW, diary UCL, 5 Dec. 1911.
29. ASW to DW, 26 Oct. 1911, UCL.
30. MAW to ASW, 24 Dec. 1911, PH.

27. Calamities, 1912–1914

1. DW, diary UCL, 16 Feb. 1912.

2. In her notebook (Honn 20) Ward recorded the inspiration for the novel as being the account of a rich old recluse called Dering, which she read in *The Times*, 1911. She combined this with a remembered episode from the 1840s when the mad Duke of Brunswick made a young politician, J. S. Duncombe, heir to all his wealth.
3. Ibid.
4. RS to MAW, 13 Jan. 1912, Honn.
5. DW, diary UCL, 17 Mar. 1912.
6. DW, diary UCL, 28 Mar. 1912.
7. MAW to ASW, 9 Apr. 1912, PH.
8. MAW to C. S. Olcott [Houghton Mifflin], 23 Mar. 1912, Hou bMS Am. 1925 (1860).
9. MAW to ASW, 'April 1912', PH.
10. MAW to ASW, 30 Apr. 1912, PH.
11. RS to MAW, 22 June 1912, PH.
12. DW, diary UCL, 17 July 1912.
13. RS to MAW, 13 June 1912, Honn.
14. DW, diary UCL, 19.10,1912.
15. MAW to DW, 22 Oct. 1912, PH.
16. DW, diary UCL, 28 Oct. 1912.
17. Honn notebook 24.
18. MAW to ASW, 4 Feb. 1912, PH.
19. MAW to DW, 10 Sept. 1912, PH.
20. DW, diary UCL, 2 Nov. 1912.
21. MAW to LC, 13 Nov. 1912, PH.
22. MAW to LC, 29 July 1913, PH.
23. DW, diary UCL, 27 Jan. 1913.
24. MAW to RS, 13 Oct. 1912, Honn.
25. DW, diary UCL, 5 Apr. 1913.
26. Mary Ward, for instance, totalled her lifetime total earnings from her fiction up to *The Case of Richard Meynell* as £105,250 and her prospective earnings from books over the next two years as £9,350. These notes are found on the backs of letters and the flyleaves of Honn notebook 16.
27. DW, diary UCL, 7 Apr. 1913.
28. DW, diary UCL, 9 Apr. 1913.
29. DW, diary UCL, 10 Apr. 1913.
30. DW, diary UCL, 13 Apr. 1913. DW's code means: 'It has all happened so quickly, I can hardly believe it. It has been bought in *her* name . . . . It is a great relief—none the less it almost makes me choke.'
31. Figures taken from notes, randomly jotted down in notebooks and on the backs of letters in the Honnold Library.
32. DW, diary UCL, 16 Apr. 1913.
33. My estimates from random notes, in Honn notebook 16.
34. DW, diary UCL, 17 Apr. 1913. DW's code means: 'today the first part of the debts has been paid, in which there is a certain sad relief. I have not seen Arnold myself.'
35. DW, diary UCL, 24 Apr. 1913. DW's code means: '[she] was overcome with fatigue yesterday evening and with other things as well, but my brother, who came to take supper with them, was at his best with her, good and lovable: then, afterwards, when he had received the second large ch[eque], he who never shows emotion, was worried about the condition of my father and poured out his feeling to my mother, and that made her happy although it was a sad happiness.'
36. *The Diaries of Virginia Woolf, Vol I, 1915–19*, ed. Anne Olivier Bell (New York,

1977), 61–2.
37. See, e.g. the malicious anecdote in *The Letters of Aldous Huxley*, ed. Grover Smith (London, 1969), 233 and Julian Huxley, *Memories*, p. 37.
38. Originally called 'The Conisborough Family' and 'The Coniston Family'. In her notes, Mrs Ward indicates the story was based on that of a well-known family whom she denotes only as 'C'.
39. Honn notebook 24.
40. DW, diary UCL, 1 June 1913.
41. MAW, *Lady Coryston's Family*, *Harper's Magazine*, Nov. 1913, p. 932.
42. DW, diary UCL, 5 May 1913.
43. DW, diary UCL, 15 May 1913.
44. DW, diary UCL, 8 May 1913. DW's code means 'a moving little conversation.'
45. MAW to DW, 6 May 1913, PH.
46. DW, diary UCL, 13 June 1913.
47. DW, diary UCL, 14 June 1913.
48. DW, diary UCL, 15 July 1913. DW's code means: 'No news about my brother, he did not go to Stocks after dinner … nor did he return here [GP] to sleep. Great anxiety, all day and evening. Mother so sad and pale.'
49. DW, diary UCL, 18 July 1913.
50. Honn notebook 21.
51. MAW to DW, 31 July 1913, PH.
52. Ibid.
53. MAW to DW, 7 Aug. 1913, PH.
54. MAW to DW, 'August 1913', PH.
55. DW, diary UCL, 15 Aug. 1913.
56. Honn notebook 21.
57. DW, diary UCL, 21 Sept. 1913.
58. DW, diary UCL, 11 Oct. 1913.
59. MAW to DW, 22 Jan. 1914, PH.
60. DW, diary UCL, 21 Oct. 1913.
61. DW, diary UCL, 28 Oct. 1913. DW's French means 'a very sad and difficult evening'.
62. DW, diary UCL, 19 Nov. 1913. DW's French means: 'He was upset because my brother stayed at the Portland Club overnight, although he phoned early to tell us.'
63. ASW to MAW, '1913', UCL.
64. DW, diary UCL, 25 Dec. 1913.
65. Curtis Brown to MAW, 22 Oct. 1914, Honn.
66. RS to MAW, 12 May 1914, Honn.
67. MAW to RS, 14 July 1914, Honn.
68. DW, diary UCL, 3 May 1914. DW's German means 'that we must sell this house'.
69. DW, diary UCL, 4 May 1914.
70. RS to MAW, 12 May 1914, Honn.
71. MAW to RS, 26 July 1914, Honn.
72. DW, diary UCL, 3 Aug. 1914.

28. The Wards and War: 1914–1916

1. DW, diary UCL, 11 Aug. 1914.
2. DW, diary UCL, 12 Aug. 1914.
3. DW, diary UCL, 27 Aug. 1914.
4. Aldous Huxley, *Eyeless in Gaza* (London, 1936), 79.

5. ASW to MAW, 2 Aug. 1914, UCL.
6. ASW to MAW, 'August 1914', UCL.
7. ASW to MAW, 'August(?) 1914', UCL.
8. DW, diary UCL, 4 Sept. 1914.
9. ASW to MAW, 21 Nov. 1914, UCL.
10. ASW to MAW, 8 Jan. 1915, UCL.
11. Ibid.
12. ASW to MAW, 2 Oct. 1914, UCL.
13. DW, diary UCL, 8 Jan. 1915.
14. ASW to MAW, 12 Jan. 1915, UCL.
15. Fragmentary letter, 'January 1915?', Honn.
16. DW, diary UCL, 26 Nov. 1914.
17. DW, diary UCL, 9 Nov. 1914.
18. Honn notebook 29.
19. DW, diary UCL, 31 Dec. 1914.
20. Honn notebook 26.
21. Ibid.
22. Ibid.
23. DW, diary UCL, 24 May 1915.
24. DW, diary UCL, 31 May 1915.
25. DW, diary UCL, 11 June 1915.
26. ASW to MAW, 'November 1914', UCL.
27. Arnold's memorandum, 'On the Conduct and Management of "A" Squadron' is in the archive of the Hertfordshire Yeomanry Historical Trust. I am grateful to Lieutenant-Colonel J. D. Sainsbury for showing me a copy. Ward complains principally against Wyld, Clayton, and the squadron sergeant-major as a 'triumvirate of fanatics' who have overworked officers, NCOs, and the men of the Squadron to the point of breakdown. He also complains of unfair promotions and that the interests of local 'Watford' men have been slighted. He alludes to his own 'troubled' relations with Wyld. The memorandum was directed in the first instance to the CO of the Regiment. It was passed on to the Director-General of the Territorial Force, Lieut.-General Sir Edward Bethune.
28. ASW to DW, 21 Aug. 1915, UCL.
29. DW, diary UCL, 26 Oct. 1915.
30. MAW to Sir Ian Hamilton, 31 Oct. 1915, PH.
31. DW, diary UCL, 11 Dec. 1915.
32. DW, diary UCL, 15 Jan. 1916.
33. Appleton's to MAW, 22 Oct. 1914, Honn.
34. Honn notebook 32.
35. Ibid.
36. Ibid.
37. Esther M. G. Smith, *Mrs Humphry Ward* (Boston, 1980), 110.

29. Soldier in Skirts: 1916–1917

1. JPT 293.
2. Theodore Roosevelt to MAW, 27 Dec. 1915, PH.
3. See William Roscoe Thayer, *Theodore Roosevelt* (New York, 1919), p. 326.
4. JPT 270.
5. Mrs Ward's diary notes, Honn notebook 34.
6. Ibid.
7. Ibid.

8. Ibid.
9. Ibid.
10. JPT 273.
11. Mrs Ward's diary notes, Honn notebook 35.
12. MAW to THW, 29 Feb. 1916, quoted in JPT 276.
13. Honn notebook 35.
14. DW, diary UCL, 2 Mar. 1916.
15. JPT 279.
16. Honn notebook 35.
17. *WR* ii. 203–4.
18. DW, diary UCL, 18 May 1916.
19. MAW, *England's Effort* (New York, 1916), p. vi.
20. MAW, *England's Effort*, p. 175.
21. MAW to Theodore Roosevelt, 'December 1916', Honn.
22. DW, diary UCL, 11 May 1916. DW's French means: '[that] if things didn't go better when he came back to England—it might be better for him to resign his Parliamentary seat in advance [of an election, presumably]'.
23. DW, diary UCL, 22 June 1916.
24. DW, diary UCL, 28 June 1916. DW's French means: 'There is nothing [against you]'.
25. Financial notes, UCL.
26. DW, diary UCL, 29 June 1916. DW's French means: 'She has been so gentle and so good to [Jan and me]'.
27. DW, diary UCL, 13 July 1916. DW's (ungrammatical) French means: 'with regard to business matters'.
28. DW, diary UCL, 14 July 1916. DW's Italian means 'payments'.
29. MAW to DW, 12 Sept. 1916, PH.
30. DW, diary UCL, 3 Aug. 1916.
31. DW, diary UCL, 23 Sept. 1916.
32. MAW to DW, 6 Oct. 1916, PH.
33. DW, diary UCL, 25 Oct. 1916.
34. DW, diary UCL, 5 Nov. 1916.
35. DW, diary UCL, 29 Oct. 1916. DW's German means 'done right'.
36. DW, diary UCL, 3 Dec. 1916.
37. DW, diary UCL, 12 Dec. 1916.
38. DW, diary UCL, 28 Dec. 1916.
39. DW, diary UCL, 30 Dec. 1916.
40. JPT 283.
41. Mrs Ward made a record of the conversation, which took place at Carlton Gardens on 15 Feb. 1917. Her notes are held at the Honnold Library.
42. Honn notebook 38.
43. DW, diary UCL, 26 Feb. 1917.
44. DW, diary UCL, 14 Mar. 1917.
45. Honn notebook 39.
46. Ibid.
47. JPT 283.
48. DW, diary UCL, 14 Mar. 1917.
49. DW, diary UCL, 16 Mar. 1917.
50. MAW, *Towards the Goal* (New York, 1917), p. vii.
51. Honn notebook 37.
52. DW, diary UCL, 23 Mar. 1917.
53. MAW to LC, 12 Sept. 1917, PH.

54. DW, diary UCL, 11 Feb. 1917.
55. DW, diary UCL, 18 Apr. 1917.
56. DW, diary UCL, 12 Aug. 1917. DW's code means: 'He told me this morning that he has not been able to keep his promise—despite his resolve, that it would not be possible to keep it [in future] . . . . [I left them] alone. More conversations this evening but we could not change his mind—and he thinks that he has right [on his side]. I have never seen my mother so upset and in truth it is intolerable, this state of affairs.'
57. DW, diary UCL, 9 Nov. 1917. DW's code means: 'the intolerable thought that Arnold could not rejoin the army until his debts were paid. It was that which weighed most heavily on her heart. And the expression with which she looked at me! [I feel entirely with her] but at what price she must free my brother! At the price of her own health and her best energies.'
58. DW, diary UCL, 10 Jan. 1918.
59. JPT 294.
60. DW, diary UCL, 1 Apr. 1918. DW's French means 'lack of consideration for my mother'.
61. DW, diary UCL, 1 Nov. 1918.
62. DW, diary UCL, 25 Dec. 1918.
63. MAW, *Harvest* (New York, 1919), 355. Although this Dodd, Mead edition is dated 1919, it seems that it was not released until 1920.

30. The End: 1918–1920

1. DW, diary UCL, 12 July 1919.
2. Honn notebook 35.
3. MAW to JPT, 30 Aug. 1919, quoted JPT 304.
4. Honn notebook 39.
5. Ibid.
6. JPT 298.
7. MAW to Barrett Wendell, 13 Feb. 1919, Hou bMS Am. 1907.1 (1331).
8. Honn notebook 39.
9. Ibid.
10. DW, diary UCL, 5 Feb. 1919.
11. MAW to Barrett Wendell, 13 Feb. 1919, Hou bMS Am. 1907.1 (1331).
12. Honn notebook 40.
13. JPT 303.
14. JPT 306.
15. DW, diary UCL, 17 Mar. 1920. Some of the entries recording Mary Ward's last hours are apparently made by JPT.
16. DW, diary UCL, 20 Mar. 1920.
17. DW, diary UCL, 22 Mar. 1920.
18. JPT, 307; DW, diary UCL, 28 Mar. 1920.
19. JPT 307.
20. DW, diary UCL, 27 Mar. 1920.
21. *The Diary of Virginia Woolf* ed. Anne Olivier Bell assisted by Andrew McNeillie (London, 1978), ii. 29.
22. See *VH* 1.
23. DW, diary UCL, 23 Apr. 1920.
24. These notes are held at UCL Library.
25. THW, notes, UCL.
26. DW, notes, UCL.

27. DW, diary UCL,13 Feb. 1922. MAW's gravestone is emblematic of her life. It lacks Christian symbol or ornament. At its head is inscribed the motto of Oxford University, *Dominus illuminatio mea*. It records her relationship to Dr Arnold of Rugby, to Tom Arnold of Oxford, to THW. It gives her dates and records that she was born in Tasmania and died in London. The text concludes with the same verse from Clough's 'Come Poet, Come!' which is found at the end of *Robert Elsmere*—'Others, I doubt not, if not we . . .'
28. DW, diary UCL, 6 May 1926.
29. DW, notes, UCL.

Family Tree of Mary Ward

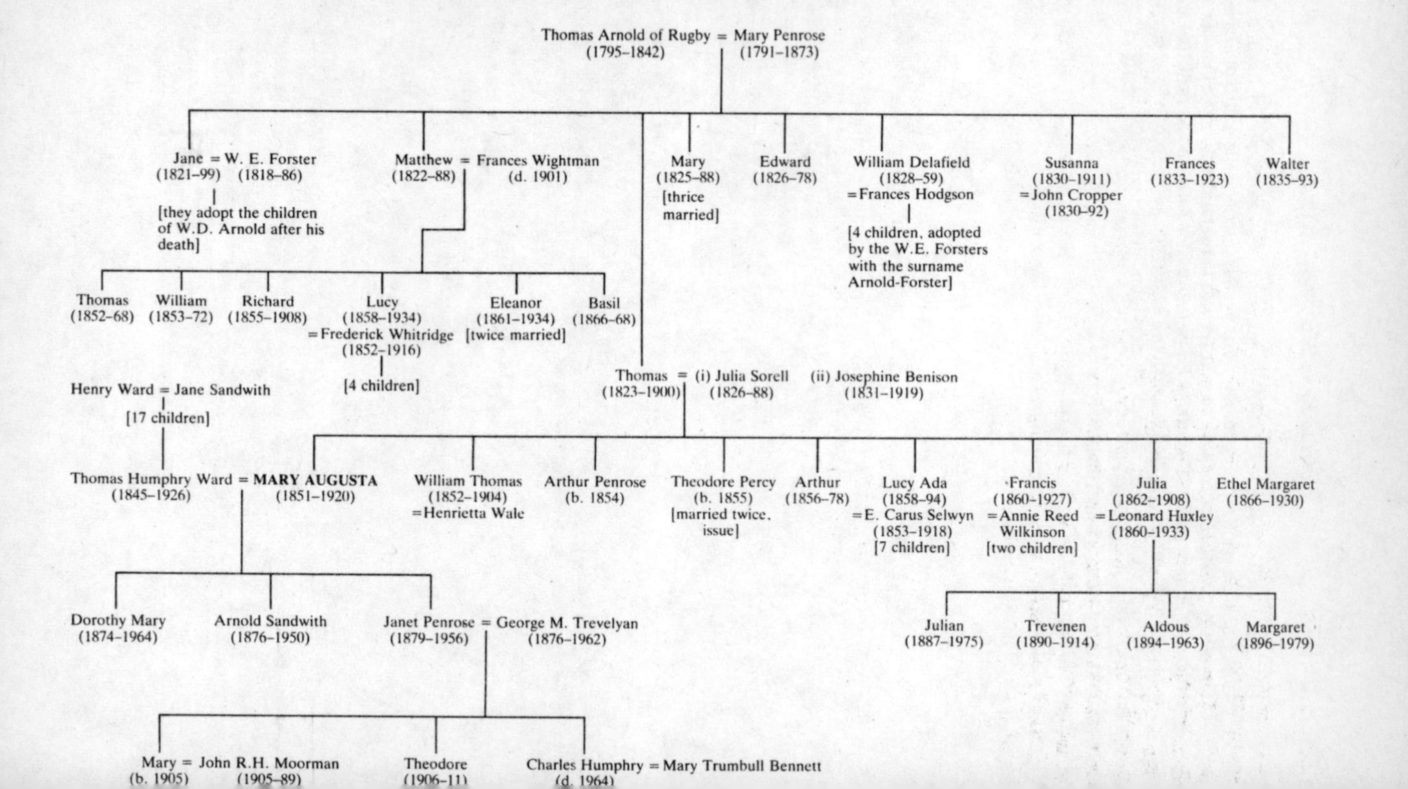

A Chronology of Mary Ward

1851 Mary Augusta Arnold born on 11 June in Hobart, Tasmania, the first child of Thomas Arnold and of Julia Arnold née Sorell.

1852 18 September: William (Thomas) Arnold born.

1854 May: Arthur (Penrose) Arnold born, dies within twenty-four hours. In autumn, Thomas Arnold announces that he will convert to Roman Catholicism. Julia persuades him to take no decision for six months.

1855 June: Julia Arnold threatens to leave her husband if he persists in his intention to convert. September: Theodore Percy Arnold born.

1856 January: Thomas Arnold received into the Catholic faith. July: Thomas Arnold resigns his position as Education Secretary at Hobart and the family sails for England. 17 October: the Arnolds arrive in London. They travel to Fox How at Ambleside where Mary and her pregnant mother remain with Thomas's mother while he goes to Dublin where J. H. Newman has secured him a tutorship (later termed a 'professorship') at the Catholic University. It pays £200 p.a. 15 December: Arthur Arnold is born at Fox How. Soon after Julia leaves for Dublin. Mary remains behind alone at Fox How, in the care of her grandmother and aunt.

1858 Autumn: Mary goes as a boarder to Anne Jemima Clough's school for girls at Eller How, Ambleside. Lucy (Ada) Arnold born in Dublin.

1860 Francis ('Frank' Sorell) Arnold born in Dublin.

1861 January: Mary goes as a boarder to Miss Davies's Rock Terrace School for Young Ladies at Shifnal, Shropshire.

1862 January: Thomas Arnold takes up the post of classics master at the Oratory School, Birmingham. His *Manual of English Literature* is published. The Arnolds (less Mary) join Thomas in March and move into a house in Edgbaston. Julia (Frances) Arnold is born.

1864 October: Mary completes her first story, 'A Tale of the Moors' (unpublished). December: Mary leaves Rock Terrace School.

1865 February (?): Mary goes to Miss May's boarding school at Clifton, near Bristol. June: Thomas Arnold having resigned his position at the Oratory school, converts back to Anglicanism. The family moves to Oxford, where he lectures and takes in private pupils. Mary is at Oxford over the Christmas vacation and hears her uncle Matthew Arnold give his Poetry Professor lectures.

1866 August: Mary Arnold completes a draft of 'Lansdale Manor' (unpublished), which she began over the summer vacation in Oxford. Ethel (Margaret), the last of the Arnold children, is born at Oxford.

1867 July: Mary Arnold leaves school, and comes home to Oxford where Thomas has had built a large house. December: Mary meets the Pattisons at Lincoln College, and by the summer of 1868 has come strongly under Mark Pattison's influence.

He arranges for her to work in the Bodleian Library where she devotes herself to the study of early Spain and its literature.

1869 July: Mary writes the story, 'A Gay Life' (unpublished). At this period she also writes a novel, 'Ailie', which is rejected by Smith, Elder and Co. in October and never published.

1870 Spring: Mary meets George Eliot at a dinner party with the Pattisons. She hears T. H. Green's lay sermon, 'The Witness of God,' which strongly affects her. Her first publication, *A Westmoreland Story*, appears in the *Churchman's Companion*. She resolves to give up fiction for 'history and criticism' leaving unfinished a novel, 'Vittoria'.

1871 January: Mary Arnold becomes intimately acquainted with T. Humphry Ward, a young Brasenose Fellow. 16 June: Humphry Ward proposes to Mary Arnold and is accepted. July: at Fox How, Mary and Humphry write 'A Morning in the Bodleian', which is printed privately at Windermere. October: Mary Arnold's article on El Cid is published in *Macmillan's Magazine*. William Arnold enters University College, Oxford, having been head boy at Rugby. Winter: E. A. Freeman invites her to write a book on Spain. She declines, on the grounds of her impending marriage.

1872 6 April: Mary Arnold marries Humphry Ward at Oxford. He is obliged to resign his fellowship and takes up the post of tutor at Brasenose. The Wards move to 5 Bradmore Road, Oxford. June: *Macmillan's Magazine* publishes Mary Ward's article on 'Alfonso the Wise'.

1873 Mary Ward is active with Louise Creighton and Mrs T. H. Green in in 'the Lectures for Women Committee' at Oxford. The first lectures take place in spring 1874.

1874 Mary Ward experiences the first twinges of the rheumatism which is to plague her through life. June: she agrees to write 'A Primer of English Poetry' for Macmillan. The contract is revoked on the basis of the sample she gives J. R. Green later in 1874. 22 July: Dorothy (Mary) Ward born. Mary Ward publishes her 'Plain Facts about Infant Feeding.' Christmas: the Wards visit Paris.

1876 February: Thomas Arnold indicates his intention of converting back to Catholicism. July: William Arnold takes a second-class degree. October: Thomas Arnold converts back to Catholicism, on the eve of his almost certainly getting the Rawlinsonian chair in Anglo-Saxon. He sells his house and leaves Oxford. Julia Arnold remains. 8 November: Arnold (Sandwith) Ward born.

1877 Mary Ward elected secretary of the newly-formed 'Association for the Education of Women'. June: William Arnold marries Henrietta Wale. October(?): Henry Wace invites Mary Ward to contribute to subsequent volumes of his *The Dictionary of Christian Biography*. November: Julia Arnold undergoes a surgical operation for breast cancer.

1878 Mary Ward is elected joint secretary of the Somerville Committee. July: her younger brother Theodore graduates from Oxford and subsequently emigrates to Tasmania to farm. September: Mary Ward's younger brother Arthur Arnold killed fighting in the Basuto War, aged 21.

1879 April: Humphry Ward contracts with Macmillan to produce an anthology, *The*

English Poets. The first edition comes out in 1880 and is a great success. Summer: the Wards vacation in the Lake District where Mary has first ideas for *Milly and Olly*. William Arnold takes up a post on the *Manchester Guardian*. October: Somerville Hall opens. 6 November: Janet (Penrose) Ward born.

1880 June: Mary Ward's personal maid, Elizabeth ('Lizzie' Harriet) Smith, enters her mistress's service in which she will remain forty years. Christmas: the Wards visit Mary's uncle W. E. Forster in Dublin, where he is Chief Secretary for Ireland.

1881 January; Humphry goes by himself to London, to try for a position as leader writer on *The Times*. February (?): Julia Arnold has a second operation for cancer. 6 March: Mary attends Revd John Wordsworth's Bampton Lecture in Oxford, and publishes a pamphlet critical of it, 'Unbelief and Sin.' July: Humphry is offered a permanent position on *The Times*. November: the Ward family move into 61 Russell Square, Bloomsbury. December: Mary Ward is commissioned by the *Manchester Guardian* to write a series of occasional papers called 'Foreign Table-Talk'. Over the next few years she writes extensively for *The Times*, the *Saturday Review*, the *Pall Mall Gazette*, *Macmillan's Magazine*, the *Oxford Chronicle*. Christmas: Macmillan publish Mary Ward's book for children, *Milly and Olly*.

1882 J. H. Newman secures a professorship for Thomas Arnold at the Royal University of Ireland which supports him until his death. July: Julia ('Judy') Arnold wins a first-class degree in English from Somerville. Autumn: Gertrude Ward joins the Russell Square household as Mary's companion. She remains eight years. November: Humphry appointed art critic on *The Times*. At this period, Mary has ominous attacks of cramp in her right hand, arm and side. She meets Henry James at the end of November.

1883 March: J. R. Green and Arnold Toynbee die in the same month. Mary Ward examines in modern languages for the Taylor Institute in Oxford. July: she proposes a book on the French Romantic Movement to Macmillan. They advance £250. The book is not written. The Wards rent Borough Farm, near Hindhead, for some weeks in the early summer, as they are to do on many vacations over the next six years. Late summer: the Wards holiday in Switzerland, where on Mark Pattison's recommendation Mary reads Amiel's *Journal Intime*. On her return in September she proposes a translation to Macmillan. November: Canon Samuel Barnett reads his paper 'Settlements of Working Men in London' at St John's College, Oxford.

1884 January: Mary Ward and Henry James attend a performance by Mary Anderson in *Comedy and Tragedy*. It inspires Ward to write *Miss Bretherton*. May: Mary Ward meets Laura Lyttelton and through her is introduced to the 'Souls'. August: Mark Pattison dies. November: Macmillan publishes *Miss Bretherton*. Lucy Arnold marries the Reverend Edward Carus Selwyn (appointed headmaster of Uppingham School, 1887).

1885 February; Mary Ward offers the unwritten *Robert Elsmere* to Macmillan's, who will not meet her demanded advance of £250. She promptly sells the project to George Smith, of Smith, Elder and Co. Spring: Toynbee Hall opened in Whitechapel, with Canon Barnett as warden. Julia Arnold marries Leonard

Huxley, who has taken up a post teaching classics at Charterhouse. December: Mary Ward's translation of Amiel's *Journal Intime* published by Macmillan.

1886 24 April: Laura Lyttleton (*née* Tennant) dies in childbirth. In the same month W. E. Forster dies. At this period Mary Ward moves across to Liberal Unionism in politics. August–October: Mary Ward suffers a physical breakdown under the strain of writing *Robert Elsmere* and goes to convalesce on the Isle of Wight and in the Lake District. December: Mary and Gertrude Ward work on *Robert Elsmere* at Borough Farm with the aim of finishing in January 1887.

1887 January: Frank Arnold qualifies as a doctor. 9 March: Mary Ward 'finishes' *Robert Elsmere*. May: the manuscript of *Robert Elsmere* is found to need over 1,300 printed pages. It is returned to Mary Ward for shortening. The Wards postpone a planned trip to the Mediterranean. July: Julia Arnold falls seriously ill. Mary Ward goes to Oxford where she nurses her mother over the summer and revises *Robert Elsmere*. 9 November: unable to complete her corrections to *Robert Elsmere* and in a state of physical breakdown Mary Ward goes to the Riviera to recover.

1888 27 January: the last corrected sheets of *Robert Elsmere* are sent to Smith. Mary Ward goes to Oxford, where her mother is dying. The novel is published on 24 February 1888. 7 April: Julia Arnold dies. *The Times* reviews *Robert Elsmere*. The following day, 8 April, Mary Ward has the first of her two discussions with Gladstone at Keble College on the theology of her novel. 15 April: Matthew Arnold dies. May: Gladstone's '*Robert Elsmere* and the Battle of Belief' published in *Nineteenth Century*. Mary Ward replies with 'The New Reformation' in the magazine's March 1889 issue. Mary Ward has her first ideas for her settlement and visits Toynbee Hall. Summer: Humphry Ward buys a plot of land at Grayswood Hill, near Haslemere, on which to build a new house. September: Mary Ward begins work on *The History of David Grieve*. As the acclaimed author of *Robert Elsmere* she visits a number of great people in their country houses. October: an estimated 100,000 copies of pirated editions of *Robert Elsmere* are sold in the US.

1889 January: Arnold Ward goes to Uppingham School to prepare for Eton entrance. April; Mary Ward makes her first trip to Rome. June: Mary Ward publishes her influential 'Appeal' against women's suffrage in *Nineteenth Century*. October: Thomas Arnold announces he will marry Josephine Benison in January 1890. November: Mary Ward explores the possibility of founding a 'settlement' on the lines of Toynbee Hall.

1890 January: Arnold Ward enters Eton as a King's Scholar. March: Mary Ward and her committee issue a circular outlining a proposed settlement, University Hall, in Gordon Square. March–April: Mary Ward visits Paris. July: the Wards move into their newly-built house at Grayswood Hill, Haslemere. 29 November: University Hall formally opened. Mary Ward gives her first public speech. December: the Residents of University Hall form a club, Marchmont Hall, for the social welfare of the area's poor.

1891 1 January: Gertrude Ward leaves Russell Square to become a district nurse in London. 17 January: the Wards move from 61 Russell Square to their new town house at 25 Grosvenor Place. April: the American international copyright bill passes into law. June: Macmillan's offer a record-breaking £7,000 for the American rights to *David Grieve*, sight unseen. September: Mary Ward finishes *David Grieve* in a state of physical prostration. Publication of the novel is held back until 22 January 1892 to facilitate simultaneous American issue. Mary Ward winters in Italy. Humphry Ward is made a trustee of the Tate Gallery.

1892 February: the Wards engage to rent Stocks, a large country house in Hertfordshire, for seven years. Mary Ward begins work on *Marcella* when they move there on 14 July. May: the Wards sell their Haslemere house for £9,000. Humphry sinks most of the profit on a Cuyp picture. August: Mary Ward suffers crippling attack of 'side' (diagnosed variously as 'neuralgia', 'internal catarrh' or 'floating kidney', revealed in August 1919 to be gallstones). November: Mary Ward gives a series of lectures in the north of England to raise funds for University Hall. Frank Arnold marries Annie Reed Wilkinson.

1893 Mary Ward decides on a new settlement to combine the best features of University Hall and Marchmont Hall. 10 October: Benjamin Jowett dies. Two years later Mary Ward sets up a memorial lecture programme at her settlement.

1894 Mary Neal begins her Saturday morning 'playrooms' at Marchmont Hall for the children of working families. 1 February: Mary Ward finishes writing *Marcella*. The novel is published in three volumes on 3 April 1894. May: Passmore Edwards agrees to finance the new settlement with a gift of £4,000 (eventually raised by another £10,000). July: the cheap (6s, one-volume) edition of *Marcella* is published by Smith, Elder, precipitating the demise of the three-volume novel. The American *Century Magazine* offers to serialize Mrs Ward's next novel. September: Lucy Selwyn dies suddenly of a blood clot in the brain. November: Mary Ward begins work on *Sir George Tressady*, a sequel to *Marcella*. Winter: the Duke of Bedford leases Mary Ward a site on the corner of Tavistock Square for her new settlement.

1895 January: Mary Ward writes *The Story of Bessie Costrell* for serialization in *Cornhill Magazine*, May–July 1895. January–March: Humphry Ward lectures on art in America, accompanied by Dorothy. March: A. Dunbar-Smith and Cecil Brewer win the competition for the architectural design of the new settlement, to be called after Passmore Edwards. Arnold Ward wins a Newcastle scholarship. He plays for Eton at cricket. He goes up to Balliol College, Oxford, in October 1895.

1896 January: Mary Ward has surgery for an internal ailment. The operation is unsuccessful. March: the Wards buy Stocks on long mortgage. William Arnold falls seriously ill with a mysterious ailment. 22 March: Mary Ward finishes writing *Sir George Tressady*, which is published 25 September 1896. May: Mary Ward taken violently ill with 'side' on a train journey from Cadenabbia to Switzerland. August: Bessie Churcher enters Mary Ward's service as secretary. October: Mary Ward has her first ideas for *Helbeck of Bannisdale*. December: Mary Ward's brother Frank Arnold incurs debts of £3,000 for which she makes herself responsible.

1897 February: Mary Ward lectures on 'The Peasant in Literature' at Glasgow. She records taking cocaine to help her through the ordeal. Mary Ward pledges £1,000 of her own money towards the building of the Passmore Edwards Settlement. William Arnold forced by ill-health to give up his position on the *Manchester*

Guardian. He and his wife come to London, where they are partly dependent on Mary Ward. March: Mary Ward writes the first chapter of *Helbeck* at Levens Hall. She remains there until June 1897 when her health suddenly collapses. July: Arnold Ward takes a first in Mods. and wins the Chancellor's Latin Verse Prize. 10 October: Passmore Edwards Settlement is formally opened (there is another ceremony on 11 February 1898). The first Saturday 'play centre' for children at Passmore Edwards Settlement is organized by Mary Neal, Bessie Churcher, and the Ward daughters. By the end of 1897 the play centre facilities are augmented by weekday evening sessions.

1898 January: Mary Ward undertakes to 'rewrite' *Helbeck* so as to protect her father's Catholic sensibilities. 28 March: Mary Ward having finished writing *Helbeck* leaves for Italy where she remains until May. There she has the first ideas for *Eleanor*. *Helbeck* is published on 10 June 1898. May: William Arnold is diagnosed as having locomotor ataxia. Gladstone dies. November: Mary Ward plans out *Eleanor*. December: Arnold Ward elected a Craven scholar.

1899 Mary Ward's prefaces to the Haworth Edition of the Brontës' work are published by Smith, Elder, 1899–1900. 28 February: a school for handicapped ('physically defective') children opens at PES. By 1906 there are twenty-three such 'special' schools in the London region caring for nearly 1,800 children. March: the Wards take an apartment near Rome. There she writes *Eleanor*. Over the summer they are visited by Henry James, and by Thomas Arnold and his wife. August: Arnold Ward gets a first in Greats. September: Mary Ward writes to *The Times* about the 'Crisis in the Church', arguing that the sacraments should not be denied to those who do not believe in biblical miracles. October: Mary Ward's aunt and godmother Jane Forster dies. November: Arnold Ward sails to Egypt as a special correspondent for *The Times*.

1900 March: *Eleanor* finished. Smith, Elder publish it on 1 November. It achieves record advance sales in Britain and in America. July: Arnold Ward returns from Egypt. 12 November: Tom Arnold dies in Dublin. Arnold Ward fails to win a fellowship at Oxford, and resolves to read for the bar.

1901 January: Mandell Creighton dies. Arnold Ward travels to India as special correspondent for *The Times*. 6 April: George Smith dies. Mary Ward—who is in Rapallo—works on a dramatized version of *Eleanor* with the American playwright Julian Sturgis. Leonard Huxley appointed to a post at Smith, Elder (eventually to the editorship of the *Cornhill Magazine*) through Mary Ward's good offices.

1902 January: Julia Huxley's boarding school for girls, Prior's Field, opens. April: Arnold Ward returns from India. August: PES opens its first Vacation School for local children during their summer holidays. October: Mary Ward's play *Eleanor* is performed at the Court Theatre, from 30 October to 15 November. It is a failure.

1903 Arnold Ward called to the bar. January: Mary Ward finishes writing *Lady Rose's Daughter*. The novel is published on 5 March 1903. In a state of physical exhaustion Mary Ward goes to Cadenabbia to recuperate. March: She experiences the onset of serious eczema. Between March and July 1903, Mary Ward writes her second play, *Agatha*. May: G. M. Trevelyan and Janet Ward become engaged.

Summer: Arnold Ward fails to get the position of private secretary to the Indian Viceroy.

1904 February: Mary Ward writes the opening chapters of *The Marriage of William Ashe*. She acquires Robin Ghyll cottage in the Lake District for her daughter Dorothy. March: Janet Ward and G. M. Trevelyan married in Oxford. 29 May: William Arnold dies of cerebral haemorrhage. July: Mary Ward has first ideas for *Fenwick's Career*, her 'Romney novel'. Humphry Ward publishes his study of Romney. November: Mary Ward has her first ideas for *The Case of Richard Meynell*.

1905 February: Mary Ward's first grandchild, Mary Trevelyan, is born. PES opens eight new play centres. March: *The Marriage of William Ashe* is published and is a great success. Ward's play, *Agatha*, is performed at His Majesty's Theatre and is another failure. May: Mary Ward begins writing *Fenwick's Career* and finishes the novel in December. September: Moberly Bell launches the *Times* Book Club which Mary Ward supports in defiance of her publisher, Reginald Smith.

1906 Mary Ward has extensive work done on her teeth, which relieves her chronic facial neuralgia. January: Arnold Ward contests the Cricklade Division of Wiltshire as a Liberal Unionist candidate, unsuccessfully. February: Julia Huxley discovered to have cancer. Mary Ward's health collapses. She is taken to the Continent, remaining there until May. May: *Fenwick's Career* is published. June: Mary Ward's mental and physical health collapses still further; in August it is feared she will die. October: she partially recovers her health. She is visited at Stocks by Mackenzie King, who enthuses her with his vision of Canada. November: Mary Ward tries unsuccessfully to mediate in the Book War between the publishers and the *Times* Book Club. The dispute lingers on until 1908. 28 December: Mary Ward begins writing *The Testing of Diana Mallory*.

1907 February: Mary Ward publishes her memoir of William T. Arnold. February–March: she goes to the Isle of Wight for a month to work on her new novel. April: the Wards celebrate their 35th wedding anniversary in Paris. May: Mary Ward experiences the first symptoms of heart disease. July: major improvements to Stocks are begun. Originally costed at £3,900 they eventually run to over three times this amount. October: Arnold discusses marriage (to Violet Asquith?) with his mother. His plans fall through. Mary Ward finishes writing *The Testing of Diana Mallory*. The novel is published on 17 September 1908.

1908 11 March: Mary, Humphry and Dorothy Ward leave for a lecture tour in North America. 19 March: the Wards arrive in New York where they stay with their relative, F. W. Whitridge. April: Arnold Ward leaves his post in Sir Charles Mathews's law firm. 4 April: Mary and Dorothy Ward travel to Philadelphia. 12 April: the Wards travel to Washington where Mary Ward meets Theodore Roosevelt and various Cabinet ministers. 20 April: the Wards go to Boston. Humphry Ward, meanwhile, has returned to England. 24 April: Mary Ward's play *William Ashe* has its first performance in London. It is judged a failure and withdrawn at the end of May. Humphry Ward receives £4,000 from the sale of a Rembrandt. May: Asquith informs a deputation of Liberals that he has no objection 'in principle' to the extension of the vote to women. 5 May: the Wards travel north to Montreal via Vermont. Mary Ward is received by the Governor General of Canada, Earl Grey. 11 May: the Wards travel to Ottawa, where they stay at Government House. 15 May: the Wards travel to Toronto. 20 May: the Wards view Niagara. The following day Sir William Van Horne makes available to them a private car on the Canadian Pacific Railway to travel across the continent. They set off on 23 May. The journey inspires the novel *Canadian Born*. Late May: the Wards reach Vancouver. Early June: the Wards return to New York and embark for England. 12 June: the 'Women's Anti-Suffrage Association' announced, with Mrs Humphry Ward at its head. 21 July: the Women's National Anti-Suffrage League formed. 27 July: Mary Ward begins writing her anti-suffrage novel, *Daphne*. 8 November: Arnold Ward gets a commission in the Hertfordshire Yeomanry. 30 November: Julia Huxley dies. December: Mary Ward launches the *Anti-Suffrage Review*.

1909 19 January: Men's League for Opposing Women's Suffrage is formed; Arnold Ward is an officer in it. Reginald Smith agrees to pay £10,000 for 'Robert Elsmere II' (eventually called *The Case of Richard Meynell*). February: Mary Ward completes the writing of *Daphne*. The novel is published in May by Cassell and sells badly. 26 February: Mary Ward debates women's suffrage with Mrs Fawcett at PES. The antis are soundly defeated. 26 March: Mary Ward speaks at a large Anti Suffrage Rally at Queen's Hall. Much heckling and disturbance. April–May: Mary Ward recuperates in Cadenabbia. On her return she is persuaded by Humphry and Arnold Ward to purchase extensive land property around Stocks. September: Mary Ward argues the anti-suffrage case at Newnham and Girton colleges. She is coldly received. Around the same period Mary Ward's formal relationship with Somerville College is broken off on account of her political activities against suffrage. October: Mary Ward finishes the writing of *Canadian Born*. The novel is published in April 1910 and sells poorly. Mary Ward is increasingly anxious about Arnold's gambling at the Portland Club in London. At the end of October, she undertakes a lecture tour in the north of England and Scotland on behalf of the Anti-Suffrage League.

1910 28 January: Arnold Ward elected Liberal Unionist MP for West Herts. March: Mary Ward petitions LCC candidates on behalf of her play centres. Mid-March: Mary Ward suffers from eczema, 'side', eyestrain, and insomnia. At the end of March she goes to Valescure to recuperate, returning in early May. 30 May: Theodore Roosevelt visits the Wards at Grosvenor Place. June: Mary Ward leads an anti-suffrage deputation to Asquith, who receives them sympathetically. July: she is active in meetings against the Conciliation Bill. August: she begins writing *The Case of Richard Meynell*. Arnold Ward promoted lieutenant in the Herts. Yeomanry. December: the Women's Anti-Suffrage League amalgamates with the Men's League for Opposing Women's Suffrage. At the General Election provoked by the failure of the Constitutional Conference Arnold Ward is re-elected with a much reduced majority.

1911 18 April: Mary Ward is in Italy where news of the death of her grandson Theo Trevelyan prostrates her. 9 August: Mary Ward finishes writing *The Case of Richard Meynell*. The novel is published on 26 October and sells poorly. 8 November: Asquith announces that his government will bring in women's suffrage. 14 December: Mrs Ward leads a deputation to Asquith and elicits from him the admission that women's suffrage would be 'a political mistake of a disastrous kind'. Suffragette militancy ensues.

1912 January: Mary Ward begins writing *The Mating of Lydia*. March–May: Mary Ward (holidaying on the Continent) learns that the sixteen-volume collective reissue of her work (called the Autograph Edition in America, the Westmoreland Edition in Britain) has failed disastrously. Summer: Mary Ward is active on her Local Government Advancement Committee. 17 July: Mary Ward examined by a gynaecologist for an unspecified condition. She has an operation on 16 October 1914, which is successful. 22 October: Mary Ward finishes the writing of *The Mating of Lydia*. The novel is not published until 15 March 1913 and sells poorly. November: Humphry Ward suffers a serious attack of amnesia. The attacks recur at intervals and with increasing intensity for the rest of his life, rendering him increasingly invalid. Mary Ward is forced off the National Union of Women Workers on account of her political views. She later (in 1914) forms her own 'Joint Advisory Committee' to lobby Parliament on social matters. 17 December: Mary Ward agrees to write *The Coryston Family*.

1913 27 January: despite the efforts of the anti-suffragists Asquith pledges to bring in a suffrage bill next session. 5 April: Arnold Ward confesses to his family that he has incurred huge losses gambling (possibly as much as £20,000). The Wards sell extensive portions of their property and of Humphry's art collection. May: Mary Ward finishes writing *The Coryston Family*. 5 May: Arnold Ward moves the rejection of the Suffrage Bill in the House. His motion succeeds. 18 July: Mary and Humphry Ward (who is now completely invalid) travel to the Tyrol in the hope of repairing his health. They remain abroad (mainly in the Alps) until 28 September. Mary Ward begins writing her anti-suffragette novel *Delia Blanchflower* during the trip. 11 October: *The Coryston Family* published to poor reviews and sales.

1914 2 March: Mary Ward finishes writing *Delia Blanchflower*. Reginald Smith has great difficulty finding an American co-publisher and the novel does not appear in print until January 1915, when it is scathingly reviewed and sells poorly. April: Mary Ward has her first ideas for *Eltham House*. May: the Wards contemplate selling Stocks. July–August: the Wards spend the summer in Scotland for economy's sake. In early August Humphry has repeated 'attacks', and arteriosclerosis is diagnosed on 4 August by doctors in Edinburgh. War is declared against Germany. Arnold Ward joins up with the Hertfordshire Yeomanry as a lieutenant. 24 August: suicide of Trevenen Huxley. 31 August: Arnold Ward's squadron is ordered to Egypt. He departs on 4 September. The Wards let out Stocks and live in Grosvenor Place or (when that is later rented out) in Stocks Cottage and in a small flat at Queen Anne's Mansions in London. 31 December: Mary Ward finishes writing *Eltham House*. The novel is published October 1915, and sells poorly. She starts writing *A Great Success* immediately.

1915 January: reports of Arnold's gambling debts in Cairo worry Mary Ward. March: women residents installed in PES for the duration of the War. May: following a visit to Oxford, Mary Ward has the idea for *Lady Connie*, her finest late novel. June: the Wards spend their summer holiday at Kelbarrow, in the Lake District. July: Arnold Ward is at odds with his Commanding Officer in Egypt, and sends a memorandum explaining his grievance to his regimental commander. He is ordered to Cyprus, where he takes command of a convalescent camp in the Troodos Mountains. August: the Herts. Yeomanry are engaged in battle in Gallipoli. Arnold's request to rejoin his regiment is declined. He continues to run up gambling debts. August: the Wards repossess Stocks. Dorothy takes charge of a corps of land girls and farms the land for the war effort. October: Arnold returns to England to present his grievances in person to the War Office. Mary Ward unsuccessfully tries to procure him a commission in the cavalry. 11 November: Mary Ward finishes writing *Lady Connie*. The novel is published in October 1916, and is a success. 27 December: Theodore Roosevelt writes to Mary Ward, suggesting she write a series of articles on the war for American readers.

1916 15 January: Arnold Ward is posted back to Egypt. 20 January: Mary Ward has an interview with Sir Edward Grey, who agrees to let her visit munitions factories, the Fleet and the Front. 31 January–15 February: Mary Ward tours munitions works in the north of England, accompanied by Lizzie Smith and Bessie Churcher. 16 February: she visits the Fleet at Invergordon. 28 February: Mary and Dorothy Ward go to the war zone in France. Henry James dies. 2 March: Mary and Dorothy Ward visit the Front, near Ypres. *A Great Success* published; not a great success. April–May: the *England's Effort* 'letters' are published in American newspapers and in book form. The work is very successful. 26 May: the Wards lay up their car for the duration of the War. 21 June: Arnold Ward returns to England to answer possible charges of misconduct. He is cleared, but there are large gambling debts for Mary Ward to settle. It is agreed that he will resign his seat in Parliament at the next election. From mid-July to mid-September the Wards vacation in the Lake District. Mary Ward works on her autobiography, *A Writer's Recollections*, eventually published in October 1918. 18 August: Arnold Ward is transferred to the the Territorial Force, and has various miscellaneous postings around Britain. 6 November: Mary Ward applies for permission to revisit the Front. Winter: Bessie Churcher suffers a heart attack. 28 December: Reginald Smith falls from the second floor window of his house and dies.

1917 After much lobbying, the Treasury agrees to fund 50 per cent of Mary Ward's play centres. 28 February: Mary and Dorothy Ward travel to France. They spend two days at the Front, before returning on 16 March. 15 July: after serialization in American newspapers, *Towards the Goal* is published and enjoys a huge success. August: until his gambling debts are settled Arnold Ward is prohibited from rejoining his regiment. He devotes himself to arguing the anti-suffrage cause in the House of Commons. 11 October: the third of Mary Ward's nephews, Frank Arnold's son Tom, is killed in War. Mrs Ward's first war novel, *Missing*, is published and is a success. 24 December: Mary Ward finishes writing *A Writer's Recollections*. It is published in October 1918.

1918 10 January: the clause guaranteeing women's suffrage is passed in the House of Lords. Spring: by energetic lobbying Mary Ward's Joint Advisory Committee succeeds in inserting a clause on behalf of physically defective children into the Education Bill as it passes through the House of Commons. April: Mary Ward finishes writing *The War and Elizabeth*. The novel is published in December 1918 and is unsuccessful. August: the Wards receive cripplingly high supertax demands. They sell much of Humphry's collection of art over the next two years to meet them. 11 November: Armistice Day. 28 December: Arnold Ward ceases to be an MP. 31 December: Mary Ward finishes *Harvest*. The novel is not published until April 1920.

1919 3 January: Mary and Dorothy Ward undertake a strenuous 900-mile motor trip through the battlegrounds of France and Belgium, returning on 30 January. Mary Ward's health collapses. March: Mary Ward receives a CBE. July: *Fields of Victory* published. 29 August: the Wards give up their London house at Grosvenor Place. Humphry continues to sell off his art collection and much of the Stocks furniture. September: Mary Ward finishes writing *Cousin Philip*. The novel is published in November 1919 and is generally ignored. From mid-September to mid-October the Wards are in the Lake District. December: Mary Ward's last public campaign, directed against the Church Assembly Act and its exclusionary line on Communion and free-thinking Christians. Christmas: Mary Ward severely troubled by neuritis.

1920 January; Mary Ward, now invalid, is moved to London. February: Mary Ward is invited by the Lord Chancellor to be one of the first women magistrates in England. March: Edinburgh University awards Mary Ward an honorary degree. 11 March: Humphry Ward falls suddenly ill and has an emergency operation. 14 March: Mary Ward bedridden with symptoms of bronchitis and heart disease. 24 March: Mary Ward dies. 27 March: she is buried at St John the Baptist Church, Aldbury.

1921 January: Lizzie Smith falls into senile dementia and is put into care. December: Stocks is sold.

1926 6 May: Humphry Ward dies and is buried at Aldbury.

1950 1 January: Arnold Ward dies and is buried at Aldbury

1964 29 March: Dorothy Ward dies and her ashes are interred at Aldbury in her mother's grave.

Select Bibliography

1. Mary Ward's Books

Milly and Olly: Or, A Holiday among the Mountains, London: Macmillan, 1881. New York: Doubleday, 1907.

Miss Bretherton, London: Macmillan, 1884. New York: J. W. Lovell, 1888.

Robert Elsmere, London: Smith, Elder, 1888. New York: Macmillan, 1888.

The History of David Grieve, London: Smith, Elder, 1892. New York: Macmillan, 1892.

Marcella, London: Smith, Elder, 1894. New York: Macmillan, 1894.

The Story of Bessie Costrell, London: Smith, Elder, 1895. New York: Macmillan, 1895. Serialized (Britain) *Cornhill Magazine*, May–July 1895. Serialized (USA) *Scribner's Magazine*, May–July 1895.

Sir George Tressady, London: Smith, Elder, 1896. New York: Macmillan, 1896. Serialized (USA) *Century Magazine*, Nov. 1895–Oct. 1896.

Helbeck of Bannisdale, London: Smith, Elder, 1898. New York: Macmillan, 1898.

Eleanor, London: Smith, Elder, 1900. New York: Harper, 1900. Serialized (USA) *Harper's Magazine*, Jan.–Dec. 1900.

Lady Rose's Daughter, London: Smith, Elder, 1903. New York: Harper, 1903. Serialized (USA) *Harper's Magazine*, May 1902–April 1903.

The Marriage of William Ashe, London: Smith, Elder, 1905. New York: Harper, 1905. Serialized (USA) *Harper's Magazine*, June 1904–May 1905.

Fenwick's Career, London: Smith, Elder, 1906. New York: Harper, 1906. Serialized (USA) *Century Magazine*, Nov. 1905–June 1906.

William Thomas Arnold, Journalist and Historian, (Mrs Humphry Ward and C. E. Montague), Manchester: Manchester University Press, 1907.

Diana Mallory (in USA *The Testing of Diana Mallory*), London: Smith, Elder, 1908. New York: Harper, 1908. Serialized (USA) *Harper's Magazine*, Nov. 1907–Oct. 1908.

Daphne (in USA *Marriage à la Mode*), London: Cassell, 1909. New York, Doubleday, Page, 1909. Serialized (USA) *McClure's Magazine*, Jan.–June 1909.

Canadian Born (in USA *Lady Merton, Colonist*), London: Smith, Elder, 1910. New York: Doubleday, Page, 1910. Serialized (Britain) *Cornhill Magazine*, Oct. 1909–May 1910. Serialized (USA) *Ladies' Home Journal*, Oct. 1909–May 1910.

The Case of Richard Meynell, London: Smith, Elder, 1911. New York: Doubleday, Page, 1911. Serialized (Britain) *Cornhill Magazine*, Jan.–Dec. 1911. Serialized (USA) *McClure's Magazine*, Jan.–Dec. 1911.

The Mating of Lydia, London: Smith, Elder, 1913. New York: Doubleday, Page, 1913. Serialized (USA) *Good Housekeeping* Nov. 1912–Nov. 1913.

The Coryston Family, London: Smith, Elder, 1913. New York: Harper, 1913. Serialized (USA) *Harper's Magazine*, May–Nov. 1913.

Delia Blanchflower, London: Ward, Lock, 1915. New York, Hearst, 1914.

Eltham House, London: Cassell, 1915. New York: Hearst, 1915.

A Great Success, London: Smith, Elder, 1916. New York: Hearst, 1916. Serialized (Britain) *Cornhill Magazine*, May–July 1915.

England's Effort, London: Smith, Elder, 1916. New York: Scribner's, 1916.
Lady Connie, London: Smith, Elder, 1916. New York: Hearst, 1916. Serialized (Britain) *Cornhill Magazine*, Dec. 1915–Oct. 1916.
Towards the Goal, London: Murray, 1917. New York: Scribner's, 1917.
Missing, London: Collins, 1917. New York: Dodd, Mead, 1917.
The War and Elizabeth (in USA *Elizabeth's Campaign*), London: Collins, 1918. New York: Dodd, Mead, 1918.
A Writer's Recollections, London: Collins, 1918. New York: Harper, 1918. Serialized (Britain) *Cornhill Magazine*, Mar.–July 1918 in shorter form. Serialized (USA) *Harper's Magazine*, Jan.–Oct. 1918.
Cousin Philip (in USA *Helena*), London: Collins, 1919. New York: Dodd, Mead, 1919.
Fields of Victory, London: Hutchinson, 1919. New York: Scribner's, 1919.
Harvest, London: Collins, 1920. New York: Dodd, Mead, 1920 (dated 1919).

2. Secondary Reading

THERE are three excellent comprehensive bibliographies to which the reader is referred. William B. Thesing and Stephen Pulsford's *Mrs Humphry Ward*, published as Victorian Fiction Research Guides XIII (St Lucia: Fiction Research Unit, Department of English, University of Queensland, 1987), contains a biographical introduction, a full listing of Mary Ward's works in printed form, a listing of reviews of the principal works when they first appeared, and a comprehensive list of article and book-length critical commentary from the 1880s to the 1980s. The second volume of Françoise Rives's *Mrs Humphry Ward, Romancière* (Lille: University of Lille, 1982) largely consists of a bibliography of exceptional thoroughness and usefulness. Rives—among much else—describes the manuscript remains of Mary Ward's work and gives their locations, compiles a wide-ranging list of historically relevant reading for the understanding of the author's cultural context, and has assembled a comprehensive and fascinating checklist of Mrs. Humphry Ward's books in translation. William S. Peterson's *Victorian Heretic: Mrs Humphry Ward's 'Robert Elsmere'* (Leicester: Leicester University Press, 1976) contains another full bibliography and is particularly valuable in its account of work written in reaction to *Robert Elsmere*. The texts of Françoise Rives's, Anne Bindslev's and William Peterson's books are recommended as the best introductory commentaries on Mary Ward's fiction and non-fiction.

ARNOLD, Thomas, *The New Zealand Letters of Thomas Arnold the Younger, 1847–1851*, ed. James Bertram, Auckland: University of Auckland Press; London: Oxford University Press, 1966.
ARNOLD, Thomas, *The Letters of Thomas Arnold the Younger, 1850–1900*, ed. James Bertram, Auckland: University of Auckland Press; London: Oxford University Press, 1980.
BELLRINGER, Alan W., 'Mrs Humphry Ward's Autobiographical Tactics', *Prose Studies* 8 (1985): 40–50.
BINDSLEV, Anne M., *Mrs Humphry Ward: A Study in Late-Victorian Feminine Consciousness and Creative Expression*, Stockholm: Almqvist and Wiksell International, 1985.
COLBY, Vineta, *The Singular Anomaly: Women Novelists of the Nineteenth Century*, New York: New York University Press, 1970.

COLLISTER, Peter, 'Mrs Humphry Ward, Vernon Lee, and Henry James', *Review of English Studies* 31 (1980): 315–21.

COLLISTER, Peter, 'Portraits of "Audacious Youth": George Eliot and Mrs Humphry Ward', *English Studies* 64 (1983): 296–317.

GWYNN, Stephen, *Mrs Humphry Ward*, London: Nisbet, 1917.

JONES, Enid Huws, *Mrs Humphry Ward*, London: Heinemann, 1973.

KNOEPFLMACHER, U. C., 'The Rival Ladies: Mrs Ward's *Lady Connie* and Lawrence's *Lady Chatterley's Lover*', *Victorian Studies* 4 (1960): 141–58.

MAISON, Margaret M., *Search Your Soul, Eustace*, London: Sheed and Ward, 1961.

NORTON-SMITH, J., 'An Introduction to Mrs Humphry Ward, Novelist', *Essays in Criticism*, 18 (1968): 420–8.

PETERSON, William S., 'Mrs Humphry Ward on *Robert Elsmere*: Six New Letters', *Bulletin of the New York Public Library* 74 (1970): 587–97.

PETERSON, William S., 'Gladstone's Review of *Robert Elsmere*: Some Unpublished Correspondence', *Review of English Studies* 21 (1970): 442–61.

PHELPS, William L., *Essays on Modern Novelists*, New York: Macmillan, 1910.

RIVES, Françoise, 'The Marcellas, Lauras, Dianas, of Mrs Humphry Ward', *Caliban* 17 (1980): 69–79.

RIVES, Françoise, 'Une Romancière victorienne face à la Grande Guerre', *Caliban* 19 (1982): 59–71.

SMITH, Esther M. G., *Mrs Humphry Ward*, Boston: Twayne, 1980.

SMITH, Esther M. G., 'Mrs Humphry Ward', in Ira B. Nadel and William E. Fredeman (eds.), *Victorian Novelists after 1885*, Detroit: Gale, 1983, 297–303.

TREVELYAN, Janet P., *The Life of Mrs Humphry Ward*, London: Constable, 1923.

WARD, Mary, *Marcella*, ed. Tamie Watters, London: Virago, 1985.

WARD, Mary, *Robert Elsmere*, ed. Rosemary Ashton, Oxford: Oxford University Press, 1987.

WOLFF, Robert Lee, *Gains and Losses: Novels of Faith and Doubt in Victorian England*, London and New York: Garland, 1977.

Index

MORE OXFORD PAPERBACKS

THE OXFORD AUTHORS

General Editor: Frank Kermode

The Oxford Authors is a series of authoritative editions of the major English writers for the student and the general reader. Drawing on the best texts available, each volume contains a generous selection from the writings—poetry and prose, including letters—to give the essence of a writer's work and thinking. Where appropriate, texts have been tactfully modernized and all are complemented by essential Notes, an Introduction, Chronology, and suggestions for Further Reading.

'The Oxford Authors series can always be relied upon to be splendid—with good plain texts and helpful notes.' Robert Nye, *Scotsman*

OSCAR WILDE

Edited by Isobel Murray

The drama of Oscar Wilde's life has for years overshadowed his achievement in literature. This is the first large-scale edition of his work to provide unobtrusive guidance to the wealth of knowledge and allusion upon which his writing stands.

Wilde had studied Greek and Latin and was familiar with American literature, while he was as well read in French as he was in English, following Gautier and Flaubert as well as Pater and Ruskin. Through her Notes Isobel Murray enables the modern reader for the first time to read Wilde as such admiring contemporaries as Pater, Yeats, and Symons read him, in a rich, shared culture of literary and visual arts.

This edition underlines the range of his achievement in many genres, including *The Picture of Dorian Gray, Salome, The Importance of Being Earnest, The Decay of Lying,* and *The Ballad of Reading Gaol.* The text is that of the last printed edition overseen by Wilde.

Also in the Oxford Authors:

Sir Philip Sidney
Ben Jonson
Byron
Thomas Hardy

OXFORD LETTERS & MEMOIRS

This popular series offers fascinating personal records of the lives of famous men and women from all walks of life.

JOURNEY CONTINUED

Alan Paton

'an extraordinary last testament, told in simple and pungent style . . . for anyone new to the period and to Paton, it will be a revelation' *Independent*

This concluding volume of autobiography (the sequel to *Towards the Mountain*) begins in 1948, the year in which Paton's bestselling novel, *Cry, the Beloved Country*, was published, and the Nationalist Party of South Africa came to power. Both events were to have a profound effect on Paton's life, and they represent two major themes in this book, literature and politics.

With characteristic resonance and trenchancy, Paton describes his career as a writer of books, which were received with extreme hostility by his fellow South Africans, and also covers his political life, notably the founding—and later Chairmanship—of the Liberal Party of South Africa, the multi-racial centre party opposed to apartheid.

'required reading for anyone who wants to understand, compassionately, the full tragedy of South Africa' *Daily Express*

Also in Oxford Letters & Memoirs:

Memories and Adventures Arthur Conan Doyle
Echoes of the Great War Andrew Clark
A Local Habitation: Life and Times 1918–1940
Richard Hoggart
Pack My Bag Henry Green

PAST MASTERS

General Editor: Keith Thomas

Past Masters is a series of concise and authoritative introductions to the life and works of men and women whose ideas still influence the way we think today.

'Put end to end, this series will constitute a noble encyclopaedia of the history of ideas.' Mary Warnock

SHAKESPEARE

Germaine Greer

'At the core of a coherent social structure as he viewed it lay marriage, which for Shakespeare is no mere comic convention but a crucial and complex ideal. He rejected the stereotype of the passive, sexless, unresponsive female and its inevitable concommitant, the misogynist conviction that all women were whores at heart. Instead he created a series of female characters who were both passionate and pure, who gave their hearts spontaneously into the keeping of the men they loved and remained true to the bargain in the face of tremendous odds.'

Germaine Greer's short book on Shakespeare brings a completely new eye to a subject about whom more has been written than on any other English figure. She is especially concerned with discovering why Shakespeare 'was and is a popular artist', who remains a central figure in English cultural life four centuries after his death.

'eminently trenchant and sensible . . . a genuine exploration in its own right' John Bayley, *Listener*

'the clearest and simplest explanation of Shakespare's thought I have yet read' Auberon Waugh, *Daily Mail*

Also available in Past Masters:

Paine Mark Philp
Dante George Holmes
The Buddha Michael Carrithers
Confucius Raymond Dawson

ANTHONY TROLLOPE IN THE WORLD'S CLASSICS

Anthony Trollope (1815–1882), one of the most popular English novelists of the nineteenth century, produced 47 novels and several biographies, travel books, and collections of short stories. The World's Classics series offers the best critical editions of his work available.

THE THREE CLERKS

Anthony Trollope
Edited with an Introduction by Graham Handley

The Three Clerks is Trollope's first important and incisive commentary on the contemporary scene. Set in the 1850s, it satirizes the recently instituted Civil Service examinations and financial corruption in dealings on the stock market.

The story of the three clerks and the three sisters who become their wives shows Trollope probing and exposing relationships with natural sympathy and insight before the fuller triumphs of Barchester, the political novels, and *The Way We Live Now*. The novel is imbued with autobiographical warmth and immediacy, the ironic appraisal of politics and society deftly balanced by romantic and domestic pathos and tribulation. The unscrupulous wheeling and dealing of Undy Scott is colourfully offset by the first appearance in Trollope's fiction of the bullying, eccentric, and compelling lawyer Mr Chaffanbrass.

The text is that of the single-volume edition of 1859, and an appendix gives the most important cuts that Trollope made for that edition.

Also in the World's Classics:

The Chronicles of Barsetshire
The Palliser Novels
Ralph the Heir
The Macdermots of Ballycloran

ARMADALE

Wilkie Collins

Edited with an Introduction by Catherine Peters

'it has the immense—and nowadays more and more rare—merit of never being dull'

T. S. Eliot's appreciation of *Armadale* still stands. The third of Wilkie Collins's four great novels of the 1860s, coming after *The Woman in White* and *No Name*, and immediately before *The Moonstone* (all available in World's Classics), *Armadale* is quintessentially a novel of its decade. It deals with the emergence of the autonomous, sexually active woman from the dichotomies of Madonna and Magdalen; with the legal tangles of the unsatisfactory marriage laws; with the perception of the growing role of scientific intrusion into the privacy of the individual psyche. Above all, it explores the divided self, and the need to acknowledge the darker side of the personality: a modern theme grafted on to a traditional melodrama, and worked out with all Collins's skill in handling a complex and exciting plot.

First published in 1866, the text of this World's Classics edition is that of the one-volume 1869 edition, checked and corrected against both the first impression and the magazine serialization which preceded it.

THE WORLD'S CLASSICS

The World's Classics series offers readers the opportunity to rediscover the great works of literature they enjoyed as children. Amongst the favourites available are such classic adventures as *Treasure Island* and *King Solomon's Mines,* fantasy tales like *Alice in Wonderland* and *The Secret Garden,* and animal stories, including *The Wind in the Willows* and *The Jungle Book*.

THE WIND IN THE WILLOWS

Kenneth Grahame

The Wind in the Willows (1908) is a book for those 'who keep the spirit of youth alive in them; of life, sunshine, running water, woodlands, dusty roads, winter firesides'. So wrote Kenneth Grahame of his timeless tale of Toad, Mole, Badger, and Rat in their beautiful and benevolently ordered world. But it is also a world under siege, threatened by dark and unnamed forces—'the Terror of the Wild Wood' with its 'wicked little faces' and 'glances of malice and hatred'—and defended by the mysterious Piper at the Gates of Dawn. *The Wind in the Willows* has achieved an enduring place in our literature: it succeeds at once in arousing our anxieties and in calming them by giving perfect shape to our desire for peace and escape.

The World's Classics edition has been prepared by Peter Green, author of the standard biography of Kenneth Grahame.

'It is a Household Book; a book which everybody in the household loves, and quotes continually; a book which is read aloud to every new guest and is regarded as the touchstone of his worth.' *A. A. Milne*